CRC HANDBOOK SERIES IN NUTRITION AND FOOD

Miloslav Rechcigl, Jr.
Editor-in-Chief

SECTION OUTLINE

SECTION A: Nutrition and Food Science
Minimum of 1 volume projected.
Nomenclature, Nutrition Literature, Nutrition Societies, Foundations, and Historical Milestones in Nutrition.

SECTION B: The Living Organisms, Their Chemical Constitution, Feeding and Digestive System, and Ecological Aspects
Minimum of 3 volumes projected.
Taxonomy, Distribution of Organisms, Ecology — Symbiosis, Feeding and Digestive System, Chemical Constitution of Organisms, and Biological Productivity.

SECTION C: The Nutrients and Their Metabolism
Minimum of 1 volume projected.
Nutrients and Growth Regulators, Antinutrients and Antimetabolites, Naturally Occurring Food Toxicants, Regulatory Aspects of Nutrition, and Availability of Nutrients.

SECTION D: Nutritional Requirements
Minimum of 4 volumes projected.
Comparative Requirements of Organisms, Qualitative Requirements of Specific Organisms, Quantitative Requirements (Nutritional Standards), and Nutritional Requirements for Specific Processes and Functions: Animals, Microorganisms, Plants.

SECTION E: Nutritional Disorders
Minimum of 2 volumes projected.
Nutritional Disorders in Living Organisms, Effect of Specific Nutrient Deficiencies and Toxicities, and Nutritional Disorders in Specific Tissues.

SECTION F: Food Composition
Minimum of 1 volume projected.
Nutrient Content and Energy Value of Food, Factors Affecting Nutrient Composition of Food, and Utilization and Biological Value of Food.

SECTION G: Diets, Culture Media, and Food Supplements
Minimum of 4 volumes projected.
Diets, Culture Media, and Food and Feed Supplements.

SECTION H: The State of World Food and Nutrition
Minimum of 1 volume projected.
World Population, Natural and Food Resources, Food Production, Food Losses, World Food Usage and Consumption, Geographic Distribution of Nutritional Diseases, Nutritional Requirements — Current and Projected, Agricultural Inputs — Current and Projected, Food Aid, Food Marketing and Distribution, Socioeconomic, Cultural, and Psychological Factors Affecting Nutrition.

SECTION I: Food Safety, Food Processing, Food Preservation
Minimum of 1 volume projected.
Food Contamination, Food Spoilage, Food Wastes, Food Laws, and Nutrition Labeling.

SECTION J: Nutrition and Food Methodology
Minimum of 1 volume projected.
Assessment of Nutritional Status, and Measuring Nutritive Value of Food.

CRC Handbook Series in Nutrition and Food

Miloslav Rechcigl, Jr., Editor-in-Chief

Nutrition Advisor and Director
Interregional Research Staff
Agency for International Development
U.S. Department of State

Section D: Nutritional Requirements
Volume I

Comparative and Qualitative Requirements

CRC PRESS, Inc.
18901 Cranwood Parkway · Cleveland, Ohio 44128

Library of Congress Cataloging in Publication Data

Main entry under title:

Nutritional requirements.

    (CRC handbook series in nutrition and food; section D)
    Bibliography: p.
    Includes index.
    1. Animal nutrition — Collected works. 2. Plants —
Nutrition — Collected works. 3. Culture media (Biology) —
Collected works. I. Rechcigl, Miloslav. II. Series.
QH519.N88   574.1'3   77-6286
ISBN 0-8493-2700-8 (Set)
    0-8493-2721-0 (Volume I)

International Standard Book Number 0-8493-2700-8 (Complete Set)
International Standard Book Number 0-8493-2721-0 (Volume I)

Library of Congress Card Number 77-6286
Printed in the United States

PUBLISHER'S PREFACE

In 1913, when the First Edition of the *Handbook of Chemistry and Physics* appeared, scientific progress, particularly in chemistry and physics, had produced an extensive literature but its utility was seriously handicapped because it was fragmented and unorganized. The simple but invaluable contribution of the *Handbook of Chemistry and Physics* was to provide a systematic compilation of the most useful and reliable scientific data within the covers of a single volume. Referred to as the "bible," the Handbook soon became a universal and essential reference source for the scientific community. The 57th Edition, published in the bicentennial year of 1976, represents more than 63 years of continuous service to millions of professional scientists and students throughout the world.

In the years following World War II, scientific information expanded at an explosive rate due to the tremendous growth of research facilities and sophisticated analytical instrumentation. The single-volume Handbook concept, although providing a high level of convenience, was not adequate for the reference requirements of many of the newer scientific disciplines. Due to the sheer quantity of useful and reliable data being generated, it was no longer feasible or desirable to select only that information which could be contained in a single volume and arbitrarily to reject the remainder. **Comprehensiveness** had become as essential as **convenience**.

By the late 1960's, it was apparent that the solution to the problem was the development of the multi-volume Handbook. This answer arose out of necessity during the editorial processing of the *Handbook of Environmental Control*. A hybrid discipline or, to be more precise, an interdisciplinary field such as Environmental Science could be logically structured into major subject areas. This permitted individual volumes to be developed for each major subject. The individual volumes, published either simultaneously or by some predetermined sequence, collectively became a multi-volume Handbook series.

The logic of this new approach was irrefutable and the concept was promptly accepted by both the scientist and science librarian. It became the format of a growing number of CRC Handbook Series in fields such as Materials Science, Laboratory Animal Science, and Marine Science.

Within a few years, however, it was clear that even the multi-volume Handbook concept was not sufficient. It was necessary to create an information structure more compatible with the dynamic character of scientific information, and flexible enough to accommodate continuous but unpredictable growth, regardless of quantity or direction. This became the objective of a "third generation" Handbook concept.

This latest concept utilizes each major subject within an information field as a "Section" rather than the equivalent of a single volume. Each Section, therefore, may include as many volumes as the quantity and quality of available information will justify. The structure achieves permanent flexibility because it can, in effect, expand "vertically" and "horizontally." Any section can continue to grow (vertically) in number of volumes, and new sections can be added (horizontally) as and when required by the information field itself. A key innovation which makes this massive and complex information base almost as convenient to use as a single-volume Handbook is the utilization of computer technology to produce up-dated, cumulative index volumes.

The *Handbook Series in Nutrition and Food* is a notable example of the "sectionalized, multi-volume Handbook series." Currently underway are additional information programs based on the same organizational design. These include information fields such as Energy and Agricultural Science which are of critical importance not only to scientific progress but to the advancement of the total quality of life.

We are confident that the "third generation" CRC Handbook comprises a worthy contribution to both information science and the scientific community. We are equally certain that it does not represent the ultimate reference source. We predict that the most dramatic progress in the management of scientific information remains to be achieved.

B. J. Starkoff
President
CRC Press, Inc.

PREFACE
CRC HANDBOOK SERIES IN NUTRITION AND FOOD

Nutrition means different things to different people, and no other field of endeavor crosses the boundaries of so many different disciplines and abounds with such diverse dimensions. The growth of the field of nutrition, particularly in the last two decades, has been phenomenal, the nutritional data being scattered literally in thousands and thousands of not always accessible periodicals and monographs, many of which, furthermore, are not normally identified with nutrition.

To remedy this situation, we have undertaken an ambitious and monumental task of assembling in one publication all the critical data relevant in the field of nutrition.

The *CRC Handbook Series in Nutrition and Food* is intended to serve as a ready reference source of current information on experimental and applied human, animal, microbial, and plant nutrition presented in concise tabular, graphical, or narrative form and indexed for ease of use. It is hoped that this projected open-ended multivolume set will become for the nutritionist what the *CRC Handbook of Chemistry and Physics* has become for the chemist and physicist.

Apart from supplying specific data, the comprehensive, interdisciplinary, and comparative nature of the *CRC Handbook Series in Nutrition and Food* will provide the user with an easy overview of the state of the art, pinpointing the gaps in nutritional knowledge and providing a basis for further research. In addition, the *Handbook* will enable the researcher to analyze the data in various living systems for commonality or basic differences. On the other hand, an applied scientist or technician will be afforded the opportunity of evaluating a given problem and its solutions from the broadest possible point of view, including the aspects of agronomy, crop science, animal husbandry, aquaculture and fisheries, veterinary medicine, clinical medicine, pathology, parasitology, toxicology, pharmacology, therapeutics, dietetics, food science and technology, physiology, zoology, botany, biochemistry, developmental and cell biology, microbiology, sanitation, pest control, economics, marketing, sociology, anthropology, natural resources, ecology, environmental science, population, law, politics, nutritional and food methodology, and others.

To make more facile use of the *Handbook,* the publication has been divided into sections of one or more volumes each. In this manner the particular sections of the *Handbook* can be continuously updated by publishing additional volumes of new data as they become available.

The Editor wishes to thank the numerous contributors, many of whom have undertaken their assignment in pioneering spirit, and the Advisory Board members for their continuous counsel and cooperation. Last but not least, he wishes to express his sincere appreciation to the members of the CRC editorial and production staffs, particularly President Bernard J. Starkoff, Mr. Gerald A. Becker, Mrs. Kathryn H. Harter, Mrs. Karen G. Ketchaver, and Mr. Paul R. Gottehrer, for their encouragement and support.

We invite comments and criticism regarding format and selection of subject matter, as well as specific suggestions for new data (and additional contributors) which might be included in subsequent editions. We should also appreciate it if the readers would bring to the attention of the Editor any errors or omissions that might appear in the publication.

Miloslav Rechcigl, Jr.
Editor-in-Chief
October 1976

PREFACE
SECTION D: NUTRITIONAL REQUIREMENTS

The section of the *CRC Handbook Series in Nutrition and Food* on nutritional requirements is projected into several volumes.

The first volume contains information relating to the qualitative requirements and utilization of nutrients in major classes of organisms, covering both plant and animal kingdoms and including microorganisms. Pertinent material is also presented on the nutritional requirements of cells and tissues.

Whenever possible, we have tried to indicate the essentiality or degree of utilization of nutrients by appropriate symbols such as R, required; Ɍ, not required; U, utilized; u, poorly utilized; Ʉ, not utilized; S, stimulatory, supportive (growth); s, moderately supportive; $, not supportive; E, essential; Ɇ, nonessential, etc.

During the preparation of this section, especially this volume, it soon became apparent that the information for many classes of organisms was either negligible or nonexistent. This was particularly true in the field of invertebrates and plants. Our knowledge of cellular requirements is also in rudimentary stages. Nevertheless, for the sake of completeness we have made an earnest effort to cover the nutritional needs of as many groups of organisms as possible, with the full realization that some of the conclusions will have to be modified as new experimental data become available. This is to be expected, especially for those systems in which the conclusions are based on indirect evidence, such as composition of typical food sources on which such organisms subsist.

The subsequent volumes will contain (1) additional nutritional data of the qualitative nature, (2) information relating to quantitative requirements (nutritional standards) of selected species of organisms for which the data are available, and (3) information on nutritional requirements for specific processes and functions.

Miloslav Rechcigl, Jr.
Editor
January 1977

MILOSLAV RECHCIGL, JR., EDITOR

Miloslav Rechcigl, Jr. is Nutrition Advisor and Director of the Interregional Research Staff in the Agency for International Development, U.S. Department of State.

He has a B.S. in Biochemistry (1954), a Master of Nutritional Science degree (1955), and a Ph.D. in nutrition, biochemistry, and physiology (1958), all from Cornell University. He was formerly a Research Biochemist in the National Cancer Institute, National Institutes of Health and subsequently served as Special Assistant for Nutrition and Health in the Health Services and Mental Health Administration, U.S. Department of Health, Education, and Welfare.

Dr. Rechcigl is a member of some 30 scientific and professional societies, including being a Fellow of the American Association for the Advancement of Science, Fellow of the Washington Academy of Sciences, Fellow of the American Institute of Chemists, and Fellow of the International College of Applied Nutrition. He holds membership in the Cosmos Club, the Honorary Society of Phi Kappa Pi, and the Society of Sigma Xi, and is recipient of numerous honors, including an honorary membership certificate from the International Social Science Honor Society Delta Tau Kappa. In 1969, he was a delegate to the White House Conference on Food, Nutrition, and Health and in the last two years served as President of the District of Columbia Institute of Chemists and a Councilor of the American Institute of Chemists.

His bibliography extends over 100 publications, including contributions to books, articles in periodicals, and monographs in the fields of nutrition, biochemistry, physiology, pathology, enzymology, and molecular biology. Most recently he authored and edited *World Food Problem: A Selective Bibliography of Reviews* (CRC Press, 1975), *Man, Food, and Nutrition: Strategies and Technological Measures for Alleviating the World Food Problem* (CRC Press, 1973), *Food, Nutrition and Health: A Multidisciplinary Treatise Addressed to the Major Nutrition Problems from a World Wide Perspective* (Karger, 1973), following his earlier pioneering treatise on *Enzyme Synthesis and Degradation in Mammalian Systems* (Karger, 1971), and that on *Microbodies and Related Particles. Morphology, Biochemistry and Physiology* (Academic Press, 1969). Dr. Rechcigl also has initiated and edits a new series on Comparative Animal Nutrition and is Associated Editor of *Nutrition Reports International.*

CONTRIBUTORS
SECTION D: NUTRITIONAL REQUIREMENTS
VOLUME I

Vernon Ahmadjian
Department of Biology
Clark University
Worcester, Massachusetts

Daniel K. Baker
College of Fisheries
University of Washington
Seattle, Washington

Michael Balls
Department of Human Morphology
University of Nottingham Medical School
Nottingham NG7 2UH, England

George W. Brown, Jr.
College of Fisheries
University of Washington
Seattle, Washington

Thomas E. Bucsko
College of Fisheries
University of Washington
Seattle, Washington

Irvine L. Burger
College of Fisheries
University of Washington
Seattle, Washington

Brendan J. Coffey
College of Fisheries
University of Washington
Seattle, Washington

Walter A. Cooke
College of Fisheries
University of Washington
Seattle, Washington

Gary D. Cortner
College of Fisheries
University of Washington
Seattle, Washington

Roland A. Coulson
Department of Biochemistry
Louisiana State University Medical
    Center
New Orleans, Louisiana

David W. T. Crompton
The Molteno Institute of Biology
    and Parasitology
University of Cambridge
Cambridge, England

R. H. Dadd
Division of Entomology and Parasitology
Agricultural Experiment Station
College of Agricultural Sciences
University of California, Berkeley
Berkeley, California

Gary Dalton
College of Fisheries
University of Washington
Seattle, Washington

David Dean
Ira C. Darling Center for Research,
    Teaching and Service (The Marine
    Laboratory)
University of Maine at Orono
Walpole, Maine

A. E. DeMaggio
Department of Biological Sciences
Dartmouth College
Hanover, New Hampshire

Kathleen D. Edwards
College of Fisheries
University of Washington
Seattle, Washington

Peter Fay
Department of Botany and Biochemistry
Westfield College (University of London)
London, England

CONTRIBUTORS
SECTION D: NUTRITIONAL REQUIREMENTS
VOLUME I

Lucienne Fenaux
 Station Zoologique
 University of Paris
 Villefrance-sur-Mer, France

Harry Gooder
 Department of Bacteriology and
   Immunology
 University of North Carolina at Chapel
   Hill
 Chapel Hill, North Carolina

William D. Gray
 512 North High Street, Apartment E
 Lancaster, Ohio

M. Gross
 College of Fisheries
 University of Washington
 Seattle, Washington

Ralph B. L. Gwatkin
 Merck Institute for Therapeutic
   Research
 Rahway, New Jersey

Timothy Hansen
 College of Fisheries
 University of Washington
 Seattle, Washington

Janis L. Hastings
 Scripps Institute of Oceanography
 University of California, San Diego
 La Jolla, California

Wyrta Heagy
 Department of Molecular Biology and
   Biochemistry
 School of Biological Sciences
 University of California, Irvine
 Irvine, California

George D. Hegeman
 Department of Microbiology
 Indiana University
 Bloomington, Indiana

W. F. Hieb
 Division of Cell and Molecular Biology
 Faculty of Natural Sciences and Mathe-
   matics
 State University of New York at Buffalo
 Buffalo, New York

Kiyoshi Higuchi
 Microbiological Associates
 Biggs Ford Road
 Walkersville, Maryland

James C. Hoeman
 College of Fisheries
 University of Washington
 Seattle, Washington

Douglas J. Jackson
 Department of Zoology
 Ramsay Wright Zoological Laboratories
 University of Toronto
 Toronto, Ontario, Canada

Michel Jangoux
 Department of Zoology
 Free University of Brussels
 Brussels, Belgium

H. George Ketola
 U.S. Department of the Interior
 Fish and Wildlife Service
 Tunison Laboratory of Fish Nutrition
 Cortland, New York

Margo Krasnoff
 Department of Biological Sciences
 Dartmouth College
 Hanover, New Hampshire

John M. Lawrence
 Department of Biology
 University of South Florida
 Tampa, Florida

CONTRIBUTORS
SECTION D: NUTRITIONAL REQUIREMENTS
VOLUME I

Howard M. Lenhoff
Department of Developmental and Cell
Biology
School of Biological Sciences
University of California, Irvine
Irvine, California

J. F. Loneragan
School of Environmental and Life Sciences
Murdoch University
Murdoch, Western Australia

T. D. Luckey
Biochemistry Department
College of Agriculture and School of
Medicine
University of Missouri-Columbia
Columbia, Missouri

Barbara A. Manz
College of Fisheries
University of Washington
Seattle, Washington

D. F. Mettrick
Department of Zoology
Ramsay Wright Zoological Laboratories
University of Toronto
Toronto, Ontario, Canada

Marjorie A. Monnickendam
Department of Clinical Ophthalmology
Institute of Ophthalmology
London, England

Alison A. Newton
Department of Biochemistry
University of Cambridge
Tennis Court Road
Cambridge, England

D. J. D. Nicholas
Department of Agricultural Biochemistry
Waite Agricultural Research Institute
University of Adelaide
Glen Osmond, South Australia

L. C. Norris
Department of Avian Sciences
Agricultural Experiment Station
College of Agricultural and Environmental
Sciences
University of California, Davis
Davis, California

Richard A. Ormsbee
U.S. Department of Health, Education, and
Welfare
Public Health Service
National Institutes of Health
National Institute of Allergy and Infectious
Diseases
Rocky Mountain Laboratory
Hamilton, Montana

Erkki Oura
Research Laboratories of the State
Alcohol Monopoly (Alko)
Helsinki, Finland

L. A. Page
U.S. Department of Agriculture
Agricultural Research Service
North Central Region
National Animal Disease Center
P.O. Box 70
Ames, Iowa

C. O. Patterson
Department of Microbiology
Indiana University
Bloomington, Indiana

Henry M. Reiswig
Redpath Museum
McGill University
Montreal, Quebec, Canada

CONTRIBUTORS
SECTION D: NUTRITIONAL REQUIREMENTS
VOLUME I

Alan W. Rodwell
Commonwealth Scientific and Industrial
    Research Organization
Division of Animal Health — Animal Health
    Research Laboratory
Parkville, Victoria, Australia

Melissa Millam Stanley
Biology Department
George Mason University
Fairfax, Virginia

Heikki Suomalainen
Research Laboratories of the State
    Alcohol Monopoly (Alko)
Helsinki, Finland

William H. Thomas
Scripps Institute of Oceanography
University of California, San Diego
La Jolla, California

Indra K. Vasil
Department of Botany
University of Florida
Gainesville, Florida

Pran Vohra
Department of Avian Sciences
Agricultural Experiment Station
College of Agricultural and Environmental
    Sciences
University of California, Davis
Davis, California

Alan A. Wright
Postgraduate School of Studies in Biological
    Sciences
University of Bradford
Bradford, West Yorkshire, England

TABLE OF CONTENTS
SECTION D: NUTRITIONAL REQUIREMENTS
VOLUME I

To my inspiring teachers at Cornell: Harold H. Williams, John K. Loosli, Richard H. Barnes, the late Clive M. McCay, and the late Leonard A. Maynard.
And to my supportive and beloved family: Eva, Jack, and Karen.

Comparative Requirements of Organisms

COMPARATIVE NUTRITION

T. D. Luckey

What is nutrition? This question is best answered from a philosophic viewpoint that integrates all aspects of nutrition, and incidentally highlights the inseparability of nutrients, metabolite, and toxicant. Nature (the genetic base) and nurture (the environment) are the two interacting forces that form each biologic unit (organelle, virus, cell, organ, organism, society, or biota) of any geographic area (Figure 1). Nurture, nature, and time are the trinity of life. Although nature can be manipulated directly and indirectly by environmental factors, it will not be discussed here. An understanding of all components of nurture places nutrition in philosophic perspective and helps define nutrition from a cellular viewpoint. Materials are taken into cells without regard to categories such as nutrient, waste product, drug, or toxicant. This concept defines comparative nutrition and provides a basis for specific nutritional considerations. Qualitative nutrient requirements will be emphasized.

Nurture is an interacting triad of physical, chemical, and biological components. The physical component of the environment includes mass, energy, gravity, density, magnetism, neutral and electromagnetic radiation, and chemical composition. The biological component of nurture is composed of self (intraorganism) and nonself (interorganism), and includes interactions with both psychological and sociological forces. The chemical component of nurture encompasses nutrition. Nutrition is that part of the chemical component of the environment taken into the organism. This includes nonnutrients and other organisms (e.g., bacteria) as well as nutrients. Nonnutrients include growth promotants (and other hormetins), drugs, semiochemicals (including pheromones), aromas, carcinogens, mutagens, teratogens, toxicants, toxins, and inert substances. Nutrients are utilizable substances taken into an organism. Essential nutrients satisfy a need for elements and metabolites that cannot be synthesized by the organism at a rate commensurate with health and optimum physiological, psychological, and sociological functions. Food is that material ingested intentionally to fulfill life functions, and consists of gases, liquids, and solids. Authors writing about vertebrate nutrition rarely discuss oxygen, nitrogen, hydrogen, or water as nutrients. Diverse government agencies regulate carbonated, alcoholic, or caffeine containing beverages. Chewing gum, snuff, ointments, smoke, and other pollutants are generally ignored by nutritionists. However, a given quantity of a chemical affects our cells in the same way, whether it entered the organism by inhalation, ingestion, or inunction (Figure 2); some chemicals are synthesized by microbes living in the host. The concentration and physiochemical properties of each compound are important; whether the compound is considered to be a nutrient, a metabolite, or a toxicant indicates a man-made compartmentalization.

The organism is an integrated unit; physical, chemical, and intraorganism components of the environment (Figure 1) dramatically affect the utilization of a nutrient or the response to a drug. They rarely alter the qualitative nutrient requirements of the organism. Examples include the effect of temperature upon the quantitative requirements for energy and thiamine, the effect of light upon the qualitative requirements of *Euglena,* the effect of oxygen or an organic carbon source for facultative anaerobes or lithotrophic organisms, and the effect of stress upon quantitative food utilization. Interorganism relationships include those with the myriad microbes that contact the surfaces of most organisms. Microbes provide a governing influence in the environment in which the metazoans evolve, and they often contribute a final insult to dying organisms. The appearance of metazoans with alimentary tracts provide new microbic habitats in the world.[16,17] The nurture of the host organism may include more microbes than host cells.

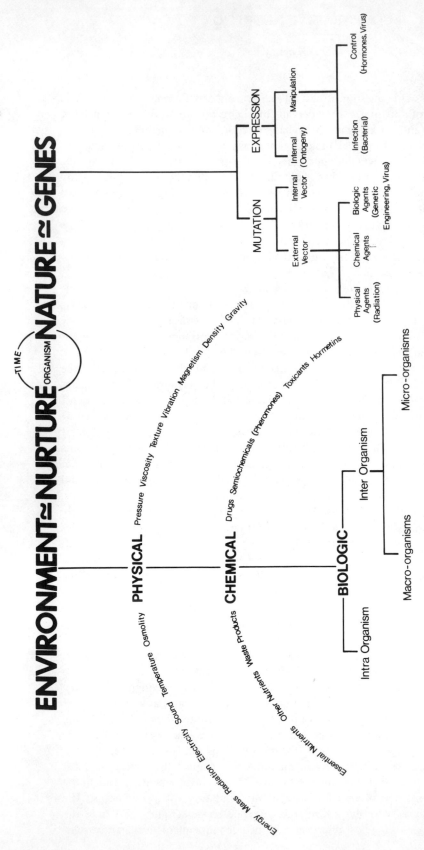

FIGURE 1.   Nutrition as the chemical component of nurture. (From Luckey, T. D. and Venugopal, B., *Metal Toxicity in Mammals, Physiologic and Chemical Basis for Toxicity*, Vol. 1, Plenum, New York, 1976. With permission.)

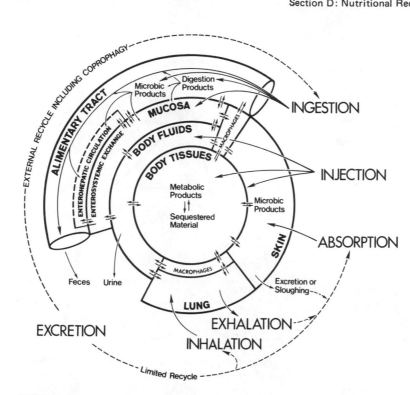

FIGURE 2.   Organism-environment interchange showing routes for nutrient entry. (From Luckey, T. D. and Venugopal, B., *Metal Toxicity in Mammals. Physiologic and Chemical Basis for Toxicity,* Vol. 1, Plenum, New York, 1976. With permission.)

Intimate relations include parasitism, e.g., infections; commensalism, e.g., microbes in the alimentary tract; and mutualism, e.g., in ruminants and in root nodules. Mutualism often reduces qualitative nutrient requirements of the host species. The synthesis of vitamins by the alimentary flora negate some dietary requirements of the host.

Vitamins and the trace metals are required in only catalytic amounts. However, depending upon their concentration, essential nutrients may also act as stimulants, chemicals undergoing stoichiometric reactions, or toxicants. Excess quantities react in an organism or cell quite differently than satisfying a nutritional need. The complete biologic activity spectrum for different compounds is categorized in Figure 3. The different responses depend upon the inherent character of the compound and the condition of the organism as well as dosage and period of observation. For example, vitamin C is an essential nutrient for certain animals and a stimulant for others. It stimulates growth when fed to other animals that can synthesize it in adequate quantities because it is a hormetin, a stimulant.[4,7,10] It also is a reducing compound that can affect reactions stoichiometrically. High doses of vitamin C are toxic, as stated by Doft[10] and confirmed by Luckey.[23] Rats fed 2 to 3% vitamin C show diarrhea, and some die; at 4 to 6% of the diet, vitamin C is clearly toxic, and most rats die within a few weeks.

When is a nutrient harmful? Luckey and Venugopal[18,21] suggested an arbitrary means to set limits (Figure 4). The contribution of the compound is balanced against the harm at different concentrations, with as full a knowledge of interactions as possible. The question can be reversed. When is a toxicant acceptable in the diet? Answers to these are clear from examples: the well-known growth stimulus from dietary antibiotics[12] and the little-understood growth stimulation in insects fed diluted insecticides.[15] When the concentration of a toxicant provides a stimulus to increase homeostasis or to provide increased health under stressful conditions, that material is not toxic at that dosage for

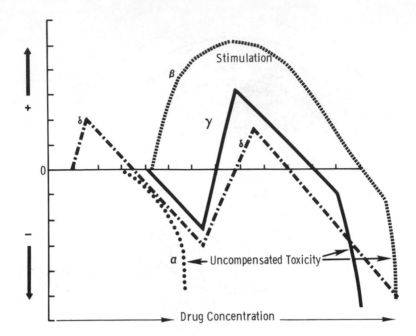

FIGURE 3.   The four types of dose-response relationships found for all drugs in three pharmacology texts. (From Townsend, J. F. and Luckey, T. D., *JAMA*, 173, 44–48, 1960. With permission. Copyright 1960, American Medical Association.)

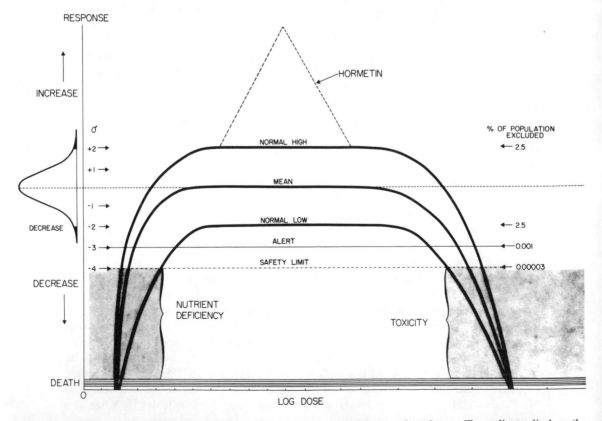

FIGURE 4.   Definition of limits of alert and safety for essential nutrients and toxicants. The ordinate displays the distribution of the population and the proportion excluded by 1,2,3, and 4 standard deviations from the mean. (From Luckey, T. D. and Venugopal, B., *Metal Toxicity in Mammals. Physiologic and Chemical Basis for Toxicity*, Vol. 1, Plenum, New York, 1976. With permission.)

the parameters measured. Obviously, meaningful parameters must be used. The highest concentration that gives reactions equivalent to controls is the zero equivalent point (ZEP). This dose should be determined in the presence of a variety of modern stimulants and stresses to suggest safety limits for toxicants and drugs in foods for jurisdictional and regulatory purposes.

The general rationale for nutrient requirements is simple. First, all living organisms require 20 to 30 inorganic elements. Those considered to be essential or stimulatory to vertebrates are indicated in Figure 5. Second, the major metabolic machinery is similar for all living organisms; qualitative nutritional requirements for organic molecules are inversely proportional to the biosynthetic capabilities of the organism. Some organisms (lithotrophs) can synthesize all their metabolites from inorganic substrates. Others cannot synthesize 1 to 20 different metabolites, which then are nutritional requirements for these organisms. Examples illustrate the relation between biosynthetic potential and nutritional requirements. Available information regarding ascorbate synthesis correlates well with the requirement for vitamin C in different groups of organisms (Table 1). Chatterjee[6] found vitamin C is synthesized in the liver by mammals and in the kidney by other animals. When fruit-eating birds and animals lost the enzyme that converts gulonic acid to vitamin C, they became restricted to a diet containing this nutrient. Another example of the inverse relationship between biosynthetic potential and nutrient requirement relates the utilization of precursors to nutritional need (Table 2): Some bacteria can make niacin from inorganic sources while other organisms cannot; some require different organic precursors; some can make no part of the vitamin; and *Hemophilus* can utilize only the coenzyme forms.

The knowledge that small amounts of many compounds are stimulatory reopens the question of what is an essential nutrient. Observations of increased growth, enhanced

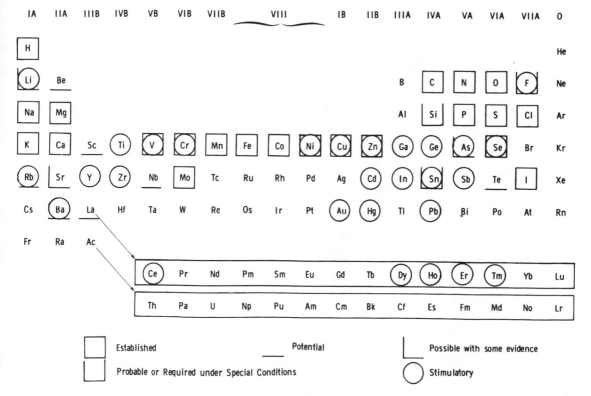

FIGURE 5.   Essential and stimulatory elements in nutrition of animals. (From Luckey, T. D., *Environ. Qual. Saf.*, Suppl., 81–103, 1975. With permission.)

Table 1
PHYLOGENY OF VITAMIN C REQUIREMENT

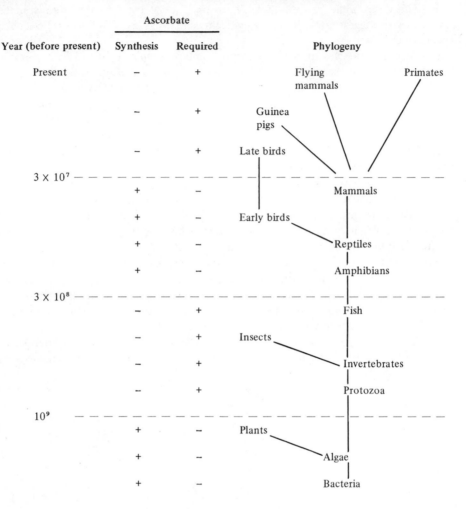

Year (before present)	Ascorbate Synthesis	Ascorbate Required	Phylogeny
Present	–	+	Flying mammals Primates
	–	+	Guinea pigs
	–	+	Late birds
3×10^7	+	–	Mammals
	+	–	Early birds
	+	–	Reptiles
	+	–	Amphibians
3×10^8	–	+	Fish
	–	+	Insects
	–	+	Invertebrates
	–	+	Protozoa
10^9	+	–	Plants
	+	–	Algae
	+	–	Bacteria

physiologic funtions, or simple changes in metabolic processes do not necessarily make the compound an essential nutrient. Criteria to evaluate the acceptance of a nutrient as essential are expressed below, with increased importance related to the magnitude of the number. Deletion or omission of the material from the diet leads to:

1. Increased requirement for another nutrient,
2. Biochemically significant change in tissue composition,
3. Decreased growth, or increased morbidity and/or mortality under special conditions,
4. Subclinical disability,
5. Functional disability under special conditions,
6. Decreased appetite or clinical disability,
7. Increased morbidity,
8. Retarded growth,
9. Inability to function normally (e.g., reproduction), or
10. Death.

A rough estimate of credibility for an essential nutrient may be obtained by adding the

Table 2

COMPARATIVE NIACIN REQUIREMENT

Compound	Azotobacter sp.	Neurospora crassa	Proteus vulgaris	Staphylococcus aureus	Dog	Chick	Lactobacillus arabinocis	Man	Hemophilus parainfluenza
N₂	+	−	−	−	−	−	−	−	−
NO₃⁻	+	+	−	−	−	−	−	−	−
Pyridine	+	+	−	−	−	−	−	−	−
Trigonellin	+	+	−	−	−	−	−	−	−
4-Pyridinecarboxylic acid	+	+	+	−	−	−	−	−	−
Me-nicotinate	+	+	+	+	+	−	+	0[a]	−
Nicotinuric acid	+	+	+	+	+	0	+	0	−
Tryptophan	+	+	+	+	+	+	−	+	−
Nicotinic acid	+	+	+	+	+	+	+	+	−
NAD or NADP	+	+	+	+	+	+	+	+	+

[a] Not tested.

numbers for which any substance has shown specificity for alleviating or preventing each criterion. The quantitative values obtained would be high for essential nutrients.

An overview of comparative nutrition can best be obtained from the perspective of evolution (Table 3). No individual and no phylum exists separately; all are related to each other and the earth, and their development is a continum through the earth to the evolution of the solar system and stars. Each individual interrelates with the contemporary occupants of his habitat and has a meaningful phylogenetic relationship to past occupants. This continuity is reflected in the unity of biochemistry[2,11] and provides a basis for generalizations in nutrition. The unity of life begins with a common supply of building material from the earth's crust and primordial soups. Common metabolic pathways for a variety of purposes, the universal DNA code, and the restricted use of D amino acids and L sugars are salient testimony for a strong concept of unity in comparative nutrition.[3]

Current concepts of the origin of life and biochemical unity throughout evolution indicate that changes in qualitative nutrient requirements are confined to very few major periods. The basic structure of living organisims was established by the first living organisms that exhibited genetic continuity (Table 3). Cell structures include compounds comparable to those found in contemporary organisms of all phyla: soluble salts and ions, simple organic compounds, complex lipids, polysaccharides, proteins, nucleic acids, high energy phosphates, and metal complexes.

The first living organism probably had the least sophisticated biosynthetic machinery, and thus had extensive nutritional requirements. Other than parasites, some heterotrophic bacteria are the most fastidious organisms known. They need most of the elements required by vertebrates (Figure 5): a simple carbon source for energy, most B vitamins, 18 to 20 amino acids, and often purines, pyrimidines, and fatty acid. The first revolutionary change in nutritional requirements (Table 3) was the replacement of the need for organic compounds by enzyme systems, as exhibited by autotrophic bacteria and plants. Basic cell structures and common metabolic pathways were well developed by the time of the appearance of the photosynthetic, oxygen-producing blue-green alagae and the aerobic bacteria.[5] Interorganism communication of Protista was developed and finely tuned during metazoan evolution to include chemical, audible, and sometimes visual

Table 3
TIME PERSPECTIVE OF EVOLUTION

Years (before present)	Begin evolution
2×10^{10}	Atoms
1×10^{10}	Stars and galaxies
8×10^9	Earth
7×10^9	Modern geology
6×10^9	Prebiotic biochemistry
5×10^9	Coacervates and unstable living systems
4×10^9	Anaerobic bacteria, procaryotes
3×10^9	Blue-green algae
2×10^9	Aerobic bacteria
1×10^9	Eucaryotes, green algae
6×10^8	Marine invertebrates
5×10^8	Land plants and fish
4×10^8	Vascular plants and land invertebrates
3.5×10^8	Gymnosperms, ferns, and amphibia
3×10^8	Reptiles
2×10^8	Mammals
1×10^8	Birds
5×10^7	Primates

systems for effective social communication (Table 3); no new nutrients are required for these functions. Most morphologic and physiologic differences that lead to speciation and the uniqueness of genera, families, and phyla are not reflected in different qualitative nutrient requirements.

Basic energy utilization and transformation systems were developed by the anaerobic bacteria and the blue-green algae; refinements of mitochondria, chloroplasts, and nucleus developed in the green algae (Table 3). This was capped by the capability of the aerobic bacteria to utilize most organic compounds for energy, using both glycolysis and the tricarboxylic acid cycle. Similar adaptability is noted in anaerobic bacteria for both energy source material and electron and proton acceptors. Some anaerobic microbes are highly sensitive to oxygen toxicity; they must either produce molecular hydrogen or their fermentation demands a substitute substrate for molecular oxygen to be reduced (Table 4). When this moiety is lacking, growth and active metabolism stops quickly; thus, electron and proton acceptors are stoichiometric nutritional requirements. These accepters are special excretion nutrients, equivalent to oxygen for aerobic organisms, that can quickly limit growth of anaerobic bacteria in a variety of ecologically interesting habitats. This proton acceptor is often a nonspecific nutrient, probably depending more upon the oxidation-reduction potential than upon chemical composition. Thus, while aerobic organisms have a nonspecific requirement for the material being oxidized, with oxygen being reduced, both sides of the energy reaction are nonspecific for many anaerobic bacteria. Doetsch and Cook[9] summarize the utilization of inorganic materials for electron acceptors by different bacteria (Table 4). Note that the compound may release oxygen which accepts the proton to form H_2O in the absence of molecular oxygen. Under certain conditions, these acceptors are specific nutrient requirements; they are probably the limiting factor for microbic growth in the intestine.

The acquisition of structural rigidity in animals suggests the possible imposition of either quantitative or qualitative metal requirements. Examples in animals were described by Luckey:[19] Si may be required as an initiator for bone formation in vertebrates; Ca in

Table 4
INORGANIC COMPOUNDS AND IONS AS
FINAL ELECTRON ACCEPTORS FOR BACTERIA

Terminal electron acceptor	End product(s)	Bacterium
NO_3^-	NH_4^+, NO_2^-	*Escherichia coli*
NO_3^-	N_2	*Pseudomonas dentrificans*
NO_3^-	$N_2 + N_2O$	*P. aeruginosa*
NO_3^- or NO	N_2O	*Corynebacterium nephridii*
SO_4^{2-}	H_2S	*Desulfovibrio desulfuricans*
CO_2	CH_4	*Methanosarcina barkeri*
CO	CH_4	*M. barkeri*
$HTeO_3^-$	Te^0	*Staphylococcus aureus*
$HSeO_3^-$	Se^0	*Salmonella typhosa*
$S_4O_6^{2-}$	$S_2O_3^{2-}$	*S. typhosa*
H_2VO_4	$VO(OH)_2$	*Veillonella alcalescens* (Micrococcus lactilyticus)
$UO_2(OH)_2$	$U(OH)_4$	*V. alcalescens*
Se^0	Se^{2-}	*V. alcalescens*
S^0	S^{2-}	*V. alcalescens*
Au^{3+}	Au^0	*V. alcalescens*

From Doetsch, R. N. and Cook, T. M., *Introduction to Bacteria and Their Ecobiology*, University Park Press, Baltimore, © 1973. With permission.

Coelenterata; Sr, Ba, and Ca in Mollusca, Ca and Mg in Echinodermata; and Ca and P in Arthropoda and Vertebrata. The metals in respiratory pigments could be viewed in like manner: Fe in Brachiopoda, Sipunula, Echiura, Annelida, and Vertebrata; Cu in Mollusca, Crustacea, and Arachinida; Va in Cephalachorda; and Mg in plants. The acquisition of other physiologic functions such as circulation of blood and lymph, complex nervous and endocrine control systems, specialized skin, and complex memory do not impose requirements for specific nutrients.

In some respects, animals resemble the original life forms more than do plants; they lack photosynthesis and exhibit heterotrophic nutritional requirements. These and mobility are associated with major differences in qualitative nutritional requirements between plants and animals. All animals require dietary organic energy sources, preformed vitamins, and amino acids; plants do not. The number of amino acids required depends upon species and individual ontogeny.[25] The metabolic efficiency gained by not synthesizing ten essential amino acids (about 70 enzymes) and all the vitamins (roughly 100 enzymes) may be utilized to develop mobility and complex sensing and response mechanisms. The constant presence of such organic compounds in the diet allowed animals to develop without the metabolic machinery to synthesize them. The example of vitamin C suggests how this principle functions in evolution (Table 1). The occasional loss of enzymes to synthesize or to process specific metabolites is normal genetic mutation in the continual probing of phylogenetic development to adapt to the ever-changing environment. Sometimes a known compound can compensate for an enzymatic loss by individuals; that compound is a unique qualitative nutrient requirement for individuals exhibiting that metabolic abberation.

The nutritional requirement for fat-soluble vitamins is phylogenetically a late acquisition. Few Protista or invertebrates and no plants require these vitamins from extraneous sources. In the Paleozoic era, about 0.5 billion years before present,[28] animals changed dramatically to a radiant evolution of animals that had extensive nervous and endocrine systems, energy store as phosphocreatine, and a requirement for fat-soluble vitamins.

Within the unity of building material and major metabolic pathways among all living organisms is diversity. The basis for nutritional diversity within the bounds of biochemical unity was explored phylogenetically.[19] This is discussed with examples to show the integration between organic nutrient requirements and biosynthetic potential of organisms.

The diversity of nutritional requirements, e.g., that exhibited by different organisms for niacin and vitamin C, remains within the metabolic unity of life. Certain compounds are needed to catalyze or control metabolic reactions; whether they are synthesized or eaten is less pertinent to metabolism than the fact that they are present in the cell. Other than the essential elements and energy sources, the compounds required in the diet are active metabolites or their precursors. Consequently, it is possible to devise a single diet that should be satisfactory for organisms in any phyla. The diet given in Table 5 was satisfactory for microbes, plants, invertebrates, and vertebrates.[13] Failures were attributed to peculiar eating habits, one or more compounds being present in toxic amounts, or deficiency of some nutrient not then appreciated. An updated version is given in Table 6.

Ontogeny provides further diversity of nutritional requirements within the frame of metabolic unity. The fertilized cell is literally a parasite of the mother, or her produce, i.e., the egg or endosperm. The egg cell has the genetic potential to carry out all the reactions of the mature individual, but lacks the differentiated maturity or capacity for many reactions or functions; it must be nurtured to develop its potential. Consequently, the nutritional requirements of immature organisms are frequently more numerous than those of mature individuals; examples are choline in the rat and arginine in man. Further,

Table 5
COMPOSITION OF DIETS

Ingredient (g/kg)	Diet number		
	U-1	U-2	U-3
Casein (Labco)	300	300	300
Glucose			480
Starch (corn)	300	200	
Alpha cell	120	60	30
Cellophane spangles		60	30
Sucrose	120	120	
Oil (corn)	80	80	80
$Na_2MoO_4 \cdot 2H_2O$		0.30	0.30
Vitamin A (IU)	10,000	10,000	10,000
Vitamin D (IU)	2,000	2,000	2,000
Vitamin E	0.30	0.30	0.30
Vitamin K (menadione)	0.010	0.010	0.010
Vitamin C	10	10	10
Thiamine	0.020	0.020	0.020
Riboflavin	0.020	0.020	0.020
Pyridoxine	0.020	0.020	0.020
Ca pantothenate	0.050	0.050	0.050
Niacin	0.100	0.100	0.100
Biotin	0.001	0.001	0.001
Folic acid	0.020	0.020	0.020
Choline chloride	2	2	2
Vitamin B_{12}	0.0005	0.0005	0.0005
i-Inositol	2	2	2
Starch (corn) as vitamin carrier	5	5	5
Salts L-II[a]	60	60	60

[a] Salts composition:

$CaCO_3$	18.0
$CaHPO_4$	3.3
K_2HPO_4	13.5
$NaHPO_4$	12.0
NaCl	3.0
KI	0.045
$MgSO_4 \cdot 7H_2O$	4.5
$MnSO_4 \cdot 4H_2O$	0.75
$Fe(C_6H_5O_7)_2$	4.5
$CuSO_4 \cdot 5H_2O$	0.23
$CoCl_2 \cdot 6H_2O$	0.03
$ZnSO_4 \cdot 7H_2O$	0.06
$Na_2B_4O_7 \cdot 10H_2O$	0.03
$AlK(SO_4)_2 \cdot 12H_2O$	0.045

In addition to Salts L-II:

MgO	4
K acetate	20

From Luckey, T. D., *Comp. Biochem. Physiol.*, 2, 100–124, 1961. With permission.

Table 6
SYNTHETIC DIET

Ingredient	%	Ingredient	mg/100 g
Glucose	6.0	Carotene	0.6
Tripalmitin	3.0	Vitamin D_3	0.05
Ethyl linolate	1.3	α-Tocopherol	33
l-Arginine	1.8	Ascorbic acid	100
d,l-Histidine	0.2	2-Me-1,4-napthoquinone	2.0
d,l-Isoleucine	2.0	Thiamine	0.5
l-Leucine	2.0	Riboflavin	1.0
l-Lysine	1.8	Niacin	10.0
d,l-Methionine	2.0	Ca pantothenate	2.0
d,l-Phenylalanine	2.0	Pyridoxine	1.0
d,l-Threonine	2.0	Choline	100
l-Tryptophan	0.5	Inositol	100
d,l-Valine	2.0	Biotin	0.05
Other amino acids	6.4	Folic acid	0.05
$NaHCO_3$	1.0	Vitamin B_{12}	0.01
NaCl	0.5	$Cr_2(SO_4)_3$	0.02
Na_2HPO_4	1.5	Na_2SeO_3	0.02
K_2HPO_4	1.4	Na_2MoO_3	0.02
$CaHPO_4$	0.3	Na_4SiO_4	0.02
$CaCO_3$	1.8	Na_3VO_4	0.02
$MgSO_4$	0.5		
$MnSO_4$	0.1		
Fe citrate	0.5		
$CuSO_4$	0.02		
$CoCl_2$	0.01		
$ZnSO_4$	0.01		
KI	0.005		

the acquisition of the alimentary microflora following birth normally reduces qualitative requirements for certain vitamins, e.g., vitamin K, biotin, and others.[14]

The diversity provided by conditioning factors of nutrition, e.g., heat, cold, excess of nutrients, drugs or stress, and biologic variation, is less significant for qualitative than for quantitative requirements. Differentiation of cell types changes qualitative nutrient requirements; this is well illustrated by comparing the nutrient requirements of roots and leaves from a single plant.

Comparative nutrition may be described with different limits of knowledge which range from descriptive enumeration of foods eaten,[8] quantitative analysis of all ingredients,[24] or the feeding of a chemically defined diet[22] to the qualitative determination of specific nutrients required for each species or groups of organisms (Table 7). This compilation of qualitative nutritional requirements was taken from Beerstecher,[3] Altman and Dittmer,[1] Florkin and Scheer,[11a] NRC Allowances,[26,27] and Luckey.[19]

Table 7
QUALITATIVE DIETARY REQUIREMENTS[a]

	Vertebrate					Invertebrate			Protista				Plant	
	Man	Ruminant (cow)	Rodent (rat/mouse)	Bird (chick)	Fish (Salmon)	Diptera (house fly)	Coleoptera (flour beetle)	Orthoptera (cockroach)	Tetrahymena pyriformis	Leuconostoc mesenteroides	Salmonella typhosa	Yeast	Spermatophytes (seed plants)	Algae
Energy Source														
Light	N	N	N	N	N	N	N	N	N	N	N	N	R	R
Inorganic	N	N	N	N	N	N	N	N	N	N	N	N	N	N
Organic	R	R	R	R	R	R	R	R	R	R	R	R	N	N
Metals														
Group I														
Na	R	R	R	R	R	R	R	R	R	R	R	R	R	R
K	R	R	R	R	R	R	R	R	R	R	R	R	R	R
Cu	R	R	R	R	R	R	R	R	R	R	R	R	R	R
Group II														
Mg	R	R	R	R	R	R	R	R	R	R	R	R	R	R
Ca	R	R	R	R	R	R	R	R	R	R	R	R	R	R
Zn	R	R	R	R	R	R	R	R	R	R	+	R	R	+
Group V														
V			+		+								R	R
Group VI														
Cr	+	+	+	+	+									
Mo	R	R	R	R	+	+	+	+	+	+	+	+	R	R
Group VII														
Mn	R	R	R	R	R	+	+	R	R	R	R	R	R	R
Group VIII														
Fe	R	R	R	R	R	R	R	R	R	R	R	R	R	+
Co	R	R	R	R	+	+	+	+	+	+	+	R	+	R
Ni	+	+	+	+	+	+	+	+	+					
Nonmetals														
H	R	R	R	R	R	R	R	R	R	R	R	R	R	R
B	-	-	-	-	-	-	-	-	-	-	-		R	R
C	R	R	R	R	R	R	R	R	R	R	R	R	R	R
Si	+	+	+	+	+								R	+
N	R	R	R	R	R	R	R	R	R	R	R	R	R	R
P	R	R	R	R	R	R	R	R	R	R	R	R	R	R
O	R	R	R	R	R	R	R	R	R	R	R	R	R	R
S	R	R	R	R	R	R	R	R	R	R	R	R	R	R
Se	+	R	R	R	+	+	+	+	+		s	s		

[a] Periphery material of comparative nutrition is not compiled. This includes special requirements within special groups, as the many supplementary compounds used for bacteria, or isolated requirements: carnitine for larvae of certain beetles (e.g., Tenebrionidae), glutathione for Hydra, and hematin for bloodsucking insects. Symbols indicate the following: upper case letter = generally required by the group; lower case letter = one or more members of the group require it, but not all; R, r = required for growth; N, n = not required; u = utilized; s = stimulatory; * = supplied by alimentary microbes; + = probably required according to conservative estimate of compiler; − = probably not required according to conservative estimate of compiler.

Table 7 (continued)
QUALITATIVE DIETARY REQUIREMENTS[a]

	Vertebrate					Invertebrate			Protista				Plant	
	Man	Ruminant (cow)	Rodent (rat/mouse)	Bird (chick)	Fish (salmon)	Diptera (house fly)	Coleoptera (flour beetle)	Orthoptera (cockroach)	*Tetrahymena pyriformis*	*Leuconostoc mesenteroides*	*Salmonella typhosa*	Yeast	Spermatophytes (seed plants)	Algae
Nonmetals (continued)														
F	u	–	u	–	–	–	–	–	–	–	–	–	–	–
Cl	R	R	R	R	R	+	+	+	+	+	R	+	+	+
Br	–	–	–	–	–	–	–	–	–	–	–	–	–	–
I	R	R	R	R	+	s					s	s		
Amino Acids[b]														
Glycine	N	N	N	R	N	N	N	N	N	R	N	N	N	N
Alanine	N	N	N	N	N	N	N	R?	N	N	N	N	N	N
Valine	R	R	R	R	R	R	R	R	R	R	N	N	N	N
Leucine	R	R	R	R	R	R	R	R	R	R	N	N	N	N
Isoleucine	R	R	R	R	R	R	R	R	R	R	N	N	N	N
Serine	N	N	N	N	N	N	R	N	R	N	N	N	N	N
Threonine	R	R	R	R	R	R	R	N	R	R	N	N	N	N
Aspartate	N	N	N	N	N	N	N	N	N	R	N	N	N	N
Glutamate	N	N	N	N	N	N	N	N	N	R	N	N	N	N
Cyst(e)ine	N	R	N	N	N	r	N	R?	N	R	N	N	N	N
Methionine	R	R	R	R	R	r	R	R	R	R	N	N	N	N
Lysine	R	R	R	R	R	R	R	R	R	R	N	N	N	N
Arginine	R	R	R	R	R	R	R	R	R	R	N	N	N	N
Histidine	R	R	R	R	R	R	R	R	R	R	N	N	N	N
Tryptophan	R	R	R	R	R	R	R	R	R	R	R	N	N	N
Phenylalinine	R	R	R	R	R	R	R	N	R	R	N	N	N	N
Tyrosine	N	R	N	N	N	N	N	N	N	R	N	N	N	N
Proline	N	N	N	N	N	r	N	R	N	R	N	N	N	N
Lipid														
Linolenate	R	+	R	R	+							–	–	–
Vitamin A	R	R	R	R	R	R	u	N	N	N	N	N	N	N
Vitamin D	R	R	R	R	R	R	–	–	–	N	N	N	N	N
Vitamin E	R	R	R	R	r				–	–	–	–	–	
Vitamin K	*	*	*	R					–	–	–	–	–	–
B Vitamins														
Biotin	*	*	R	R	R	R	R	R	N	R			N	N
Deoxybiotin			u	u	u									
Choline		*	R	R	R		R	R	N		N	N	N	N
Cobalamine	R	*	R	R	R	N?		R					R	r
Folate	R	*	R	R	R	R	R	R	R	R			N	N
Inositol	–	–	–	–	–	–	N	R						
Lipoic acid	N	N	N	N	N	N	n	n	R	–	N	N	N	N
Niacin	R	N	R	R	+	R	R		R	R	N	N	N	N
Pantothenate			R	R		R	R		R	R	N	N	N	N

b Ruminants may obtain all of their amino acids from their micro flora.

Table 7 (continued)
QUALITATIVE DIETARY REQUIREMENTS[a]

	Vertebrate					Invertebrate			Protista				Plant	
	Man	Ruminant (cow)	Rodent (rat/mouse)	Bird (chick)	Fish (salmon)	Diptera (house fly)	Coleoptera (flour beetle)	Orthoptera (cockroach)	Tetrahymena pyriformis	Leuconostoc mesenteroides	Salmonella typhosa	Yeast	Spermatophytes (seed plants)	Algae
B Vitamins (continued)														
Pyridoxine	R	*	R	R	R	R	R	R	R	N	N	N	N	N
Pyridoxal	u	u	u	u	u	u	u	u	u					
Pyridoxamine	u	u	u	u	u	u	u	u	u					
Riboflavin	R	*	R	R	R	R	R	R	R	N		N	N	n
Thiamine	R	*	R	R	R	R	R	R		N			N	N
Other														
Carbohydrate	+	+	R	R	+	+	+	+	+	+		–	–	–
Ascorbate	R	–	–	–	–	+	–	–	–	–	–	–	N	N

SUMMARY

Nutrition is that part of the chemical component of nurture, the environment, which is taken into the organism. Nutrition interacts with the physical and biologic components of nurture to direct the ontogeny of each individual. All materials taken into the cell constitute the total nutrition of that cell. Thus, nutrition includes essential nutrients, other nutrients, and nonnutrients taken into the organism by all routes. The response to each compound is a function of the character and dosage of the compound. Excess of any nutrient is harmful, while minute quantities of many nonnutrients are stimulatory and help maintain adequate homeostasis for health and survival.

Comparative nutrition is locked into the unity of biochemistry. All cells exhibit a remarkable metabolic sameness to provide morphological structures and physiological functions essential for life. The essential elements utilized by organisms are those abundant in the earth's crust that have inherent characteristics suitable for the continuity of life. The organic moiety of cells has common basic building blocks, metabolites, polymers, and complexes. Some bacteria and most plants can utilize inorganic compounds to synthesize all their metabolic machinery. Some require an organic energy source plus inorganic material. Other bacteria and all animals require organic energy sources, vitamins, and amino acids; they have inadequate enzymes to synthesize these essential metabolites. Those biosynthetic lesions are the basis of qualitative nutritional requirements for certain organic compounds. These essential organic nutrients include compounds requiring many enzymes for their biosynthesis. This loss of enzymatic capacity imposes a dietary requirement to supply certain essential structures. Vitamins and the essential amino acids typify the exchange of biosynthesis potential for nutrient requirement.

Within the frame of biochemical unity, diversity in nutrition derives from phylogenetic, ontogenetic, and individual differences in the character of organisms. The genetic potential of each species has evolved in time to fit the ever-changing environment. The physical, chemical, and biological environment is a major factor in ontogenetic development and individual variations. The circle is complete: nutrition is a part of nurture that affects the genetic nature of the organism, which in turn determines qualitative nutritional requirements.

REFERENCES

1. **Altman, P. L. and Dittmer, D. S.,** *Biological Data Book,* Vol. 1, Federation of American Societies for Experimental Biology, Bethesda, Md., 1972.
2. **Baldwin, E.,** *An Introduction to Comparative Biochemistry,* 4th ed., Cambridge University Press, Cambridge, Engl., 1964.
3. **Beerstecher, E., Jr.,** The biochemical basis of chemical needs, in *Comparative Biochemistry,* Vol. 6, Florkin, M. and Mason, H. S., Eds., Academic Press, New York, 1964, 119–220.
4. **Briggs, G. M., Luckey, T. D., Elvhjem, C. A., and Hart, E. B.,** Effect of ascorbic acid on chick growth when added to purified rations, *Proc. Soc. Exp. Biol. Med.,* 55, 130–133, 1944.
5. **Brock, T. D.,** *Biology of Microorganisms,* Prentice-Hall, Englewood Cliffs, N.J., 1973.
6. **Chatterjee, I. B.,** Evolution and the biosynthesis of ascorbic acid, *Science,* 182, 1271–1272, 1973.
7. **Combs, G. F. and Pesti, G. M.,** Influence of ascorbic acid on selenium nutrition in the chick, *J. Nutr.,* 106, 958–966, 1976.
8. **Crawford, M. A., Ed.,** *Comparative Nutrition of Wild Animals,* Academic Press, New York, 1968.
9. **Doetsch, R. N. and Cook, T. M.,** *Introduction to Bacteria and Their Ecobiology,* University Park Press, Baltimore, 1973.
10. **Doft, F. S.,** Role of bacteria in prevention of pantothenic acid deficiency in rats fed penicillin and/or ascorbic acid, in *Proc. 5th Int. Congr. Nutr.,* American Institute of Nutrition, Washington, D.C., 1960, 5.
11. **Florkin, M. and Mason, H. S.,** *Comparative Biochemistry: A Comprehensive Treatise,* Vol. 1–7, Academic Press, New York, 1960–1965.
11a. **Florkin, M. and Scheer, B. T.,** *Chemical Zoology,* Vol. 1–6, Academic Press, New York, 1968–1973.
12. **Luckey, T. D.,** Antibiotics in nutrition, in *Antibiotics, Their Chemistry and Nonmedical Uses,* Goldberg, H. S., Ed., D. Van Nostrand, Princeton, N.J., 1959.
13. **Luckey, T. D.,** A study in comparative nutrition, *Comp. Biochem. Physiol.,* 2, 100–124, 1961.
14. **Luckey, T. D.,** Effects of microbes on germfree animals, *Adv. Appl. Microbiol.,* 7, 169–223, 1965.
15. **Luckey, T. D.,** Insecticide hormoligosis, *J. Econ. Entomol.,* 61, 7–12, 1968.
16. **Luckey, T. D.,** Introduction to the ecology of the intestinal flora, *Am. J. Clin. Nutr.,* 23, 1430–1432, 1970.
17. **Luckey, T. D.,** Introduction to intestinal microecology, *Am. J. Clin. Nutr.,* 25, 1292–1294, 1972.
18. **Luckey, T. D.,** Hormology with inorganic compounds, *Environ. Qual. Saf.,* Suppl., 81–103, 1975.
19. **Luckey, T. D.,** Introduction to comparative animal nutrition, in *Comparative Animal Nutrition,* Rechcigl, M., Ed., S. Karger, Basel, 1976, chap. 1.
20. **Luckey, T. D.,** Bicentennial overview of intestinal microecology, *Am. J. Clin. Nutr.,* 28, 1–8, 1977.
21. **Luckey, T. D. and Venugopal, B.,** *Metal Toxicity in Mammals. Physiologic and Chemical Basis for Toxicity,* Vol. 1, Plenum, New York, 1976.
22. **Luckey, T. D., Moore, P. R., Elvehjem, C. A., and Hart, E. B.,** Growth of chicks on purified and synthetic diets containing amino acids, *Proc. Soc. Exp. Biol. Med.,* 64, 348–351, 1947.
23. **Luckey, T. D.,** unpublished.
24. **Morrison, F. B.,** *Feeds and Feeding,* Morrison Publishing, Clinton, Iowa, 1959.
25. **Munro, H. N.,** *Mammalian Protein Metabolism,* Vol. 4, Academic Press, New York, 1969.
26. National Research Council, *Nutrient Requirements of Laboratory Animals,* National Academy of Sciences, Washington, D.C., 1972.
27. National Research Council, *Recommended Dietary Allowances,* National Academy of Sciences, Washington, D.C., 1974.
28. **Silverstein, A.,** *The Biological Sciences,* Rinehart, San Francisco, 1974.
29. **Townsend, J. F. and Luckey, T. D.,** Hormologosis in pharmacology, *JAMA,* 173, 44–48, 1960.

Microorganisms

CELLULAR FUNCTIONS REQUIRED FOR REPLICATION OF VIRUSES

A. A. Newton

Viruses can only replicate inside living cells, and they are entirely dependent on the host cell not only for all low molecular weight precursors but also for the supply of energy for virus-directed biosynthetic processes. In addition, viruses make use of many of the organelles and enzyme systems of the host cell in the processes of viral replication. Thus, all known viruses use the host cell system of ribosomes and enzymes of protein synthesis for the synthesis of viral proteins. The extent to which a virus relies on host cell functions varies and may be investigated by the use of specific inhibitors of the host cell enzyme or by employing mutants of the host cell deficient in the relevant function. Because few mutants of animal cells are available, little is known about such requirements for animal viruses.

Much of the specificity of a virus for a particular cell type results from the initial virus-cell interaction through recognition by the virus of specific receptors in the cell surface. These are listed in Table 1.

Table 1
FEATURES OF BACTERIAL CELLS REQUIRED FOR ATTACHMENT AND PENETRATION OF BACTERIOPHAGES

Host cell	Bacteriophage	Feature required	Ref.
		Outer membrane (OM)	1
Escherichia coli	ϕX174, P1, P2, S13,	OM, core lipopolysaccharide	2
	Ω8	OM, side chain of polysaccharide	1
	T_2, T_6	OM, protein	1
	T_5	OM, lipoprotein	1
	T_3, T_4, T_7	OM, lipopolysaccharide	1
	λ	OM, outer membrane component	3
Salmonella	C, ϵ^{15}, ϵ^{24}, P_{22}	OM, O side chain of polysaccharide	1
	F0	OM, amino sugar in core polysaccharide	1
Bacillus sphaericus	P1	OM, surface protein	4
Staphylococcus	3C, 71, 77, 52A	OM, teichoic acid-peptidoglycan	1
B. subtilis	ϕ, ϕ25, ϕ29	OM, teichoic acid-peptidoglycan	1
	SP3, SP10, SP02		
Streptococcus	A, C, G	OM, C-carbohydrate peptidoglycan	1
B. brevis	B. brevis phages	T layer protein	1
Streptococcus lactis	m 13	Cytoplasmic membrane	5
Salmonella, E. coli, Citrobacter	Vi-phage II	Vi-antigen	1
Salmonella arizona, E. coli, Proteus mirabilis	$\chi_1 - \chi_7$	Flagella	6
B. subtilis	PBS 1	Flagella	7
B. pumilis	PBP 1	Flagella	8
Caulobacter	ϕCp 34	Flagella, swarmer cells only	9
E. coli	R17, M12, Qβ, f_2, f_4	F pili	1, 10
	Fd, M13, Ec9, HR	F pili	11, 12
	If1, If2	I pili	13
Pseudomonas aeruginosa	7S, PP7	Pili	14
Caulobacter crescentus	ϕcb5	Pili, swarmer cells only	15
E. coli	$T_1 - T_7$, ϕX174	Membrane adhesions	16

Table 2

FEATURES OF BACTERIAL CELLS REQUIRED FOR REPLICATION OF BACTERIOPHAGES

Host cell	Bacteriophage	Host system required	Function	Ref.
Escherichia coli	ϕX174, f1, fd, M13	Membrane sites	Viral DNA $\rightarrow$ RFI RFI $\rightarrow$ RFII	17-21
	T_4, T_7, λ, ϕII, P2	Membrane sites	DNA replication	18, 22—24
Salmonella typhimurium	P22	Membrane sites	DNA replication	25
E. coli	λ	Integration site in DNA	Integration	26
	M13	dnaA gene product	DNA replication	17, 27
	ϕX174	dnaB gene product	DNA replication	17a
	ϕX174	dnaC/D gene product	DNA replication	17a
	ϕX174, M13, G4	DNA polymerase III	DNA replication	17a
	ϕX174, M13, G4	dnaG gene product	Initiation of DNA replication	17, 28
	ϕX174	dnaH gene product	DNA replication	17
	ϕX174, fd	dnaZ gene product	DNA replication	17
	ϕX174, G4, M13	DNA ligase, DNA polymerase I	DNA replication	17a
	ϕX174, G4, M13	Unwinding protein	DNA replication	17a
	ϕX174, G4	Host factors I and II	DNA replication	29
	ϕX174	Host factors i and n	DNA replication	30
	ϕX174, P2, 186	rep gene	DNA replication	31
	T_7	Thioredoxin	Viral DNA polymerase	32
	λ	gro P gene product	DNA replication	33
	λ	Ligase	Circularization of DNA	34
	ϕX174, M13, λ	RNA polymerase	RNA synthesis	35, 36
	M13	RNA polymerase III	DNA synthesis	17a
	T_1, T_3, T_5, T_7, ϕII	RNA polymerase	Early RNA synthesis	37
	λ, ϕ80, 21, 381	gro N gene product	RNA synthesis	38
	λ, T_7, ϕX174, T_4, T_5	Rho factor, kappa factor	Termination of RNA synthesis	39, 40
	T_7	RNAase III	Processing mRNA	41
	T_4	RNAase P	Processing tRNA	42
	Qβ, f2	Ribosomal protein S1	RNA synthesis	43
	Qβ, f2	Elongation factors Ts, Tu	RNA synthesis	44, 45
	Qβ	Host factor I	-ve strand synthesis	46
	R17	RNA polymerase	RNA synthesis	47
Mycoplasma	Mu L51	RNA polymerase	RNA synthesis	48
E. coli	T_5	Protease	Cleavage of viral proteins	49
	T_4, T_5, λ	gro E gene function	Assembly	50, 51
	T even, λ, ϕ80	mop gene function	Assembly	52
	M13	Host membrane function	Release of virus	53
Pseudomonas	PM2	Host membrane function	Assembly and release of virus	54, 55

Table 3
REQUIREMENTS FOR MULTIPLICATION OF VIRUSES IN ANIMAL CELLS

Virus	Host system required	Function	Ref.
Picornaviruses	Membrane receptors	Adsorption of virus	56
Polyoma virus	Neuraminic-acid-containing receptor	Adsorption of virus	57
Myxo- and paramyxoviruses	Neuraminic-acid-containing receptor	Adsorption of virus	58
Adenovirus	Receptors in plasma membrane	Adsorption of virus	59
Epstein Barr virus	Receptor on B lymphocyte membrane	Adsorption of virus	60
Avian tumor viruses	Strain-specific receptors on cell membrane	Penetration of virus	61, 62
Reoviruses	Lysosomal enzymes	Uncoating of virus	63, 64
Polyoma virus	Ligase	DNA integration	65
	DNA polymerase C	DNA replication	66
	Unwinding protein	Relaxation of DNA	67
	Histones	Viral structure	68, 69
Adenovirus	RNA polymerase II	Viral mRNA synthesis	70
	RNA polymerase III	5S RNA synthesis	71
SV40	RNA polymerase II	Viral mRNA synthesis	72
Herpes virus	RNA polymerase II	Viral mRNA synthesis	73, 74
Influenza virus	RNA polymerase II	Unknown	75
Many viruses	RNA modification systems	Adenylation, capping and shortening of viral mRNA	76 76
Influenza virus	Nuclease function	Initiation of transcription	77
Parvovirus	S phase of cell cycle	DNA replication	78, 79
Oncornaviruses	S phase of cell cycle	Establishment of infection	81
	Mitosis	Establishment of infection	81
Poliovirus	Protease	Polypeptide cleavage	82, 83
Myxoviruses	Protease	Polypeptide cleavage	84
Paramyxoviruses	Protease	Polypeptide cleavage	85
Rhabdoviruses	Cellular phospholipids and glycolipids	Viral assembly	86
Myxoviruses	Glycosylating enzymes	Synthesis of viral glycoproteins	87, 88
Paramyxoviruses	Glycosylating enzymes	Synthesis of viral glycoproteins	89
Poliovirus	Cell membranes	Assembly	90

REFERENCES

1. Lindberg, A. A., *Annu. Rev. Microbiol.*, 27, 205–241, 1973.
2. Jazwinski, S. M., Lindberg, A. A., and Kornberg, A., *Virology*, 66, 268–282, 1975.
3. Randall-Hazelbauer, L. and Schwartz, M., *J. Bacteriol.*, 116, 1436–1446, 1973.
4. Howard, L. and Tipper, D. J., *J. Bacteriol.*, 113, 1491–1504, 1973.
5. Oram, J. D., *J. Gen. Virol.*, 13, 59–71, 1971.
6. Edwards, S. and Meynell, G. G., *J. Gen. Virol.*, 2, 443–444, 1968.
7. Joys, T. M., *J. Bacteriol.*, 90, 1575–1577, 1965.
8. Lovett, P. S., *Virology*, 47, 743–752, 1972.
9. Fukuda, A., Miyakawa, K., Iba, H., and Okada, Y., *Virology*, 71, 583–592, 1976.
10. Crawford, E. M. and Gesteland, R. F., *Virology*, 22, 165–167, 1964.
11. Marvin, D. A. and Hohn, B., *Bacteriol. Rev.*, 33, 172–209, 1969.
12. Marco, R., Jazwinski, S. M., and Kornberg, A., *Virology*, 62, 209–223, 1974.
13. Meynell, G. G. and Lawn, A. M., *Nature*, 217, 1184–1186, 1968.
14. Holloway, B. W., Krishnapillai, V., and Stanisich, V., *Annu. Rev. Genet.*, 5, 425–446, 1971.
15. Schmidt, J. M., *J. Gen. Microbiol.*, 45, 347–353, 1966.
16. Bayer, M. E. and Starkey, T. W., *Virology*, 49, 236–256, 1972.
17. Schekman, R., Weiner, J. H., Weiner, A., and Kornberg, A., *J. Biol. Chem.*, 250, 5859–5865, 1975.
17a. Schekman, R., Weiner, A., and Kornberg, A., *Science*, 186, 987–993, 1974.
18. Siegel, P. J. and Schaechter, M., *Annu. Rev. Microbiol.*, 27, 261–282, 1973.

19. Staudenbauer, W. L. and Hofschneider, P. H., *Biochem. Biophys. Res. Commun.*, 42, 1035–1041, 1971.
20. Forsheit, A. B. and Ray, D. S., *Virology*, 43, 647–664, 1971.
21. Grandis, A. S. and Webster, R. E., *Virology*, 55, 39–52, 1973.
22. Shah, D. B. and Berger, H., *J. Mol. Biol.*, 57, 17–34, 1971.
23. Hallick, L. M., Boyce, R. P., and Echols, H., *Nature*, 223, 1239–1242, 1969.
24. Ljungquist, E., *Virology*, 52, 120–129, 1973.
25. Levine, M., *Curr. Top. Microbiol. Immunol.*, 58, 135–156, 1972.
26. Rothman, J., *J. Mol. Biol.*, 12, 892–912, 1965.
27. Mitra, S. and Stallions, D. R., *Eur. J. Biochem.*, 67, 37–45, 1976.
28. Dasgupta, S. and Mitra, S., *Eur. J. Biochem.*, 67, 47–51, 1976.
29. Wickner, S. and Hurwitz, J., *Proc. Natl. Acad. Sci. U.S.A.*, 73, 1953–1957, 1976.
30. Weiner, J. H., McMacken, R., and Kornberg, A., *Proc. Natl. Acad. Sci. U.S.A.*, 73, 752–756, 1976.
31. Eisenberg, S., Scott, J. F., and Kornberg, A., *Proc. Natl. Acad. Sci. U.S.A.*, 73, 1594–1597, 1976.
32. Mark, D. and Modrich, P., *Fed. Proc. Fed. Am. Soc. Exp. Biol.*, 34, 639, 1975.
33. Georgopolous, C. P. and Herskowitz, I., in *The Bacteriophage* λ, Hershey, A. D., Ed., Cold Spring Harbor Laboratory, Cold Spring Harbor, N.Y., 1971, 553–564.
34. Tomizawa, J. and Ogawa, T., *Cold Spring Harbor. Symp. Quant. Biol.*, 33, 533–551, 1968.
35. Clements, J. B. and Sinsheimer, R. L., *J. Virol.*, 15, 151–160, 1975.
36. Sternberg, N., *Virology*, 73, 139–154, 1976.
37. Bautz, E. K. F., *Prog. Nucleic Acid Res. Mol. Biol.*, 12, 129–160, 1972.
38. Georgopolous, C. P., in *The Bacteriophage* λ, Hershey, A. D., Ed., Cold Spring Harbor Laboratory, Cold Spring Harbor, N.Y., 1971, 639–645.
39. Schafer, R. and Zillig, N., *Eur. J. Biochem.*, 33, 201–206, 1973.
40. Hayashi, M. N. and Hayashi, M., *J. Virol.*, 9, 207–215, 1972.
41. Dunn, J. J. and Studier, F. W., *Proc. Natl. Acad. Sci. U.S.A.*, 70, 3296–3300, 1973.
42. Schedl, P. and Primakoff, P., *Proc. Natl. Acad. Sci. U.S.A.*, 70, 2091–2095, 1973.
43. Inouye, H., Pollack, Y., and Petre, J., *Eur. J. Biochem.*, 45, 109–117, 1974.
44. Weissman, C., Feix, G., Slor, H., and Pollet, R., *Proc. Natl. Acad. Sci. U.S.A.*, 57, 1870–1877, 1967.
45. Blumenthal, T., Landers, T. A., and Weber, K., *Proc. Natl. Acad. Sci. U.S.A.*, 69, 1313–1317, 1972.
46. Franze de Fernandez, M. T., Eoyang, L., and August, J. T., *Nature*, 219, 588–590, 1968.
47. Igarishi, S. J. and Bissonette, R. P., *J. Biochem.* (Tokyo), 70, 845-854, 1971.
48. Das, S. and Maniloff, J., *J. Virol.*, 18, 969–976, 1976.
49. Zweig, M. and Cummings, D. J., *J. Mol. Biol.*, 80, 505–518, 1973.
50. Georgopolous, C. P., Hendrix, R. W., Kaiser, A. D., and Wood, W. B., *Nature New Biol.*, 239, 38–41, 1972.
51. Sternberg, N., *J. Mol. Biol.*, 76, 25–44, 1973.
52. Takano, T. and Kakefuda, T., *Nature New Biol.*, 239, 34–37, 1972.
53. Staudenbauer, W. L. and Hofschneider, P. H., *Mol. Gen. Genet.*, 138, 203–212, 1973.
54. Cota Robles, E., Espejo, R. T., and Haywood, P. W., *J. Virol.*, 2, 56–68, 1968.
55. Dahlberg, J. E. and Franklin, R. M., *Virology*, 42, 1073–1086, 1970.
56. Holland, J. J. and Maclaren, L. C., *J. Exp. Med.*, 109, 487–504, 1959.
57. Eddy, B. E., Rowe, W. P., Hartley, J. W., Stewart, S. E., and Huebner, R. J., *Virology*, 6, 290–298, 1958.
58. Choppin, P. W. and Compans, R. W., in *Comprehensive Virology*, Vol. 4, Fraenkel-Conrat, H. and Wagner, R. R., Eds., Plenum, New York, 1975, 122–125.
59. Philipson, L., *J. Virol.*, 1, 868–875, 1967.
60. Jondal, M. and Klein, G., *J. Exp. Med.*, 138, 1365–1378, 1973.
61. Piraino, F., *Virology*, 32, 700–707, 1967.
62. Crittenden, L. B., *J. Natl. Cancer Inst.*, 41, 145–153, 1968.
63. Silverstein, S. C., Astell, C., Levin, D. H., Schonberg, M., and Acs, G., *Virology*, 47, 797–806, 1972.
64. Chang, C.-T. and Zweerink, H. J., *Virology*, 46, 544–555, 1971.
65. Beard, P., *Biochim. Biophys. Acta*, 269, 385–396, 1972.
66. Pigiet, V., Eliasson, R., and Reichard, P., *J. Mol. Biol.*, 84, 197–216, 1974.
67. Champoux, J. J. and Dulbecco, R., *Proc. Natl. Acad. Sci. U.S.A.*, 69, 143–146, 1972.
68. Frearson, P. M. and Crawford, L. V., *J. Gen. Virol.*, 14, 141–155, 1972.
69. Seebeck, T. and Weil, R., *J. Virol.*, 13, 567–576, 1974.
70. Ledinko, N., *Nature New Biol.*, 233, 247–248, 1971.

71. Weinman, R., Raskas, H. J., and Roeder, R. G., *Proc. Natl. Acad. Sci. U.S.A.,* 71, 3426–3430, 1974.
72. Jackson, A. H. and Sugden, W., *J. Virol.,* 10, 1086–1089, 1972.
73. Ben Zeev, A., Asher, Y., and Becker, Y., *Virology,* 71, 302–311, 1976.
74. Preston, C. M. and Newton, A. A., *J. Gen. Virol.,* in press, 1976.
75. Mahy, B. W. J., Hastie, N. D., and Armstrong, S. J., *Proc. Natl. Acad. Sci. U.S.A.,* 69, 1421–1424, 1972.
76. Weinberg, R. A., *Annu. Rev. Biochem.,* 42, 329–354, 1973.
77. Kelly, D. C., Avery, R. J., and Dimmock, N. J., *J. Virol.,* 13, 1155–1161, 1974.
78. Siegl, G. and Gautschi, M., *Arch. Gesamte Virusforsch.,* 40, 105–118, 1973.
79. Tennant, R. W. and Hand, R. E., *Virology,* 42, 1054–1063, 1970.
80. Humphries, E. H. and Temin, H., *J. Virol.,* 10, 82–87, 1972.
81. Temin, H., *J. Cell. Physiol.,* 69, 53–58, 1967.
82. Korant, B. D., *J. Virol.,* 10, 751–759, 1972.
83. Villa Komaroff, L., Guttman, N., Baltimore, D., and Lodish, H. F., *Proc. Natl. Acad. Sci. U.S.A.,* 72, 4157–4161, 1975.
84. Stanley, P., Gandhi, S., and White, D., *Virology,* 53, 92–106, 1973.
85. Scheid, A. and Choppin, P., *Virology,* 57, 475–490, 1974.
86. Wagner, R., in *Comprehensive Virology,* Vol. 4, Fraenkel-Conrat, H. and Wagner, R. R., Eds., Plenum, New York, 1975, 26.
87. Laver, W. G. and Webster, R. G., *Virology,* 30, 104–115, 1966.
88. Burge, B.-W. and Huang, A. S., *J. Virol.,* 6, 176–182, 1970.
89. Klenk, H. D. and Choppin, P. W., *Proc. Natl. Acad. Sci. U.S.A.,* 66, 57–64, 1970.
90. Perlin, M. and Phillips, B. A., *Virology,* 53, 107–114, 1973.

QUALITATIVE REQUIREMENTS AND UTILIZATION OF NUTRIENTS: *RICKETTSIALES*

R. A. Ormsbee

The order *Rickettsiales* is made up of a group of small coccoid or rod-shaped microorganisms, most of which grow in arthropods and many of which parasitize man or other mammals. The majority of the species in this order are obligate intracellular parasites in both host and vector. They usually are transmitted from vertebrate host to vertebrate host by blood-sucking arthropods including ticks, mites, and insects.

The 8th edition of *Bergey's Manual of Determinative Bacteriology*[1] lists three families in the order *Rickettsiales: Rickettsiaceae, Bartonellaceae,* and *Anaplasmataceae.* Most of the data presented in this chapter concern members of the family *Rickettsiaceae* because this family has been the most extensively studied. The family is divided into the tribes *Rickettsieae, Ehrlichieae,* and *Wolbachieae.* Most of our knowledge of this family is a product of studies of members of the genera *Rickettsia* and *Coxiella* of the tribe *Rickettsieae.* Relatively little is known about the biology of organisms in the other two tribes.

Nutrition of *Rickettsiales* cannot be discussed on the same basis as that of *Escherichia coli,* for example, because most *Rickettsiales* are obligate intracellular parasites. If they have been grown on cell-free media, as is the case with *Rochalimaea quintana, Wolbachia malophagi,* and members of the family *Bartonellaceae,* the media have been complex mixtures of organic and inorganic substances, the essential components of which have not been defined.

It is now clear that the *Rickettsieae* are essentially fastidious bacteria. They contain both DNA and RNA, and can synthesize proteins, lipids, and complex polysaccharides. They possess cell walls similar in appearance to those of free-living Gram-negative bacteria and usually are categorized as Gram-negative microorganisms. Members of the genus *Rickettsia* are relatively closely related to one another but are only remotely related to the genus *Coxiella,* as indicated by the great difference between the two genera in DNA base composition. Nutritional needs of the *Rickettsieae,* insofar as they can be characterized in terms of rickettsial growth in host cell cultures or on complex, noncellular, organic media, will be covered in Section 7.

This chapter will be confined to a review of substrates that are involved in the metabolism of those members of the tribe *Rickettsieae* that are obligate intracellular parasites. These data are summarized in Table 1. It probably can be assumed that a metabolic capability found in one species of the genus *Rickettsia* is present also in other members of that genus. Thus, from the combined data on *R. rickettsii, R. typhi,* and *R. prowazekii,* the conclusion can be drawn that members of the genus *Rickettsia* characteristically possess all or most of the essential enzymatic components of the citric acid cycle of intermediary carbohydrate metabolism and ability to synthesize lipids, proteins, and complex polysaccharides. It must be stated, however, that very little is known about the metabolic capabilities of *R. tsutsugamushi.* The genus *Coxiella,* however, possesses not only these capabilities, but also the ability to utilize glucose via an Embden-Meyerhof pathway. No evidence that any member of the genus *Rickettsia* can metabolize glucose or any of its 6- or 3-carbon phosphorylated derivates has yet been adduced. Whether members of the genus *Rickettsia* truly lack the intrinsic ability to metabolize glucose or whether the failure to detect this ability is due to an artefact of cell permeability is unknown at present. Glucose metabolism in *Coxiella* has been demonstrated only with cell-free extracts of rickettsiae, not with suspensions of intact rickettsial cells. Comparable cell-free extracts made from *Rickettsia* cells have not yet been tested.

Additional information on the metabolism of these organisms is furnished in reviews by Weiss,[25,26] Paretsky,[27] and Wisseman.[28]

Table 1
SUBSTRATES REPORTEDLY INVOLVED IN ENZYME-CATALYZED METABOLIC REACTIONS OF THE GENERA *RICKETTSIA* AND *COXIELLA* OF THE *RICKETTSIEAE*

Numbers Denote Bibliographic References

Substrate	*R. rickettsii*	*R. typhi*	*R. akari*	*R. tsutsugamushi*	*C. burnetii*	Metabolic test system[a]
Acetate		21[d]				WRCS
Acetyl phosphate					3	CFE
Adenine	17	19	19	17	3	HCC, CFE
Adenine diphosphate					3	HCC, CFE
Amino acids[b]	17	19, 24	19	17		HCC, WRCS
Asparagine		13				WRCS
Aspartate					7	CFE
Adenosine triphosphate					3	CFE
Carbamyl phosphate					7	CFE
Citrate					3	CFE
Iso-citrate					4	CFE
Citrulline					7	CFE
Formaldehyde					6	CFE
Fructose-6-phosphate					10	CFE
Fructose-1,6-diphosphate					10	CFE
Fumarate	23	15			2	WRCS
Glucose					9	CFE
Glucose-6-phosphate					4	CFE
Glutamate	16, 23	12,[e] 14,[e] 15, 18			2, 3	WRCS, CFE
Glutamine	16	13			2	WRCS
Glyceraldehyde-3-phosphate					10	CFE
Glycine		20[d]			6	WRCS, CFE
α-Ketoglutarate	23	15			2	WRCS
L-Leucine					8	WRCS
Methionine		22,[d] 24				WRCS
Malate	23	15, 18			2, 3	WRCS, CFE
Ornithine					7	CFE
Oxaloacetate	23	12,[e] 15			2, 3	WRCS, CFE
L-Phenylalanine					8	CFE
Phosphoenolpyruvate					10	CFE
6-Phosphogluconate					5	CFE
Pyruvate	16, 23	15			2	WRCS
Ribonucleotides[c]					11	CFE
L-Serine					2, 6	WRCS, CFE
Succinate	23	15			2	WRCS
Ureidosuccinate					7	CFE

[a] WRCS = whole rickettsial cell suspension; CFE = cell-free extract; HCC = host cell culture.
[b] Uptake or incorporation from a mixture of 15 radiolabeled amino acids.
[c] These included the triphosphoribonucleosides of adenine, guanine, uracil and cytosine.
[d] Demonstrated with *R. prowazekii* only.
[e] Demonstrated with *R. prowazekii* also.

REFERENCES

1. Moulder, J. W., Weiss, E., Philip, C. B., Brooks, M. A., Weinman, D., Ristic, M., and Kreier, J. P., The Rickettsias, in *Bergey's Manual of Determinative Bacteriology,* 8th ed., Buchanan, R. E. and Gibbons, N. E., Eds., Williams and Wilkins, Baltimore, 1974, 882–914.
2. Ormsbee, R. A. and Peacock, M. G., Metabolic activity in *Coxiella burnetii, J. Bacteriol.,* 88, 1205, 1964.
3. Paretsky, D., Downs, C. M., Consigli, R. A., and Joyce, B. K., Studies on the physiology of rickettsiae. I. Some enzyme systems of *Coxiella burnetii, J. Infect. Dis.,* 103, 6, 1958.
4. Consigli, R. A. and Paretsky, D., Oxidation of glucose-6-phosphate and isocitrate by *Coxiella burnetii, J. Bacteriol.,* 83, 206, 1962.
5. McDonald, T. L. and Mallavia, L., Biochemistry of *Coxiella burnetii:* 6-phosphogluconic acid dehydrogenase, *J. Bacteriol.,* 102, 1, 1970.
6. Myers, W. F. and Paretsky, D., Synthesis of serine by *Coxiella burnetii, J. Bacteriol.,* 82, 761, 1961.
7. Mallavia, L. P. and Paretsky, D., Studies on the physiology of rickettsiae. V. Metabolism of carbamyl phosphate by *Coxiella burnetii, J. Bacteriol.,* 86, 232, 1963.
8. Mallavia, L. P. and Paretsky, D., Physiology of rickettsiae. VII. Amino acid incorporation by *Coxiella burnetii* and by infected hosts, *J. Bacteriol.,* 93, 1479, 1967.
9. Paretsky, D., Consigli, R. A., and Downs, C. M., Studies on the physiology of rickettsiae. III. Glucose phosphorylation and hexokinase activity in *Coxiella burnetii, J. Bacteriol.,* 83, 538, 1962.
10. McDonald, T. L. and Mallavia, L., Biochemistry of *Coxiella burnetii:* Embden Meyerhof pathway, *J. Bacteriol.,* 107, 864, 1971.
11. Jones, F., Jr. and Paretsky, D., Physiology of rickettsiae. VI. Host-independent synthesis of polyribonucleotides by *Coxiella burnetii, J. Bacteriol.,* 93, 1063, 1967.
12. Bovarnick, M. R. and Snyder, J. C., Respiration of typhus rickettsiae, *J. Exp. Med.,* 89, 561, 1949.
13. Hahn, F. E., Cohn, Z. A., and Bozeman, F. M., Metabolic studies of rickettsiae. V. Metabolism of glutamine and asparagine in *Rickettsia mooseri, J. Bacteriol.,* 80, 400, 1960.
14. Bovarnick, M. R. and Miller, J. C., Oxidation and transamination of glutamate by typhus rickettsiae, *J. Biol. Chem.,* 184, 661, 1950.
15. Wisseman, C. L., Jr., Hahn, F. E., Jackson, E. B., Bozeman, F. M., and Smadel, J. E., Metabolic studies of rickettsiae. II. Studies on the pathway of glutamate oxidation by purified suspensions of *Rickettsia mooseri, J. Immunol.,* 68, 251, 1952.
16. Weiss, E., Rees, H. B., Jr., and Hayes, J. R., Metabolic activity of purified suspensions of *Rickettsia rickettsii, Nature,* 213, 1020, 1967.
17. Weiss, E., Green, A. E., Grays, R., and Newman, L. W., Metabolism of *Rickettsia tsutsugamushi* and *Rickettsia rickettsi* in irradiated host cells, *Infect. Immun.,* 8, 4, 1973.
18. Weiss, E., Coolbaugh, J. C., and Williams, J. C., Separation of viable *Rickettsia typhi* from yolk sac and L cell host components by renografin density gradient centrifugation, *Appl. Microbiol.,* 30, 456, 1975.
19. Weiss, E., Newman, L. W., Grays, R., and Green, A. E., Metabolism of *Rickettsia typhi* and *Rickettsia akari* in irradiated L cells, *Infect. Immun.,* 6, 50, 1972.
20. Bovarnick, M. R. and Schneider, L., The incorporation of glycine-1-C^{14} by typhus rickettsiae, *J. Biol. Chem.,* 235, 1727, 1960.
21. Bovarnick, M. R., Incorporation of acetate-1-C^{14} into lipid by typhus rickettsiae, *J. Bacteriol.,* 80, 508, 1960.
22. Bovarnick, M. R., Schneider, L., and Walter, H., The incorporation of labelled methionine by typhus rickettsiae, *Biochim. Biophys. Acta,* 33, 414, 1959.
23. Price, W. H., The epidemiology of Rocky Mountain spotted fever. I. The characterization of strain virulence of *Rickettsia rickettsii, Am. J. Hyg.,* 58, 248, 1953.
24. Kohno, S., Natsume, Y., and Shishido, A., Studies on metabolism of rickettsiae. II. Incorporation of ^{14}C amino acid mixture and ^{35}S methionine by *Rickettsia mooseri, Jpn. J. Med. Sci. Biol.,* 14, 213, 1961.
25. Weiss, E., Comparative metabolism of rickettsiae and other host dependent bacteria, *Zentralbl. Bakteriol. Parasitenk. Infektionskr. Hyg. Abt. Orig.,* 206, 292, 1968.
26. Weiss, E., Growth and physiology of rickettsiae, *Bacteriol. Rev.,* 37, 259, 1973.
27. Paretsky, D., Biochemistry of rickettsiae and their infected hosts with special reference to *Coxiella burnetii, Zentralbl. Bakteriol. Parasitenk. Infektionskr. Hyg. Abt. Orig.,* 206, 283, 1968.
28. Wisseman, C. L., Jr., Some biological properties of rickettsiae pathogenic for man, *Zentralbl. Bakteriol. Parasitenk. Infektionskr. Hyg. Abt. Orig.,* 206, 299, 1968.

QUALITATIVE REQUIREMENTS AND UTILIZATION OF NUTRIENTS: CHLAMYDIAE

L. A. Page

The morphology, chemical composition, and mode of multiplication of chlamydiae qualify them as bacteria, but to this date it has not been possible to propagate them outside of living host cells; hence, they are obligate intracellular parasites. They may be propagated readily in mammalian cell cultures or in laboratory animals. Examination of their biochemical activities in suspensions free of host cells has consistently demonstrated only a few enzymatic capabilities (as summarized in Figure 1) along with their principal metabolic failure, i.e., production and storage of high energy compounds such as ADP or ATP. This failure has earned chlamydiae the label "energy parasites"[1,2] and underlines their dependence on host cells for multiplication. Chlamydiae survive outside of host cells in a hardy, sporelike form (elementary body) which is released from infected host cells as the end product of a unique intracellular developmental cycle. There are two species: *Chlamydia trachomatis,* which causes ocular and genital diseases of man, and *C. psittaci,* which is widely distributed in birds and mammals, where it causes pneumonitis, arthritis, placentitis, enteritis, and/or generalized fatal infection.

REFERENCES

1. **Moulder, J. W.,** *The Psittacosis Group as Bacteria,* CIBA Lectures, John Wiley & Sons, New York, 1964.
2. **Moulder, J. W.,** *BioScience,* 19, 875–881, 1969.

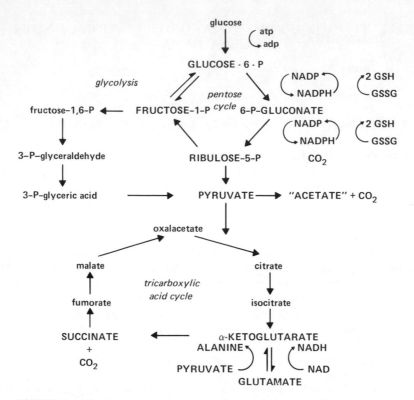

FIGURE 1.   Metabolism of glucose and related substances in chlamydiae. Intermediates that are either produced or attacked by chlamydiae are shown in upper case characters; those that are neither produced nor attacked are shown in lower case. Reactions known to occur in chlamydiae are indicated by solid arrows. *Chlamydia psittaci* converts glutamate to α-ketoglutarate almost exclusively by transamination; *C. trachomatis* converts by NAD-mediated oxidation. Abbreviations used: NAD and NADH, oxidized and reduced nicotinamide-adenine dinucleotide; NADP and NADPH, oxidized and reduced nicotinamide-adenine dinucleotide phosphate; GSSG and GSH, oxidized and reduced glutathione. (Reprinted with permission, from Moulder, J. W., in the October 1969 *BioScience* published by the American Institute of Biological Sciences.)

Table 1
DEMONSTRABLE ENZYMATIC ACTIVITIES OF CHLAMYDIAE

Species	Strain[a]	Enzyme	Ref.
C. trachomatis	TE 55	Glucose-6-phosphate dehydrogenase	2
	TE 55	6-Phosphogluconate dehydrogenase	2
	TE 55	Glutamate-pyruvate transaminase	3, 4
	TE 55	Pyruvate dehydrogenase	3, 4
	TW3, 440L	Glycogen synthetase	5
C. psittaci	MN	Glucose-6-phosphate dehydrogenase	1
	MN	6-Phosphogluconate dehydrogenase	1
	MN, NJT, UPA	Glutamate-pyruvate transaminase	3, 4
	MN, NJT, UPA	Glutamate dehydrogenase	3, 4
	MN	α-Ketoglutarate dehydrogenase	3
	MN	Pyruvate decarboxylase	3

[a] TE 55: Tang, 1955; TW3: Taiwan 3; 440L: lymphogranuloma venereum 440; MN: meningopneumonitis; NJT: New Jersey turkey; UPA: Utah sheep polyarthritis.

REFERENCES

1. **Moulder, J. W., Grisso, D. L., and Brubaker, R. R.,** *J. Bacteriol.,* 89, 810–812, 1965.
2. **Weiss, E.,** *J. Bacteriol.,* 90, 243–253, 1965.
3. **Weiss, E.,** *J. Bacteriol.,* 93, 177–184, 1967.
4. **Weiss, E.,** *Am. J. Ophthalmol.,* 63, 1098–1101, 1967.
5. **Jenkin, H. M. and Fan, V. S. C.,** *Excerpta Med. Int. Congr. Ser.,* 223, 52–59, 1971.

Table 2
CULTURE MEDIA: NATURAL AND SYNTHETIC FOR CHLAMYDIAE

A. C. Trachomatis

System	Cell type/species	Base Medium	Additions	Antibiotics	pH	Incubation temperature (°C)	Ref.
Cell culture	Line cells						
	Human						
	HeLa 229	Eagle's MEM[a]	5% FBS[b]	SM[c] and VAN[d] (100 μg/ml)	7.4	35	1
	FL	Eagles's BSS[e]	20% human serum	SM (100 μg/ml)	NS[f]	35	2
	Mouse						
	L cell (plaques)	Eagle's MEM	10% FBS	NS	NS	NS	3
	L cell (spinner)	Eagle's MEM	10% FBS	NS	NS	NS	4
	McCoy (irradiated)	Eagle's MEM	10% FBS	NS	NS	35	5
	Monkey						
	MK-2	Eagle's MEM	10% FBS 0.27% glucose	SM (200 μg/ml)	7.2	37	6
	BSC-1	Eagle's MEM	10% FBS	SM (300 μg/culture)	NS	35	7
	Bovine						
	BK-21	Eagle's MEM	10% FBS, 0.035% NaHCO₃, 0.36% glucose	NS	7.5	35—37	8
Organ culture	Monkey						
	Conjunctiva						9

a Minimum essential medium.
b Fetal bovine serum.
c Streptomycin sulfate.
d Vancomycin sulfate.
e Basic salt solution.
f Not stated by author.

Table 2 (continued)
CULTURE MEDIA: NATURAL AND SYNTHETIC FOR CHLAMYDIAE

A. C. Trachomatis (continued)

System	Species	Age	Inoculation route	Ref.
Laboratory animals	Embryonating chicken embryos	6–7 days	Yolk sac, 35C	10
	Mice	3 weeks	Lung (intranasal)	11
	Guinea pigs	12 weeks	Skin (intradermal)	12
	Primates			
	Rhesus (*Macacca* sp.)	NS	Conjunctiva (swab)	13
	Baboon (*Papio* sp.)	NS	Conjunctiva (swab)	14

B. C. Psittaci

System	Species cell code	Base medium	Additions	Antibiotics	pH	Incubation temperature (°C)	Ref.
Cell culture	Line cells						
	Human						
	HeLa 229	Eagle's MEM[a]	5% FBS[b]	SM[c] and VAN[d]	7.4	37	15
	Mouse						
	L cell 929 (plaques)	Eagle's MEM	10% FBS	SM (50μg/ml)	NS[f]	37	3
	L cell (spinner)	Waymouth MB752	0.2% BSA,[g] 0.005% oleate	SM (100 μg/ml), VAN (100 μg/ml)	7.2	37	16
	McCoy	Eagle's MEM	10% horse serum		7.2	37	5, 10
	Monkey						
	MK-2	Eagle's MEM	5% FBS	NS	NS	NS	17
		Waymouth	0.2% BSA, 0.005% oleate, 0.5% FBS	NS	NS	NS	17
Primary cell culture	Chicken embryo						
	Cells	M 199	None	NS	7.4	37	18
	Explants	Hank's BSS	25% chicken serum	NS	NS	35	19

g Bovine serum albumin.

Table 2 (continued)
CULTURE MEDIA: NATURAL AND SYNTHETIC FOR CHLAMYDIAE

System	Host species	Chlamydial strains	Host age	Inoculation route	Incubation temperature (°C)	Ref.
		B. C. Psittaci (continued)				
Laboratory animals	Embryonating Chicken Embryos	All	6—7 days	Yolk sac	37—39	20
	Mice	Avian	3 weeks	Intraperitoneal	—	21
	Mice	Mammalian	3 weeks	Intranasal, intracerebral	—	21
	Guinea pigs	Avian and mammalian	8 weeks	Intraperitoneal	—	21

REFERENCES

1. Jenkin, H. M., J. Infect. Dis., 116, 390—399, 1966.
2. Bernkopf, H., Mashiah, P., and Becker, Y., Ann. N.Y. Acad. Sci., 98, 62—81, 1962.
3. Banks, J., Eddie, B., Schachter, J., and Meyer, K. F., Infect. Immunol., 1, 259—262, 1970.
4. Schechter, E. M., J. Bacteriol., 91, 2069—2080, 1966.
5. Gordon, F. B., Dressler, H. R., Quan, A. L., McQuillein, W. T., and Thomas, J. I., Appl. Microbiol.. 23, 123—129, 1972.
6. Fan, V. S. C. and Jenkin, H. M., Infect. Immunol. 10, 464—470, 1974.
7. Bernkopf, H., Am. J. Ophthalmol., 63, 1206—1207, 1967.
8. Blythe, W. A., Taverne, J., and Garrett, A. J., Excerpta Med. Int. Congr. Ser., 223, 79—87, 1971.
9. Barron, A. L., Mount, D. T., Cornell, D. S., and Brody, H., Excerpta Med. Int. Congr. Ser., 223, 71—78, 1971.
10. Gordon, F. B., Dressler, H. R., and Quan, A. L., Am. J. Ophthalmol., 63, 1044—1048, 1967.
11. Graham, D. M., Am. J. Ophthalmol., 63, 1173—1190, 1967.
12. Blythe, W. A., Am. J. Ophthalmol., 63, 1153—1161, 1967.
13. Dawson, C. R., Mordhorst, C. H., and Thygeson, P., Ann. N.Y. Acad. Sci., 98, 167—176, 1962.
14. Collier, L. H., Ann. N.Y. Acad. Sci., 98, 188—196, 1962.
15. Jenkin, H. M. and Fan, V. S. C., Excerpta Med. Int. Congr. Ser., 223, 52—59, 1971.
16. Morrison, S. J. and Jenkin, H. M., In Vitro, 8, 94—100, 1972.
17. Makino, S., Jenkin, H. M., Yu, H. M., and Townsend, D., J. Bacteriol., 103, 62—70, 1970.
18. Piraino, F., J. Bacteriol., 98, 475—480, 1969.
19. Weiss, E. and Huang, J. S., J. Infect. Dis, 94, 107—125, 1954.
20. Page, L. A., Excerpta Med. Int. Congr. Ser., 223, 40—51, 1971.
21. Meyer, K. F., in Diseases of Poultry, 5th ed., Biester, H. E. and Schwarte, L. H., Eds., Iowa State University Press. Ames, 1965, 692.

QUALITATIVE REQUIREMENTS AND UTILIZATION OF NUTRIENTS: WALL-DEFECTIVE MICROBIAL VARIANTS (WDMV)

H. Gooder

INTRODUCTION

Terminology employed for wall-defective microbial variants (protoplasts, spheroplasts, transitional phase variants, L-forms, and L-phase variants) has been defined previously,[1] and extensive reviews of these organisms are available.[2-4] Briefly, the wall-defective microbial variants (WDMV) are bacteria or fungi that lack a major structural component from the cell wall material that lies external to the cytoplasmic membrane. Such growth forms can be induced in vitro or isolated from natural sources in vivo.

NUTRITION

In general, complex media have been used for either the induction or isolation of WDMV, so detailed studies of their nutritional requirements or comparative studies to the parent organism are rare. Undoubtedly, nutritional requirements vary depending on the species from which the WDMV was derived. It would be expected that these requirements would be the same as those of the parent species, and these are summarized elsewhere in this volume in the chapters on the qualitative requirements of bacteria (by G. D. Hegeman and C. O. Patterson) and yeasts (by E. Oura and H. Suomalainen). There are few investigations in which a chemically defined medium was employed, and in these reports no significant nutrient differences between the WDMV and parent strain were noted.[5-8] Minor differences can be explained by the selection of mutants during adaptation of the WDMV to culture in liquid medium.[3] In a recent extensive comparative investigation of *Streptococcus faecium* and a derived L-phase variant, the same 14 amino acids, a carbon (energy) source (glucose), vitamins, purines, pyrimidines, and salts were essential, while an additional 7 amino acids were stimulatory for both organisms.[9] Possibilities for nutritional differences between the WDMV and parent organism involve the hypothesis that the WDMV may have a decreased or absent requirement for a nutrient that is uniquely a cell wall constituent, e.g., diaminopimelic acid in some Gram-negative bacterial species. One example has been reported.[10]

UTILIZATION OF NUTRIENTS

Utilization of nutrients by WDMV for metabolism or growth requires a very different environment from that employed for the parent strain. Absence of a cell wall polymer and exposure of the cytoplasmic membrane to the surroundings can affect, for example, transport mechanisms, the access of toxic materials to the membrane, and osmotic stabilization. For initial isolation and growth of L-forms, solidification of medium by low concentrations (0.7 to 1.2%) of agar is usually required. Similarly, serum (5 to 10%) or protein fractions obtained from serum are employed, probably to bind unknown toxic materials present in culture media.

The osmolarity of the culture or suspension medium can be increased by the addition of a variety of solutes, commonly salts (ions) or sugars. Osmotic protection usually occurs within a limited osmolarity range and is often dependent on the nature of the solute and species of organism. Synergistic and antagonistic interactions between added solutes are found. Estimated intracellular osmotic pressure of Gram-positive bacteria 1500 to 2000 mOsm and Gram-negative bacteria 150 to 600 mOsm is reflected in the reported osmotic stabilization of their WDMV for utilization of nutrients (Table 1).

Table 1

SOLUTES FOUND SUITABLE FOR OSMOTIC STABILIZATION
OF WALL-DEFECTIVE MICROBIAL VARIANTS

A. Suitable for Growth

Salts: NaCl, NH_4Cl, $CaCl_2$, $MgCl_2$, NaH_2PO_4 (*Streptococcus pyogenes*, 0.43—0.51 M[13])
KCl (*Proteus mirabilis*, 0.2—0.5 M[14]; Salmonella, Hemophilus[13]) NaCl, NH_4Cl (*Strepto-coccus faecium*, 0.43 M[15])
Carbohydrates: Sucrose (*Escherichia coli*, 0.3 M[10]; *Neisseria gonorrhoeae*, 0.3 M[16]; *P. mirabilis*, 0.2—0.5 M[17])
Miscellaneous: Polyvinylpyrrolidone 40,000 (*N. gonorrhoeae*, 4—5%[18])

B. Suitable for Metabolic Studies

Salts: As in A above plus $CoSO_4$[19]
Carbohydrates: As in A above plus cellobiose, lactose, melibiose, raffinose,[20] rhamnose[21]
Polyamines: Spermine, Spermidine, protamine, polylysine, cadaverine, putrescine[22]

REFERENCES

1. **McGee, Z. A., Wittler, R. G., Gooder, H., and Charache, P.,** *J. Infect. Dis.* 123, 433—438, 1971.
2. **Guze, L. B., Ed.,** *Microbial Protoplasts, Spheroplasts and L-forms,* Williams and Wilkins, Baltimore, 1967.
3. **Hijmans, W., van Boven, C. P. A., and Clasener, H. A. L.,** in *The Mycoplamatales and the L-phase of Bacteria,* Hayflick, L., Ed., Appleton-Century-Crofts, New York, 1969, 67—143.
4. **Mattman, L. H.,** *Cell Wall Deficient Forms,* CRC Press, Cleveland, 1974.
5. **Medill, M. A. and O'Kane, D. J.,** *J. Bacteriol.,* 68, 530—533, 1954.
6. **Minck, R., Kirn, A., Galleron, T.,** *Ann. Inst. Pasteur* (Paris), 92, 138—141, 1957.
7. **Panos, C. and Hymes, L. M.,** *Arch. Biochem.,* 105, 326—328, 1964.
8. **van Boven, C. P. A., Kastelein, M. J. W., and Hijmans, W.,** *Ann. N.Y. Acad. Sci.,* 143, 749—754, 1967.
9. **Gregory, W. W. and Gooder, H.,** Abstracts of the Annual Meeting of the American Society for Microbiology, ASM, Washington, D.C., 1973, G 241, p. 66.
10. **Lederberg, J. and St. Clair, J.,** *J. Bacteriol.,* 75, 143—160, 1958.
11. **Ward, J. B.,** *J. Bacteriol.,* 124, 668—678, 1975.
12. **Gregory, W. W. and Gooder, H.,** Abstracts of the Annual Meeting of the American Society for Microbiology, ASM, Washington, D.C., 1976, D 36, p. 57.
13. **Dienes, L. and Sharp, J. T.,** *J. Bacteriol.,* 71, 208—213, 1956.
14. **Landman, O. E., Altenbern, R. A., and Ginoza, H. S.,** *J. Bacteriol.,* 567—576, 1958.
15. **Kind, J. R. and Gooder, H.,** *J. Bacteriol.,* 103, 686—691, 1970.
16. **Roberts, R. B.,** *J. Bacteriol.,* 92, 1609—1614, 1966.
17. **Altenbern, R. A. and Landman, O. E.,** *J. Bacteriol.,* 79, 510—518, 1960.
18. **Lawson, J. W. and Douglas, J. T.,** *Can. J. Microbiol.,* 19, 1145—1151, 1973.
19. **Dark, F. A. and Strange, R. E.,** *Nature,* 180, 759—760, 1957.
20. **Abrams, A.,** *J. Biol. Chem.,* 234, 383—388, 1959.
21. **Eddy, A. A. and Williamson, D. H.,** *Nature,* 179, 1252—1253, 1957.
22. **Harold, F. M.,** *J. Bacteriol.,* 88, 1416—1420, 1964.

QUALITATIVE REQUIREMENTS
AND UTILIZATION OF NUTRIENTS: BACTERIA*

G. D. Hegeman and C. O. Patterson

The following table summarizes the qualitative nutritional requirements, so far as they are known, of 245 genera of bacteria. Taxonomy and arrangement follow that of the 8th edition of *Bergey's Manual.*[1] In the interest of brevity, only a few typical carbon and energy sources are given for most heterotrophs. In few cases are the minimal nutritional requirements clearly known. Laboratory cultures, when available, are usually maintained on general purpose, "complex" media unless the organism has demonstrated its inability to grow without further supplementation. Strains of marine origin may require 2 to 3% NaCl in media for growth, and typically exhibit an absolute requirement for the sodium cation; this is usually satisfied by inclusion in growth media of at least 1% (w/v) NaCl.

Although carbon dioxide is listed as sole carbon source for some genera, it is clear that almost all "autotrophs" are capable of considerable assimilation of preformed organic compounds from the medium (eg., acetate), and should most probably be regarded as "mixotrophs" in the sense of Rittenberg,[2] although they may be capable of growth in an entirely inorganic medium. Blanks in the table indicate that no good information was found.

Numerous specific recipes for commonly used media are presented in Appendix VI of the ATCC Catalog,[3] and for certain groups in Aaronson,[4] Schlegel,[5] and in *Bergey's Manual.*[1] Koser[6] has extensively reviewed the vitamin requirements of bacteria. Mineral nutrition of microorganisms is treated by Weinberg.[7]

When no specific reference is cited, the source of the information was either *Bergey's Manual* 1 or one of a number of widely used current texts that treat the biology of the bacteria, e.g., Reference 8. *Bergey's Manual* in particular has such a complete list of references that the reader will find this a particularly useful companion to this table. The user is directed to older editions of *Bergey's Manual* and *Index Bergeyana*[9] for synonyms and generic names not in current use.

Key to Symbols Used

Nitrogen sources	Relation to oxygen	
1. Ammonia or ammonium salts	A	— aerobic
2. Nitrates	aA	— anaerobic
3. Amino acids	mA	— microaerophilic
4. Urea	fa	— facultatively anaerobic
5. Fixes atmospheric N_2	U	— unknown
6. Complex and/or undetermined		
7. Peptone		

In the case of many organisms for which ammonia (or ammonium ion) serves as sole nitrogen source, amino acids, urea, or peptone may serve as well. If this specific information was not available, ammonia is listed alone.

Growth Factor and Carbon and Energy Source Requirements

C — complex
U — undetermined
N — not examined (organisms not available in pure culture or never cultivated in defined medium)

Note: Numbers in parentheses in table are references.

*This work was supported in part by U.S. Public Health Service Research Grant HD-07314 from the National Institute of Child Health and Human Development and a research grant from the Petroleum Research Fund of the American Chemical Society.

1. Phototrophic Bacteria[a]

Genus	Relation to oxygen	Nitrogen source	Growth factors	Carbon sources	Energy sources
Rhodospirillaceae					
Rhodospirillum	aA–phototrophic; fa to mA–dark heterotrophic; some strict aA	1,3,4, 5,7	None, or biotin, or p-ABA (10)	CO_2, fatty acids, some sugars, acetate, pyruvate, most TCA cycle intermediates, some amino acids	Light, H_2, simple organic compounds
Rhodopseudomonas	aA–phototrophic; fa (with NO_3^-)– dark heterotrophic	1,2,3, 4,5,7	None, or thiamine, biotin, nicotinate, p-ABA, pantothenate (10,11)	CO_2, alcohols, fatty acids, amino acids, pyruvate, TCA cycle intermediates, some sugars	Light, H_2, $S^=$, $S_2O_3^=$, simple organic compounds
Rhodomicrobium	aA–phototrophic; fa to mA–dark heterotrophic	1,5,7	None	CO_2, some alcohols, most TCA cycle intermediates	Light, H_2, simple organic compounds
Chromatiaceae					
Chromatium	aA	1,4,5, 7	None, or B_{12} (11,12)	CO_2, acetate, pyruvate, glucose (some strains)	Light, H_2, S^0, $S^=$, $S_2O_3^=$, simple organic compounds
Thiocystis	aA	1,2,4, 5	None	CO_2, acetate, pyruvate	Light, H_2, S^0, $S^=$, $S_2O_3^=$
Thiosarcina	aA	1,5	None	CO_2, acetate, pyruvate	Light, H_2, S^0, $S^=$
Thiospirillum	aA	1,5	B_{12}	CO_2, acetate, pyruvate	Light, H_2, S^0, $S^=$
Thiocapsa	aA, one species mA in darkness	1,5	None	CO_2, acetate, pyruvate, fructose, glycerol	Light, H_2, S^0, S, $S_2O_3^=$, simple organic compounds
Lamprocystis	aA	1,5	None	CO_2, acetate, pyruvate	Light, H_2, S^0, $S^=$
Thiodictyon	aA	1,5	None	CO_2, acetate, pyruvate	Light, H_2, S^0, $S^=$
Thiopedia	aA	1,5?	N	CO_2, N	Light, H_2, S^0, $S^=$
Amoebobacter	aA	1,5	B_{12}	CO_2, pyruvate	Light, H_2, S^0, $S^=$, $S_2O_3^=$, sulfite
Ectothiorhodospira	aA	1,5	B_{12}, elevated NaCl levels	CO_2, U	Light, H_2, S^0, $S^=$, $S_2O_3^=$, acetate

[a] For phototrophic bacteria H_2 is not a true source except for the *Rhodospirillaceae* growing in the dark, aerobically. It is rather an electron donor for photosynthetic growth.

1. Phototrophic Bacteria[a] (continued)

Genus	Relation to oxygen	Nitrogen source	Growth factors	Carbon sources	Energy sources
			Chlorobiaceae		
Chlorobium	aA	1,5	B_{12} (11)	CO_2, acetate (in presence of $S^=$)	Light, S^O, $S^=$, $S_2O_3^=$, H_2 in presence of H_2S
Prosthecochloris	aA	1,5	B_{12}	CO_2, acetate (in presence of $S^=$)	Light, S^O, $S^=$, H_2 in presence of H_2S
Pelodictyon	aA	1,5	None	CO_2, acetate (in presence of $S^=$)	Light, S^O, $S^=$, H_2 in presence of H_2S
Clathrochloris	aA	1	N	CO_2, N	Light, S^O, $S^=$, H_2?

2. Gliding Bacteria

Genus	Relation to oxygen	Nitrogen source	Growth factors	Carbon sources	Energy sources
			Myxococcaceae		
Myxococcus	A	6	C, U, including amino acids; no definite vitamin requirements shown (13–16)	Peptides and amino acids	
			Archangiaceae		
Archangium	A	6	C, U	Peptides and amino acids	
			Cystobacteraceae		
Cystobacter	A	6	C, U	Peptides and amino acids	
Melittangium	A	6	C, U, but apparently including amino acid mixtures	Peptides and amino acids	
Stigmatella	A	6	C, U, but apparently including amino acid mixtures and thiamine (16)	Peptides and amino acids	
			Polyangiaceae		
Polyangium	A	6,2	U or N, or none	Peptides and amino acids; one strain uses cellulose	
Nannocystis	A	6	U, apparently includes B_{12}	Peptides and amino acids	
Chondromyces	A	6	C, U	Peptides and amino acids	
			Cytophagaceae		
Cytophaga	A or fa	1,2,7,3	None or U	Polysaccharides such as cellulose, agar, chitin	
Flexibacter	A, one speices fa	1,2,3,7 (17)	U, but including various amino acids and thiamine, may depend on culture temperature (17)	Glucose, galactose, sucrose, starch (17)	

2. Gliding Bacteria (continued)

Genus	Relation to oxygen	Nitrogen source	Growth factors	Carbon source	Energy source
			Cytophagaceae (continued)		
Herpetosiphon	A	2,7	U	Glucose, galactose, sucrose; cellulose for some strains	
Flexithrix	A	2,7	None	Glucose, galactose, sucrose	
Saprospira	A	7	U, but including various amino acids and vitamins	Glucose, galactose, sucrose, acetate	
Sporocytophaga	A	1,2,4,7	None	Glucose, cellobiose, cellulose, mannose	
			Beggiatoaceae		
Beggiatoa	A—mA	3	(Some sulfhydryl compounds; catalase stimulates growth)	CO_2, reduced sulfur compounds ($S^=$, S^O, $S_2O_3^=$, etc.)	
Vitreoscilla	A?	6	C, U	U	
Thioploca	mA	6	N	U	
			Simonsiellaceae		
Simonsiella	A	6	N, grows on blood and serum agar	U	
Alysiella	A	6	N, grows on blood and serum agar	U	
			Leucotrichaceae		
Leucothrix	A	6	None, or U NaCl	Many simple carbon compounds are oxidized (18)	
Thiothrix	mA?	6	N	N (CO_2 and reduced sulfur compounds?)	
		Families and Genera of Gliding Bacteria of Uncertain Affiliation			
Toxothrix	mA?	6	N	N	
			Achromatiaceae		
Achromatium	mA?	6	N	N (CO_2 and reduced sulfur compounds?)	
			Pelonemataceae		
Pelonema	mA?	6	N	N	
Achroonema	mA—A?	6	N	N	
Peloploca	mA?	6	N	N	
Desmanthos	mA?	6	N	N	

3. The Sheathed Bacteria

Genus	Relation to oxygen	Nitrogen source	Growth factors	Carbon source	Energy source
Sphaerotilus	A	1,3,7,2 with vitamin B_{12} or methionine	None, if grown on organic nitrogen; B_{12} or methionine if grown on nitrate	Some alcohols, organic acids, and sugars (e.g., ethanol, acetate and sorbitol); varies with strain.	
Leptothrix	A	3,7	None, or biotin and thiamine	Organic acids and sugars	
Streptothrix	A	1,2,7 glutamate	B_{12}, thiamine	Glucose, lactose, and sucrose	
Lieskeella	A?	6	N	N. Not in pure culture; metabolism virtually unknown	
Phragmidiothrix	A?	6	N	N. Not in pure culture; not cultivated on artificial medium.	
Crenothrix	A?	6	N	N. Not in pure culture; not cultivated on artificial medium.	
Clonothrix	A?	6	N	N. Not in pure culture; not cultivated on artificial medium.	

4. Budding and/or Appendaged Bacteria

Genus	Relation to oxygen	Nitrogen source	Growth factors	Carbon source	Energy source
Hyphomicrobium	A, or fa in presence of NO_3	1,2	None	Formate, acetate, propionate, etc.	
Hyphomonas	A	1, 7	CO_2	Isolated only once, grew on rich media such as coagulated serum, blood agar, etc.	
Pedomicrobium	A	1, 7			
Caulobacter	A		None or riboflavin or biotin or B_{12} (19)	Wide variety of sugars, amino acids, and organic acids (19)	
Asticcacaulis	A		Biotin	Wide variety of sugars, amino acids, and organic acids (19)	
Ancalomicrobium	fa	1	Pantothenate, possibly others		
Prosthecomicrobium	A	1	B_{12}, biotin, thiamine	Sugars	
Thiodendron	A?		$S^=$		
Pasteuria	A		U, satisfied by peptone, yeast extract		
Blastobacter	?U			Never obtained in pure culture	
Seliberia	fa		None		
Gallionella	A		None?	CO_2	Fe^{++} (FeS)
Nevskia	A			? has never been grown in pure culture	
Planctomyces	A?			Has not been grown in artificial media	
Metallogenium	A		N; satisfied by serum or growth in mixed culture with living fungi		
Caulococcus	mA	6	N	Has not been grown in pure culture or on artificial media	
Kusnezovia	A?			Has not been grown in pure culture or on artificial media	

5. Spirochetes

Genus	Relation to oxygen	Nitrogen source	Growth factors	Carbon source	Energy source
Spirochaeta	aA or fa	3?	U	Growth reported only on complex media; minimal requirements unknown	
Cristispira	U	6	N	Has never been grown in pure culture; nutritional require-ments unknown.	
Treponema	aA	6		Several species never cultured *in vitro*; of species cultured *in vitro*, all require peptone-yeast extract-serum media. Minimal requirements unknown	
Borrelia	aA	6	N	Minimal nutritional requirements unknown; complex media needed for growth	
Leptospira	A	1	B_{12}, thiamine, possibly others for some strains		

6. Spiral and Curved Bacteria

Spirillaceae

Genus	Relation to oxygen	Nitrogen source	Growth factors	Carbon source	Energy source
Spirillum	A, or fa in presence of NO_3^-	1	None, or C	C or sugars (e.g., glucose), organic acids, alcohols.	
Campylobacter	mA–aA	6	C, U usually satisified by serum	C. Pyruvate and lactate utilized; amino acids (?).	

Spiral and Curved Genera of Uncertain Affiliation

Genus	Relation to oxygen	Nitrogen source	Growth factors	Carbon source	Energy source
Bdellovibrio	A		Complex and/or undetermined; many strains obligately parasitic on other bacteria	U	
Microcyclus	A	1	C, U (biotin?)	U	
Pelosigma	aA	6	U	U	
Brachyarcus	aA or A	6	N	N	

7. Gram-negative Aerobic Rods and Cocci

Pseudomonadaceae

Genus	Relation to oxygen	Nitrogen source	Growth factors	Carbon source	Energy source
Pseudomonas	A, or fa with NO_3^-	1	None, except *P. maltophilia* (methio-nine), *P. vesicularis* (biotin, pantothenate, and cyanocobalamin), and *P. diminuta* (biotin, pantothenate, cyanocobalamin, and cysteine)	Wide variety of organic compound of all classes of which lactate and succinate seem nearly universal (20)	
Xanthomonas	A	1	C, including nico-tinate, some amino acids (especially gluta-mate and methionine)	Glucose and a variety of other sugars	

7. Gram-negative Aerobic Rods and Cocci (continued)

Genus	Relation to oxygen	Nitrogen source	Growth factors	Carbon source	Energy source
		Pseudomonadaceae (continued)			
Zoogloea	A	1 (if biotin and B_{12} provided), 3,6	B_{12}	Xylose, fructose, sucrose, mannose, ethanol and glycerol, and some amino acids (aspartate, glutamate, arginine, histidine, among others)	
Gluconobacter	A	1	C, including pABA	Amino acids, ethanol, glucose, and other alcohols including polyols (incompletely oxidized) from the medium by assimilation	
		Azotobacteracae			
Azotobacter	A	1,2,3,5	None	Carbohydrates, alcohols, organic acids, some use starch and rhamnose	
Azomonas	A	1,2,3,5	None	Carbohydrates, alcohols, organic acids, *none* use starch or rhamnose	
Beijerinckia	A	1,2,3,5	None	Hexoses and sucrose used by all; pattern of use of other compounds besides sugars is species specific	
		Rhizobiaceae			
Rhizobium	A	1,2,3,5 in specific symbioses with leguminous plants	U or biotin or none	*Monomeric* carbohydrates, organic acids, or amino acids; yeast extract commonly used; mannitol; pentoses preferred	
Agrobacterium	A	1,2,3	None or U (21)	Wide variety of monomeric carbohydrates, organic acids, amino acids; polymers of these not used	
		Methylomonadaceae			
Methylomonas	A	2	None	Only methane or methanol	
Methylococcus	A	2	None	Only methane or methanol	
		Halobacteriaceae (22)			
				All require 2 *M* (about 12%) NaCl in growth medium	
Halobacterium	A	6	None	Amino acids almost exclusively	
Halococcus	A	6,2 (some strains)	None	Amino acids almost exclusively; some hydrolyze starch	
		Gram-negative Rods and Cocci of Uncertain Affiliation			
Alcaligenes	A, aA with NO_3 respiration	1,2,3	None	Organic acids, amino acids, some use carbohydrates	
Acetobacter	A	1,3	None, or pABA, pantothenate, and nicotinate	Ability to metabolize acetate or its precursors species specific. Ethanol and lactate are good carbon sources. Usually grown on complex medium.	

7. Gram-negative Aerobic Rods and Cocci (continued)

Genus	Relation to oxygen	Nitrogen source	Growth factors	Carbon source	Energy source

Gram-negative Rods and Cocci of Uncertain Affiliation (continued)

Genus	Relation to oxygen	Nitrogen source	Growth factors	Carbon source
Brucella	A	1	Thiamin, niacin, biotin (pantothenate may stimulate)	Amino acids (23) (some may require 5–10% CO_2 gas)
Bordetella	A	1,4	Nicotinate, cysteine, methionine	Amino acids, usually grown on complex media
Francisella	A		Undefined, but no growth on ordinary media unless enriched as with blood or animal tissues	Usually grown on complex media
Thermus	A	1,3	None	Amino acids, sugars, organic acids

8. Gram-negative Facultatively Anaerobic Rods

Enterobacteriaceae

Genus	Relation to oxygen	Nitrogen source	Growth factors	Carbon source
Escherichia	fa	1	None	Sugars and sugar alcohols, amino acids and fatty acids
Edwardsiella	fa	1	None	Maltose fermentatively
Citrobacter	fa	1	None	Like *Escherichia* but also organic polycarboxylic acids
Salmonella	fa	1	None (*S. typhimurium* requires tryptophan)	Like *Citrobacter* (24)
Shigella	fa	1	None	Like *E. coli*, but a smaller range of compounds
Klebsiella	fa	1	None	Sugars and sugar alcohols, organic and some amino acids
Enterobacter	fa	1	None	Sugars, sugar alcohols, organic acids
Hafnia	fa	1	None	Arabinose, maltose, mannitol, xylose, citrate, and gluconate among others
Serratia	fa	1	None	Like *Hafnia* but does not use arabinose
Proteus	fa	1	None, or nicotinate and pantothenate	Some sugars, sugar alcohols, amino acids
Yersinia	fa	1	None	Glucose, fructose, maltose, mannitol, glycerol, and mannose as well as some amino acids
Erwinia	fa	1	None (some may require organic N)	Fructose, glucose, galactose, and some amino acids

Vibrionaceae

Genus	Relation to oxygen	Nitrogen source	Growth factors	Carbon source
Vibrio	fa	1	None	Glutamate, succinate aerobically, glucose anaerobically by all
Aeromonas	fa	1	None	Glucose, or certain amino acids, fructose, maltose

Gram-negative Rods and Cocci of Uncertain Affiliation (continued)

Genus	Relation to oxygen	Nitrogen source	Growth factors	Carbon source	Energy source
			Vibrionaceae (continued)		
Plesiomonas	fa	1	None	Glucose, inositol (less versatile than other *Vibrionaceae*)	
Photobacterium	fa	1	None	Glucose, glycerol, some amino acids,	
Lucibacterium	fa	1	None	Glucose, glycerol, some amino acids,	
Zymomonas	fa	1	Pantothenate	Glucose, fructose, and sucrose	
Chromobacterium	fa	1,3	None	Glucose, fructose, other sugars and some amino acids	
Falvobacterium	fa	6	None, or thiamine, perhaps other B vitamins	Amino acids	
Haemophilus	fa	Poorly defined, but grow well only on complex media, as chocolate agar or blood agar; niacin adenine dinucleotide (NAD) "V requirement," "X requirement" (hemin) (25,26,27) required for some.		Glucose and sugars		
Pasteurella	fa	Complex and poorly defined, usually grown on blood agar		Glucose and fructose		
Actinobacillus	fa	Minimal requirements unknown, growth reported only on complex media		Glucose, fructose, and xylose		
Cardiobacterium	fa	3	Pantothenate, nicotinamide, pyridoxine, thiamine, biotin	Glucose, fructose, mannose, sorbitol, and sucrose	
Streptobacillus	fa	Minimal nutritional requirements unknown; serum, ascitic fluid, or blood required for growth		Glucose, fructose, maltose, galactose, among others		
Calymmatobacterium	fa	Isolated and grown on fresh egg yolk medium		N		

9. Gram-negative Strict Anaerobes

Bacteroidaceae

Genus	Relation to oxygen	Nitrogen source	Growth factors	Carbon source	Energy source
Bacteroides	aA	6	None, or hemin, or B vitamins	Glucose, or maltose and sucrose generally used; some amino acids. Some strains don't ferment carbohydrates, only amino acids.	
Fusobacterium	aA	6	U. Rich media used for cultivation, serum or ascitic fluid supplements sometimes used.	Various carbohydrates or peptone	
Leptotrichia	aA (5% CO_2 needed)	6	C, U	Glucose and probably other carbohydrates	

9. Gram-negative Strict Anaerobes (continued)

Genus	Relation to oxygen	Nitrogen source	Growth factors	Carbon source	Energy source
		Gram-negative Strictly Anaerobic Genera of Uncertain Affiliation			
Desulfovibrio	aA	1,5	None	Lactate, pyruvate, and usually malate. Sulfate required as terminal electron acceptor.	
Butyrivibrio	aA	1	Variable, usually B vitamins, cysteine	Glucose, fructose, maltose, and other carbohydrates	
Succinivibrio	aA	6	Poorly defined	Glucose, fructose, sucrose, and other carbohydrates	
Succinomonas	aA	1	B vitamins	Glucose, maltose, dextrin, and starch	
Lachnospira	aA	1	B vitamins	Glucose, fructose, sucrose, and other carbohydrates	
Selenomonas	aA (CO_2 required)	1,3	C, U organic growth factors (especially B vitamins) required, but not adequately studied. Usually grown on complex media.	Glucose and other carbohydrates; certain amino acids; lactic acid	

10. Gram-negative Cocci and Coccobacillus

Genus	Relation to oxygen	Nitrogen source	Growth factors	Carbon source	Energy source
Neisseria	fa (CO_2 required)	3	C. Complex and variable. Glutamine, iron	Glucose and a small collection of other carbohydrates	
Branhamella	A	3	Biotin	Lactate or succinate; some amino acids	
Moraxella	A	1,3	U. Most strains more or less fastidious, but specific growth requirements usually not known. Some require oleic acid or serum.	Organic acids, alcohols, and amino acids. Acetate, lactate, pyruvate used by most.	
Acinetobacter	A	1	None	A wide range of organic compounds (28)	
Paracoccus	A or fa with NO_3^-; one species facultatively chemolithotrophic with H_2	1,2	None, or complex and undefined requirements; halophiles require at least 0.4 M NaCl.	CO_2 and H_2 (some); a wide variety of organic compounds including sucrose and maltose; acetate, pyruvate, and glycerol	
Lampropedia	A	1,3	Biotin, thiamine	Pyruvate, lactate, butyrate, and other organic acids	

11. Gram-negative Anaerobic Cocci

Genus	Relation to oxygen	Nitrogen source	Growth factors	Carbon source	Energy source
Veillonella	aA (CO$_2$ required)	3	Variable, pyridoxal, thiamine indispensable; pABA, biotin, putrescine, and cadaverine required by some strains	Pyruvate, lactate, malate, fumarate, or oxalacetate	
Acidaminococcus	aA	3	C. Complex and multiple. Tryptophan, glutamate, valine, and arginine required; cysteine, histidine, tyrosine, phenylalanine, serine usually required. B$_{12}$, pyridoxal, pantothenate, biotin, pABA required for growth on amino acid media.	Glutamic acid and other amino acids	
Megasphaera	aA	3	C, U, can be satisfied by yeast extract and thioglycolate	Fructose, glucose, and lactate	

12. Gram-negative, Chemolithotrophic Bacteria

Nitrobacteraceae

Genus	Relation to oxygen	Nitrogen source	Growth factors	Carbon source	Energy source
Nitrobacter	A	NO$_2^-$	None	CO$_2$	NO$_2^-$
Nitrospina	A	NO$_2^-$	None	CO$_2$	NO$_2^-$
Nitrococcus	A	NO$_2^-$	None	CO$_2$	NO$_2^-$
Nitrosomonas	A	1	None	CO$_2$	NH$_3^+$, NH$_2$OH
Nitrosospira	A	1	None	CO$_2$	NH$_3^+$, NH$_2$OH
Nitrosococcus	A	1	None	CO$_2$	NH$_3^+$, NH$_2$OH
Nitrosolobus	A	1	None	CO$_2$	NH$_3^+$, NH$_2$OH, biuret

Organisms Metabolizing Sulfur

Genus	Relation to oxygen	Nitrogen source	Growth factors	Carbon source	Energy source
Thiobacillus	A, or fa with NO$_3^-$ present	1,2	None	CO$_2$	S^0, S$^=$, Fe^{++} S$_2$O$_3^=$, SO$^=$ polythionates. Some use organic compounds.
Sulfolobus	A	1	None	CO$_2$, glutamate, ribose	S^0, glutamate, ribose
Thiobacterium	A?	6	N	N	N
Macromonas	mA	6	N	N	N
Thiovulum	mA	6	N	N	N
Thiospira	mA?	6	N	N	N

Siderocapsaceae

Genus	Relation to oxygen	Nitrogen source	Growth factors	Carbon source	Energy source
Siderocapsa	A–mA	6	N	N	N
Naumanniella	A?	6	N	N	N
Ochrobium	A?	6	N	N	N
Siderococcus	mA	6	N	N	N

13. Methane-producing Bacteria

Genus	Relation to oxygen	Nitrogen source	Growth factors	Carbon source	Energy source
Methanobacterium	aA	1	N	CO_2, formate, acetate, methanol, CO	H_2, formate, acetate, methanol, CO
Methanosarcina	aA	1	None or N	CO_2, acetate, methanol	H_2, acetate, methanol, CO
Methanococcus	aA	1	N	CO_2, acetate, formate	H_2, acetate, butyrate

14. Gram-positive Cocci

Micrococcaceae

Genus	Relation to oxygen	Nitrogen source	Growth factors	Carbon source	Energy source
Micrococcus	A	1, some 3	None, or N, or biotin, thiamine	Pyruvate, lactate, acetate, succinate, glutamate, glucose, and other carbohydrates	
Staphylococcus	fa	3	C, amino acids, especially biotin, uracil	Glucose and many other carbohydrates	
Planococcus	A	6	None	Glucose, maltose, lactose, and sucrose, among others	

Streptococcaceae

Genus	Relation to oxygen	Nitrogen source	Growth factors	Carbon source	Energy source
Streptococcus	fa	1	C, may include amino acids, purines, pyrimidines, vitamins, and fatty acids	Glucose, maltose, lactose, sucrose, and other sugars	
Leuconostoc	fa	1	C, nicotinate, thiamine, pantothenate, biotin, amino acids	Glucose by all; maltose, fructose, lactose by most	
Pediococcus	mA (CO_2 required)	3	C, niacin, biotin, riboflavin, pyridoxin, pantothenate, amino acids, mevalonate, etc.	Glucose, maltose, and some other carbohydrates	
Aerococcus	mA–fa		C, pantothenate, nicotinate, biotin, purines, amino acids	Glucose, maltose, and some other carbohydrates	
Gemella	fa	6	None	Carbohydrates	

Peptococcaceae

Genus	Relation to oxygen	Nitrogen source	Growth factors	Carbon source	Energy source
Peptococcus	aA	3,7	None	Peptides and amino acids, some sugars	
Peptostreptococcus	aA	6,7	U	Glucose, other carbohydrates, pyruvate	
Ruminococcus	aA	1	C, including B vitamins, cellobiose, branched-chain acids, etc.	Cellobiose, other carbohydrates	
Sarcina	aA	1,3	C, including B vitamins and amino acids	Glucose and few other carbohydrates	

15. Endospore-forming Rods and Cocci

Genus	Relation to oxygen	Nitrogen source	Growth factors	Carbon source	Energy source
			Bacillaceae		
Bacillus	A or fa (by fermentation or with NO_3^-)	1,2,3,5	None to C to U	A variety of compounds including carbohydrates, organic acids and alcohols, amino acids, and purines	
Sporolactobacillus	mA	1,7	C, leucine, valine, biotin, pantothenate	Glucose and other sugars and sugar alcohols	
Clostridium	A or aA	1,5,6,7	None or biotin or C and U	Glucose and other carbohydrates or amino acids singly or in combinations or purines, H_2, and CO_2; acetate and ethanol pairwise	
Desulfotomaculum	aA	6	None	Lactate, pyruvate (reducible sulfur compound required)	
Sporosarcina	A	1,7	U (free ammonia?)	Peptides (amino acids), and perhaps organic acids	
			Endospore-forming Rods of Uncertain Affiliation		
Oscillospira	aA?	6	N	N	

16. Gram-positive, Asporogenous Rod-shaped Bacteria

			Lactobacillaceae		
Lactobacillus	A or aA	3	C, strain specific	Glucose, fructose, and other sugars and sugar alcohols	
			Genera of Uncertain Affiliation		
Listeria	A—mA (may need CO_2)	3	C, biotin, riboflavin, thiamine, thioctate, amino acids	Glucose and a few other carbohydrates	
Erysipelothrix	A—mA (may need CO_2)	3	Riboflavin, oleate	Glucose and other sugars and sugar alcohols	
Caryophanon	A	1, glutamate	U, but includes thiamine, biotin	Acetate and glutamate, among others	

17. Actinomycetes and Related Organisms

			Coryneform Group		
Corynebacterium	A—fa	1	C, U, but includes nicotinate, pimelate, sometimes thiamine, some amino acids; β-alanine or pantothenate, sometimes pABA + purines	Sugars (glucose, maltose), some amino acids	
Arthrobacter	A	1,2,6	U, sometimes none, or biotin, *terregens* factor (29,30)	Sugars (e.g., glucose, mannose), organic acids, amino acids (31)	
Cellulomonas	fa?	1,2	Biotin, thiamine	Cellulose, other carbohydrates, acetate	
Kurthia	A	6	N	N	

17. Actinomycetes and Related Organisms (continued)

Genus	Relation to oxygen	Nitrogen source	Growth factors	Carbon source	Energy source
Propionibacteriaceae					
Propionibacterium	aA–mA		Biotin, sometimes pantothenate, thiamine, pABA	Glucose, glycerol, lactate, fructose, other carbohydrates variable	
Eubacterium	aA		C, U	Glucose, other carbohydrates variable	
Actinomycetales					
Actinomycetes	aA–fa (CO₂ may be required)		C, U	Glucose, other carbohydrates variable	
Arachnia	fa	6	C, U	Glucose, other carbohydrates variable	
Bifidobacterium	aA	6,7,1	Riboflavin, thiamine, pantothenate, sometimes folate, some strains U	Glucose, galactose, fructose; other carbohydrates variable	
Bacterionema	fa		Riboflavin, thiamine, nicotinate, pantothenate, cysteine; hemin if anaerobic	Glucose, fructose, sucrose; other carbohydrates variable	
Rothia	A	6,7	N	Glucose, maltose, sucrose, and other carbohydrates	
Mycobacteriaceae					
Mycobacterium	A–mA (CO₂ may be stimulating)	6	None (saprophytes) or N or *mycobactin* (some) and U (30)	Saprophytes may grow on simple media with one of a variety of organic compounds. Some of the parasitic types have never been cultivated outside living cells.	
Frankiaceae					
Frankia	mA?	6,5 (symbiotically)	N	U	
Actinoplanaceae					
Actinoplanes	A	6,7	N	U (amino acid mixtures will serve)	
Spirillospora	A	6,7	N	U (peptides and amino acids in mixture)	
Streptosporangium	A	6,7	N	U (peptides)	
Amorphosporangium	A	6,7	None	U (amino acids)	
Ampullariella	A	6,7	N	U (amino acids)	
Pilimelia	A	6,7	N	U (peptides)	
Planomonospora	A	6,7	U	U (a variety of substrates)	
Planobispora	A	6	N	Arabinose, glucose, xylose, and other carbohydrates	
Dactylosporangium	A	6	N	Arabinose, glucose, and other carbohydrates; amino acids; other substrates	
Kitasatoa	A	6	None	Carbohydrates including sucrose and glycerol; other compounds	

17. Actinomycetes and Related Organisms (continued)

Genus	Relation to oxygen	Nitrogen source	Growth factors	Carbon source	Energy source
Dermatophilaceae					
Dermatophilus	A,fa	6	N	U	
Geodermatophilus	A	6	N	Glucose, fructose, mannitol; other carbohydrates variable	
Nocardiaceae					
Nocardia	A	1,2?	U or none	U (parasites); saprophytes may use a wide variety of hydrophobic and hydrophilic organic compounds. Two species are facultative chemo-autotrophs.	
Pseudonocardia	A	6,7	U or none	U. Probably a variety of carbohydrates and amino acids, singly and in combination	
Streptomycetaceae					
Streptomyces	A	6	None	Glucose, other carbohydrates and other organic compounds	
Streptoverticillium	A	2,6,7	None or U	Glucose, other carbohydrates; amino acids	
Sporichthya	fa	2,6,7	None	Sucrose, fructose	
Microellobospira	A	2,6,7	None	Glucose, other carbohydrates	
Micromonosporaceae					
Micromonospora	A or aA	2,6,7	None	Glucose, sucrose, other carbohydrates	
Thermoactinomyces	A	6,7	None	Glucose, other carbohydrates	
Actinobifida	A	6,7	U or none	Glucose, sucrose, asparagine	
Thermomonospora	fa	6,7	U	U	
Microbispora	A–fa	6,7	Thiamine; other B vitamins	U	
Micropolyspora	A	6,3,7	U	U (apparently prefers amino acids to carbohydrates)	

18. Rickettsias

Genus	Relation to oxygen	Nitrogen source	Growth factors	Carbon source	Energy source
Rickettsiaceae					
Rickettsia	A?	6	N	Cultivated only within eukaryotic host cells	
Rochalimaea	A	6	N, requires blood + serum + erythrocytes	U	
Coxiella	A?	6	N	Cultivated only within eukaryotic host cells and chick embryo yolk sacs	
Ehrlichia	A?	6	N	Cultivated only within eukaryotic host cells, especially leukocytes	

18. Rickettsias (continued)

Genus	Relation to oxygen	Nitrogen source	Growth factors	Carbon source	Energy source
		Rickettsiaceae (continued)			
Cowdria	A?	6	N	Cultivated only within eukaryotic host cells	
Neorickettsia	A?	6	N	Cultivated only within eukaryotic host cells, expecially lymphoid tissue of *Canidae*	
Wolbachia	A	6	N	Typically cultivated within eukaryotic host cells, especially arthropods, and chick yolk sac. One species reported (1924) cultivable on glucose-blood-bouillon agar. (32)	
Symbiotes	A?	6	N	Observed only within cells of bedbugs (*Cimex*)	
Blattabacterium	A?	6	N	Observed only within cells of cockroaches (*Blatta*)	
Rickettsiella	A?	6	N	Typically cultivated within cells of insect larvae	
		Bartonellaceae			
Bartonella	A	6	N	Cultivable on cell-free media containing fresh blood, or fresh serum and hemoglobin	
Grahamella	A	6	N	Cultivable on cell-free complex media; addition of hemoglobin favors growth	
		Anaplasmataceae			
Anaplasma	A?	6	N	Have not been cultivated; observed in ruminant erythrocytes	
Paranaplasma	A?	6	N	Have not been cultivated; observed in erythrocytes of cattle, possibly sheep	
Aegyptianella	A?	6	N	Have not been cultivated; observed in erythrocytes of cattle, sheep, birds	
Haemobartonella	A?	6	N	Not cultivated; observed within erythrocytes of whole blood of vertebrates	
Eperythrozoon	A?	6	N	Not cultivated; observed within whole blood of vertebrates	
		Chlamydiaceae			
Chlamydia	A?	6	N	Not cultivated. If separated from host cells, and supplied with extensive organic and inorganic cofactors (especially ATP), will carry out certain metabolic activities, but will not divide and cannot be propagated.	

19. The Mycoplasmas

Genus	Relation to oxygen	Nitrogen source	Growth factors	Carbon source	Energy source
Mycoplasma	A–aA	6	C, U, including sterols (especially cholesterol), long-chain fatty acids. (33,34) "T" strains require urea and cholesterol.	U, most species utilize either glucose or arginine, or rarely both. Glucose utilizers usually also ferment a small range of other carbohydrates such as fructose, galactose, starch, etc.	

Acholeplasmataceae

Genus	Relation to oxygen	Nitrogen source	Growth factors	Carbon source	Energy source
Acholesplasma	fa	6	C, U, including nicotinate, riboflavin, folate, pyridoxine, pyridoxol, thiamine, long-chain fatty acids, amino acids, nucleic acid precursors	U, a small range of carbohydrates such as glucose, maltose, starch known to be utilized; possibly others	

Genera of Uncertain Affiliation

Genus	Relation to oxygen	Nitrogen source	Growth factors	Carbon source	Energy source
Thermoplasma	A?	6	U, can be supplied by yeast extract	U, apparently uses glucose	
Spiroplasma	fa	6	C, U, including sterols	U, including glucose	

20. Cyanobacteria (Blue-green Algae)

Genus	Relation to oxygen	Nitrogen source	Growth factors	Carbon source	Energy source
All genera	A	2, some 1, some 5, some 4	Usually none, some marine isolates require B_{12} (35)	Long thought to be strict photoautotrophs, using light as sole energy source, CO_2 as sole carbon source. Some strains now known to be facultative autotrophs, growing heterotrophically using glucose, sucrose, and possibly other sugars. Some strains known to photoassimilate acetate, glucose, glycolate, and possibly other organic carbon sources. (36, 37, 38)	

REFERENCES

1. **Buchanan, R. E. and Gibbons, N. E.**, Eds., *Bergey's Manual of Determinative Bacteriology,* 8th ed., Williams & Wilkins, Baltimore, 1974.
2. **Rittenberg, S. C.**, *Adv. Microb. Physiol.,* 3, 159–196, 1969.
3. *Catalogue of Strains,* The American Type Culture Collection, 11th ed., 12301 Parklawn Drive, Rockville, Md . 20852, 1974.
4. **Aaronson, S.**, *Experimental Microbial Ecology,* Academic Press, New York, 1970.
5. **Schlegel, H.**, Ed., *Anreicherungskultur und Mutantenauslese,* Gustav Fischer Verlag, Stuttgart, 1965.
6. **Koser, S. A.**, *Vitamin Requirements of Bacteria and Yeasts,* Charles C Thomas, Springfield, Ill., 1968.
7. **Weinberg, E. D.**, Ed., *Microorganisms and Minerals,* Marcel Dekker, New York, 1977.
8. **Stanier, R. Y., Adelberg, E. A., and Ingraham, J. L.**, *The Microbial World,* 4th ed., Prentice-Hall, Englewood Cliffs, N.J., 1976.
9. **Buchanan, R. E., Holt, J. G., and Lessel, E. F., Jr.**, *Index Bergeyana,* Williams and Wilkins, Baltimore, 1966.
10. **Hutner, S.**, *J. Bacteriol.,* 52, 213-221, 1946.
11. **Uspenskaya, V. E. and Kondrat'eva, E. N.**, *Mikrobiologiya,* 31, 325–328 (original pagination 396–401), 1962.
12. **Pfennig, N.**, *Naturwissenschaften,* 48, 136, 1961.
13. **Dworkin, M.**, *Ann. Rev. Microbiol.,* 20, 75–106, 1966.
14. **Hemphill, H. E. and Zahler, S. A.**, *J. Bacteriol.,* 95, 1011–1023, 1968.
15. **Mayer, D.**, *Arch. Mikrobiol.,* 58, 186–200, 1967.
16. **McCurdy, H. D. and Khouw, B. T.**, *Can. J. Microbiol.,* 15, 731–738, 1969.
17. **Simon, G. D. and White, D.**, *Arch. Mikrobiol.,* 78, 1-16, 1971.
18. **Harold, R. and Stanier, R. Y.**, *Bacteriol. Rev.,* 19, 49-64, 1955.
19. **Poindexter, J. S.**, *Bacteriol. Rev.,* 28, 231-295, 1964.
20. **Stanier, R. Y., Palleroni, N. J., and Doudoroff, M.**, *J. Gen. Microbiol.,* 43, 1-159, 1966.
21. **Starr, M. P.**, *J. Bacteriol.,* 52, 187-194, 1946.
22. **Gibbons, N. E.**, in *Methods in Microbiology,* Norris, J. R. and Ribbons, D. W., Eds., Academic Press, New York, 1969, 169–183.
23. **Gerhardt, P.**, *Bacteriol. Rev.,* 22, 81–98, 1958.
24. **Gutnick, D., Calvo, J. M., Klopotowski, T., and Ames, B. N.**, *J. Bacteriol.,* 100, 215-219, 1969.
25. **Zinneman, K.**, *Ergeb. Mikrobiol. Immunitaetsforsch. Exp. Ther.,* 33, 307-368, 1960.
26. **Zinneman, K., Rogers, K. B., Frazer, J., and Boyce, J. M. H.**, *J. Pathol. Bacteriol.,* 96, 413–419, 1968.
27. **White, D. C., Leidy, G., Jamieson, J. D., and Shope, R. E.**, *J. Exp. Med.,* 120, 1-12, 1964.
28. **Bauman, P., Doudoroff, M., and Stanier, R. Y.**, *J. Bacteriol.,* 95, 1520-1541, 1968.
29. **Morrison, N. E., Antoine, A. E., and Dewbrey, E. E.**, *J. Bacteriol.,* 89, 1630, 1965.
30. **Snow, G. A.**, *Bacteriol. Rev.,* 99-125, 1970.
31. **Ensign, J. C. and Wolfe, R. S.**, *J. Bacteriol.,* 87, 924-932, 1964.
32. **Hertig, M. and Wolbach, S. B.**, *J. Med. Res.,* 44, 329-374, 1924.
33. **Aluotto, B. B., Wittler, R. G., Williams, C. O., and Faber, J. E.**, *Int. J. Syst. Bacteriol.,* 20, 35-58, 1970.
34. **Hayflick, L. and Chanock, R. M.**, *Bacteriol. Rev.,* 29, 185-221, 1965.
35. **Van Baalen, C.**, *Bot. Mar.,* 4, 129-139, 1962.
36. **Grodzinski, B. and Colman, B.**, *Planta,* 124, 125-133, 1975.
37. **Kenyon, C. N., Rippka, R., and Stanier, R. Y.**, *Arch. Mikrobiol.,* 83, 216–236, 1972.
38. **Khoja, T. M. and Whitton, B. A.**, *Br. Phycol. J.,* 10, 139–148, 1975.

QUALITATIVE REQUIREMENTS AND UTILIZATION OF NUTRIENTS: MYCOPLASMA

A. W. Rodwell

Table 1
TAXONOMY OF THE CLASS MOLLICUTES

The term mycoplasma is used in the trivial sense for a heterogeneous group of pro-karyotic microorganisms belonging to the class Mollicutes containing a single order, Mycoplasmatales. This order is separated into three families on the basis of their growth requirement for sterol, size of genome, location of NADH oxidase, and on the occur-rence in family III of helical organisms. Mycoplasmas are the smallest microorganisms capable of autonomous growth. They have limited biosynthetic capabilities, and, with the probable exception of *Thermoplasma,* are nutritionally exacting. Their complex require-ments for lipids and lipid precursors for the synthesis of the plasma membrane, together with the sensitivity of the exposed plasma membrane to damage by surface-active com-pounds, has complicated nutritional studies. Most partly defined media have contained proteins as lipid carriers or binders, and the presence of these precludes the determina-tion of the amino acid requirements since some mycoplasmas at least can obtain amino acids by the degradation of protein.

Class: Mollicutes
   Order: Mycoplasmatales
      Family I: Mycoplasmataceae
          1. Sterol required for growth
          2. Genome size about 5.0×10^8 daltons
          3. NADH oxidase localized in cytoplasm
        Genus I: *Mycoplasma* (about 50 species current)
            1. Do not hydrolyze urea
        Genus II: *Ureaplasma* (single species with serotypes)
            1. Hydrolyzes urea
      Family II: Acholeplasmataceae
          1. Sterol not required for growth
          2. Genome size about 1.0×10^9 daltons
          3. NADH oxidase localized in the membrane
        Genus I: *Acholeplasma* (six species current)
      Family III: Spiroplasmataceae (proposed)
          1. Helical organisms during some phase of growth
          2. Sterol required for growth
          3. Genome size about 1.0×10^9 daltons
          4. NADH oxidase localized in cytoplasm
        Genus I: *Spiroplasma* (one species current)
        Genera of uncertain taxonomic position: *Thermoplasma* (single species);
            Anaeroplasma (two species)

Compiled by S. Razin and J. G. Tully from recommendations of the International Association of Microbiological Societies' Subcommittee on the Taxonomy of the Mycoplasmatales.

Table 2
ENERGY SOURCES FOR GROWTH

Distribution of Arginine Dihydrolase
and Glycolysis Pathways Among Mycoplasmas

The genus *Mycoplasma* contains both fermentative and nonfermentative species. Fermentative species require a metabolizable carbohydrate, usually supplied as glucose. The energy source for the nonfermentative species is less well understood. Most have been shown to possess one or more of the three enzymes of the arginine dihydrolase pathway.[1,2] Arginine supplementation of undefined growth media resulted in a marked increase in growth yield of *M. arthritidis*[1] and *M. hominis*.[3] *M. arthritidis* can use a variety of substrates in addition to arginine.[4,5]

The energy source for *Ureaplasma* is not known. All known species of *Acholeplasma*, *Spiroplasma*, *Thermoplasma*, and *Anaeroplasma* catabolize glucose, while none hydrolyze arginine.

Genus and species	Glucose catabolism	Arginine hydrolysis
Mycoplasma		
M. agalactiae subsp. *agalactiae*	–	–
M. agalactiae subsp. *bovis* (*M. bovimastiditis*)	–	–
M. alkalescens	–	+
M. anatis	+	–
M. arginini	–	+
M. arthritidis	–	+
M. bovigenitalium	–	–
M. bovirhinis	+	–
M. bovoculi	+	+ or –
M. buccale (formerly orale 2)	–	+
M. canis	+	–
M. canadensis	–	+
M. capricolum	+	+ or –
M. caviae	+	–
M. conjunctivae	+	–
M. cynos	+	–
M. dispar	+	–
M. edwardii	+	–
M. equirhinis	–	+
M. faucium (formerly orale 3)	–	+
M. feliminutum	+	–
M. felis	+	–
M. fermentans	+	+
M. flocculare	+	–
M. gallinarum	–	+
M. gallisepticum	+	–
M. gateae	–	+
M. hominis	–	+
M. hyopneumoniae/*M. suipneumoniae*	+	–
M. hyorhinis	+	–
M. hyosynoviae	–	+
M. iners	–	+
M. lipophilum	–	+
M. maculosum	–	+
M. meleagridis	–	+

Table 2 (continued)
ENERGY SOURCES FOR GROWTH

Genus and species	Glucose catabolism	Arginine hydrolysis
Mycoplasma (continued)		
M. moatsii	+	+
M. molare	+	−
M. mycoides subsp. capri	+	−
M. mycoides subsp. mycoides	+	−
M. neurolyticum	+	−
M. opalescens	−	+
M. orale (formerly orale 1)	−	+
M. ovipneumoniae	+	−
M. pneumoniae	+	−
M. primatum	−	+
M. pulmonis	+	−
M. putrefaciens	+	−
M. salivarium	−	+
M. spumans	−	+
M. synoviae	+	NT
M. verecundum	−	−
M. gallisepticum	+	−
Ureaplasma		
U. ureaplasma	−	−
Acholeplasma		
A. axanthum	+	−
A. equifoetale	+	−
A. granularum	+	−
A. laidlawii	+	−
A. modicum	+	−
A. oculi	+	−
Spiroplasma		
S. citri	+	−
Thermoplasma		
T. acidophilum	+	−
Anaeroplasma		
A. abactoclasticum	+	−

From data compiled by J. G. Tully and S. Razin.

REFERENCES

1. Barile, M. F., Schimke, R. T., and Riggs, D. B., *J. Bacteriol.*, 91, 189–192, 1966.
2. Schimke, R. T., Berlin, C. M., Sweeney, E. W., and Carroll, W. R., *J. Biol. Chem.*, 241, 2228–2236, 1966.
3. Razin, S., Gottfried, L., and Rottem, S., *J. Bacteriol.*, 95, 1685–1691, 1968.
4. VanDemark, P. J. and Smith, P. F., *J. Bacteriol.*, 88, 1602–1607, 1964.
5. VanDemark, P. J. and Smith, P. F., *J. Bacteriol.*, 89, 373–377, 1965.

Table 3
NUTRITIONAL REQUIREMENTS OF *MYCOPLASMA MYCOIDES* AND *ACHOLEPLASMA LAIDLAWII*

Only two completely defined media have been described for the growth of mycoplasmas: one for *M. mycoides* subspecies *mycoides* strain Y,[1] and the other for *A. laidlawii* strain B.[2-4] The nutritional requirements of other strains of *M. mycoides* subspecies *mycoides* and *capri* are probably very similar to those for strain Y. A partly defined medium[5] permitted the definition of most of the requirements, apart from those for amino acids, of *A. laidlawii* strain B. The requirements of *A. laidlawii* strains A and B are probably very similar and the data have been combined. *Abbreviations:* E = essential; R = required but essentiality uncertain, may be either E or S; Ɍ = not required; S = not essential for growth, growth stimulated in medium believed to be uncontaminated by factor; U = utilized; u = poorly utilized; ᵾ = not utilized; — = no data available.

Nutrient	*M. mycoides* strain Y	Ref.	*A. laidlawii* strains A and B	Ref.
Inorganic				
Cations				
K^+	E[a]	6	E[a]	5, 7
Mg^{2+}	E	6	E	2, 5
NH_4^+	—		U[b]	5, 8
Anions				
PO_4^{3-}	E	6	R	2, 5
Amino acids				
Alanine	S[d]	1	(Ɍ, E)[c]	4
Arginine	E	1	E	4
Asparagine	E	1	(Ɍ, U)[c]	3, 4
Aspartic acid	Ɍ	1	(E, u)[c]	3, 4
Cysteine	E	1	Ɍ	4
Cystine	Ɍ	1	E	2, 4
Glutamic acid	Ɍ	1	ᵾ[b]	4, 8
Glutamine	E	1	E	3, 4
Glycine	E	1	E	4
Histidine	E	1	E	4

[a] K^+ was inhibitory at concentrations greater than 0.04 M.[5,6]

[b] The addition of NH_4^+ stimulated growth of strain A in a medium deficient in certain amino acids.[5] Strain A possesses a glutamate dehydrogenase for the synthesis of glutamic from α-ketoglutarate and NH_4^+.[8]

[c] The amino acid requirements of strain B, but not those of strain A, have been defined.[3,4] Strain A probably possesses some synthetic ability for amino acids; see Footnote b. Strain B was unable to incorporate glutamic acid, only the amide. Aspartic acid was poorly incorporated, but asparagine was not essential for growth in the presence of aspartic acid. Cysteine was not required in the presence of cystine. Phenylalanine and tyrosine could replace alanine, but neither was required in the presence of alanine. Evidence for the operation of the shikimic acid pathway for the synthesis of aromatic amino acids was obtained by the incorporation of ^{14}C shikimic acid. All other amino acids were essential.

[d] A nutritional antagonism was found between alanine and glycine. At high concentrations of glycine the growth rate was increased when alanine was supplied as the tri- or tetrapeptide (L-ala$_3$ or L-ala$_4$). Alanyl-alanine was inhibitory.[1]

Table 3 (continued)
NUTRITIONAL REQUIREMENTS OF
MYCOPLASMA MYCOIDES AND
ACHOLEPLASMA LAIDLAWII

Nutrient	M. mycoides strain Y	Ref.	A. laidlawii strains A and B	Ref.
Amino Acids (continued)				
Isoleucine	E	1	E	4
Leucine	E	1	E	4
Lysine	E	1	E	4
Methionine	E	1	E	4
Phenylalanine	E	1	(E, R̶)[c]	4
Proline	E	1	E	4
Serine	E	1	E	4
Threonine	E	1	E	4
Tryptophan	E	1	E	4
Tyrosine	E	1	(E, R̶)[c]	4
Valine	E	1	E	4
Carbohydrates				
Glucose	E[e]	6	E[f]	2, 5
Glycerol	E[g]	6, 9	R̶	2, 5
Acetate	U̶		U[h]	10–12, 13–15
Nucleic Acid Precursors				
			[i]	2, 16, 17
Adenine	(U, R̶)[j]	18	u	
Guanine	E	19	u	
Uracil	E	19	u	
Thymine	E	19	U̶	17
Cytosine	U̶		R̶, E[i]	2, 17
Adenosine	U		E	2, 16
Guanosine	U		E	2, 16
Uridine	U		R̶, E[i]	17
Thymidine	U		E, R̶[i,m]	2, 16
Cytidine	U		E, R̶[i]	2, 17

[e] Replaceable by maltose and less effectively by mannose or fructose, but not by galactose.

[f] Glucose could be replaced by maltose, but not by galactose, fructose, mannose, lactose, or sucrose for growth of strain A.[5]

[g] Required for glyceride synthesis; M. mycoides is unable to obtain L-glycerol-3-phosphate from glucose.[9]

[h] A. laidlawii strains A and B can synthesize straight chain saturated fatty acids[10-13] and carotenoids[14,15] from acetate. Neither strain can synthesize unsaturated fatty acids, which must be provided exogenously, but both strains can elongate shorter chain monoenoic acid precursors.[11,13] Unsaturated fatty acids could be replaced by a cyclopropane fatty acid for the growth of strain A.[26]

[i] Adenosine, guanosine, cytidine, and thymidine can supply the nucleic acid precursor requirements of both strains.[2,16] Thymidine is not required by strain A in the presence of folinic acid.[16] A. laidlawii can synthesize uridine monophosphate, the precursor of all four pyrimidine triphosphates, from uridine or cytosine.[17] The free bases are apparently poorly utilized.

[j] Adenine is efficiently utilized for the synthesis of adenosine monophosphate,[18] but does not stimulate growth of strain Y in the presence of a nonlimiting concentration of guanine.[27]

Table 3 (continued)
NUTRITIONAL REQUIREMENTS OF MYCOPLASMA MYCOIDES AND ACHOLEPLASMA LAIDLAWII

Nutrient	M. mycoides strain Y	Ref.	A. laidlawii strains A and B	Ref.
Vitamins and Coenzymes				
Coenzyme A	E	1	E	2
Pantetheine	U	1	U	20
Pantothenate	U	1	U	2
Riboflavin	E	19	R	2, 5
Nicotinic acid	E	19	R	2, 5
Thiamin	E	19	R	2, 5
Pyridoxin or pyridoxal	—		R[k]	2, 5
Pyridoxamine	(R, S)[l]	1	—	
Biotin	R	1	R	2
Folinic acid	R	1	(E, R)[m]	16
Folic acid	—		U	16
α-Lipoic acid	S	1	R	2, 5
Lipids				
Sterols	E[n]	21	R[o]	14, 22
Fatty acids	E[p]	23	E[h]	10—13, 26
Polyamines				
Spermine or spermidine	S	6	R[q]	2, 5

[k] The requirement for these was not tested individually.

[l] Pyridoxamine stimulated growth in the absence of alanine or alanyl peptides, but was not required in their presence.

[m] Folinic acid was essential in the absence of thymidine, but not required in its presence.[16]

[n] Cholesterol is incorporated into the membrane without esterification or alteration of structure. It may be replaced for growth of strain Y and other sterol-requiring mycoplasmas by 3β-cholestanol, 7-cholesten-3β-ol, β-sitosterol, stigmasterol, and ergosterol.[21,24,25] 3α-Cholesterol, 3α-cholestanol, coprostan-3β-ol, coprostan-3α-ol, 4-cholesten-3-one, and 5-cholesten-3-one inhibited growth in the presence of cholesterol.[21,24]

[o] Although acholeplasmas do not require or synthesize sterols, they incorporate exogenous sterol into the membrane.[14,22]

[p] Strain Y is unable to synthesize or alter the chain length of either saturated or unsaturated fatty acids. Elaidate, transvaccenate, cis-12-octadecenoate, isopalmitate, isostearate, and anteisoheptadecanoate supported optimum growth when they were the only fatty acid supplied, and the membrane lipids then contained only one fatty acid. With most cis-monoenoic acids there was little or no growth unless a straight-chain saturated acid was supplied in addition, and both the chain length and position of the double bond of the monoenoic acid had a marked effect on the chain length of the saturated acids optimal for growth.

[q] Spermine at a concentration of 10 μg/ml inhibited growth of strain A.[5]

Table 3 (continued)

REFERENCES

1. Rodwell, A. W., *J. Gen. Microbiol.,* 58, 39–47, 1969.
2. Tourtellotte, M. E., Morowitz, H. J., and Kasimer, P., *J. Bacteriol.,* 88, 11–15, 1964.
3. Tourtellotte, M. E., in *The Mycoplasmatales and the L-phase of Bacteria,* Hayflick, L., Ed., Appleton-Century-Crofts, New York, 1969, 451–468.
4. Tourtellotte, M. E., Dept. of Animal Diseases, University of Connecticut College of Agriculture, Storrs, Conn., unpublished.
5. Razin, S. and Cohen, A., *J. Gen. Microbiol.,* 30, 141–154, 1963.
6. Rodwell, A. W., *Ann. N.Y. Acad. Sci.,* 143, 88–109, 1967.
7. Cho, H. W. and Morowitz, H. J., *Biochim. Biophys. Acta,* 183, 295–303, 1969.
8. Yarrison, G., Young, D. W., and Choules, G. L., *J. Bacteriol.,* 110, 494–503, 1972.
9. Plackett, P. and Rodwell, A. W., *Biochim. Biophys. Acta,* 210, 230–240, 1970.
10. Pollack, J. D. and Tourtellotte, M. E., *J. Bacteriol.,* 93, 636–641, 1967.
11. Rottem, S. and Panos, C., *Biochemistry,* 9, 57–63, 1970.
12. Rottem, S. and Razin, S., *J. Gen. Microbiol.,* 48, 53–63, 1967.
13. Romijn, J. C., Golde, L. M. G. van, McElhaney, R. N., and Deenen, L. L. M. van, *Biochim. Biophys. Acta,* 280, 22–32, 1972.
14. Smith, P. F. and Rothblat, G. H., *J. Bacteriol.,* 83, 500–506, 1962.
15. Smith, P. F., *J. Gen. Microbiol.,* 32, 307–319, 1963.
16. Razin, S., *J. Gen. Microbiol.,* 28, 243–250, 1962.
17. Smith, D. W. and Hanawalt, P. C., *J. Bacteriol.,* 96, 2066–2076, 1968.
18. Sin, I. L. and Finch, L. R., *J. Bacteriol.,* 112, 439–444, 1972.
19. Rodwell, A. W., *Ann. N.Y. Acad. Sci.,* 79, 499–507, 1960.
20. Rottem, S., Muhsam-Peled, O., and Razin, S., *J. Bacteriol.,* 113, 586–591, 1973.
21. Rodwell, A. W., *J. Gen. Microbiol.,* 32, 91–101, 1963.
22. Argaman, M. and Razin, S., *J. Gen. Microbiol.,* 38, 153–160, 1965.
23. Rodwell, A. W. and Peterson, J. E., *J. Gen. Microbiol.,* 68, 173–186, 1971.
24. Smith, P. F., *J. Lipid Res.,* 5, 121–125, 1964.
25. Archer, D. B., *J. Gen. Microbiol.,* 88, 329–338, 1975.
26. Panos, C. and Leon, O., *J. Gen. Microbiol.,* 80, 93–100, 1974.
27. Rodwell, A. W., unpublished data.

Table 4

NUTRITIONAL REQUIREMENTS OF MYCOPLASMAS
OTHER THAN *MYCOPLASMA MYCOIDES* AND
ACHOLEPLASMA LAIDLAWII

Information from studies utilizing undefined or partly defined media on the nutritional requirements of mycoplasmas other than *M. mycoides* and *A. laidlawii* is summarized.

Mycoplasma	Nutrient	Comment	Ref.
Strain J (classification uncertain)	Fatty acids	A party defined medium was described in which fatty acids were supplied by diacetyl tartrate esters of monoglycerides	1
	Coenzyme A	Stimulated growth; it was replaceable by pantetheine or pantethine, but not by pantothenate.	1
	Biotin	Stimulated growth	1
	α-Lipoic acid	Stimulated growth	2
M. arthritidis strain 07		A partly defined medium was described in which lipids were supplied by a serum lipoprotein fraction. The growth response to the individual components was not reported.	3
M. synoviae	Nicotinamide adenine dinucleotide	Required for growth in undefined medium	4
	Cysteine	Appears to be a specific requirement not replaceable by thioglycollate	5
Ureaplasma ("T-mycoplasmas")	Urea	Required for growth in undefined media.	6
		Urea is hydrolyzed.	7
		The function of urea in the nutrition of ureaplasmas is not understood. It was replaceable for growth by putrescine plus allantoin.	8
Thermoplasma acidophilum	Inorganic salts, glucose, polypeptide(s)	Optimum pH for growth between 1 and 2. Optimum temperature for growth about 60°C. Appears to be relatively nonexacting in its nutritional requirements.	9
		Growth was obtained in an inorganic salts medium and glucose with the addition of polypeptide fraction(s) isolated from yeast. The function of the polypeptide(s) is not known. Sterol is not required.	10

Table 4 (continued)
NUTRITIONAL REQUIREMENTS OF MYCOPLASMS OTHER THAN *MYCOPLASMA MYCOIDES* AND *ACHOLEPLASMA LAIDLAWII*

Mycoplasma	Nutrient	Comment	Ref.
Spiroplasma		Only two spiroplasmas, (*S. citri* and the agent of corn stunt disease) have been cultivated in artificial media. No information is available on the nutritional requirements of these organisms, apart from the requirement for sterol by *S. citri*.	11–13

REFERENCES

1. Lund, P. G. and Shorb, M. S., *Proc. Soc. Exp. Biol. Med.*, 121, 1070–1075, 1966.
2. Lund, P. G. and Shorb, M. S., *J. Bacteriol.*, 94, 279–280, 1967.
3. Smith, P. F., *Proc. Soc. Exp. Biol. Med.*, 88, 628–631, 1955.
4. Chalquest, R. R., and Fabricant, J., *Avian Dis.*, 4, 515–539, 1960.
5. Chalquest, R. R., *Avian Dis.*, 6, 36–43, 1962.
6. Ford, D. K. and MacDonald, J., *J. Bacteriol.*, 93, 1509–1512, 1967.
7. Shepard, M. C. and Lunceford, C. D., *J. Bacteriol.*, 93, 1513–1520, 1967.
8. Masover, G. K., Benson, J. R., and Hayflick, L., *J. Bacteriol.*, 117, 765–774, 1974.
9. Darland, G., Brock, T. D., Samsonoff, W., and Conti, S. F., *Science*, 170, 1416–1418, 1970.
10. Smith, P. F., Langworthy, T. A., and Smith, M. R., *J. Bacteriol.*, 124, 884–892, 1975.
11. Saglio, P., Lafleche, D., Bonissol, C., and Bové, J. M., *Physiol. Veg.*, 9, 569–592, 1971.
12. Chen, T. A. and Liao, C. H., *Science*, 188, 1015–1017, 1975.
13. Williamson, D. L. and Whitcomb, R. F., *Science*, 188, 1018–1020, 1975.

QUALITATIVE REQUIREMENTS AND UTILIZATION OF NUTRIENTS: BLUE-GREEN ALGAE

P. Fay

I. MINERAL ELEMENTS

All species require the principal constituents of cell material (carbon, hydrogen, oxygen, and nitrogen) for growth. For main sources of carbon and nitrogen, see Parts II and III. Sulfate is the main source of sulfur, and orthophosphate of phosphorus. Only a single report[26] supports the requirement of boron, and there is no available information on copper and zinc requirement for growth of blue-green algae. Nevertheless, these and various other elements (e.g., vanadium, chromium, tungsten, titanium) are included in trace amounts in the currently used culture media for blue-green algae, as they are known to be components (cofactors) of essential plant enzymes.

In the table below, R = required, R̸ = not required, S = supportive.

I. MINERAL ELEMENTS

Species	Major elements						Minor elements						Ref.
	P	S	K	Na	Mg	Ca	Fe	Mn	B	Mo[a]	Co	V	
Agmenellum quadruplicatum				R	R	R							1
Anabaena cylindrica	R	R	R	R[b]	R	R		R		R			2—6
A. doliolum										R			7[c]
A. flos-aquae			R	R									8
A. spiroides	R	R	R	R	R	R	R						9
A. variabilis[d]			R	R	R[e]								10, 11
Anacystis nidulans[f]	R		R	R		R	R	R					10, 12—16
Aphanizomenon flos-aquae	R	R	R				R						17
Calothrix parietina											R	R̸	18, 19
Chlorogloea fritschii[g]	R	R	R	R̸	R	R	R			R			20
Chroococcus turgidus			R	R									21
Coccochloris elabens				R	R	R							1
C. peniocystis	R	R	R		R	R̸	R				R		18, 22
Microcoleus chthonoplastes		S											1
Microcystis aeruginosa		R	R	R		R[h]					R		17, 18, 23, 24
Nostoc commune						R			R		R		25
N. muscorum			R			R[i]		R	R		R	R̸	10, 18, 19, 26, 27
Oscillatoria limosa					S[j]	S[j]							28
Phormidium luridum						R		R					29, 30
P. persicinum	R[k]		R	R	R	R							32

[a] Molybdenum is not required when blue-green algae are grown in the presence of ammonium-nitrogen.
[b] Required both when grown with nitrate or on elemental nitrogen.
[c] Also reports stimulation of growth by sodium tungstate.
[d] Nonheterocystous strain.
[e] Stimulates Hill reaction.
[f] Synechococcus sp.[31]
[g] Chlorogloepsis sp.[31]
[h] In trace amounts only.
[i] Especially under nitrogen-fixing conditions.
[j] Stimulates sheath formation.
[k] Can utilize both inorganic and organic phosphate as a source of phosphorus.

I. MINERAL ELEMENTS (continued)

REFERENCES

1. Van Baalen, C., *Bot. Mar.,* 4, 129, 1962.
2. Fogg, G. E., *Ann. Bot.* (London), 13, 241, 1949.
3. Wolfe, M., *Ann. Bot.* (London), 18, 299, 1954.
4. Allen, M. B. and Arnon, D. I., *Plant Physiol.,* 30, 366, 1955.
5. Allen, M. B. and Arnon, D. I., *Physiol. Plant.,* 8, 653, 1955.
6. Allen, M. B., *Sci. Mon.,* 83, 100, 1956.
7. Tyagi, V. V. S., *Ann. Bot.* (London), 38, 485, 1974.
8. Bostwick, C. D., Brown, L. R., and Tischer, R. G., *Physiol. Plant.,* 21, 466, 1968.
9. Volk, S. L. and Phinney, H. K., *Can. J. Bot.,* 46, 619, 1968.
10. Kratz, W. A. and Myers, J., *Am. J. Bot.,* 42, 282, 1955.
11. Susor, W. A. and Krogmann, D. W., *Biochim. Biophys. Acta,* 88, 11, 1964.
12. Richter, G., *Planta,* 57, 202, 1961.
13. Fredricks, W. W. and Jagendorf, A. T., *Arch. Biochem. Biophys.,* 104, 39, 1964.
14. Batterton, J. C. and Van Baalen, C., *Can. J. Microbiol.,* 14, 341, 1968.
15. Ingram, L. O. and Thurston, E. L., *J. Bacteriol.,* 125, 369, 1976.
16. Öquist, G., *Physiol. Plant.,* 25, 188, 1971.
17. McLachlan, J., Some Aspects of the Autecology of *Aphanizomenon flos-aquae* Born. et Flah. Studied Under Culture Conditions, Ph.D. thesis, Oregon State University, Corvallis, 1957.
18. Holm-Hansen, O., Gerloff, G. C., and Skoog, F., *Physiol. Plant.,* 7, 665, 1954.
19. Holm-Hansen, O., A Study of Major and Minor Element Requirements in the Nutrition of Blue-green Algae, Ph.D. thesis, University of Wisconsin, Madison, 1954.
20. Fay, P., Studies on Growth and Nitrogen Fixation in *Chlorogloea fritschii* Mitra, Ph.D. thesis, University of London, 1962.
21. Emerson, R. and Lewis, C. M., *J. Gen. Physiol.,* 25, 579, 1942.
22. Gerloff, G. C., Fitzgerald, G. P., and Skoog, F., *Am. J. Bot.,* 37, 835, 1950.
23. Gerloff, G. C., Fitzgerald, G. P., and Skoog, F., *Am. J. Bot.,* 39, 26, 1952.
24. Zehnder, A. and Gorham, P. R., *Can. J. Microbiol.,* 6, 645, 1960.
25. Taha, E. E. M. and Elrefai, A. E. M. H., *Arch. Mikrobiol.,* 43, 67, 1962.
26. Eyster, C., *Nature,* 170, 755, 1952.
27. Eyster, C., in *Proc. 9th Int. Bot. Congr. Montreal,* Vol. 2, Runge Press, Ottawa, 1959, 109.
28. Foerster, J. W., *Trans. Am. Microsc. Soc.,* 83, 420, 1964.
29. Piccioni, R. G. and Mauzerall, D. C., *Biochim. Biophys. Acta,* 423, 605, 1976.
30. Tel-Or, E. and Stewart, W. D. P., *Nature,* 258, 715, 1975.
31. Stanier, R. Y., Kunisawa, R., Mandel, M., and Cohen-Bazire, G., *Bacteriol. Rev.,* 35, 171, 1971.
32. Pintner, I. J. and Provasoli, L., *J. Gen. Microbiol.,* 18, 190, 1958.

II. CARBON SOURCES

All species can grow in light fixing carbon dioxide in photosynthesis. (Most blue-green algae can utilize bicarbonate as a source of carbon dioxide.) For certain species, carbon dioxide appears to be the only source of carbon able to support growth, though they may use external organic compounds as an energy source. The potential of organotrophic nutrition varies a great deal among the blue-green algae, ranging from a limited utilization of organic compounds in the light, through light-dependent assimilation of organic substrates (facultative photoheterotrophy), to an ability of growth on organic nutrients in the dark (facultative chemoheterotrophy). Sugars are the principal substrates that can support heterotrophic growth. The assimilation of organic acids and other organic compounds appears to be light dependent, and these can only serve as an accessory source of carbon. Growth rates in the dark, as compared with those in the light, are generally low. It is not possible to compare directly the available information on stimulation by, utilization of, or growth on organic compounds, since the test conditions applied may have differed considerably. The reader is advised to consult the relevant references. More detailed general discussions on the utilization of organic compounds by blue-green algae are given in two recently published books (Carr, N. G. and Whitton, B. A., Eds., *The Biology of Blue-green Algae,* Blackwell, Oxford, 1973; Fogg, G. E., Stewart, W. D. P., Fay, P., and Walsby, A. E., *The Blue-green Algae,* Academic Press, London, 1973) and in a review (Stanier, R. Y., *Biochem. Soc. Trans.,* 3, 352, 1975).

In the tables, U = utilized, u = poorly utilized, Ụ = not utilized; S = supportive, s = moderately supportive, Ṣ = not supportive.

Table II.A
ALCOHOLS AND SUGARS

Species	Alcohols				Sugars														Ref.
	Ethanol	Erythritol	Glycerol	Mannitol	Arabinose	Fructose	Fructose-6-phosphate	Galactose	Glucose	Lactose	Maltose	Mannose	Rhamnose	Ribose	Sorbose	Sucrose	Trehalose	Xylose	
Agmenellum quadruplicatum			U[a]						S[b]										1, 2
Anabaena cycadeae[c]						U			U[d]			U[d]							3
A. cylindrica									ψ							U[a]			4
A. flos-aquae						U[a]			U[a]							U[a]			5
A. spiroides									U[e]										6
A. variabilis			g[e]	g[e]		g[e]	g[e]	g[e]	ψ	g[e]	g[e]			g[e]		g[e]		g[e]	7—9
A. variabilis[f]			g[e]	g[e]		g[e]	g[e]	g[e]	S[a]	g[e]	g[e]								6
Anabaena sp.[g]		g[h]	g[h]	g[h]	g[h]		g[h]	g[h]	U[e]	g[h]	u[h]	g[h]	g[h]	g[h]	g[h]	U[e]	g[h]	g[h]	6, 10, 11
Anabaenopsis circularis				g[e]	g[h]	U[e]	g[e]	g[h]	U[e]	g[h]	g[e]		g[h]	g[h]	g[h]	g[e]		g[h]	12
Anacystis marina			g[e]	g[e]	g[h]	S[j]	g[e]	g[e]	g[h]	g[e]	g[e]			g[e]				g[e]	7, 13—15
A. nidulans[i]									U[a]										16
Aphanocapsa sp.[k]						U[h]			U[h]		u[h]					U[h]			11
Calothrix brevissima									ψ[e]										6
C. desertica						u[h]			U[h]		u[h]					U[h]			11
C. membranaceae									ψ[e]		U[h]					U[e]			6
C. parietina				u[d]		ψ[h]			U[e]										6, 11, 16, 17
Chlorogloea fritschii[l]									U[e]										16
Chlorogloea sp.									ψ										4
Chroococcus turgidus																			

a   In the light.
b   In dim light.
c   Endophytic, isolated from *Cycas.*
d   In the presence of combined nitrogen only.
e   Both in the light and in the dark.
f   Non heterocystous strain.
g   Strain 1551 (Indiana Culture Collection).
h   In the dark. (All species that utilize an organic compound in the dark were found, when tested, to also assimilate the same compound in the light.)
i   *Synechococcus* sp.[29]
j   Increased substrate respiration in the dark.
k   Several strains.
l   *Chlorogloeopsis* sp.[29]

Table II.A (continued)
ALCOHOLS AND SUGARS

Species	Ethanol	Erythritol	Glycerol	Mannitol	Arabinose	Fructose	Fructose-6-phosphate	Galactose	Glucose	Lactose	Maltose	Mannose	Rhamnose	Ribose	Sorbose	Sucrose	Trehalose	Xylose	Ref.
Coccochloris elabans			u[h]						U[h]										12
C. peniocystis						u[a]			U[h]										14
Fremyella diplosiphon								U[h]	U[h]							U[h]			18
Gloeocapsa alpicola						U[a]				u[h]									14
Lyngbya aestuarii									g[h]										12
L. lagerheimii									U[e]										1, 6
Lyngbya sp.[k]									U?[m]										6, 19
Microchaete sp.									U[a]										6
Microcoleus chthonoplastes									g[h]										12
M. tenerrimus									g[h]										12
M. vaginatus									U[e]										6
Nodularia sphaerocarpa									U[e]										6
Nostoc ellipsosporum			g[e]	g[e]		U[h]	g[e]		u[h]		U[h]					U[h]			11
N. muscorum[n]					g[e]	U[h]					U[h]	U[h]							7, 11, 20
N. muscorum[o]			g[e]	g[e]	g[e]	g[e]	g[e]	g[e]	U[e]	g[e]	g[e]			g[e]		g[e]		g[e]	7
N. punctiforme[p]						U[h]				U[h]	u[h]	U[h]				U[e]			3, 21
Nostoc sp.[r]			U[a]	g[h]				s[h]	U[a]	g[h]	U[h]	g[h]		g[h]		U[a]		g[h]	22, 23
Oscillatoria amphibia									g[h]										12
O. lagerheimii									g[h]										12
O. subtilissima									g[h]										12
Oscillatoria sp.									U[e]										6
Phormidium foveolarum						U[h]			U[s]										4
P. luridum	g								u[h]		u[h]					U[h]			11
P. persicinum		g	g													U[a]			24
Phormidium sp.[t]									U[a]										19

m   Utilized by strains 487 (Indiana Culture Collection) and 6404[6] but not by strain 6703.[6]
n   Gerloff strain.
o   Different strains.
p   Endophytic, isolated from Gunnera.
r   Endophytic, isolated from Macrozamia.
s   In the presence of yeast extract.
t   Strain 485 (Indiana Culture Collection).

Table II.A (continued)
ALCOHOLS AND SUGARS

Species	Alcohols				Sugars														Ref.
	Ethanol	Erythritol	Glycerol	Mannitol	Arabinose	Fructose	Fructose-6-phosphate	Galactose	Glucose	Lactose	Maltose	Mannose	Rhamnose	Ribose	Sorbose	Sucrose	Trehalose	Xylose	
Plectonema boryanum				U^h		U^h			U^h		u^h			U^h		U^h			11, 25, 26
P. calothricoides						U^h			U^h		ψ^h					U^h			11
P. notatum									U^a										4, 6
P. terebrans									ψ^h										12
Plectonema sp.									U^a										6
Schizothrix calcicola									U^e										6
Spirulina sp.									ψ^e										6
Synechococcus cedrorum								U^h	ψ		ψ			U^h					4
Tolypothrix tenuis						U^h			U^e							u^h			6, 11, 27, 28

Table II.A (continued)
ALCOHOLS AND SUGARS

REFERENCES

1. Van Baalen, C., Hoare, D. S., and Brandt, E., *J. Bacteriol.,* 105, 685, 1971.
2. Ingram, L. O., Van Baalen, C., and Calder, J. A., *J. Bacteriol.,* 114, 701, 1973.
3. Winter, G., *Beitr. Biol. Pflanz.,* 23, 295, 1935.
4. Allen, M. B., *Arch. Mikrobiol.,* 17, 34, 1952.
5. Moore, B. G. and Tischer, R. G., *Can. J. Microbiol.,* 11, 877, 1965.
6. Kenyon, C. N., Rippka, R., and Stanier, R. Y., *Arch. Mikrobiol.,* 83, 216, 1972.
7. Kratz, W. A. and Myers, J., *Am. J. Bot.,* 42, 282, 1955.
8. Pearce, J. and Carr, N. G., *J. Gen. Microbiol.,* 54, 451, 1969.
9. Ohki, K. and Katoh, T., *Plant Cell Physiol.,* 16, 53, 1975.
10. Watanabe, A. and Yamamoto, Y., *Nature,* 214, 738, 1967.
11. Khoja, T. and Whitton, B. A., *Br. Phycol. J.,* 10, 139, 1975.
12. Van Baalen, C., *Bot. Mar.,* 4, 129, 1962.
13. Carr, N. G. and Pearce, J., *Biochem. J.,* 99, 28P, 1966.
14. Smith, A. J., London, J., and Stanier, R. Y., *J. Bacteriol.,* 94, 972, 1967.
15. Peschek, G. A. and Broda, E., *Naturwissenschaften,* 60, 479, 1973.
16. Rippka, R., *Arch. Mikrobiol.,* 87, 93, 1972.
17. Fay, P., *J. Gen. Microbiol.,* 39, 11, 1965.
18. Diakoff, S. and Scheibe, J., *Physiol. Plant.,* 34, 125, 1975.
19. Pope, D. H., *Can. J. Bot.,* 52, 2369, 1974.
20. Allison, F. E., Hoover, S. R., and Morris, H. J., *Bot. Gaz.,* 98, 433, 1937.
21. Harder, R., *Z. Bot.,* 9, 145, 1917.
22. Hoare, D. S., Ingram, L. O., Thurston, E. L., and Walkup, R., *Arch. Mikrobiol.,* 78, 310, 1971.
23. Ingram, L. O., Calder, J. A., Van Baalen, C., Plucker, F. E., and Parker, P. L., *J. Bacteriol.,* 114, 695, 1973.
24. Pintner, I. J. and Provasoli, L., *J. Gen. Microbiol.,* 18, 190, 1958.
25. Pan, P., *Can. J. Microbiol.,* 18, 275, 1972.
26. White, A. W. and Shilo, M., *Arch. Microbiol.,* 102, 123, 1975.
27. Kiyohara, T., Fujita, Y., Hattori, A., and Watanabe, A., *J. Gen. Appl. Microbiol.,* 6, 176, 1960.
28. Cheung, W. Y. and Gibbs, M., *Plant Physiol.,* 41, 731, 1966.
29. Stanier, R. Y., Kunisawa, R., Mandel, M., and Cohen-Bazire, G., *Bacteriol. Rev.,* 35, 171, 1971.

Table II.B
ORGANIC ACIDS

Species	Acetate	Aconitate	Butyrate	Citrate	Formate	Fumarate	Glycolate	Lactate	Malate	Oleate	Oxaloacetate	α-Oxoglutarate	Palmitate	Propionate	Pyruvate	Stearate	Succinate	Ref.
Agmenellum quadruplicatum	U[a]							g[b]							U[a]		U[a]	1, 2
Anabaena cylindrica	U[c]																	3, 4
A. flos-aquae	U[a]			U[a]		U[a]	U	U[a]							U[a]		U[a]	5—8
A. variabilis[e]	U[a]	g[b]	g[d]	U[d]		g[d]	U[d]	g[d]	g[d]					u[b]	g[d]		g[d]	6, 9—12
Anabaenopsis circularis	U[a]		g[b]	g[b]		g[b]	g[d]	g[b]	g[b]		g[b]				g[b]		g[b]	13, 14
Anacystis marina	g[b]				U[a]			g[b]	s	u[a]				g[b]	s[a]			1
A. nidulans[f]	U[a]		g[d]	U[d]		g[d]	U[d]	g[d]	g[b]		g[b]	g[b]	u[a]	u[a]	g[b]	u[a]	g[d]	6, 9—11, 15—17
Chlorogloea fritschii[g]	U[d]			g[b]		g[b]	g[b]					g[b]						6, 13, 18, 19
Coccochloris elabens	g[b]							g[b]										1
C. peniocystis	U[a]					u[b]			u			g[b]						16
Fremyella diplosiphon	U[a]														U		u	20
Gloeocapsa alpicola	g[b]							g[b]	u[a]					U[a]	u[a]		u[a]	16, 21
Lyngbya aestuarii	u[a]							g[b]							u[a]	u[b]		1
L. lagerheimii	g[b]							g[b]										1, 2
Microcoleus chthonoplastes	g[b]							g[b]										1
M. tenerrimus	u[b]			u[b]														1
Nostoc muscorum[h]	U[a]		g[d]	g[b]		g[d]	g[d]	g[d]	g[d]					g[d]	g[d]		g[d]	6, 9, 10
N. muscorum[i]																		13, 22
N. punctiforme[j]																		23
Nostoc sp.[k]	U[a]														g[b]			24, 25

a In the light.
b In the dark.
c In the light; poorly utilized in the dark.
d Both in the light and in the dark.
e Nonheterocystous strain.
f *Synechococcus* sp.[29]
g *Chlorogloeopsis* sp.[29]
h Gerloff strain.
i Allison strain.
j Endophytic, isolated from *Gunnera*.
k Endophytic, isolated from *Macrozamia*.

Table II.B (continued)
ORGANIC ACIDS

Species	Acetate	Aconitate	Butyrate	Citrate	Formate	Fumarate	Glycolate	Lactate	Malate	Oleate	Oxaloacetate	α-Oxoglutarate	Palmitate	Propionate	Pyruvate	Stearate	Succinate	Ref.
Oscillatoria amphibia	$b							$b										1
O. lagerheimii	$b							$b										1
O. subtilissima	$b							$b										1
Oscillatoria sp.	$						U[a]											7
Phormidium persicinum	U[a]					$												26
Phormidium sp.[1]	$b																$	17
Plectonema terebrans	U							$b										1
Spirulina platensis																		27
Tolypothrix tenuis	U[a]														ψ[b]			13, 28

[1] Strain 485 (Indiana Culture Collection).

Table II.B (continued)
ORGANIC ACIDS

REFERENCES

1. Van Baalen, C., *Bot. Mar.*, 4, 129, 1962.
2. Van Baalen, C., Hoare, D. S., and Brandt, E., *J. Bacteriol.*, 105, 685, 1971.
3. Codd, G. A. and Stewart, W. D. P., *Arch. Mikrobiol.*, 94, 11, 1973.
4. Codd, G. A., and Stewart, W. D. P., *Plant Sci. Lett.*, 3, 199, 1974.
5. Moore, B. G. and Tischer, R. G., *Can. J. Microbiol.*, 49, 351, 1967.
6. Hoare, D. S., Hoare, S. L., and Moore, R. B., *J. Gen. Microbiol.*, 49, 351, 1967.
7. Miller, A. G., Cheng, K. H., and Colman, B., *J. Phycol.*, 7, 97, 1971.
8. Tarant, A. A. and Colman, B., *Can. J. Bot.*, 50, 2067, 1972.
9. Kratz, W. A. and Myers, J., *Am. J. Bot.*, 42, 282, 1955.
10. Hoare, D. S. and Moore, R. B., *Biochim. Biophys. Acta*, 109, 622, 1965.
11. Pearce, J. and Carr, N. G., *J. Gen. Microbiol.*, 49, 301, 1967.
12. Butler, M. and Capindale, J. B., *Can. J. Microbiol.*, 21, 1372, 1975.
13. Hoare, D. S., Hoare, S. L., and Smith, A. J., in *Progress in Photosynthesis Research*, Vol. 3, Metzner, H., Ed., Tübingen, 1969, 1570.
14. Watanabe, A. and Yamamoto, Y., *Nature*, 214, 738, 1967.
15. Nichols, B. W., Harris, R. V., and James, A. T., *Biochem. Biophys. Res. Commun.*, 20, 256, 1965.
16. Smith, A. J., London, J., and Stanier, R. Y., *J. Bacteriol.*, 94, 972, 1967.
17. Pope, D. H., *Can. J. Bot.*, 52, 2369, 1974.
18. Fay, P., *J. Gen. Microbiol.*, 39, 11, 1965.
19. Miller, S. J. and Allen, M. M., *Arch. Mikrobiol.*, 86, 1, 1972.
20. Diakoff, S. and Scheibe, J., *Physiol. Plant.*, 34, 125, 1975.
21. Smith, A. J. and Lucas, C., *Biochem. J.*, 124, 23P, 1971.
22. Allison, R. K., Skipper, H. E., Reid, M. R., Short, W. A., and Hogan, G. L., *J. Biol. Chem.*, 204, 197, 1953.
23. Harder, R., *Z. Bot.*, 9, 145, 1917.
24. Ingram, L. O., Calder, J. A., Van Baalen, C., Plucker, F. E., and Parker, R. L., *J. Bacteriol.*, 114, 695, 1973.
25. Hoare, D. S., Ingram, L. O., Thurston, E. L., and Walkup, R., *Arch. Mikrobiol.*, 78, 310, 1971.
26. Pintner, I. J. and Provasoli, L., *J. Gen. Microbiol.*, 18, 190, 1958.
27. Baron, C., Crance, J. M., and Forin, M. C., *C. R. Soc. Biol.*, 168, 501, 1974.
28. Cheung, W. Y. and Gibbs, M., *Plant Physiol.*, 41, 731, 1966.
29. Stanier, R. Y., Kunisawa, R., Mandel, M., and Cohen-Bazire, G., *Bacteriol. Rev.*, 35, 171, 1971.

III. NITROGEN SOURCES, VITAMINS, AND GROWTH REGULATORS

Most blue-green algae can utilize a variety of nitrogen sources for growth. Among the inorganic sources of nitrogen, ammonia and nitrate are the most readily assimilated. Many, but apparently not all, species can fix elemental nitrogen, some only at low oxygen concentrations, when the above sources of combined nitrogen are scarce or absent in their environment. Certain species exhibit a marked proteolytic activity, and/or can assimilate amino acids and other organic nitrogenous compounds. A number of marine species show a requirement for vitamin B_{12}, and the growth of some fresh-water blue-green algae appears to be affected by the presence of a plant growth regulator. (For a more detailed discussion of nitrogen metabolism in blue-green algae, see Fogg, G. E., Stewart, W. D. P., Fay, P., and Walsby, A. E., *The Blue-green Algae*, Academic Press, London, 1973.)

In the tables, U = utilized, u = poorly utilized, Ʉ = not utilized; S = supportive, s = moderately supportive, $ = not supportive; R = required, Ɍ = not required.

Table III.A

INORGANIC NITROGEN SOURCES AND ORGANIC NITROGENOUS COMPOUNDS

Species	Inorganic N sources					Organic N sources					Purines				Pyrimidines					
	Ammonia	Dinitrogen (N₂)	Hydroxylamine	Nitrate	Nitrite	Diethylamine	N-Acetylglucosamine	Methylamine	Urea	Uric acid	Adenine	Guanine	Hypoxanthine	Xanthine	Cytosine	Orotic acid	Thiouracil	Thymine	Uracil	Ref.
Agmenellum quadruplicatum	U[a]	ψ[a]	ψ[a]	U[a]	U[a]	ψ[a]	ψ[a]	ψ[a]	u[a]	u[a]	ψ[a]		U[a]	U[a]	ψ[a]	ψ[a]	ψ[a]	ψ[a]	ψ[a]	1–4
Anabaena ambigua		U																		5
A. azollae[b]		U																		6
A. cycadeae[c]	U	U[d]	U	U	U				U[e]											7, 8
A. cylindrica	U	U	U	U	U															9–14
A. doliolum	U	U																		15
A. fertilissima		U																		5
A. flos-aquae	U	U		U	U				U											16–18
A. gelatinosa		U																		19
A. humicola		U																		9
A. levanderi		U																		20
A. naviculoides		U																		19
A. spiroides		U		U					S											21, 22
A. variabilis		ψ																		9, 19, 23, 24
A. variabilis[f]	U	U		U					U[e]											25
Anabaena sp.[g]		U																		22
Anabaenopsis circularis											u[a]	u[a]								22, 26, 30
Anacystis nidulans[h]	U	ψ		U					ψ	ψ[i]						u[a]		u[a]	u[a]	25, 27, 28

a   In light.
b   Endophytic, isolated from *Azolla*.
c   Endophytic, isolated from *Cycas*.
d   In the dark in the presence of fructose, glucose, or sucrose.
e   Can be sole source of nitrogen.
f   Nonheterocystous strain.
g   Several strains.
h   *Synechococcus* sp.[60]
i   Decomposes uric acid to allantoin.

Table III.A (continued)
INORGANIC NITROGEN SOURCES AND ORGANIC NITROGENOUS COMPOUNDS

Species	Ammonia	Dinitrogen (N₂)	Hydroxylamine	Nitrate	Nitrite	Diethylamine	N-Acetylglucosamine	Methylamine	Urea	Uric acid	Adenine	Guanine	Hypoxanthine	Xanthine	Cytosine	Orotic acid	Thiouracil	Thymine	Uracil	Ref.
Aphanzomenon flos-aquae		U[j]																		29
Aulosira fertilissima		U																		5
Calothrix brevissima		U																		26, 30
C. deserica		U																		22
C. elenkinii	U	U																		31
C. parietina	u	U		U					∅											32, 37
C. scopulorum	U	U		U																33, 34
Chlorogloea fritschii[k]	U	∅		U					U[a]											22, 35, 36
Coccochloris elabens		∅		U						u[a]										1
C. peniocystis		U		U																37, 38
Cylindrospermum gorakhporense		U																		5
C. licheniforme		U																		9
C. majus		U																		9
C. sphaerica		U																		39
Fischerella major		U																		40, 41
F. muscicola		U																		37
Gloeocapsa dimidiata		∅																		37
G. membranea		∅																		42, 43
Gloeocapsa sp.[l]		U																		31
Hapalosiphon fontinalis	U	U							∅											1, 32
Lyngbya aestuarii		?		U																1
L. lagerheimii		∅																		22
Lyngbya sp.[m]		∅[n]																		

j Result of field test.

k *Chlorogloeopsis* sp.[60]

l Strains 795 (Indiana Culture Collection) and 6501 (Berkeley Culture Collection), probably independent isolates of the same species.[43]

m Strain 6703.[22]

n At low pO_2.

Table III.A (continued)
INORGANIC NITROGEN SOURCES AND ORGANIC NITROGENOUS COMPOUNDS

Species	Inorganic N sources					Organic N sources					Purines				Pyrimidines					Ref.
	Ammonia	Dinitrogen (N_2)	Hydroxylamine	Nitrate	Nitrite	Diethylamine	N-Acetylglucosamine	Methylamine	Urea	Uric acid	Adenine	Guanine	Hypoxanthine	Xanthine	Cytosine	Orotic acid	Thiouracil	Thymine	Uracil	
Mastigocladus laminosus	U	U																		44
Microcheate sp.[o]	U	U		U																22
Microcoleus chthonoplastes	U	∅[a]		U					∅[a]											1
M. tenerrimus		∅[a]																		1
M. vaginatus		∅[a]																		22
Microcystis aeruginosa	U[a]	∅[a]	§[a]	U[a]					U[a]											37, 45, 46
Nodularia sphaerocarpa	U	U																		22
N. spumigena		U		U	U				U											47
Nostoc calcicola		U																		9
N. commune		U																		48
N. cycadeae[p]	U	U		U																49
N. entophytum		U																		33, 34
N. muscorum[r]	U	U		U																50
N. muscorum[s]		U							U[e]											25, 37, 51—53
N. paludosum		U																		9
N. punctiforme[t]		U																		23, 24
N. sphaericum[u]		U																		54
Oscillatoria amphibia		∅																		1
O. brevis		∅																		33
O. subtilissima		∅																		1
Oscillatoria sp.[v]		∅																		22
Oscillatoria sp.[w]		U[n]																		22

o   Strains 1544 (Indiana Culture Collection) and 6601 (Berkeley Culture Collection).

p   Endophytic, isolated from Cycas, Encephalartos, and Macrozamia species.

r   Allison strain.

s   Gerloff strain.

t   Endophytic, isolated from Gunnera.

u   Endophytic, isolated from Blasia.

v   Strain 6401 (Berkeley Culture Collection).

w   Strains 6407, 6412, 6506, and 6602 (Berkeley Culture Collection).

Table III.A (continued)
INORGANIC NITROGEN SOURCES AND ORGANIC NITROGENOUS COMPOUNDS

Species	Inorganic N sources					Organic N sources					Purines				Pyramidines					Ref.
	Ammonia	Dinitrogen (N_2)	Hydroxylamine	Nitrate	Nitrite	Diethylamine	N-Acetylglucosamine	Methylamine	Urea	Uric acid	Adenine	Guanine	Hypoxanthine	Xanthine	Cytosine	Orotic acid	Thiouracil	Thymine	Uracil	
Phormidium foveolarum	U	Ψ		U																19, 32
P. luridum	U	Ψ		U																32
P. persicinum	u	Ψ		U																55
P. tenue		U[n]																		37
Phormidium sp.[x]		U[n]																		22
Plectonema boryanum[y]		Ψ																		56
P. nostocorum	U	U[n]							U											37
P. notatum[z]	U	Ψ		U					U[a]	u[a]										22, 32
P. terebrars		U[n]		U																1
Plectonema sp.[aa]		Ψ[bb]																		22
Schizothrix calcicola		U																		22
Scytonema arcangelii		U																		20
S. hofmanni		U																		20
Stigonema dendroideum		Ψ[n]																		57
Spirulina sp.[cc]		Ψ							Ψ	Ψ										22
Synechococcus cedrorum	U	U		U					U											27
Tolypothrix tenuis	U	U		U	U															26, 30, 58
Westiellopsis prolifica		U																		59

x   Identical with *Lyngbya* sp., strain 6409.[22]
y   Strain 594 (Indiana Culture Collection).
z   Strain 6306.[22]
aa   Strain 1541 (Indiana Culture Collection).
bb   Examined only under aerobic conditions.
cc   Strain 6313 (Berkeley Culture Collection).

Table III.A (continued)
INORGANIC NITROGEN SOURCES AND ORGANIC NITROGENOUS COMPOUNDS

REFERENCES

1. Van Baalen, C., *Bot. Mar.,* 4, 129, 1962.
2. Van Baalen, C. and Marler, J. E., *J. Gen. Microbiol.,* 32, 457, 1963.
3. Stevens, S. E. and Van Baalen, C., *Arch. Mikrobiol.,* 72, 1, 1970.
4. Kapp, R., Stevens, S. E., and Fox, J. L., *Arch. Microbiol.,* 104, 135, 1975.
5. Singh, R. N., *Indian J. Agric. Sci.,* 12, 743, 1942.
6. Venkataraman, G. S. *Indian J. Agric. Sci.,* 32, 22, 1962.
7. Winter, G., *Beitr. Biol. Pflanz.,* 23, 295, 1935.
8. Douin, R., *C. R. Acad, Sci.,* 236, 956, 1953.
9. Bortels, H., *Arch. Mikrobiol.,* 11, 155, 1940.
10. Fogg, G. E., *J. Exp. Biol.,* 19, 78, 1942.
11. Fogg, G. E., *Ann. Bot.* (London), 13, 241, 1949.
12. Allen, M. B. and Arnon, D. I., *Plant Physiol.,* 30, 366, 1955.
13. Hattori, A., *Plant Cell Physiol.,* 3, 355, 1962.
14. Cobb, H. D. and Myers, J., *Am. J. Bot.,* 51, 753, 1964.
15. Singh, H. N. and Shrivastava, B. S., *Can. J. Microbiol.,* 14, 1341, 1968.
16. Davis, E. B., Tischer, R. G., and Brown, L. R., *Physiol. Plant.,* 19, 823, 1966.
17. Bone, D. H., *Arch. Mikrobiol.,* 80, 234 1971.
18. Bone, D. H., *Arch. Mikrobiol.,* 80, 241, 1971.
19. De, P. K., *Proc. R. Soc. London Ser. B,* 127, 121, 1939.
20. Cameron, R. E. and Fuller, W. H., *Soil Sci. Soc. Am. Proc.,* 24, 353, 1960.
21. Volk, S. L. and Phinney, H. K., *Can. J. Bot.,* 46, 619, 1968.
22. Kenyon, C. N., Rippka, R., and Stanier, R. Y., *Arch. Mikrobiol.,* 83, 216, 1972.
23. Drewes, K., *Zentralbl. Bakteriol. Parasitenkd.* (Abt. II), 76, 88, 1928.
24. Allison, F. E. and Morris, H. J., *Science,* 71, 221, 1930.
25. Kratz, W. A. and Myers, J., *Am. J. Bot.,* 42, 282, 1955.
26. Watanabe, A., *Arch. Biochem. Biophys.,* 34, 50, 1951.
27. Birdsey, E. C. and Lynch, V. H., *Science,* 137, 763, 1962.
28. Pigott, G. H. and Carr, N. G., *Arch. Mikrobiol.,* 79, 1, 1971.
29. Stewart, W. D. P., Fitzgerald, G. P., and Burris, R. H., *Arch. Mikrobiol.,* 62, 336, 1968.
30. Watanabe, A., *J. Gen. Appl. Microbiol.,* 5, 21, 1959.
31. Taha, M. S., *Mikrobiologiya,* 32, 421, 1963.

Table III.A (continued)
INORGANIC NITROGEN SOURCES AND ORGANIC NITROGENOUS COMPOUNDS

32. Allen, M. B., *Arch. Mikrobiol.,* 17, 34, 1952.
33. Stewart, W. D. P., *Ann. Bot.* (London), 26, 439, 1962.
34. Stewart, W. D. P., *J. Exp. Bot.,* 15, 138, 1964.
35. Fay, P. and Fogg, G. E., *Arch. Mikrobiol.,* 42, 310, 1962.
36. Fay, P., *J. Gen. Microbiol.,* 39, 11, 1965.
37. Williams, A. E. and Burris, R. H., *Am. J. Bot.,* 39, 340. 1952.
38. Gerloff, G. C., Fitzgerald, G. P., and Skoog, F., *Am. J. Bot.,* 37, 835, 1950.
39. Venkataraman, G. S., *Proc. Natl. Acad. Sci. India,* 31, 100, 1961.
40. Pankow, H., *Naturwissenschaften,* 51, 274, 1964.
41. Mitra, A. K., *Proc. Natl. Acad. Sci. India,* 31, 98, 1961.
42. Wyatt, J. T. and Silvey, J. K. G., *Science,* 165, 908, 1969.
43. Rippka, R., Neilson, A., Kunisawa, R., and Cohen-Bazire, G., *Arch. Mikrobiol.,* 76, 341, 1971.
44. Fogg, G. E., *J. Exp. Bot.,* 2, 117, 1951.
45. Zehnder, A. and Gorham, P. R., *Can. J. Microbiol.,* 6, 645, 1960.
46. McLachlan, J. L. and Gorham, P. R., *Can. J. Microbiol.,* 8, 1, 1962.
47. Camm, E. L. and Stein, J. R., *Can. J. Bot.,* 52, 719, 1974.
48. Herisset, A., *Bull. Soc. Chim. Biol.,* 34, 532, 1952.
49. Watanabe, A. and Kiyohara, T., in *Studies on Microalgae and Photosynthetic Bacteria,* Japanese Society of Plant Physiologists, Eds., University of Tokyo Press, 1963, 189.
50. Allison, F. E., Hoover, S. R., and Morris, H. J., *Bot. Gaz.,* 98, 433, 1937.
51. Burris, R. H., Eppling, F. J., Wahlin, H. B., and Wilson, P. W., *J. Biol. Chem.,* 148, 349, 1943.
52. Magee, W. E. and Burris, R. H., *Am. J. Bot.,* 11, 777, 1954.
53. Allison, R. K., Skipper, H. E., Reid, M., Short, W. A., and Hogan, G. L., *Plant Physiol.,* 29, 164, 1954.
54. Pankow, H. and Maertens, B., *Arch. Mikrobiol.,* 48, 203, 1964.
55. Pintner, I. J. and Provasoli, L., *J. Gen. Microbiol.,* 18, 190, 1958.
56. Stewart, W. D. P. and Lex, M., *Arch. Mikrobiol.,* 73, 250, 1970.
57. Venkataraman, G. S., *Indian J. Agric. Sci.,* 31, 213, 1961.
58. Fujita, Y. and Hattori, A., *Plant Cell Physiol.,* 1, 281, 1960.
59. Pattnaik, H., *Ann. Bot.* (London), 30, 231, 1966.
60. Stanier, R. Y., Kunisawa, R., Mandel, M., and Cohen-Bazire, G., *Bacteriol. Rev.,* 35, 171, 1971.

Table III.B
AMINO ACIDS

Species	Alanine	Arginine	Asparagine	Aspartate	Citrulline	Cystein	Glutamine	Glutamate	Glycine	Histidine	Hydroxyproline	Isoleucine	Leucine	Lysine	Methionine	Ornithine	Phenylalanine	Proline	Serine	Threonine	Tryptophan	Tyrosine	Valine	Ref.
Agmenellum quadruplicatum	U[a]	u[a]	U[a]	U[a]	ψ[a]	ψ[a]	U[a]	U[a]	u[a]	U[a]		ψ[a]	ψ[a]	U[a]	ψ[a]		u[a]	u[a]	U[a]	u[a]	ψ[a]	ψ[a]	ψ[a]	1, 2
Anabaena cylindrica	U[a]	U[a]	U[e]	U[e]	ψ[a]	ψ[a]	U[a]	U[a]	u[e]	U[a]		ψ[a]	ψ[a]	ψ[a]	ψ[a]		u[a]	u[a]	U[a]	u[a]	ψ[a]	ψ[a]	ψ[a]	3
A. spiroides		U							S															4
A. variabilis[b]			u[a]	u[a]				U[a]	ψ[a]				u[a]							u[a]				5–7
Anacystis nidulans[c]		ψ[a]		ψ[e]			u[e]	ψ[e]					u[a]	ψ[a]										8, 9
Chlorogloea fritschii[d]				u[a]					u[a]				u[a]											10
Coccochloris elabens																								1
C. peniocystis	U[e]		U[e]	U[e]	U[e]		u[e]	u[e]				U[e]	U[e]	u[e]	u[e]		U[e]	u[e]	u[e]	U[e]		u[e]	U[e]	8
Fremyella diplosiphon				U[a]				u[a]	ψ[a]				U[a]	ψ[a]			u[e]	u[e]						11, 12
Gloeocapsa alpicola								u[a]																8
Gloeotrichia pisum									ψ[a]					ψ[a]		u[a]			u[a]					13
Microcoleus chthonoplastes	ψ[a]	ψ[a]	u[a]	ψ[a]	U[a]	ψ[a]	U[a]	ψ[a]	ψ[a]			u[a]	u[a]	ψ[a]	ψ[a]		ψ[a]		u[a]	ψ[a]			ψ[a]	1
Microcystis aeruginosa					ψ[e]												U		u[a]					13, 14[f]
Nostoc sp.[g]									u															15
Phormidium luridum	U		U[a]	ψ[a]			U[e]	ψ[a]		ψ			U	ψ[a]	U		U					U	U	11
P. persicinum		u[a]	u[a]					U[a]						u[a]										16
Phormidium sp.[h]		U[e]	u[e]																					17
Plectonema terebrans	U[e]	u[a]	ψ[e]	ψ[e]		ψ[e]	U[e]	ψ[e]	ψ[a]	u[e]			U[e]		u[e]		U[e]	U[e]	U[e]	ψ[e]	ψ[e]		ψ[e]	1
Tolypothrix tenuis	U[e]	U[e]							u[e]															18

a In the light.
b Nonheterocystous strain.
c *Synechococcus* sp.[19]
d *Chlorogloeopsis* sp.[19]
e In the dark.
f Data from tests with cultures that were not bacteria free.
g Endophytic, isolated from *Macrozamia*.
h Strain 485 (Indiana Culture Collection).

Table III.B (continued)
AMINO ACIDS

REFERENCES

1.   Van Baalen, C., *Bot. Mar.,* 4, 129, 1962.
2.   Kapp, R., Stevens, S. E., and Fox, J. L., *Arch. Microbiol.,* 104, 135, 1975.
3.   Brownell, P. F. and Nicholas, D. J. D., *Plant Physiol.,* 42, 915, 1967.
4.   Volk, S. L. and Phinney, H. K., *Can. J. Bot.,* 46, 619, 1968.
5.   Kratz, W. A. and Myers, J., *Am. J. Bot.,* 42, 282, 1955.
6.   Hood, W., Leaver, A. G., and Carr, N. G., *Biochem. J.,* 114, 12P, 1969.
7.   Hood, W. and Carr, N. G., *J. Bacteriol.,* 107, 365, 1971.
8.   Smith, A. J., London, J., and Stanier, R. Y., *J. Bacteriol.,* 94, 972, 1967.
9.   Maclean, F. J., Forrest, H. S., and Myers, J., *Biochem. Biophys. Res. Commun.,* 18, 623, 1965.
10. Fay, P., *J. Gen. Microbiol.,* 39, 11, 1965.
11. Crespi, H. L., Daboll, H. F., and Katz, J. J., *Biochim. Biophys. Acta,* 200, 26, 1970.
12. Diakoff, S. and Scheibe, J., *Physiol. Plant.,* 34, 125, 1975.
13. Stephens, G. S., Vaidya, B. S., and Saxena, O. P., *Indian J. Exp. Biol.,* 7, 43, 1969.
14. McLachlan, J. L. and Gorham, P. R., *Can. J. Microbiol.,* 8, 1, 1962.
15. Hoare, D. S., Ingram, L. O., Thurston, E. L., and Walkup, R., *Arch. Microbiol.,* 78, 310, 1971.
16. Pintner, I. J. and Provasoli, L., *J. Gen. Microbiol.,* 18, 190, 1958.
17. Pope, D. H., *Can. J. Bot.,* 52, 2369, 1974.
18. Kiyohara, T., Fujita, Y., Hattori, A., and Watanabe, A., *J. Gen. Appl. Microbiol.,* 6, 176, 1960.
19. Stanier, R. Y., Kunisawa, R., Mandel, M., and Cohen-Bazire, G., *Bacteriol. Rev.,* 35, 171, 1971.

Table III.C
AMINO ACID MIXTURES, VITAMINS, AND GROWTH REGULATORS

Species	Casein hydrolysate	Peptone	Tryptone	Yeast extract	Biotin	Thiamine	Vitamin B_{12}	β-Indolylacetic acid (IAA)	Ref.
Agmenellum quadruplicatum	u[a]				Ɍ	Ɍ	R		1
Anabaena cylindrica					Ɍ	Ɍ	Ɍ	s	2, 3
A. variabilis[b]	S[a]	g[c]	g[c]	g[c]	Ɍ	Ɍ	Ɍ		4
Anacystis marina	g[d]				Ɍ	Ɍ	Ɍ		1
A. nidulans[e]	g[c]	g[c]	g[c]	g[c]	Ɍ	Ɍ	Ɍ	s	3, 4
Aphanizomenon flos-aquae					Ɍ	Ɍ	Ɍ		5
Calothrix brevissima					Ɍ	Ɍ	Ɍ		6
C. parietina	U				Ɍ	Ɍ	Ɍ		2
C. scopulorum					Ɍ	Ɍ	Ɍ		7
Chlorogloea fritschii[f]								s	3
Chroococcus turgidus	U				Ɍ	Ɍ	Ɍ		2
Coccochloris elabens	u[a]				Ɍ	Ɍ	R		1, 8, 9
C. peniocystis					Ɍ	Ɍ	Ɍ		5
Fremyella diplosiphon	U[d]								10
Gloeocapsa dimidiata					Ɍ	Ɍ	Ɍ		5
G. membranina					Ɍ	Ɍ	Ɍ		11
Lyngbya aestuarii	g[d]				Ɍ	Ɍ	Ɍ		1
L. lagerheimii	g[d]				Ɍ	Ɍ	R		1
Microcoleus chthonoplastes	u[a]				Ɍ	Ɍ	Ɍ		1
M. tenerrimus	g[d]				Ɍ	Ɍ	Ɍ		1
M. vaginatus[g]					Ɍ	Ɍ	Ɍ		12
Microcystis aeruginosa					Ɍ	Ɍ	Ɍ		5
Nostoc entophytum					Ɍ	Ɍ	Ɍ		7
N. muscorum					Ɍ	Ɍ	Ɍ	s	2—4
Nostoc sp.								R	13
Oscillatoria amphibia	g[d]				Ɍ	Ɍ	R		1
O. brevis					Ɍ	Ɍ	Ɍ		7
O. lagerheimii	g[d]				Ɍ	Ɍ	R		1
O. lutea					Ɍ	Ɍ	Ɍ[h]		12
O. rubescens					Ɍ	Ɍ	Ɍ		14
O. subtilissima	g[d]				Ɍ	Ɍ	R		1
Phormidium faveolarum	U				Ɍ	Ɍ	Ɍ	s	2, 3
P. luridum	U				Ɍ	Ɍ	Ɍ		2
P. persicinum					Ɍ	Ɍ	R		15
P. tenue					Ɍ	Ɍ	Ɍ		5

[a] In the light.
[b] Nonheterocystous strain.
[c] In the light and in the dark.
[d] In the dark.
[e] Synechococcus sp.[9]
[f] Chlorogloeopsis sp.[9]
[g] Var. fuscus.
[h] However, Oscillatoria lutea var. auxotrophica and var. contorta require vitamin B_{12} for growth.

Table III.C (continued)
AMINO ACID MIXTURES, VITAMINS, AND GROWTH REGULATORS

Species	Casein hydrolysate	Peptone	Tryptone	Yeast extract	Biotin	Thiamine	Vitamin B_{12}	β-Indolylacetic acid (IAA)	Ref.
Phormidium sp.[i]		U[j]							16
Plectonema boryanum	S[k]								17
P. nostocorum					Ř	Ř	Ř		5
P. notatum	U				Ř	Ř	Ř		2
P. terebrans	U[a]				Ř	Ř	Ř		1
Schizothrix calcicola					Ř	Ř	Ř		12
Synechococcus cedrorum	U				Ř	Ř	Ř		2
Tolypothrix tenuis	U[d]				g[d]	g[d]	g[d]	s	3, 18

[i] Strain 485 (Indiana Culture Collection).
[j] Reconstituted algal protein hydrolysate, utilized also in the dark but better in the light.
[k] Supports growth on glucose in the dark.

REFERENCES

1.   Van Baalen, C., *Bot. Mar.*, 4, 129, 1962.
2.   Allen, M. B., *Arch. Mikrobiol.*, 17, 34, 1952.
3.   Ahmad, M. R. and Winter, A., *Planta*, 78, 277, 1968.
4.   Kratz, W. A. and Myers, J., *Am. J. Bot.*, 42, 282, 1955.
5.   Gerloff, G. C., Fitzgerald, G. P., and Skoog, F., *Am. J. Bot.*, 37, 216, 1950.
6.   Watanabe, A., *Arch. Biochem. Biophys.*, 34, 50, 1951.
7.   Stewart, W. D. P., *Ann. Bot.* (London), 26, 439, 1962.
8.   Burkholder, P. R., in *Symposium on Marine Microbiology*, Oppenheimer, C. H., Ed., Charles C Thomas, Springfield, Ill., 1963, 133.
9.   Stanier, R. Y., Kunisawa, R., Mandel, M., and Cohen-Bazire, G., *Bacteriol. Rev.*, 35, 171, 1971.
10. Diakoff, S. and Scheibe, J., *Physiol. Plant.*, 34, 125, 1975.
11. Williams, A. E. and Burris, R. H., *Am. J. Bot.*, 39, 340, 1952.
12. Baker, A. F. and Bold, H. C., Phycological Studies. X. Taxonomic Studies on Oscillatoriaceae, University of Texas Publ. No. 7004, Austin, 1970.
13. Bunt, J. S., *Nature*, 192, 1274, 1961.
14. Staub, R., *Schweiz. Z. Hydrol.*, 23, 82, 1961.
15. Pintner, I. J., and Provasoli, L., *J. Gen. Microbiol.*, 18, 190, 1958.
16. Pope, D. H., *Can. J. Bot.*, 52, 2369, 1974.
17. White, A. W. and Shilo, M., *Arch. Microbiol.*, 102, 123, 1975.
18. Kiyohara, T., Fujita, Y., Hattori, A., and Watanabe, A., *J. Gen. Appl. Microbiol.*, 6, 176, 1960.

QUALITATIVE REQUIREMENTS AND UTILIZATION OF NUTRIENTS: ALGAE*

J. L. Hastings and W. H. Thomas

INTRODUCTION

Most algae are photoautotrophs; they are photosynthetic and utilize CO_2 as a carbon source. Some have heterotrophic capabilities and grow on organic carbon sources. The criterion listed in the tables for determining utilization of such sources is that a given alga will grow in darkness on a given substrate (facultative heterotrophy). Stimulation of growth by a substance in the light does not indicate utilization.

Inorganic requirements include all those known for all plants; major and trace inorganic elements are required. It should also be noted that the Bacillariophyceae and certain Chrysophyceae require Si for growth. Media for algae that include these inorganics are given in Stein[†] for both fresh-water and marine algae.

Many algae require certain vitamins. These are listed as "required" in the tables if the alga does not grow well in a purely mineral salts medium.

Nitrate and ammonium are the most common nitrogen sources for algae. Utilization of other nitrogen sources is shown in the tables.

Abbreviations: R = required, ᴿ̶ = not required; U = utilized, U̶ = not utilized; u = poorly utilized.

* The compilation of this information was supported by Grant DES 74-23972 from the Oceanography Section, U.S. National Science Foundation.
[†] Stein, J., Ed., *Handbook of Phycological Methods: Culture Methods and Growth Measurements,* Cambridge University Press, New York, 1973.

Table 1
VITAMINS, SUGARS, AND ALCOHOLS

Vitamins are listed as "required" if the alga does not grow well in a purely mineral salts medium. Criterion for determining utilization of sugars and alcohols as a carbon source: growth of the alga in darkness on a given substrate (facultative heterotrophy). Stimulation of growth by a substance in the light does not indicate utilization.

Species (synonym)	Vitamins				Carbon sources — Sugars										Carbon sources — Alcohols					Ref.
	Thiamine	Biotin	B12	Other	Arabinose	Fructose	Galactose	Glucose	Lactose	Maltose	Mannose	Raffinose	Sucrose	Xylose	n-Butanol	Ethanol	Glycerol	Methanol	Other	
Cyanophyceae																				
Agmenellum quadruplicatum	R	R	R					U					U[a]							152, 166
Anabaena cylindrica	R	R	R					U					U[a]							4, 82
A. gelatinosa		R	R																	26
A. inaequalis	R	R	R																	82
A. naviculoides		R	R																	26
A. spiroides	R	R	R			U		U		U			U[b]							80
A. variabilis						U	U	U	U	U			U[b]				U			80, 82, 86
Anabaena sp.						U		U		u			U[b]							80, 82, 83
Anabaenopsis circularis								U		u			U[c]							80, 83, 168
Anabaenopsis sp.																				82
Anacystis marina	R	R	R			U	U	U	U	U			U[b]				U			152, 166
A. nidulans	R	R	R					U												82, 86, 150, 153
Aphanocapsa sp.	R	R	R					U[d,e]												117, 141
Aphanizomenon flos-aquae	R	R	R																	46
Calothrix brevissima	R	R	R			U		U		u			U[a]							82, 83, 167
C. desertica								U												80
C. membranacea						U		U		u			U[b]							59, 130
C. parietina	R	R	R																	4, 80
C. scopulorum	R	R																		155
Chlorogloea fritschii					U	U		u, fU[87]		U			U						u[g]	40, 80, 82, 83, 87, 109

a   Combined N absent.
b   Combined N present.
c   Combined N absent; viable after 3 months in the dark.
d   Two strains showed slow growth, one strain showed poor growth, and two more failed to grow.[141]
e   One strain can utilize glucose in the dark; another strain cannot.[117]
f   Only with NO_3 added.
g   Mannitol.

Table 1 (continued)
VITAMINS, SUGARS, AND ALCOHOLS

Cyanophyceae (continued)

Species (synonym)	Thiamine	Biotin	B_{12}	Other	Arabinose	Fructose	Galactose	Glucose	Lactose	Maltose	Mannose	Raffinose	Sucrose	Xylose	n-Butanol	Ethanol	Glycerol	Methanol	Other	Ref.
Chlorogloea sp.	R	R	R																	141
Chroococcus turgidus	R	R	R					U												4
Coccochloris elabens	R	R	R					U												152, 166
C. peniocystis	R	R	R					U												46, 152
Fischerella epiphytica								U					U[h]							103
F. musicola								U					U[h]							103
F. tisserantii								U					U[h]							103
Fremyella diplosiphon				U[i]				U?					U							177
Gloeocapsa alpicola		R	R			U		U											U[i]	150, 152
G. dimidiata	R	R	R																	46
G. membranina	R	R	R																	172
Lyngbya aestuarii	R	R	R					U												166
L. lagerheimii	R	R	R					U,U[80]												80, 166
Lyngbya sp.													U[j]							82
Microcoleus chthonoplastes	R	R	R					U												166
M. tenerrimus	R	R	R					U												166
M. vaginatus		R	R																	80
M. vaginatus fuscus	R	R	R					U												9
Microcystis aeruginosa (*Diplocystis aeruginosa*)	R	R	R					U												46, 152
Nodularia sphaerocarpa								U												80
Nostoc commune						U		U					U[b]							80
N. ellipsosporum		R	R					u		U			U[b]							82, 83
N. entophytum	R	R	R			U,U[83]		U[k],U[83,86]	U	U,U[83]			U[k],U[a,83,86]				U			155
N. muscorum	R	R	R	U,[m] U[g]		U	U	U	U	U	U		U	U			U		U,[m] U[g]	6, 83, 86
N. punctiforme					u	U	U	U[n],U[o]	U				U							55, 82
Nostoc sp.	R	R	R	U[t,g]				U	U				U				U		U[t,g]	62, 70, 80
Oscillatoria amphibia	R	R	R								U			U			U			166

h Growth was often, but not always, better with NO_3 present.
i Inositol.
j Combined N present; viable after 3 months in dark.
k By one strain; utilized by one strain.
l Growth is slow. Also, a buffer must be used or the pH drops and kills the cells.
m Dextrin, insulin.
n Growth may just be *very* slow since it does occur in the light with 10^{-5} M DCMU inhibiting photosynthesis.
o Fastest growth when casamino acids added.

Table 1 (continued)
VITAMINS, SUGARS, AND ALCOHOLS

Column groups: **Vitamins** (Thiamine, Biotin, B$_{12}$, Other); **Carbon sources — Sugars** (Arabinose, Fructose, Galactose, Glucose, Lactose, Maltose, Mannose, Raffinose, Sucrose, Xylose); **Alcohols** (n-Butanol, Ethanol, Glycerol, Methanol); Other; Ref.

Cyanophyceae (continued)

Species (synonym)	Thiamine	Biotin	B$_{12}$	Other	Arabinose	Fructose	Galactose	Glucose	Lactose	Maltose	Mannose	Raffinose	Sucrose	Xylose	n-Butanol	Ethanol	Glycerol	Methanol	Other	Ref.
O. brevis	R	R	R																	155
O. lutea	R	R	R																	9
O. lutea auxotrophica	R	R	R																	9
O. lutea contorta	R	R	R																	9
O. rubescens	R	R	R																	153
O. subtillisima	R	R	R					U					U[a]							166
O. tenuis								U												82
Oscillatoria sp.																				80
Phormidium foveolarum	R	R	R					U, U[q,4,u]					U[a]							4
P. luridum	R	R	R			U		U		u										4, 82, 83
P. persicinum[p]	R	R	R																	121
P. tenue	R	R	R																	46
Phormidium sp.						U		U[q,r]		u			U							80, 83
Plectonema boryanum						U		U[q,r,114]		U			U[a]							80, 114
P. calothricoides						U		U		U			U[a]							82, 83
P. nostocorum	R	R	R					U[s]												46
P. notatum	R	R	R					U												4
P. terebrans	R	R	R					U											U[t]	166
Schizothrix calcicola	R	R	R																	9, 80
S. calcicola glomerata	R	R	R																	9
S. calcicola mucosa	R	R	R																	9
S. calcicola spiralis	R	R	R																	9
S. calcicola vermiformis	R	R	R																	82
Scytonema sp.													U[a]							4, 152
Synechococcus cedrorum	R	R	R					U												152
S. elongatus								U												152
S. lividus								U[u]		U										
Tolypothrix tenuis	R?	R?	R?			U, U[s3]		U[u]					u, U[s2,s3]							80, 82—84

[p] Carbon sources tested for heterotrophic growth are not individually listed.

[q] Very slow growth rate. Growth with glucose does occur in the light with photosynthesis inhibited by 10^{-5} M DCMU.

[r] Growth in the dark occurred only if a continuous flow-through system was used.

[s] In presence of yeast autolysate.

[t] Ribose.

[u] In presence of casein hydrolysate.

Table 1 (continued)
VITAMINS, SUGARS, AND ALCOHOLS

Species (synonym)	Vitamins				Carbon sources															Ref.
					Sugars										Alcohols					
	Thiamine	Biotin	B_{12}	Other	Arabinose	Fructose	Galactose	Glucose	Lactose	Maltose	Mannose	Raffinose	Sucrose	Xylose	n-Butanol	Ethanol	Glycerol	Methanol	Other	
Rhodophyceae																				
Antithamnion glanduliferum	R	R	R																	160
A. spirographidis (sarniense)	R	R	R																	160
Antithamnion sp.	R	R	R																	160
Asterocytis ramosa	R	R	R		U	U	U						U							45
Bangia fuscopurpurea	R	R	R																	133
Erythrotrichia carnea	R	R	R																	44
Goniotrichum alsidii (elegans)	R	R	R																	42
Nemalion helminthoides (multifidum)	R	R	R																	43
Polysiphonia urceolata	R	R	R																	44
Porphyra tenera (Conchocelis phase)	R	R	R																	72
Porphyridium (=purpureum) cruentum	R	R	R					U												27, 29
Rhodella maculata	R	R	R					U												165
Cryptophyceae																				
Chilomonas paramecium[a]	R	R	R	R				U							U	U		U	U[b]	2, 23, 64, 134
Chroomonas salina	R	R	R																	7
Cryptomonas borealis	R	R						U									U[c]			175
C. marssonii	R	R						U												175
C. ovata	R	R	R																	175
Cryptomonas sp.	R	R						U												137
Cyanophora paradoxa[d]	R	R	R					U												137
Hemiselmis virescens	R	R																		29, 32
Rhodomonas lens	R	R	?																	134
R. ovalis	R	R	R																	77

[a] Colorless.

[b] *i*-Butanol, *i*-propanol, and *n*-propanol; however, *n*-hexanol is utilized.

[c] Grows only with high concentrations.

[d] Genus of uncertain taxonomic position.

Table 1 (continued)
VITAMINS, SUGARS, AND ALCOHOLS

Dinophyceae

Species (synonym)	Vitamins				Carbon sources — Sugars										Carbon sources — Alcohols					Ref.
	Thiamine	Biotin	B$_{12}$	Other	Arabinose	Fructose	Galactose	Glucose	Lactose	Maltose	Mannose	Raffinose	Sucrose	Xylose	n-Butanol	Ethanol	Glycerol	Methanol	Other	
Amphidinium carteri	R	R	R																	106
A. rhynchocephalum	R	R	R																	106
Cachonina niei	R	R	R																	99
Entomosigma sp. (Heterosigma akashivo)	R	R	R																	76
Exuviaella cassubica	R	R	R																	136
Exuviaella sp.	R	R	R																	74
Glenodinium foliaceum	R	R	R																	32
G. halli	?	R	R																	48
G. monotis	R	R	R	?	ψ	ψ	ψ	ψ											U[a]	118
Gonyaulax polyedra	R	?	R	?	ψ	ψ	ψ	ψ	ψ	ψ	ψ		ψ	ψ	ψ	ψ	ψ		ψ[b]	58, 134, 162
Gymnodinium brevis	R	R	R		ψ	ψ	ψ	ψ	ψ	ψ	ψ	ψ	ψ	ψ	ψ	ψ	ψ			3, 173
G. foliacium		R						U												29
G. inversum	R	R	R	?																176
G. simplex	?	?	?																	163
G. splendens	R	R	R																	157
Gyrodinium californicum	?	R	R					U								U			U[c]	137
G. cohnii	R	R	R				U	U					U							71, 132, 135
G. resplendens	R	R	R																	136
G. uncatenum	R	R	R																	136
Heterosigma inlandica	R	R	R	R[d,e]																78
Oxyrrhis marina	R	R	R																	30, 32, 33
Peridinium balticum	R	R	R																	136
P. chattoni	R	R	R			ψ	ψ	ψ			ψ		ψ						ψ[f]	136
P. cinctum var. ovuplanum	R	R	R			ψ	ψ	ψ												19
P. hangoei	R	R	R																	73

[a] Ribose.

[b] Ribose and many other carbohydrates; also dulcitol, ethylene glycol, mannitol, and sorbitol.

[c] Hexanol.

[d] A quinone – must be a fully formed, substituted benzoquinone with an unsaturated side chain of at least 30 carbons. Ubiquinone or plastoquinone will suffice.

[e] Steroid – several common steroids such as cholesterol will satisfy this requirement, which probably is not absolute since some growth can be obtained without steroids.

[f] Rhamnose.

Table 1 (continued)
VITAMINS, SUGARS, AND ALCOHOLS

Species (synonym)	Vitamins				Carbon sources															Ref.
					Sugars										Alcohols					
	Thiamine	Biotin	B₁₂	Other	Arabinose	Fructose	Galactose	Glucose	Lactose	Maltose	Mannose	Raffinose	Sucrose	Xylose	n-Butanol	Ethanol	Glycerol	Methanol	Other	
Dinophyceae (continued)																				
P. trochoideum	R	R	R																	32
Polykrikos schwartzii	R	R	R																	75
Prorocentrum micans	R	R	R																	27, 29, 79
Symbiodinium micro-adriaticum[g]	R	R	R					U												105, 107
Woloszynskia limnetica (Peridinium sp.)	R	R	R																	137

g The taxonomic position of zoochlorellae is uncertain.

Species (synonym)	Thiamine	Biotin	B₁₂	Other	Arabinose	Fructose	Galactose	Glucose	Lactose	Maltose	Mannose	Raffinose	Sucrose	Xylose	n-Butanol	Ethanol	Glycerol	Methanol	Other	Ref.
Bacillariophyceae																				
Achnanthes brevipes	R	R	R					U												90
Amphipleura rutilans	R	R	R					U												90
Amphiprora paludosa	R[a]	R	R,[b] R[s,9]					U												90
Amphora coffaeiformis	R		R					U[c]												54, 59, 90
A. lineolata	R		R					U												90
A. paludosa duplex	R	R	R																	90
A. perpusilla	R	R	R																	68
Asterionella formosa	R[d]	R	R[d]																	134
Bellerochea sp. (polymorpha?)	R[d]	R	R[d]																	56
Bellerochea sp. (spinifera?)	R	R	R					U												56
Chaetoceros ceratosporum	R	R	R																	54, 93
C. gracilis	R	R	R																	163
C. lorenzianus	R	R	R																	53, 54
C. pelagicus	R	R	R					U												53, 54, 93
C. pseudocrinitus	R	R	R					U												54, 93
C. simplex	R	R	R																	54
Coscinodiscus asteromphalus	R	R	R					U												53, 54, 93

a By seven strains; required by three strains.
b By six strains; required by four strains.
c By eight strains; not utilized by two strains.
d Required by one strain; not required by another.

Table 1 (continued)
VITAMINS, SUGARS, AND ALCOHOLS

Bacillariophyceae (continued)

Species (synonym)	Thiamine	Biotin	B₁₂	Other	Arabinose	Fructose	Galactose	Glucose	Lactose	Maltose	Mannose	Raffinose	Sucrose	Xylose	n-Butanol	Ethanol	Glycerol	Methanol	Other	Ref.
Coscinodiscus sp.	R		R		U	U								U	U[e]				U[f]	171
Cyclotella caspia	R[g]	R[g]	R		U	U	u				U		U	U	U[e]		U		U[f]	53, 54, 93
C. cryptica			R[g]		U	U	U	U			U		U	U			U			61, 90, 149, 171
C. nana			R[g]					U[h]												52, 54, 93
Cylindrotheca closterium (*Nitzschia closterium*)	R	R	R																	54
C. fusiformis (*Nitzschia closterium*)	R[j]	R	R[i]					U[k], U[89]	U				U			U	U			54, 89, 90, 94
Detonula confervacea	R	R	R																	52
Ditylum brightwellii	R	R	R																	54
Fragilaria brevistrata	R	R	R																	98
F. capucina	R	R	R																	134
Fragilaria sp. (*pinnata?*)	R	R	R																	56
Fragilaria sp. (*rotundissima?*)	R	R	R																	56
Gomphonema parvulum								U												92
Hantzschia amphionys								U												92
Licmophora hyalina		R	R																	98
Melosira juergensii	R	R	R					U												158
M. lineata	R	R	R					U												164
M. nummuloides	R	R	R			U		U[60,93] U									U		U[l]	29, 60, 93,
Melosira sp. (*nummuloides*)	R	R	R																	164
Navicula bottnica	R	R	R																	51
Navicula bottnica	R	R	R																	90
N. corymbosa	R		R																	90
N. inceria	R		R					U												90

e   Rhamnose, ribose.
f   Myo-inositol, mannitol, sorbitol, dulcitol.
g   By five strains.
h   By three strains.
i   B₁₂ not required by three strains; required by one.
j   By four strains.
k   By three strains; not utilized by one strain.
l   Mannitol.

Table 1 (continued)
VITAMINS, SUGARS, AND ALCOHOLS

Bacillariophyceae (continued)

Species (synonym)	Thiamine	Biotin	B$_{12}$	Other	Arabinose	Fructose	Galactose	Glucose	Lactose	Maltose	Mannose	Raffinose	Sucrose	Xylose	n-Butanol	Ethanol	Glycerol	Methanol	Other	Ref.
N. meniscus	R																			90
N. minima			R																	92
N. pelliculosa	R	R	R			U		U												56, 92
Nitzschia alba[n]	R		R		U	U	U	U	U	U		U	U	U		U	U		U[m]	91
N. angularis affinis	R	R	R				U	U												90
N. brevirostris	R		R																	54
N. curvilineata	R		R					U												90
N. filiformis	R		R					U												90
N. fonticola								U												92
N. frustulum	R[o]		R[p]					U[q]												90
N. hybridaeformis	R		R					U												90
N. laevis	R[r]	R	R[r]					U												54, 90
N. lanceolata	R	R	R					U												90
N. leucosigma[n]	R		R																	91
N. marginata	R		R					U												90
N. obtusa scalpelliformis	R		R					U												90
N. ovalis	R		R					U												90
N. palea[s]								U												92
N. punctata	R		R					U, u[127]												90
N. putrida[n]	R		R																	91, 127
N. seriata			R																	53
N. silicula	R		R																	90
N. tenuissima	?		?					U												90
Opephora sp.	R	R	R																	90
Phaeodactylum tricornutum	R		R					U												29, 66
Rhizosolenia setigera			R?																	53
Skeletonema costatum	R		R[t]					U												29, 31, 93

m Cellobiose, ethylene glycol, fucose, mannitol, melibiose, α-methylglucoside, rhamnose, sorbose, and trehalose.
n Colorless.
o By six strains.
p By four strains; required by two strains.
q By three strains; utilized by three strains.
r By two strains.
s 27 strains.
t Several other compounds similar to B$_{12}$ can be utilized in its place.[35]

Table 1 (continued)
VITAMINS, SUGARS, AND ALCOHOLS

Species (synonym)	Vitamins				Carbon sources															Ref.
	Thiamine	Biotin	B$_{12}$	Other	Arabinose	Fructose	Galactose	Glucose	Lactose	Maltose	Mannose	Raffinose	Sucrose	Xylose	n-Butanol	Ethanol	Glycerol	Methanol	Other	
Bacillariophyceae (continued)																				
Stauroneis amphoroides	R		R					U												90
Stephanopyxis tunis	R	R	R																	134
Synedra affinis	R		R					U												90
Synedra sp. (rotundissima?)	R	R	R																	56
Tabellaria flocculosa	R	R	R																	134
Thalassiosira fluviatilis			R					U?												53, 54, 93, 149
T. nordenskioldii			R?																	53
Haptophyceae																				
Apistonema aestuari (Pontosphaera roscoffensis)	R	R	R																	120
Chrysochromulina brevefilum	R	R	R					U			U		U				U			120
C. kappa	R	R	R					U			U		U				U			120
C. strobilus	R	R	R					U			U		U				U			120
Cricosphaera elongata			R					U												29
Dicrateria inornata	R	R	R					U			U									120
Emiliana (Coccolithus) huxleyi	R[b]	R	R,[c] R[120]					u		U							U		U[a]	53, 120, 149
Hymenomonas carterae	R	R	R																	137
H. elongata	R	R	R					U			U									31, 120
Isochrysis galbana	R	R	R																	29, 120, 134
Ochrosphaera neopolitana	R	R	R																	120, 122
Pleurochrysis scherffelii	R	R	R																	120, 134
Pyrmnesium parvum	R,[49] R	R	R,[49] R		U	U	U	U	U	U	U		U	U			U[d]		U[e,g] U[f,49]	29, 32, 49, 120, 139
Syracosphaera sp.	R	R	R																	120

a Propylene glycol.
b By two strains; not required by one strain.
c By two strains; required by one strain.
d Grows only with high concentration of glycerol.
e Rhamnose, ribose, sorbose, cellobiose, mellibiose, raffinose.
f Mannitol.
g Inositol, sorbitol, thioglycerol.

Table 1 (continued)
VITAMINS, SUGARS, AND ALCOHOLS

Chrysophyceae

Species (synonym)	Thiamine	Biotin	B$_{12}$	Other	Arabinose	Fructose	Galactose	Glucose	Lactose	Maltose	Mannose	Raffinose	Sucrose	Xylose	n-Butanol	Ethanol	Glycerol	Methanol	Other	Ref.
Mallomonas epithalattia	R	R	R																	29
Microglena arenicola	R	R	R					U												29, 32, 120
Monochrysis lutheri	R	R	R					U												29, 32, 120
Ochromonas danica	R	R	R		U	U	U	U^a	U	U	U	U?	?	U			U			1, 59, 120, 129
O. malhamensis	R	R	R		U	U	U	U^a	U	U?	U	U?	U	U			U		U?^{b}	69, 123, 129
O. minuta	R	R	R																	120
Pavlova gyrans	R	R	R																	120, 122
Poteriochromonas stipitata	R	R	R																	129
Stichochrysis immobilis	R	R	R																	134
Synerypta (Synochromonas) korschikoffi	R	R	R																	120, 130
Synura caroliniana	R	R	R																	120, 134
S. petersenii	R	R,R^{120}	R																	120, 137
S. sphagnicola	R	R	R																	120

[a] Other organic nutrients also present in a defined medium.[1]
[b] Ribose.

Table 1 (continued)
VITAMINS, SUGARS, AND ALCOHOLS

Species (synonym)	Vitamins				Carbon sources															Ref.
					Sugars										Alcohols					
	Thiamine	Biotin	B$_{12}$	Other	Arabinose	Fructose	Galactose	Glucose	Lactose	Maltose	Mannose	Raffinose	Sucrose	Xylose	n-Butanol	Ethanol	Glycerol	Methanol	Other	
Xanthophyceae																				
Botrydiopsis intercedens	R	R	R			U					U		U			U	U			13
Bumilleriopsis brevis	R	R	R			U		U			U		U			U	U			13
Chlorellidium tetrabotrys	R	R	R			U		U			U		U			U	U			13
Monodus subterraneus	R	R	R		U	U	U	U	U		U	U	U				U			107, 108
Tribonema aequale	R	R	R			U		U	U		U		U			U	U			12
T. minus	R	R	R					U	U	U			U	U		U	U			13
Eustigmatophyceae																				
Polyedriella helvetica	R	R	R			U		U			U		U			U	U			13
Phaeophyceae																				
Desmarestia ligulata	R	R	R																	111
D. viridis	R	R	R																	111
Ectocarpus confervoides	R	R				ψ					U								ψ^a	14
E. fasciculatus	R	R							U	U										116
E. parasitica	R	R																		27
Litosiphon pusillus	R	R	R																	116
Petalonia zostericola	R	R	R																	161
Pylaiella (Pilayella) littoralis	R	R	R																	116
Scytosiphon lomentaria	R	R	R																	161
Waerniella lucifuga	R	R	R																	27

a Mannitol, dextrose.

Table 1 (continued)
VITAMINS, SUGARS, AND ALCOHOLS

Euglenophyceae

| Species (synonym) | Vitamins | | | | Carbon sources | | | | | | | | | | | | | | | Ref. |
| | | | | | Sugars | | | | | | | | | | Alcohols | | | | | |
	Thiamine	Biotin	B₁₂	Other	Arabinose	Fructose	Galactose	Glucose	Lactose	Maltose	Mannose	Raffinose	Sucrose	Xylose	n-Butanol	Ethanol	Glycerol	Methanol	Other	
Astasia longa (A. chattoni, A. klebsii)	R	R	R	R				U[a]							U	U		U	U[b]	2, 10, 17, 134
A. quartana	R	R	R													U	U	U	U[c]	2
Euglena anabaena minor[d]	R	R	R	R				U					U		U	U	U	U		2
E. deses[e]		R	R[f]	R				U					U			U	U			2
E. gracilis bacillaris	R[h]	R	R[f]					U[g]												2, 25, 65
E. gracilis typica	R	R	R[f]					U							U	U		U	U[i]	2, 134
E. gracilis urophora	R	R	R					U[k]					U		U	U		U	U[j]	2, 134
E. gracilis (Vischer strain)																				25
E. klebsii[d,l]	R	R	R	?				U					U							2
E. mutabilis	R	R	R																	98
E. pisciformis	R[m]	R	R					U					U				U			2, 37, 134
E. spirogyra	R	R	R																	88
E. stellata	R	R	R																	2, 134
E. viridis	R	R	R					U									U			102, 134
Eutreptiella sp.	R	R	R																	74
Peranema trichophorum	R	R	R	R																156
Phacus pyrum	R	R	R																	134
Trachelomonas abrupta	?	R	R																	138
T. pertyi	?	R	R																	134

a A mutant strain was able to utilize glucose, whereas the parent strain could not.[10]

b i-Pentanol, n-pentanol, and i-propanol; however, n-hexanol and n-propanol are utilized.

c n-Hexanol and n-propanol.

d May require no vitamins after adaptation to media lacking vitamins, as with other species of Euglena.

e According to E. G. Pringsheim, the organism used in these studies was E. geniculata.

f Pseudocobalamin also utilized.

g pH 4.5.

h Can be replaced by pyrimidine.

i i-Butanol, i-pentanol, n-pentanol, and i-propanol; however, n-hexanol and n-propanol are utilized.

j i-Pentanol, n-pentanol, and i-propanol; however, i-butanol, n-hexanol, and n-propanol are utilized.

k According to E. G. Pringsheim, the organism used in these studies was E. mutabilis.

l pH 4.5 or 6.8.

m Can be replaced by pyrimidine and thiazole.

Table 1 (continued)
VITAMINS, SUGARS, AND ALCOHOLS

Species (synonym)	Vitamins				Carbon sources															Ref.
					Sugars										Alcohols					
	Thiamine	Biotin	B_{12}	Other	Arabinose	Fructose	Galactose	Glucose	Lactose	Maltose	Mannose	Raffinose	Sucrose	Xylose	n-Butanol	Ethanol	Glycerol	Methanol	Other	
Prasinophyceae																				
Platymonas chui								U[a]												164
P. subcordiformis		R	R					U[a]												164
P. tetrathele	R	R	R					ψ[a]												164
Pyramimonas inconstans	R	R	R																	134
Stephanoptera gracilis	R	R	R					ψ												47
Tetraselmis carteriiformis	R							U[b]												164
Tetraselmis sp.						u														164

a Requires source of reduced nitrogen such as thymine when grown in the dark.
b For good growth in the dark, a source of reduced nitrogen in the form of a purine or a pyrimidine was essential.

Table 1 (continued)
VITAMINS, SUGARS, AND ALCOHOLS

Chlorophyceae

Species (synonym)	Thiamine	Biotin	B_{12}	Other	Arabinose	Fructose	Galactose	Glucose	Lactose	Maltose	Mannose	Raffinose	Sucrose	Xylose	n-Butanol	Ethanol	Glycerol	Methanol	Other	Ref.
Acetabularia mediterranea	R	R	R																	146
Asterococcus superbus	?	R	?																	132
Astrephomene gubernaculifera	?[154] R	R	R																	16, 154
Brachiomonas submarina	R	R	R																	34
Bracteacoccus cinnabarinus	R	R	R					U												115
B. engadiensis	R	R	R					U												115
B. minor	R	R	R					U												115, 145
B. terrestris	R	R	?					U												115
Carteria crucifera	?		?																	18
Chlamydobotrys stellata	R	R	R					U												124
Chlamydomonas actinochloris	?		?																	18
C. agloeformis	R	R	R	R																2, 134
C. calyptrata	?		?																	18
C. carrosa[a]	?		?																	18
C. chlamydogama	R	R	R																	67
C. dysosmos								U									U			97
C. eugametos[b]	R	R	R																	170
C. gloeogama	?		?																	18
C. gloeopara	?		?																	18
C. inflexa	?		?																	18
C. kakosmos	?		?																	18
C. mexicana	?		?																	18
C. microsphaera acuta	?		?																	18
C. microsphaerella	?		?																	18
C. minuta	?		?																	18
C. moewusii	R	R	R	R	U			U		U			U	U	U	U	U			2, 67, 134
C. moewusii rotunda	?		?																	18
C. mundana	R	R	R					U		U						U				38, 39
C. mundana astigmata	?	R	R														U			39
C. mutabilis	?		?																	18
C. pallens	R	R	R																	125

[a] Four strains.

[b] An obligate phototroph; individual compounds tested as carbon sources are not listed.[169]

Table 1 (continued)
VITAMINS, SUGARS, AND ALCOHOLS

| Species (synonym) | Vitamins | | | | Carbon sources | | | | | | | | | | | | | | | Ref. |
| | Thiamine | Biotin | B_{12} | Other | Sugars | | | | | | | | | | Alcohols | | | | Other | |
					Arabinose	Fructose	Galactose	Glucose	Lactose	Maltose	Mannose	Raffinose	Sucrose	Xylose	n-Butanol	Ethanol	Glycerol	Methanol		
Chlorophyceae (continued)																				
C. peterfi	?	R	?																	18
C. pichinchae	R	R	R																	63
C. pulsatilla[c]	R	R	R																	29, 34
C. radiata	?	R	?					U^d												18
C. reinhardii	R	R	R^e		U	U	U	U	U	U	U		U	U		U	U		U, f, ug	98, 134, 143
C. sectilis	?		?																	18
C. spreta	?							U												29
C. typhlos	?		?																	18
Chlorella autotrophica	R	R	R		U	U	U	U	U	U	U	U	U							147
C. candida	R	R	R		U	U	U	U	U	U	U	U	U							147
C. ellipsoidea	R	R	R		U	U	U	U	U	U	U	U	U							147
C. emersonii	R	R	R		U	U	U	U	U	U	U	U	U							147
C. emersonii globosa	R	R	R		U	U, uᵍ,ⁱ	U	U	U	U	U	U	U							81, 147
C. fusca	R	R	R		U	U	U	U	U	U	U	U	U							147
C. fusca vacuolata	R	R	R		U	U	U	U^h	U	U	U	U	U							147
C. infusionum	R	R	R		U	U	U	U	U	U	U	U	U							147
C. infusionum auxenophila	R	R	R		U	U	U	U	U	U	U	U	U							81
C. kessleri					?	u	?	?	?	U	U	U	C							81
C. luteoviridis							U	?	U											147
C. miniata	R	R	R		U	U	?	U	?	U	U		U							81
C. minutissima	R	R	R		U	U	U	?	U	U	U	U	U							147
C. mutabilis	R	R	R		U	U	U	?	U	U	U	U	U							147
C. nocturna	R	R	R		U	u	U	U	U	U	U	U	U							147
C. photophila	R	R	R		U	U	u	U	U	U	U	U	U							147
C. pringsheimii	R	R	R		U	U	U	u	U	U	U	U	U							147
C. protothecoides	R	R	R		U	U	U	U	U	U	U	U	U							147
C. protothecoides communis	R	R	R		U	U	U	U	U	U	U	U	U							147

c Probably an obligate phototroph; substrates failing to support dark growth are not individually listed.
d Seven strains.
e By one strain; required by one strain.
f Ribose and mannitol.
g Acetylmethyl carbinol.
h By one strain; utilized by one strain.

Table 1 (continued)
VITAMINS, SUGARS, AND ALCOHOLS

Species (synonym)	Thiamine	Biotin	B$_{12}$	Other	Arabinose	Fructose	Galactose	Glucose	Lactose	Maltose	Mannose	Raffinose	Sucrose	Xylose	n-Butanol	Ethanol	Glycerol	Methanol	Other	Ref.
		Vitamins					Sugars									Alcohols				
Chlorophyceae (continued)																				
C. protothecoides galactophila	R	R	R		U	U	U	U	U	U	U	U	U							147
C. protothecoides mannophila	R	R	R		U	U	U	U	U	U	U	U	U							147
C. pyrenoidosa	R	R	R			U	U	U		U	U		U							147
C. regularis	R	R	R		U	U	U	U	U	U	U	U	U							147
C. regularis aprica	R	R	R		U	U	U	U	U	U	U	U	U							147
C. regularis imbricata	R	R	R		U	U	U	U	U	U	U	U	U							147
C. rubescens	R	R	R		U	U[81]	U	U	U	U	U	U	U							22
C. saccharophila	R	R	R		U	U	U	U	U	U	U	U	U							81, 147
C. simplex	R	R	R		U	U	U	U	U	U	U	U	U							147
C. sorokiniana	R	R	R		U	U	U	U	U	U	U	U	U							147
C. vannielii	R	R	R		U	U	U	U	U	U	U	U	U							147
C. variabilis	R	R	R		U	U	U	U	U	U	U	U	U							147
C. vulgaris	R	R	R		U	U,?[81]	U,[110],[81]	U		U,[110]	U	U	U							81, 110, 134, 147
C. vulgaris luteoviridis	R	R	R		U	U	U,?	U		U	U	U	U							147
C. zofingiensis		R	R			u	?	?		U	U	U	U							81
Chlorococcum aplanosporum	R	R	R					U						U							115
C. diplobionticum	R	R	R					U													115
C. echinozygotum	R	R	R					U													115
C. ellipsoideum	R	R	R					U													115
C. hypnosporum	R	R	R					U													115
C. macrostigmatum	R	R	R					U													115
C. minutum	R	R	R					U						U							8, 115
C. multinucleatum	R	R	R					U													115
C. oleofaciens	R	R	R					U													115
C. perforatum	R	R	R					U													115
C. pingaideum	R	R	R					U													115
C. punctatum	R	R	R					U													115
C. scabellum	R	R	R					U													115
C. tetrasporum	R	R	R					U													115
C. uliginosum	R	R	R					U											U[i]	8

i Ribose.

Table 1 (continued)
VITAMINS, SUGARS, AND ALCOHOLS

| Species (synonym) | Vitamins | | | | Carbon sources | | | | | | | | | | | | | | | Ref. |
	Thiamine	Biotin	B₁₂	Other	Arabinose	Fructose	Galactose	Glucose	Lactose	Maltose	Mannose	Raffinose	Sucrose	Xylose	n-Butanol	Ethanol	Glycerol	Methanol	Other	
Chlorophyceae (continued)																				
C. vacuolatum	R	R	R																	115
C. wimmeri	R	R	R					U												115
Chlorococonium elongatum					U		U	U	U	U			U						ψ,ju,k	100
Chlorosarcinopsis auxotrophica	R	R	R																	50
C. eremi	R	R	R					U											ui	50
C. pseudominor	R	R	R																	50
C. sempervirens	R	R	R																	98
Chlorosphaera consociata	R	R	R																	159
Closterium acerosum	R	R	R																	159
C. braunii	R	R	R																	159
C. ehrenbergii	R	R	R																	159
C. maculentum	R	R	R																	159
C. peracerosum	R	R	R																	159
C. pusillum	R	R	R																	159
C. siliqua	R	R	R																	159
C. strigosum	R	R	R																	159
C. turgidum	R	R	R																	95
Coelastrum morus (?)	R	R	R																	159
Cosmarium botrytis	R	R	R																	159
C. impressulum	R	R	R																	159
C. laeve	R	R	R																	159
C. lundelli	R	R	R																	159
C. meneghini	R	R	R																	159
C. turpini	R	R	R																	85
Cyanidium caldarium[l]	R	R	R,R⁶³			U	U	U	U	U			U			U	U			4,5
Cylindrocystis brebissonii	R	R	R																	63, 159
Desmidium swartzii	R	R	R																	159

j   Dextran.
k   Levulose.
l   Genus of uncertain taxonomic position.[148]

Table 1 (continued)
VITAMINS, SUGARS, AND ALCOHOLS

Chlorophyceae (continued)

Species (synonym)	Thiamine	Biotin	B₁₂	Other	Arabinose	Fructose	Galactose	Glucose	Lactose	Maltose	Mannose	Raffinose	Sucrose	Xylose	n-Butanol	Ethanol	Glycerol	Methanol	Other	Ref.
Dictyochloris fragrans		R	R																	115
Dictyococcus cinnabarinus		R	R					U												22
Diplostauron elegans		R	R			ψ	ψ	U	ψ	ψ			ψ	ψ		U	ψ	ψ	ψ[m,n]	101
Docidium manubrium		R	R																	159
Dunaliella primolecta		R	R					ψ								U	ψ	ψ		29, 134
D. salina		R	R					ψ												47, 134
D. tertiolecta		R	R					ψ[o]												87
D. viridis		R	R					ψ												47
Eremosphaera viridis		R	R																	151
Eudorina elegans		R	R																	36
Friedmannia israelensis		R	R		U	U		U						U						21, 50
Furcilla stigmatophora		R	R		ψ	U		ψ				ψ				U				11
Gloeocystis gigas	?		?			ψ	ψ	U	ψ	ψ			ψ		U		ψ			18
G. maxima	?		?																	18
Gonium multicoccum		R	R																	144
G. pectorale		R	R																	154
Haematococcus (Balticola) buetschlii		R	R					ψ												28, 29, 34
H. capensis borealis		R	R																	126
H. capensis typ.		R	R																	126
H. droebakensis[c]		R	R					ψ												28, 29, 34
H. pluvialis		R	R	R				ψ												2, 104, 126, 134
H. zimbabsiensis		R	R																	126
Hyalotheca dissiliens		R	R																	159
Lobomonas pyriformis		R	R																	112
L. rostrata		R	R																	95
L. sphaerica		R	R					u?												131
Micrasterias cruxmelitensis		R	R																	159
Nannochloris atomus		R	R																	142
N. oculata		R	R		ψ															29, 32

m Ribose, deoxyribose, rhamnose, glucose-1-P, fructose-1-6 DP.
n Propanol, inositol.
o Cell-free extracts can metabolize glucose.

Table 1 (continued)
VITAMINS, SUGARS, AND ALCOHOLS

Chlorophyceae (continued)

Species (synonym)	Vitamins				Carbon sources — Sugars										Alcohols					Ref.
	Thiamine	Biotin	B$_{12}$	Other	Arabinose	Fructose	Galactose	Glucose	Lactose	Maltose	Mannose	Raffinose	Sucrose	Xylose	n-Butanol	Ethanol	Glycerol	Methanol	Other	
Nautococcus pyriformis	R	R	R					U												115
Neochloris alveolaris	R	R	R					U												115
N. aquatica	R	R	R					U												21, 115
N. cohaerens														U						52
N. gelatinosa	R	R	R					U												115
N. minuta	R	R	R					U											U^i	21, 115
N. pseudoalveolaris	R	R	R					U												115
Oocystis naegeli	R	R	R					U												22
Pandorina morum	R	R	R					U							U	U	U		U^i	113
P. unicocca	R	R	R					U			U				U	U	U		U^p	140
Platydorina caudata	R	R	R			U		U	U	U			U	U	U	U	U			57
P. tetrathele	R	R	R												U	U				34
Pleurotaenium trabecula	R^r	R	R					U							U	U		U	U^s	159
Polytoma caudatum	R^r	R		R											U	U	U	U	U^s	2, 134
P. obtusum	R	R		R											U	U	U	U	U^t	2, 134
P. ocellatum	R^v	R	R	R	U			U						U	U	U	U	U	U^u	2, 102, 134
P. uvella	R^v	R	R	R	U			U		U			U	U	U	U	U	U	U^s	2, 134
Polytomella caeca		R	R					U		U			U	U	U	U	U	U	U, q,w	2, 102, 134, 174
Prototheca zopfii	R^v	R	R		U			U		U			U	U	U	U	U	U	U^x	2, 134
Radiosphaera dissecta	R	R	R					U							U	U		U		115
Raphidonema nivale	R	R	R							U	U									63
Scenedesmus costulatus						U	u			U	U		U							15

p   Rhamnose, propanol, inositol.
q   n-Amyl alcohol and hexanol.
r   Can be replaced by thiazole.
s   i-Butanol, n-pentanol, i-propanol, and n-propanol.
t   i-Butanol, n-pentanol, and n-propanol.
u   i-Butanol, n-hexanol, and n-propanol; however, i-propanol and i-propanol are not utilized, and n-pentanol is poorly utilized.
v   Can be replaced by pyrimidine and thiazole.
w   i-Pentanol, n-pentanol, and i-propanol; however, i-butanol, n-hexanol, and n-propanol are utilized.
x   i-Butanol, n-pentanol, and n-propanol; however, i-pentanol is poorly utilized, and i-propanol is not utilized.

Table 1 (continued)
VITAMINS, SUGARS, AND ALCOHOLS

Species (synonym)	Vitamins				Carbon sources															Ref.
					Sugars										Alcohols					
	Thiamine	Biotin	B_{12}	Other	Arabinose	Fructose	Galactose	Glucose	Lactose	Maltose	Mannose	Raffinose	Sucrose	Xylose	n-Butanol	Ethanol	Glycerol	Methanol	Other	
Chlorophyceae (continued)																				
S. obliquus	R	R	R			U				U										110, 134
S. obtusiusculus	R[y]	R	R				U	U												22
Selenastrum minutum	?	R	?				U	U												95
Sphaerobotrys fluviatilis	R		R																	18
Spongiochloris excentrica	R	R	R					U												115
S. lamellata	R	R	R					U												115
S. spongiosis	R	R	R					U												115
Spongiococcum alabamense	R	R	R					U												115
S. excentricum	R	R	R					U												115
S. multinucleatum	R	R	R					U												115
S. tetrasporum	R	R	R					ψ												115
Sporotetras pyriformis	?		?																	18
Staurastrum gladiosum	R	R	R																	159
S. paradoxum	R	R	R																	159
S. sebaldii ornatum	R	R	R																	159
Staurodesmus pachyrhynchus	R	R	R																	34
Stephanosphaera pluvialis	R	R	R[e]																	96, 142
Stichococcus cylindricus (?)	R	R	R					U												22
S. diplosphaera	R	R	R																	24
Stigeoclonium aestivale	R	R	R																	24
S. farctum	R	R	R																	24
S. helveticum	R	R	R																	24
S. pascheri	R	R	R																	24
S. subsecundum	R	R	R																	24
S. tenue	R	R	R																	24
S. variabile	R	R	R																	24
Valvox aureus	R	R	R																	128
V. globator	R, R[119]	R	R																	119, 128
V. tertius	R	R	R																	119, 128
Volvulina pringsheimii	R	R	R, R[2 o]	?																20, 154
V. steinii	R	R	R					ψ		ψ			ψ	ψ		U	ψ		ψ[z]	20, 154
Xanthidium cristatum	R	R	R																	159

y By six strains; required by one strain.
z Rhamnose, ribose, glucose-1-phosphate.

Table 1 (continued)
VITAMINS, SUGARS, AND ALCOHOLS

| | Vitamins | | | | Carbon sources | | | | | | | | | | | | | | | |
Species (synonym)	Thiamine	Biotin	B₁₂	Other	Arabinose	Fructose	Galactose	Glucose	Lactose	Maltose	Mannose	Raffinose	Sucrose	Xylose	n-Butanol	Ethanol	Glycerol	Methanol	Other	Ref.
									Sugars								Alcohols			
Charophyceae																				
Chara aspera	R	R	R																	41
C. globularis	R	R	R																	41
C. zeilanica	R	R	R																	41

Table 1 (continued)
VITAMINS, SUGARS, AND ALCOHOLS

REFERENCES

1. Aaronson, S. and Baker, H., *J. Protozool.*, 6, 282–284, 1959.
2. Albritton, E. C., *Standard Values in Nutrition and Metabolism*, W. B. Saunders, Philadelphia, 1954.
3. Aldrich, D. V., *Science*, 137, 988, 1962.
4. Allen, M. B., *Arch. Mikrobiol.*, 17, 34, 1952.
5. Allen, M. B., *Arch. Mikrobiol.*, 32, 270, 1959.
6. Allison, F. E., Hoover, F. R., and Morris, H. J., *Bot. Gaz.* (Chicago), 98, 433–463, 1937.
7. Antia, N. J., Cheng, J. Y., and Taylor, F. J. R., *Proc. VI Int. Seaweed Symp.*, Vol. 6, Direccion General Pesca Maritima, Madrid, 1969, 17–29.
8. Archibold, P. A. and Bold, H. C., Phycological Studies. XI. The Genus *Chlorococcum* Meneghini, Pub. No. 7015, University of Texas, Austin, 1970.
9. Baker, A. F. and Bold. H. C., Phycological Studies. X. Taxonomic Studies in the Oscillatoriaceae, Pub. No. 7004, University of Texas, Austin, 1970.
10. Barry, S. N. C., *J. Protozool.*, 9, 395–400, 1962.
11. Belcher, J. H., *Arch. Mikrobiol.*, 58, 181–185, 1967.
12. Belcher, J. H. and Fogg, G. E., *Arch. Mikrobiol.*, 30, 17, 1958.
13. Belcher, J. H. and Miller, J. D. A., *Arch. Mikrobiol.*, 36, 219, 1960.
14. Boalch, G. T., *J. Mar. Biol. Assoc. U.K.*, 41, 287–304, 1961.
15. Bristol Roach, B. M., *Ann. Bot.* (London), 41, 509–517, 1927.
16. Brooks, A. E., Physiology and Genetics of *Astrephomene gubernaculifera*, Ph.D. thesis, Indiana University, Bloomington, 1965.
17. Buetow, D. E. and Padilla, G. M., *J. Protozool.*, 10, 121–123, 1963.
18. Cain, J., *Can. J. Bot.*, 43, 1367, 1965.
19. Carefoot, J. R., *J. Phycol.*, 4, 129–131, 1968.
20. Carefoot, J. R., *J. Protozool.*, 14, 15–18, 1967.
21. Chantanachat, S. and Bold, H. C., Phycological Studies. II. Some Algae from Arid Soils, Pub. No. 6218, University of Texas, Austin, 1962.
22. Chodat, F. and Schopfer, J. F., *Schweiz. Z. Hydrol.*, 22, 103–110, 1960.
23. Cosgrove, W. B. and Swanson, B. K., *Physiol. Zool.*, 25, 287, 1952.
24. Cox, E. R. and Bold, H. C., Phycological studies. VII. Taxonomic investigations of *Stigeoclonium*, Pub. No. 6618, University of Texas, Austin, 1966.
25. Cramer, M. and Myers, J., *Arch. Mikrobiol.*, 17, 384–402, 1952.
26. De, P. K., *Proc. R. Soc. London Ser. B*, 127, 121, 1939.
27. Droop, M. R., *J. Gen. Microbiol.*, 16, 286–293, 1957.
28. Droop, M. R., Studies on the Phytoflagellate *Haematococcus pluvialis*, thesis, Cambridge University, England, 1954.
29. Droop, M. R., in *Algae Physiology and Biochemistry*, Stewart, W. D. P., Ed., Bot. Monogr. 10, Blackwell Scientific Publications, Oxford, Engl., 1974, 530–559.
30. Droop, M. R. and Pennock, J. F., *J. Mar. Biol. Assoc. U.K.*, 51, 455–470, 1971.
31. Droop, M. R., *J. Mar. Biol. Assoc. U.K.*, 34, 229, 1955.
32. Droop, M. R., *J. Mar. Biol. Assoc. U.K.*, 37, 323, 1958.
33. Droop, M. R., *J. Mar. Biol. Assoc. U.K.*, 38, 605, 1959.
34. Droop, M. R., *Rev. Algol.*, 5(4), 247, 1961.
35. Droop, M. R. et. al., Preprints Abstr. Papers, Int. Oceanogr. Congr., New York, 1959, 916.
36. Dusi, H., *Ann. Inst. Pasteur*, 64, 340–343, 1940.
37. Dusi, H., *C.R. Soc. Biol.*, 130, 419–421, 1939.
38. Eppley, R. W. and Maciasr, F. M., *Limnol. Oceanogr.*, 8, 411–416, 1963.
39. Eppley, R. W. and Maciasr, F. M., *Physiol. Plant.*, 15, 72, 1962.
40. Fay, P., *J. Gen. Microbiol.*, 39, 11–20, 1965.
41. Forsberg, C., *Physiol. Plant.*, 18, 275–290, 1965.
42. Fries, L., *Nature*, 183, 558–559, 1959.
43. Fries, L., *Experientia*, 17, 75–76, 1961.
44. Fries, L., *Nature*, 202, 110, 1964.
45. Fries, L. and Pettersson, H., *Br. Phycol. Bull.*, 3, 417–422, 1968.
46. Gerloff, G. C., Fitzgerald, G. P., and Skoog, F., *Am. J. Bot.*, 37, 216, 1950.
47. Gibor, A., *Biol. Bull.*, 111, 223, 1956.
48. Gold, K., *J. Protozool.*, 11, 85, 1964.
49. Gooday, G. W., *Arch. Mikrobiol.*, 72, 9–15, 1970.

Table 1 (continued)
VITAMINS, SUGARS, AND ALCOHOLS

50. Groover, R. D. and Bold, H. C., Phycological Studies. VIII. The Taxonomy and Comparative Physiology of the Chlorosarcinales and Certain Other Edaphic Algae, Pub. 6907, University of Texas, Austin, 1969.
51. Guillard, R. R. L., Organic sources of nitrogen for marine centric diatoms, in *Symposium on Marine Microbiology*, Oppenheimer, C. H., Ed., Charles C Thomas, Springfield, Ill., 1963.
52. Guillard, R. R. L. and Ryther, J. H., *Can. J. Microbiol.*, 8, 229, 1962.
53. Guillard, R. R. L., in *Symposium on Marine Microbiology*, Oppenheimer, C. H., Ed., Charles C Thomas, Springfield, Ill., 1963, 93.
54. Guillard, R. R. L., *J. Phycol.*, 4, 59–64, 1968.
55. Harder, R., *Z. Bot.*, 9, 145–242, 1917.
56. Hargraves, P. E. and Guillard, R. R. L., personal communication, from Provasoli, L. and Carlucci, A. F., in *Algae Physiology and Biochemistry*, Bot. Monogr. 4, Stewart, W. D. P., Ed., Blackwell Scientific Publications, Oxford, Engl., 1974, 741–787.
57. Harris, D. O., *J. Phycol.*, 5, 205–210, 1969.
58. Haxo, F. T. and Sweeney, B. M., Luminescence Biol. Systems Proc. Conf. Luminescence, Asilomar, Calif., 1955, 415.
59. Heinrich, H. C., *Naturwissenschaften*, 14, 418, 1955.
60. Hellebust, J. A. and Guillard, R. R. L., *J. Phycol.*, 8, 132–136, 1972.
61. Hellebust, J. A., The uptake and utilization of organic substance by marine phytoplankters, in Symp. on Org. Matter in Nat. Waters, University of Alaska, College, 1969.
62. Hoare, D. S., Ingram, L. O., Thurston, E. L., and Walkup, R., *Arch. Mikrobiol.*, 78, 310–321, 1971.
63. Hoham, R., Laboratory and Field Studies on Snow Algae of the Pacific Northwest, Ph.D. thesis, University of Washington, Seattle, 1971.
64. Holz, G. G., *J. Protozool.*, 1, 114, 1954.
65. Hurlbert, R. E. and Rittenberg, S. C., *J. Protozool.*, 9, 170–182, 1962.
66. Hutner, S. H., *Trans. N.Y. Acad. Sci.*, 10, 136, 1948.
67. Hutner, S. H. and Provasoli, L., *Biochem. Physiol. Protozoa*, 1, 27, 1951.
68. Hutner, S. H. and Provasoli, L., *Phycol. News Bull.*, 6, 7, 1953.
69. Hutner, S. H., Provasoli, L., and Filfus, J., *Ann. N.Y. Acad. Sci.*, 56, 852, 1953.
70. Ingram, L. D., Calder, J. A., Van Baalen, C., Plucker, F. E., and Parker, P. L., *J. Bacteriol.*, 114, 695–700, 1973.
71. Ishida, Y. and Katoda, H., *Mem. Res. Inst. Food Sci. Kyoto Univ.*, 26, 10–17, 1965.
72. Iwasaki, H., *Plant Cell Physiol.*, 6, 325–336, 1965.
73. Iwasaki, H., *Bull. Plankton Soc. Jpn.*, 16, 132–139, 1969.
74. Iwasaki, H., *J. Oceanogr. Soc. Jpn.*, 27, 152–157, 1971.
75. Iwasaki, H., *Bull. Jpn. Soc. Sci. Fish.*, 37, 606–609, 1971.
76. Iwasaki, H., Fujiyama, T., and Yamashita, E., *J. Fac. Fish. Anim. Husb. Hiroshima Univ.*, 7, 259–267, 1968.
77. Iwasaki, H., Okaka, Y., and Tanake, S., *Bull. Plankton Soc. Jpn.*, 16, 140–144, 1969.
78. Iwasaki, H. and Sasada, K., *Bull. Jpn. Soc. Sci. Fish.*, 35, 943–947, 1969.
79. Kain, J. M. and Fogg, G. E., *J. Mar. Biol. Assoc. U.K.*, 39, 33, 1960.
80. Kenyon, C. N., Rippka, R., and Stanier, R. Y., *Arch. Mikrobiol.*, 83, 216–236, 1972.
81. Kessler, E., *Arch. Microbiol.*, 85, 153–158, 1972.
82. Khoja, T. and Whitton, B. A., *Arch. Mikrobiol.*, 79, 280–282, 1971.
83. Khoja, T. M. and Whitton, B. A., *Br. Phycol. J.*, 10, 139–148, 1975.
84. Kiyohara, T. et al., *J. Gen. Appl. Microbiol.*, 6, 176, 1960.
85. Korn, R. W., *Physiol. Plant.*, 22, 1158–1165, 1969.
86. Kratz, W. A. and Meyers, J., *Am. J. Bot.*, 42, 282, 1955.
87. Kwon, Y. M. and Grant, B. R., *Plant Cell Physiol.*, 12, 29–39, 1971.
88. Leedale, G. F., Meeuse, B. J. D., and Pringsheim, E. G., *Arch. Microbiol.*, 50, 133–155, 1965.
89. Lewin, J. C. and Hellebust, J. A., *Can. J. Microbiol.*, 16, 1123–1129, 1970.
90. Lewin, J. C. and Lewin, R. A., *Can. J. Microbiol.*, 6, 127–134, 1960.
91. Lewin, J. and Lewin, R. A., *J. Gen. Microbiol.*, 46, 361, 1967.
92. Lewin, J., *J. Gen. Microbiol.*, 9, 305, 1953.
93. Lewin, J. C., in *Symposium on Marine Microbiology*, Oppenheimer, C. H., Ed., Charles C Thomas, Springfield, Ill., 1963, 229.
94. Lewin, J. C., *Phycologia*, 4, 142, 1965.
95. Lewin, R. A., *Phycol. News Bull.*, 5, 21, 1952.
96. Lewin, R. A., *J. Gen. Microbiol.*, 10, 93, 1954.
97. Lewin, R. A., *J. Gen. Microbiol.*, 11, 459, 1954.
98. Lewin, R. A., *Sciencaj Studoj*, p. 187, 1958.

Table 1 (continued)
VITAMINS, SUGARS, AND ALCOHOLS

99. Loeblich, A. R., III, *J. Protozool.,* 16 (suppl.), 20, 1969.
100. Loefer, J. B., *Arch. Protistenkd.,* 84, 456, 1935.
101. Lynn, R. I. and Starr, R. C., *Arch. Protistenkd.,* 112, 283–302, 1970.
102. Lwoff, A. and Dusi, H., *C.R. Soc. Biol.,* 127, 53–55, 1938.
103. Martin, T. C. and Wyatt, J. T., *J. Phycol.,* 10, 57–64, 1974.
104. McLachlan, J. and Craigie, J. S., *Can. J. Bot.,* 43, 1449–1456, 1965.
105. McLaughlin, J. J. A. and Zahl, P. A., *Ann. N.Y. Acad. Sci.,* 77, 55–72, 1959.
106. McLaughlin, J. J. A. and Provasoli, L., *J. Protozool.,* 4 (suppl.), 7, 1957.
107. Miller, J. D. A. and Fogg, G. E., *Arch. Mikrobiol.,* 28, 1–17, 1957.
108. Miller, J. D. A. and Fogg, G. E., *Arch. Mikrobiol.,* 30, 1, 1958.
109. Miller, J. S. and Allen, M. M., *Arch. Mikrobiol.,* 86, 1–12, 1972.
110. Mineeva, L. A., *Mikrobiologiya,* 30, 586–592, 1961.
111. Nakahara, H., Alternation of Generations of Some Brown Algae in Unialgal and Axenic Cultures, Ph.D. thesis, Hokkaido University, Sapporo, Japan, 1972.
112. Osterud, K. L., *Physiol. Zool.,* 19, 19, 1946.
113. Palmer, E. G. and Starr, R. C., *J. Phycol.,* 7, 85–89, 1971.
114. Pan, P., *Can. J. Microbiol.,* 18, 275–280, 1972.
115. Parker, B. C., Bold, H. C., and Deason, T. R., *Science,* 133, 761, 1961.
116. Pedersen, M., *Physiol. Plant.,* 22, 977–983, 1969.
117. Pelroy, R. A., Rippka, R., and Stanier, R. Y., *Arch. Mikrobiol.,* 87, 303–322, 1972.
118. Pincemin, J. M., Action de Facteurs Physiques, Chimiques et Biotiques sur Quelques Dinoflagelles et Diatomees en Culture, Ph.D. thesis, Universite D'Aix-Marseille, France, 1971.
119. Pintner, I. J. and Provasoli, L., *Proc. IX Int. Bot. Congr. Montreal,* University of Toronto Press, 1959, 300–301 (abstr.).
120. Pintner, I. J. and Provasoli, L., *Bull. Misaki Mar. Biol. Inst. Kyoto Univ.,* 12, 25–31, 1968.
121. Pintner, I. J. and Provasoli, L., *J. Gen. Microbiol.,* 18, 190, 1958.
122. Pintner, I. J. and Provasoli, L., in *Symposium on Marine Microbiology,* Oppenheimer, C. H., Ed., Charles C Thomas, Springfield, Ill., 1963, 114.
123. Pringsheim, E. G., *Q. J. Microsc. Sci.,* 93, 71–96, 1952.
124. Pringsheim, E. G. and Wiessner, W., *Nature,* 188, 919–921, 1960.
125. Pringsheim, E. G., *Nature,* 195, 604, 1962.
126. Pringsheim, E. G., *J. Phycol.,* 2, 1–7, 1966.
127. Pringsheim, E. G., *Arch. Mikrobiol.,* 56, 60–67, 1967.
128. Pringsheim, E. G., *Antonie van Leeuwenhoek J. Microbiol. Serol.,* 36, 33–43, 1970.
129. Pringsheim, E. G., *Arch. Mikrobiol.,* 23, 181, 1955.
130. Pringsheim, E. G., *Planta,* 52, 405, 1958.
131. Pringsheim, E. G., *Arch. Mikrobiol.,* 46, 227, 1963.
132. Provasoli, L. and Gold, K., *Arch. Mikrobiol.,* 42, 196–203, 1962.
133. Provasoli, L., in Proc. IV Int. Seaweed Symp., De Virville, A. D. and Feldman, J., Eds., Pergamon Press, Oxford, 1964, 9–17.
134. Provasoli, L., *Annu. Rev. Microbiol.,* 12, 279, 1958.
135. Provasoli, L. and Gold, K., *J. Protozool.,* 4 (suppl.), 7, 1957.
136. Provasoli, L. and McLaughlin, J. J. A., *J. Protozool.,* 2(suppl.), 10, 1955.
137. Provasoli, L. and Pintner, I. J., *Ann. N.Y. Acad. Sci.,* 56, 839, 1953.
138. Provasoli, L. and Pintner, I. J., *Phycol. News Bull.,* 8, 7, 1955.
139. Rahat, M. and Spira, Z., *J. Protozool.,* 14, 45–48, 1967.
140. Rayburn, W. R., Morphology, Nutrition and Physiology of Sexual Reproduction of *Pandorina unicocca* sp. nov., Ph.D. thesis, Indiana University, Bloomington, 1971.
141. Rippka, R., *Arch. Mikrobiol.,* 87, 93–98, 1972.
142. Ryther, J. H., *Biol. Bull.,* 106, 198, 1954.
143. Sager, R. and Granick, S., *Ann. N.Y. Acad. Sci.,* 56, 831–838, 1953.
144. Saito, S., *J. Phycol.,* 8, 169–175, 1972.
145. Sheath, R. G. and Hellebust, J., *J. Phycol.,* 10, 34–41, 1974.
146. Shepard, D. C., in *Methods in Cell Physiology,* Vol. 4, Prescott, D. M., Ed., Academic Press, New York, 1969, 46–69.
147. Shihira, I. and Krauss, R. W., *Chlorella: Physiology and Taxonomy of 41 Isolates,* University of Maryland Press, College Park, 1965.
148. Silva, P., in *Physiology and Biochemistry of Algae,* Lewin, R. A., Ed., Academic Press, New York, 1962, 827.

Table 1 (continued)
VITAMINS, SUGARS, AND ALCOHOLS

149. Sloan, P. R. and Strickland, J. D. H., *J. Phycol.*, 2, 29, 1966.
150. Smith, A. J., London, J., and Stanier, R. Y., *J. Bacteriol.*, 94, 972–983, 1967.
151. Smith, R. L. and Bold H. C., Phycological studies VI. Investigations on the Algal Genera *Eremosphaera* and *Oocystis*, Pub. No. 6612, University of Texas, Austin, 1966.
152. Stanier, R. Y., Kunisawa, R., Mandel, M., and Cohen-Bazire, G., *Bacteriol. Rev.*, 35, 171–205, 1971.
153. Staub, R., *Schweiz. Z. Hydrol.*, 23, 82–198a, 1961.
154. Stein, J. R., Ph.D. thesis, University of California, Berkeley, 1957.
155. Stewart, W. D. P., *Ann. Bot.* (London), 26, 439, 1962.
156. Storm, J. and Hunter, S. H., *Ann. N.Y. Acad. Sci.*, 56, 901, 1953.
157. Sweeney, B. M., *Am. J. Bot.*, 41, 821, 1954.
158. Swift, M. J. and McLaughlin, J. J. A., *Ann. N.Y. Acad. Sci.*, 175, 577–600, 1970.
159. Tassigny, M., *J. Phycol.*, 7, 213–215, 1971.
160. Tatewaki, M. and Provasoli, L., *Bot. Mar.*, 6, 193–203, 1964.
161. Tatewaki, M., Personal communication, from Provasoli, L. and Carlucci, A. F., in *Algae Physiology and Biochemistry*, Bot. Monogr. 4, Stewart, W. D. P., Ed., Blackwell Scientific Publications, Oxford, Engl., 1974, 741–787.
162. Thomas, W. H., *J. Protozool.*, 2(suppl.), 2, 1955.
163. Thomas, W. H., *J. Phycol.*, 2, 17, 1966.
164. Turner, M. F., in Scottish Marine Biological Association, Report of the Council for 1969/70, Oban, Scotland, 1970, 13.
165. Turner, M. F., *Br. Phycol. J.*, 5, 15–18, 1970.
166. Van Baalen, C., *Bot. Mar.*, 4, 129, 1962.
167. Watanabe, A., *Arch. Biochem. Biophys.*, 34, 50, 1951.
168. Watanabe, A. and Yamamoto, Y., Heterotrophic nitrogen fixation by the blue-green alga *Anabaenopsis circularis*, *Nature*, 214, 738, 1967.
169. Wetherell, D. F., *Physiol. Plant.*, 11, 260, 1958.
170. Wetherell, D. F. and Krauss, R. W., *Am. J. Bot.*, 44, 609, 1957.
171. White, A. W., *J. Phycol.*, 10, 292–300, 1974.
172. Williams, A. E. and Burris, R. H., *Am. J. Bot.*, 39, 340, 1952.
173. Wilson, W. B. and Collier, A., *Science*, 121, 394, 1955.
174. Wise, D. L., *J. Protozool.*, 6, 19–23, 1959.
175. Wright, R. T., *Limnol. Oceanogr.*, 9, 163–178, 1964.
176. Wright, R. T. and Hobbie, J. E., *Ecology*, 47, 447–464, 1966.
177. Wyatt, J. T., Martin, C., and Jackson, J. W., *Phycologia*, 12, 153–161, 1973.

Table 2
FATTY ACIDS AND OTHER ORGANIC ACIDS

Criterion for determining utilization of fatty acids and other organic acids as a carbon source: growth of the alga in darkness on a given substrate (facultative heterotrophy). Stimulation of growth by a substance in the light does not indicate utilization.

Species (synonym)	Fatty acids											Other organic acids[a]								Ref.
	Acetic	i-Butyric	n-Butyric	n-Caproic	n-Heptylic	n-Nonylic	n-Octylic	Propionic	i-Valeric	n-Valeric	Other	Citric	Fumaric	Lactic	Malic	Phospho-glyceric	Pyruvic	Succinic	Miscellaneous	
Cyanophyceae																				
Agmenellum quadruplicatum	U[b]																			67, 71
Anabaena flos-aquae	U													U			U	U	U[c]	29
A. variabilis	U[b]		U					U							U					37
Anacystis marina	U												U						U[c]	67, 71
A. nidulans	U		U					U						U	U		U	U	U[d]	37, 65, 67
Chlorogloea fritschii	U[b]												U	U	U		U	U	U[,e,22,67] u[f]	22, 48, 67
Coccochloris elabens	U[b]													U			U	U	U[c]	67, 71
C. peniocystis	U[b]											U			U		U	U	U[d]	65, 67, 71
Fremyella diplosiphon	U											U		U						76
Gloeocapsa alpicola	U[b]																U	U	U[d]	65, 67
Lyngbya aestuarii	U													U						71
L. lagerheimii	U													U						71
Microcoleus chthonoplastes	U													U						71
M. tenerrimus	U																			71
Microcystis aeruginosa	U[b]		U					U					U		U		U	U	U[c]	67
Nostoc muscorum														U			U	U		37
N. punctiforme												U?			u?				U[g]	26

[a] Negative results reported for organic acids may not be significant, since it has been shown that *Prototheca zopfii* and *Euglena gracilis* utilize them only at pH 3.5—5.5. Most of the negative results tabulated were obtained in media having a pH near neutrality.

[b] Solid agar medium used.[6,7]

[c] Glutamate.

[d] Glutamate, aspartate, leucine.

[e] Glycolate, α-ketoglutarate, aspartic acid, glutamic acid.

[f] Glutamine, glycine.

[g] Tannin, aspargine.

Table 2 (continued)
FATTY ACIDS AND OTHER ORGANIC ACIDS

Species (synonym)	Carbon sources																			Ref.
	Fatty acids											Other organic acids[a]								
	Acetic	i-Butyric	n-Butyric	n-Caproic	n-Heptylic	n-Nonylic	n-Octylic	Propionic	i-Valeric	n-Valeric	Other	Citric	Fumaric	Lactic	Malic	Phospho-glyceric	Pyruvic	Succinic	Miscellaneous	
Cyanophyceae (continued)																				
Nostoc sp.	⊕											⊕					⊕		⊕[h]	30, 34
Oscillatoria amphibia	⊕													⊕						71
O. subtilissima	⊕													⊕						71
Plectonema terebrans	⊕[b]													⊕						71
Synechococcus cedrorum	⊕[b]																⊕	⊕	⊕[c]	67
S. elongatus	⊕[b]																⊕	⊕	⊕[c]	67
S. lividus	⊕																⊕	⊕	⊕[c]	67

h Gluconic acid, glucuronic acid, glucosamine, citrulline.

Table 2 (continued)
FATTY ACIDS AND OTHER ORGANIC ACIDS

Species (synonym)	Carbon sources																			Ref.
	Fatty acids											Other organic acids[a]								
	Acetic	i-Butyric	n-Butyric	n-Caproic	n-Heptylic	n-Nonylic	n-Octylic	Propionic	i-Valeric	n-Valeric	Other	Citric	Fumaric	Lactic	Malic	Phospho-glyceric	Pyruvic	Succinic	Miscellaneous	
Rhodophyceae																				
Asterocytis ramosa	U																			23
Porphyridium cruentum	U																			16
Rhodella maculata	U																			69
Cryptophyceae																				
Chilomonas paramecium[b]	U			U	U															13, 31
Cryptomonas borealis	U				U			U	U	U	U[c]	U	U	U	U			U		74
C. marssonii	U							U	U	U	U[c]	U	U	U	U					74
Hemiselmis virescens	U							U	U	U			U	U					U[d]	17

b Colorless.
c n-Caprylic acid.
d Glycolic acid.

Table 2 (continued)
FATTY ACIDS AND OTHER ORGANIC ACIDS

Dinophyceae

Species (synonym)	Carbon sources — Fatty acids											Carbon sources — Other organic acids[a]								Ref.
	Acetic	i-Butyric	n-Butyric	n-Caproic	n-Heptylic	n-Nonylic	n-Octylic	Propionic	i-Valeric	n-Valeric	Other	Citric	Fumaric	Lactic	Malic	Phospho-glyceric	Pyruvic	Succinic	Miscellaneous	
Gonyaulax polyedra	U		Ψ					Ψ					Ψ	Ψ	Ψ			Ψ	Ψ[b]	68
Gymnodinium brevis	U		Ψ					Ψ		Ψ	Ψ[c]	Ψ	Ψ	Ψ	Ψ			Ψ	Ψ[c]	2
G. foliaceum	U															Ψ				16
G. inversum	U[d]																			75
Gyrodinium cohnii	U		U					U	U		u[e]									35, 58
Oxyrrhis marina	U							U	U										U[f]	17, 58
Peridinium cinctum var. *ovoplanum*	Ψ	Ψ						Ψ				Ψ		Ψ			Ψ	u	Ψ[b]	10
Prorocentrum micans	U																			16

b Malonate.
c Several other fatty acids.
d Apochlorotic, obligate heterotroph.
e Oleic.
f Glycolic acid.

Table 2 (continued)
FATTY ACIDS AND OTHER ORGANIC ACIDS

Species (synonym)	Acetic	i-Butyric	n-Butyric	n-Caproic	n-Heptylic	n-Nonylic	n-Octylic	Propionic	i-Valeric	n-Valeric	Other	Citric	Fumaric	Lactic	Malic	Phospho-glyceric	Pyruvic	Succinic	Miscellaneous	Ref.
Bacillariophyceae																				
Achnanthes brevipes	U																			42
Amphipleura rutilans	U													U						42
Amphiprora paludosa	U													U						42
Amphora coffaeiformis	U[b]													U[c]						42
A. lineolata	U													U						42
Chaetoceros ceratosporum														U						41
C. pelagicus														U						41
C. pseudocrinitus														U						41
Coscinodiscus asteromphalus														U						41
Coscinodiscus sp.												U		U				U	U[d]	72
Cyclotella caspia	U													U						41
C. cryptica	U[h]											U		U				U	U, e,f	42, 72
C. nana														U[g]						41
Cylindrotheca fusiformis (*Nitzschia closterium*)								U				U	U	U[i]	U		U	U	U,j,u, k,U[l]	38, 42

b By nine strains, utilized by one strain.
c By eight strains; utilized by two strains.
d Galacturonic acid, glycolate, glycine, alanine, glutamic acid, valine, serine, aspartic acid, arginine, lysine, glucosamine, peptone, tryptone, yeast extract, sewage effluent.
e Galacturonic acid, glycolate, glycine, alanine, glutamic acid, valine, serine, aspartic acid, arginine, lysine, glucosamine, yeast extract, sewage effluent, tryptone.
f Peptone.
g By three strains.
h By four strains.
i By three strains; not utilized by one strain.
j Tryptone, yeast extract.
k Casamino acids.
l Glycolic acid, alanine, arginine, asparagine, aspartic acid, glutamic acid, glycine, glutamine.

Table 2 (continued)
FATTY ACIDS AND OTHER ORGANIC ACIDS

Bacillariophyceae (continued)

Species (synonym)	Acetic	i-Butyric	n-Butyric	n-Caproic	n-Heptylic	n-Nonylic	n-Octylic	Propionic	i-Valeric	n-Valeric	Other	Citric	Fumaric	Lactic	Malic	Phospho-glyceric	Pyruvic	Succinic	Miscellaneous	Ref.
Melosira lineata	U[m]													U					U[n]	70
M. nummuloides	U													U						16, 28, 41
Navicula incerta	U													U						42
N. menisculus	U													U						42
N. pelliculosa	U		U					U		U	U[o]	U	U	U	U	U	U	U	U[p]	40
Nitzschia alba[q]	U									U	U	U		U	U	U	U	U	U[p]	39
N. angularis affinis	U												U	U						42
N. curvilineata	U													U						42
N. filiformis	U[s]													U						42
N. frustulum	U													U[t]						42
N. hybridaeformis	U													U						42
N. laevis	U													U				U	U[r]	39
N. marginata	U													U						42
N. obtusa scalpelliformis	U													U						42
N. ovalis	U													U						42
N. punctata	U													U[3], u			U	U	U[r]	39, 55
N. putrida[q]	U													U						42
N. tenuissima	U													U					U[n]	17
Phaeodactylum tricornutum	U													U					U[n,r]	17, 41, 64
Skeletonema costatum	U													U						42
Stauroneis amphoroides	U													U						42
Synedra affinis	U													U						42
Thalassiosira fluviatilis	u																		U[r]	41, 64

m With ammonium sulfate as a nitrogen source.
n Glycolic acid.
o Lauric, myristic, oleic, and steric acids.
p Formic, galacturonic, gluconic, glucuronic, glycolic, α-ketoglutaric, and oxalic acids.
q Colorless.
r Glutamic acid.
s By six strains.
t By five strains; utilized by one strain.

Table 2 (continued)
FATTY ACIDS AND OTHER ORGANIC ACIDS

Haptophyceae

Species (synonym)	Carbon sources																			Ref.
	Fatty acids											Other organic acids[a]								
	Acetic	i-Butyric	n-Butyric	n-Caproic	n-Heptylic	n-Nonylic	n-Octylic	Propionic	i-Valeric	n-Valeric	Other	Citric	Fumaric	Lactic	Malic	Phospho-glyceric	Pyruvic	Succinic	Miscellaneous	
Apistonema sp.	ψ																		ψ[b]	17
Chrysochromulina brevefilum	ψ																			53
C. kappa	ψ																			53
C. strobilus	ψ																			53
Cricosphaera elongata	ψ																		ψ[b]	17
Dicrateria inorata	ψ																			53
Emiliania (Coccolithus) huxleyi	u																		ψ[f]	64
Isochrysis galbana	ψ																		ψ[b]	17
Prymnesium parvum	ψ												ψ		ψ		ψ	ψ	ψ,c,d; ψ[b,e]	17, 25, 59

b Glycolic acid.

c Aminobutyrate, betaine, ethionine, histidine, isoleucine, leucine, methionine, phenylalanine, sarcosine, serine, tryptophan, valine, acetoin, glycolate, α-ketoglutarate.

d Alanine, arginine, asparagine, glutamic acid.

e Alanine.

f Glutamic acid.

Table 2 (continued)
FATTY ACIDS AND OTHER ORGANIC ACIDS

Species (synonym)	Carbon sources																			Ref.
	Fatty acids											Other organic acids[a]								
	Acetic	i-Butyric	n-Butyric	n-Caproic	n-Heptylic	n-Nonylic	n-Octylic	Propionic	i-Valeric	n-Valeric	Other	Citric	Fumaric	Lactic	Malic	Phospho-glyceric	Pyruvic	Succinic	Miscellaneous	
Chrysophyceae																				
Mallomonas epithalattia	U																			16
Microglena arenicola	U																			16
Monochrysis lutheri	U																		U[b]	17
Ochromonas malhamensis	U																		U[c]	33, 54

b Glycolic acid.
c Glutamic acid.

Table 2 (continued)
FATTY ACIDS AND OTHER ORGANIC ACIDS

Species (synonym)	Carbon sources																			Ref.
	Fatty acids											Other organic acids[a]								
	Acetic	i-Butyric	n-Butyric	n-Caproic	n-Heptylic	n-Nonylic	n-Octylic	Propionic	i-Valeric	n-Valeric	Other	Citric	Fumaric	Lactic	Malic	Phospho-glyceric	Pyruvic	Succinic	Miscellaneous	
Xanthophyceae																				
Botrydiopsis intercedens	U											U	U					U		7
Bumilleriopsis brevis	U											U	U					U		7
Chlorellidium tetrabotrys	U											U	U					U		7
Monodus subterraneus	U											a	a	a	a			a		47
Tribonema aequale	a											U	a					a		6
T. minus	a											a	a					a		7
Eustigmatophyceae																				
Polyedriella helvetica	a											a	a					a		7
Phaeophyceae																				
Ectocarpus confervoides	a																		ψ[b]	8

[b] Na-glycerophosphate.

Table 2 (continued)
FATTY ACIDS AND OTHER ORGANIC ACIDS

Euglenophyceae

Species (synonym)	Carbon sources																			Ref.
	Fatty acids											Other organic acids[a]								
	Acetic	i-Butyric	n-Butyric	n-Caproic	n-Heptylic	n-Nonylic	n-Octylic	Propionic	i-Valeric	n-Valeric	Other	Citric	Fumaric	Lactic	Malic	Phospho-glyceric	Pyruvic	Succinic	Miscellaneous	
Astasia longa (A. chattoni. A. klebsii)[b]	U	u	U	U	U	U	U	U	U	U	u[c]			U?	U?		U?	U?		1, 4, 9, 32
A. quartana	U	u	U	U	U	U	U	U	U	U	U[d]	U		U	U		u	U		1
Euglena anabaena minor	U	U	U	U[e]	U[e]	U[e]	U[e]	U	U	U		U		U	U	U	U	U		1
E. deses[f]	U	U	U	U[e]	U[e]	U[e]	U[e]	U	U	U		U	D	U	U	U	U	U	ψ,g,14 U[h],14	1, 14
E. gracilis bacillaris	U	U	U	U	U	U	U	U	U	U	U[i]	U		U	U	U	U	U		1
E. gracilis typica	U	U	U	U	u	u	U	U	u	U	u[c]	U	d	U	U		U	U		1
E. gracilis urophora	U	U	U									D	D				U	U?		14
E. gracilis (Vischer strain)	U		U									U						U	ψ,j	14
E. klebsii[k]	U		U					U	U	U		U		U	U		U	U		1
E. pisciformis	U	U	U	U[e]	U[e]	U[e]	U[e]	U	U	U					U	U	U	U		20
E. stellata	U		U					U							U	U				
E. viridis	U													U						

b Apochlorotic, obligate heterotroph.
c i-Caproic and n-decylic acids.
d i-Caproic acid.
e At toxic concentrations.
f According to E. G. Pringsheim, the organism used in these studies was E. geniculata.
g Glycolic acid.
h Alanine, aspartic acid, glutamic acid.
i n-Decylic acid; however, i-caproic acid is poorly utilized.
j Alanine, glutamic acid.
k According to E. G. Pringsheim, the organism used in these studies was E. mutabilis.

Table 2 (continued)
FATTY ACIDS AND OTHER ORGANIC ACIDS

Chlorophyceae

Species (synonym)	Acetic	i-Butyric	n-Butyric	n-Caproic	n-Heptylic	n-Nonylic	n-Octylic	Propionic	i-Valeric	n-Valeric	Other	Citric	Fumaric	Lactic	Malic	Phospho-glyceric	Pyruvic	Succinic	Miscellaneous	Ref.
Astrephomene gubernaculifera	U																		ψ[b]	66
Brachiomonas submarina	U,[19] U[c]																		U[b]	17, 19
B. submarina pulsifera	U[c]																			17
Bracteacoccus cinnabarinus	U																			52
B. engadiensis	ψ/U																			52
B. minor	U																			52
B. terrestris	ψ/U																			52
Chlamydobotrys stellata	U		U																	56
Chlamydomonas agloeformis	U	ψ		ψ				ψ		ψ		ψ		U	ψ	ψ		ψ		1
C. dysosmos	ψ/U			ψ								ψ		U	ψ	ψ	U	ψ		43
C. moewusii	ψ/U		ψ					ψ		ψ		ψ	ψ	U	ψ	ψ	U	ψ	ψ[d]	1
C. mundana	ψ/U		ψ					ψ		ψ		ψ	ψ	ψ	ψ	ψ	ψ	ψ	ψ[b]	21
C. pulsatilla[e]	U																		ψ[f], u[g]	17
C. reinhardii	U[c]	ψ	ψ					ψ				ψ	ψ	ψ	ψ		ψ	ψ	ψ[b]	60
C. spreta	?																			16, 17
Chlorella autotrophica	ψ/U																			62
C. candida	ψ/U																			62
C. ellipsoidea	ψ/U																		ψ[b]	17, 62
C. emersonii	ψ/U																			62
C. emersonii globosa	ψ/U																			62
C. fusca	ψ/U																			36, 62
C. fusca vacuolata	ψ/U																			62
C. infusionum	ψ/U																			62
C. infusionum auxenophila	ψ/U																			62
C. kessleri	U																			36

b Glycolic acid.
c Slow but repeatable growth, accompanied by loss of chlorophyll.
d Acetamide, casein hydrolysate.
e Seven strains tested.
f Formate, glycerophosphate, formaldehyde (10^{-4} M), oxalate, tartrate, trans-aconitate, α-ketoglutarate, glutamine, glutamate, asparagine, aspartate, glycine.
g Monacetin, acetylmethyl acetate.

Table 2 (continued)
FATTY ACIDS AND OTHER ORGANIC ACIDS

Chlorophyceae (continued)

Species (synonym)	Acetic	i-Butyric	n-Butyric	n-Caproic	n-Heptylic	n-Nonylic	n-Octylic	Propionic	i-Valeric	n-Valeric	Other	Citric	Fumaric	Lactic	Malic	Phospho-glyceric	Pyruvic	Succinic	Miscellaneous	Ref.
C. luteoviridis	U																			36
C. miniata	U																			62
C. minutissima	U																			36
C. mutabilis	U																			62
C. nocturna	U																			62
C. photcphila	U																			62
C. pringsheimii	U																			62
C. protothecoides	U																			62
C. protothecoides communis	U																			62
C. protothecoides galactophila	U																			62
C. protothecoides mannophila	U																			62
C. pyrenoidosa	U		U					U					U	U	U			U		61
C. regularis	U																			62
C. regularis aprica	U																			62
C. regularis imbricata	U																			62
C. rubescens	U																			12
C. saccharophila	U																			36, 62
C. simplex	U																			62
C. sorokiniana	U																			62
C. vannielii	U																			62
C. variabilis	U																			62
C. vulgaris	U																			36, 49, 62
C. vulgaris luteoviridis	U																			62
C. zofingiensis	U																			36
Chlorococcum aplanosporum	U																			52
C. diplobionticum	U																			52
C. echinozygotum	U																			52
C. ellipscideum	U																			52
C. hypnosporum	U																			52
C. macrostigmatum	U																			52
C. minutum	U																			52
C. multinucleatum	U																			52
C. oleofaciens	U																			52
C. perforatum	U																			52

Table 2 (continued)
FATTY ACIDS AND OTHER ORGANIC ACIDS

Carbon sources

Chlorophyceae (continued)

Species (synonym)	Fatty acids											Other organic acids[a]							Miscellaneous	Ref.
	Acetic	i-Butyric	n-Butyric	n-Caproic	n-Heptylic	n-Nonylic	n-Octylic	Propionic	i-Valeric	n-Valeric	Other	Citric	Fumaric	Lactic	Malic	Phospho-glyceric	Pyruvic	Succinic		
C. pinguideum	ψ																			52
C. punctatum	ψ																			52
C. scabellum	ψ																			52
C. tetrasporum	ψ																			52
C. vacuolatum	ψ																			52
C. wimmeri	ψ			ψ																52
Chlorogonium elongatum	U		U					ψ	u		ψ[i]									44
Cyanidium caldarium[h]	U	u	U									U		U		U				3
Diplostauron elegans	U			ψ				ψ				ψ	ψ	ψ	ψ		U	ψ	ψ[j]	45
Dunaliella primolecta	ψ																		U[b]	17
D. salina	ψ																			24
C. viridis	U[k]																			24
Furcilla stigmatophora	ψ	ψ[k]	u[k]					ψ[k]							ψ		u		ψ[l]	5
Haematococcus (Balticola) buetschlii	ψ[l],[19] U											ψ	ψ		ψ			ψ	ψ[b]	17, 19
H. capensis																				17
H. droebakensis[m]	U[n],ψ[n,46]	ψ	U,ψ[46]																	17, 19
H. pluvialis	ψ	ψ	U,ψ[46]	ψ				ψ		ψ		ψ	ψ	U,ψ[46]	U,ψ[46]	ψ	U,ψ[46]	U,ψ[46]	ψ[o]	1, 17, 19, 46
Lobomonas pyriformis	U																			50
L. sphaerica	U												u?	u?	u?					57
Nannochloris oculata[p]	ψ																		ψ[b]	17

h Genus of uncertain taxonomic position.[63]

i Formate, fumarate.

j Isocitrate, glyoxylate, α-ketoglutarate, alanine, arginine, asparagine, glutamic acid, tyrosine, glycine, glycylglycine, cysteine, histidine.

k Both yeast extract and peptone added.

l Formic, glycolic, oxalic, tartaric, maleic, and glycerophosphoric acids, and glycine.

m Probably an obligate phototroph; substrates failing to support dark growth are not individually listed.

n Different strains were used in these two cases.

o Glycolic, benzoic, phthalic, glutaric, adipic, pimelic, maleic, aconitic, tartaric, malonic, glycolic, glyoxylic, α-ketoglutarate, and caproic acids.

p Obligate phototroph.

Table 2 (continued)
FATTY ACIDS AND OTHER ORGANIC ACIDS

Species (synonym)	Carbon sources — Fatty acids											Other organic acids[a]								Ref.
	Acetic	i-Butyric	n-Butyric	n-Caproic	n-Heptylic	n-Nonylic	n-Octylic	Propionic	i-Valeric	n-Valeric	Other	Citric	Fumaric	Lactic	Malic	Phospho-glyceric	Pyruvic	Succinic	Miscellaneous	

Chlorophyceae (continued)

Species (synonym)	Acetic	i-Butyric	n-Butyric	n-Caproic	n-Heptylic	n-Nonylic	n-Octylic	Propionic	i-Valeric	n-Valeric	Other	Citric	Fumaric	Lactic	Malic	Phospho-glyceric	Pyruvic	Succinic	Miscellaneous	Ref.
Neochloris alveolaris	U																			52
Pandorina morum	U[q]		Ψ									Ψ	Ψ					Ψ	Ψ[r]	51
Platydorina caudata	Ψ											Ψ	Ψ		Ψ	Ψ		Ψ	Ψ[s]	27
Polytoma caudatum	U	Ψ	U	U	Ψ	Ψ	Ψ	Ψ	Ψ	Ψ	Ψ[t]									1
P. obtusum	U	U	U	U	U	U	U	U	U	U	U[t]			Ψ	Ψ		U	Ψ		1
P. ocellatum	U	U	U	U	U	Ψ	U	U	u	U	U[u]			Ψ	Ψ?		U	Ψ?		1
P. uvella	U	Ψ	U	U	U	U	u	Ψ	Ψ, U[x]	Ψ	U[u]	Ψ		U?	Ψ		u	U?		1, 73
Polytomella caeca[w]	U	U	U	U	U	U	U	U	Ψ, U[x]	U	U[v]		Ψ	Ψ?	Ψ		u,U[7,3]	u		1
Prototheca zopfii	Ψ	U	U	U	U		U	U	u	U	U[t]	Ψ	Ψ	U?			U[y]	Ψ		52
Radiosphaera dissecta	U									U										49
Scenedesmus obliquus																				24
Stephanoptera gracilis	Ψ							Ψ				Ψ							Ψ[b]	17, 19
Stephanosphaera pluvialis	Ψ?U[r,9]		Ψ											U[y]			U	Ψ	Ψ[z]	
Volvulina steinii	U													U				Ψ		11

q Only by one strain out of three.
r Glyoxylate, glutarate, glycolate, and eight amino acids.
s Isocitrate, glycine, alanine, asparagine, glutamic acid.
t i-Caproic and n-decylic acids.
u i-Caproic acid; however, n-decylic acid is not utilized.
v i-Decylic acid; however, i-decylic acid is poorly utilized.
w Apochlorotic, obligate heterotroph.
x Valerate.
y Only at pH 3.0—5.5.
z Glyoxylate, α-ketoglutarate, isocitrate, glutamate, ascorbate.

Table 2 (continued)
FATTY ACIDS AND OTHER ORGANIC ACIDS

REFERENCES

1. Albritton, E. C., *Standard Values in Nutrition and Metabolism,* W. B. Saunders, Philadelphia, 1954.
2. Aldrich, D. V., *Science,* 137, 988, 1962.
3. Allen, M. B., *Arch. Mikrobiol.,* 17, 34, 1952.
4. Barry, S. N. C., *J. Protozool.,* 9, 395–400, 1962.
5. Belcher, J. H., *Arch. Mikrobiol.,* 58, 181–185, 1967.
6. Belcher, J. H. and Fogg, G. E., *Arch. Mikrobiol.,* 30, 17, 1958.
7. Belcher, J. H. and Miller, J. D. A., *Arch. Mikrobiol.,* 36, 219, 1960.
8. Boalch, G. T., *J. Mar. Biol. Assoc. U.K.,* 41, 287–304, 1961.
9. Buetow, D. E. and Padilla, G. M., *J. Protozool.,* 10, 121–123, 1963.
10. Carefoot, J. R., *J. Phycol.,* 4, 129–131, 1968.
11. Carefoot, J. R., *J. Protozool.,* 14, 15–18, 1967.
12. Chodat, F. and Schopper, J. F., *Schweiz. Z. Hydrol.,* 22, 103–110, 1960.
13. Cosgrove, W. B. and Swanson, B. K., *Physiol. Zool.,* 25, 287, 1952.
14. Cramer, M. and Myers, J., *Arch. Mikrobiol.,* 17, 384–402, 1952.
15. Droop, M. R., *Arch. Mikrobiol.,* 20, 391–397, 1954.
16. Droop, M. R., in *Algal Physiology and Biochemistry,* Bot. Monogr. 10, Stewart, W. D. P., Ed., Blackwell Scientific Publications, Oxford, Engl., 1974, 530–559.
17. Droop, M. R. and McGill, S., *J. Mar. Biol. Assoc. U.K.,* 46, 679–684, 1966.
18. Droop, M. R., *J. Mar. Biol. Assoc. U.K.,* 37, 323, 1958.
19. Droop, M. R., *Rev. Algol.,* 5(4), 247, 1961.
20. Dust, H., *Ann. Inst. Pasteur* (Paris), 70, 311–312, 1944.
21. Eppley, R. W. and Maciasr, F. M., *Limnol. Oceanogr.,* 8, 411–416, 1963.
22. Fay, P., *J. Gen. Microbiol.,* 39, 11–20, 1965.
23. Fries, L. and Petterson, H., *Br. Phycol. Bull.,* 3, 417–422, 1968.
24. Gibor, A., *Biol. Bull.,* 111, 223, 1956.
25. Gooday, G. W., *Arch. Mikrobiol.,* 72, 9–15, 1970.
26. Harder, R., *Z. Bot.,* 9, 145–242, 1917.
27. Harris, D. O., *J. Phycol.,* 5, 205–210, 1969.
28. Hellebust, J. A. and Guillard, R. R. L., *J. Phycol.,* 8, 132–136, 1972.
29. Hoare, D. S., Hoare, S. L., and Moore, R. B., *J. Gen. Microbiol.,* 49, 351–370, 1967.
30. Hoare, D. S., Ingram, L. O., Thurston, E. L., and Walkup, R., *Arch. Mikrobiol.,* 78, 310–321, 1971.
31. Holz, G. G., *J. Protozool.,* 1, 114, 1954.
32. Hunter, F. R. and Lee, J. W., *J. Protozool.,* 9, 74–78, 1962.
33. Hutner, S. H., Provasoli, L., and Filfus, J., *Ann. N.Y. Acad. Sci.,* 56, 852, 1953.
34. Ingram, L. O., Calder, J. A., Van Baalen, C., Plucker, F. E., and Parker, P. L., *J. Bacteriol.,* 114, 695–700, 1973.
35. Ishida, Y. and Katoda, H., *Mem. Res. Inst. Food Sci. Kyoto Univ.,* 26, 10–17, 1965.
36. Kessler, E., *Arch. Mikrobiol.,* 85, 153–158, 1972.
37. Kratz, W. A. and Meyers, J., *Am. J. Bot.,* 42, 282, 1955.
38. Lewin, J. C. and Hellebust, J. A., *Can. J. Microbiol.,* 16, 1123–1129, 1970.
39. Lewin, J. and Lewin, R. A., *J. Gen. Microbiol.,* 46, 361, 1967.
40. Lewin, J. C., *J. Gen. Microbiol.,* 9, 305, 1953.
41. Lewin, J. C., in *Symposium on Marine Microbiology,* Oppenheimer, C. H., Ed., Charles C Thomas, Springfield, Ill., 1963, 229.
42. Lewin, J. C. and Lewin, R. A., *Can. J. Microbiol.,* 6, 127, 1960.
43. Lewin, R. A., *J. Gen. Microbiol.,* 11, 459, 1954.
44. Loefer, J. B., *Arch. Protistenkd.,* 84, 456, 1935.
45. Lynn, R. I. and Starr, R. C., *Arch. Protistenkd.,* 112, 283–302, 1970.
46. McLachlan, J. and Craigie, J. S., *Can. J. Bot.,* 43, 1449–1456, 1965.
47. Miller, J. D. A. and Fogg, G. E., *Arch. Mikrobiol.,* 30, 1, 1958.
48. Miller, J. S. and Allen, M. M., *Arch. Mikrobiol.,* 86, 1–12, 1972.
49. Mineeva, L. A., *Mikrobiologiya,* 30, 586–592, 1961.
50. Osterud, K. L., *Physiol. Zool.,* 19, 19, 1946.
51. Palmer, E. G. and Starr, R. C., *J. Phycol.,* 7, 85–89, 1971.
52. Parker, B. C., Bold, H. C., and Deason, T. R., *Science,* 133, 761, 1961.
53. Pinter, I. J. and Provasoli, L., *Bull. Misaki Mar. Biol. Inst. Kyoto Univ.,* 12, 25–31, 1968.
54. Pringsheim, E. G., *Q. J. Microsc. Sci.,* 93, 71–96, 1952.
55. Pringsheim, E. G., *Arch. Mikrobiol.,* 56, 60–67, 1967.

Table 2 (continued)
FATTY ACIDS AND OTHER ORGANIC ACIDS

56. Pringsheim, E. G. and Weissner, W., *Nature,* 188, 919–921, 1960.
57. Pringsheim, E. G., *Arch. Mikrobiol.,* 46, 227, 1963.
58. Provasoli, L. and Gold, K., *Arch. Mikrobiol.,* 42, 196–203, 1962.
59. Rahat, M. and Spira, Z., *J. Protozool.,* 14, 45–48, 1967.
60. Sager, R. and Granick, S., *Ann. N. Y. Acad. Sci.,* 56, 831–838, 1953.
61. Samejima, H. and Meyers, J., *J. Gen. Microbiol.,* 18, 107, 1958.
62. Shihira, I. and Krauss, R. W., *Chlorella: Physiology and Taxonomy of 41 Isolates,* University of Maryland Press, College Park, 1965.
63. Silva, P., in *Physiology and Biochemistry of Algae,* Lewin, R. A., Ed., Academic Press, New York, 1962, 827.
64. Sloan, P. R. and Strickland, J. D. H., *J. Phycol.,* 2, 29, 1966.
65. Smith, A. J., London, J., and Stanier, R. Y., *J. Bacteriol.,* 94, 972–983, 1967.
66. Stein, J. R., Ph.D. thesis, University of California, Berkeley, 1957.
67. Stanier, R. Y., Kunisawa, R., Mandel, M., and Cohen-Bazire, G., *Bacteriol. Rev.,* 35, 171–205, 1971.
68. Thomas, W. H., *J. Protozool.,* 2(suppl.), 2, 1955.
69. Turner, M. F., *Br. Phycol. J.,* 5, 15–18, 1970.
70. Turner, M. F., in Scottish Marine Biological Association, Report of the Council for 1969/70, Oban, Scotland, 1970, 13.
71. Van Baalen, C., *Bot. Mar.,* 4, 129, 1962.
72. White, A. W., *J. Phycol.,* 10, 292–300, 1974.
73. Wise, D. L., *J. Protozool.,* 6, 19–23, 1959.
74. Wright, R. T., *Limnol. Oceanogr.,* 9, 163–178, 1964.
75. Wright, R. T. and Hobbie, J. E., *Ecology,* 47, 447–464, 1966.
76. Wyatt, J. T., Martin, C., and Jackson, J. W., *Phycologia,* 12, 153–161, 1973.

Table 3
AMINO ACIDS

Criterion for determining utilization of amino acids as a nitrogen source: growth of the alga on a given substrate in the absence of another nitrogen compound.

Species (synonym)	Alanine	Arginine	Asparagine	Aspartic acid	Glutamic acid	Glutamine	Glycine	Histidine	Isoleucine	Leucine	Lysine	Ornithine	Phenylalanine	Proline	Serine	Tryptophan	Tyrosine	Valine	Other	Ref.
Cyanophyceae																				
Agmenellum quadruplicatum	ψ[b]	ψ					ψ,u[7]				ψ									7, 46
Anabaena variabilis			U				ψ													27
Anacystis marina							ψ													7
A. nidulans			u[a]				ψ													27, 41
Aphanocapsa sp.	ψ		U		ψ	u[a]	ψ													32
Coccochloris elabens		ψ					U				ψ									46
Fremyella diplosiphon		U	U				U													47
Fremyella sp.		U					ψ				ψ									47
Microcoleus chthonoplastes		ψ	U				?				ψ ψ									46
Nostoc muscorum			U				u	ψ												27
Phormidium persicinum			ψ[b]	ψ	ψ		u	u				U								36
Plectonema terebrans		U	U	ψ	ψ[b]	u[a]	U					U	U		U	ψ				46
Synechococcus sp.	ψ[b]	U	U			U	U			U			U		U	ψ			ψ[c]	7, 32
Tolypothrix tenuis																				26

a By one strain.
b Only one strain tested.
c Threonine.

Table 3 (continued)
AMINO ACIDS

Rhodophyceae

Species (synonym)	Alanine	Arginine	Asparagine	Aspartic acid	Glutamic acid	Glutamine	Glycine	Histidine	Isoleucine	Leucine	Lysine	Ornithine	Phenylalanine	Proline	Serine	Tryptophan	Tyrosine	Valine	Other	Ref.
Asterocytis ramosa		U	U		U															18
Nemalion multifidum			u				u	U				n								17
Porphyridium sp.					?a	U	U,?				?				?					7, 48
Rhodella maculata	u	U	U	u	U		u	U	u	u	u	U	u	u	u	u	u	u	U,b	44
Rhodosorus marinus		U	U		U		?	U				ub		u					U,c	17

a Contaminated, possible algal growth.
b Glycylglycine, methionine, threonine.
c Citrulline.

Table 3 (continued)
AMINO ACIDS

Cryptophyceae

Species (synonym)	Alanine	Arginine	Asparagine	Aspartic acid	Glutamic acid	Glutamine	Glycine	Histidine	Isoleucine	Leucine	Lysine	Ornithine	Phenylalanine	Proline	Serine	Tryptophan	Tyrosine	Valine	Other	Ref.
Chilomonas paramecium[a]							U												U[b]	23
Chroomonas salina					u		U												u[c]	7
Hemiselmis virescens							U									u				7, 8, 14
Rhodomonas lens																				7

[a] Colorless.
[b] Various amino acids tested as carbon sources were not utilized.
[c] Cysteine, cystine.

Table 3 (continued)
AMINO ACIDS

Dinophyceae

Species (synonym)	Alanine	Arginine	Asparagine	Aspartic acid	Glutamic acid	Glutamine	Glycine	Histidine	Isoleucine	Leucine	Lysine	Ornithine	Phenylalanine	Proline	Serine	Tryptophan	Tyrosine	Valine	Other	Ref.
Amphidinium carteri	U	U	U,[38] Ū		U,[38] Ū		U,[38] Ū				U	U			U			U	U[a,38] Ū[b]	7, 38, 48
A. rhynchocephalum	U	U	U		U		U				Ū	U			Ū			Ū	U[a]	38
Cachonia niei	Ū		Ū		Ū		Ū					Ū							Ū[b]	48
Gymnodinium brevis																			Ū[c]	2
G. simplex	U		u		u		U													43
G. splendens			U		U		U													38
Gyrodinium californicum		U	U				U													38
G. resplendens		U	U																	38
G. uncatenum		U	U																	38
Heterocapsa kollmeriana			U		U			U		U				U			U		U[d]	42
Heterosigma inlandica		u	u			U														25

a Methionine.
b Threonine.
c Several amino acids tested as carbon sources were not utilized.
d Methionine, thymine.

Table 3 (continued)
AMINO ACIDS

Bacillariophyceae

Species (synonym)	Alanine	Arginine	Asparagine	Aspartic acid	Glutamic acid	Glutamine	Glycine	Histidine	Isoleucine	Leucine	Lysine	Ornithine	Phenylalanine	Proline	Serine	Tryptophan	Tyrosine	Valine	Other	Ref.
Amphora hyalina	U[a]				ψ		U													7
Chaetoceros affinis			ψ U		u u	u u	U[a]				u	U[a]			ψ			ψ[a]	u[b]	48
C. gracilis					u		u													43
C. lorenzianus					ψ U	u U	u?													20
C. pelagicus					U	U U	U													20
Coscinodiscus asteromphalus																				20
Cyclotella caspia							u u													20
C. cryptica	ψ		ψ		ψ[c]	U[c]	U[?]ψ[d,a,s]								ψ			ψ	ψ[b]	20, 48
C. nana							u					U[a]								7
Cylindrotheca closterium					U															7
C. fusiformis (Nitzschia closterium)																				1
Detonula confervacea	U			ψ	u	u	u u								u					20
Fragilaria pinnata	U	U	U		ψ	U	U					ψ		U	ψ		U	ψ	ψ[e]	7
Melosira juergensii	ψ		U		ψ		ψ	U	ψ	U	ψ	U	ψ	U	U	ψ	U	ψ	u[b]	42
M. nummuloides	u		ψ		ψ		u	U		U	ψ				u			U		22
Melosira sp.[f]							U											U	U[b]	48
Navicula biskanteri	U		U		U		U				U				U			ψ	ψ[b]	7
N. incerta	U		ψ		ψ		ψ				ψ				ψ			ψ		48
N. salinarum[g]	U																			48

[a] Contaminated.
[b] Threonine.
[c] By two strains; poorly utilized by one strain.
[d] By two strains; not utilized by one strain.
[e] Threonine, cysteine, cystine, methionine, α-aminoisobutyric acid.
[f] These results are questionable since growth on nitrate was also very slow.
[g] These results are questionable since growth on all nitrogen sources, including nitrate, was slow.

Table 3 (continued)
AMINO ACIDS

Bacillariophyceae (continued)

Species (synonym)	Alanine	Arginine	Asparagine	Aspartic acid	Glutamic acid	Glutamine	Glycine	Histidine	Isoleucine	Leucine	Lysine	Ornithine	Phenylalanine	Proline	Serine	Tryptophan	Tyrosine	Valine	Other	Ref.
Nitzschia acicularis					U[i]		U													7
N. alba[h]			U		U															28
N. frustulum	U				U[i]		U					U			U			U	U[b]	48
N. leucosigma[h]			U		U[i]															28
N. ovalis	U				U		U					U			U			U	U[b]	28
N. putrida[h]					u															20
N. seriata	U		U		U		ψ				U	U			U			U		48
Nitzschia sp.	U		u		U		U				U	U			U			U	U, b,k,21,48 ψ[j,21]	7, 21, 48
Phaeodactylum tricornutum				U		U		ψ	U	U			U	U		U	U	U		20
Rhizosolenia setigera					ψ[i]		u,20 U													20, 48
Skeletonema costatum	ψ		ψ		ψ[i]	u	ψ				ψ	ψ			ψ			ψ	ψ[b]	48
Stephanopyxis turris	ψ		ψ		ψ[i]	u					ψ	ψ			ψ			ψ	ψ[b]	20, 48
Thalassiosira fluviatilis	ψ		ψ		U	u	u,20 ψ				ψ	U			U			ψ		20, 48
T. nordenskioldii					U	u	u					U								20

h Colorless.
i As a carbon source, see Table 2.
j Cysteine, taurine.
k Citrulline, hydroxyproline, methionine, norvaline.

Table 3 (continued)
AMINO ACIDS

Species (synonym)	Alanine	Arginine	Asparagine	Aspartic acid	Glutamic acid	Glutamine	Glycine	Histidine	Isoleucine	Leucine	Lysine	Ornithine	Phenylalanine	Proline	Serine	Tryptophan	Tyrosine	Valine	Other	Ref.
Haptophyceae																				
Cricosphaera carterae	U[a]		ψ		ψ		ψ				U	U			ψ			U	ψ[b]	48
Emiliana (Coccolithus) huxleyi	ψ	?	ψ		ψ, u[f,20]		U, u[f,20]	ψ	?	?	ψ	ψ			ψ			ψ	ψ[b]	7, 20, 48
Hymenomonas elongata				?		u[g]	?				?				ψ	ψ		?	?[c]	13
Isochrysis galbana	U	U	ψ		ψ		U	U	U	U	U		?		U			U	ψ[b]	7, 48
Prymnesium parvum[d]	U	U	U	U	U[e]		U	U	U	U	U	U	U		?	U	U	U	U[c]	13, 30

[a] Contaminated.
[b] Threonine.
[c] Methionine.
[d] Various strains differences.
[e] By three strains; not utilized by one strain.
[f] By three strains.
[g] By two strains, utilized by one strain.

Table 3 (continued)
AMINO ACIDS

Chrysophyceae

Species (synonym)	Alanine	Arginine	Asparagine	Aspartic acid	Glutamic acid	Glutamine	Glycine	Histidine	Isoleucine	Leucine	Lysine	Ornithine	Phenylalanine	Proline	Serine	Tryptophan	Tyrosine	Valine	Other	Ref.
Monochrysis lutheri		U		U	U		U,13	U	U	U	U		U		U	U		U	U[a]	7, 13
Ochromonas malhamensis							u	U	U	U										24

[a] Methionine.

Table 3 (continued)
AMINO ACIDS

Xanthophyceae

Species (synonym)	Alanine	Arginine	Asparagine	Aspartic acid	Glutamic acid	Glutamine	Glycine	Histidine	Isoleucine	Leucine	Lysine	Ornithine	Phenylalanine	Proline	Serine	Tryptophan	Tyrosine	Valine	Other	Ref.
Heterothrix sp.							u													7
Monallantus salina	V	V		V	V		V	V	V	V	V?	V	V	V	V	u	V?	V	V;?[a]	7
Monodus subterraneus	U	U	V?	V	V	U	V			U	V		U		U	U	U	U	U[b]	31
Tribonema aequale						U														10

[a] Cystine; however, threonine is not utilized.
[b] Cystine and threonine.

Table 3 (continued)
AMINO ACIDS

Euglenophyceae

Species (synonym)	Alanine	Arginine	Asparagine	Aspartic acid	Glutamic acid	Glutamine	Glycine	Histidine	Isoleucine	Leucine	Lysine	Ornithine	Phenylalanine	Proline	Serine	Tryptophan	Tyrosine	Valine	Other	Ref.
Euglena anabaena minor[a]	U	U			U					U	U		U	U	U	U	U	U		—
E. deses	U	U	U		U		U	U		U	U		U	U	u	U	U	U		—
E. gracilis bacillaris	U	U			U[b]		U	U		U					u		U	U		—
E. gracilis typica	U	U	U		U[b]		U	U		U	U		U	U	U	U	U	U		—
E. gracilis urophora		U			U															—
E. klebsii[c]	U	U	U[d]		U		U	U		U	U		U	U	U	U	U	U		—
E. pisciformis	U	U	U		U		U	U		U	U		U	U	U	U	U	U		—
E. stellata	U	U			U		U	U		U	U		U	U	U	U	U	U		—

a According to E. G. Pringsheim, the organism used in these studies was *E. geniculata*.
b Only at pH 3.0—5.5.
c According to E. G. Pringsheim, the organism used in these studies was *E. mutabilis*.
d Growth is obtained only if thiamine is present.

Table 3 (continued)
AMINO ACIDS

Prasinophyceae

Species (synonym)	Alanine	Arginine	Asparagine	Aspartic acid	Glutamic acid	Glutamine	Glycine	Histidine	Isoleucine	Leucine	Lysine	Ornithine	Phenylalanine	Proline	Serine	Tryptophan	Tyrosine	Valine	Other	Ref.
Platymonas subcordiformis	U		U		U		U				U[a]	U			U			U[a]	U[b]	33, 48
Platymonas sp.	U	U	U		U		U				U[a]	U			U			U	U[c]	48
Prasinocladus marinus							U				U	U								7
Stephanoptera gracilis			u	U	U		U													19
Tetraselmis maculata		U					U[d]													7
T. striata		U					U													7
Tetraselmis sp.			U				U								U					45

a Contaminated.
b Threonine, contaminated.
c Threonine.
d 120—200% of maximal growth relative to that on NH_4^+ (= 100%).

Table 3 (continued)
AMINO ACIDS

Chlorophyceae

Species (synonym)	Alanine	Arginine	Asparagine	Aspartic acid	Glutamic acid	Glutamine	Glycine	Histidine	Isoleucine	Leucine	Lysine	Ornithine	Phenylalanine	Proline	Serine	Tryptophan	Tyrosine	Valine	Other	Ref.
Asterococcus superbus	U			U	U	U	U				U	U			U					12
Brachiomonas submarina	U	U	U	U	U		U	U	u	u	U	U	u	u	u	u	u	u	u[a]	15
Carteria crucifera	U	U		U	U	U					U	U			u					12
Chlamydomonas actinochloris	U[b]		U	U	U	U	U[b]				U	U			U[b]					12
C. agloeformis	U			U	U	U	U				U	U			U					1
C. calyptrata	U			U[d]	U	U	U				U[e]	U			U					12
C. carrosa[c]	U[f]		U	U	U	U	U[e]				U	U			U					12
C. chlamydogama	U			U	U	U	U				U	U			U					12
C. eugametos	U			U	U	U	U				U	U			U					12
C. gloegama	U			U	U	U	U				U	U			U					12
C. gloeopara	U			U	U	U	U				U	U			U					12
C. inflexa	U			u	U	U	U				U	U			U					12
C. kakosmos	U			U	U	U[f]	U				U	U			U					12
C. mexicana	U			U	U	U	U				U	U			U					12
C. microsphaera acuta	U			U	U	U	U				U	U			U					12
C. microsphaerella	U			U	U	U	U				U	U			U					12
C. minuia	U			U	U	U	U				U	U			U					12
C. moewusii	U		U	U	U	U	U				U	U			U					12
C. moewusii rotunda	U			U	U	U	U				U	U			U					12

[a] Threonine.
[b] By one strain; not utilized by one strain.
[c] Four strains.
[d] By three strains; not utilized by one strain.
[e] By three strains; utilized by one strain.
[f] By one strain; poorly utilized by one strain.

Table 3 (continued)
AMINO ACIDS

Chlorophyceae (continued)

Species (synonym)	Alanine	Arginine	Asparagine	Aspartic acid	Glutamic acid	Glutamine	Glycine	Histidine	Isoleucine	Leucine	Lysine	Ornithine	Phenylalanine	Proline	Serine	Tryptophan	Tyrosine	Valine	Other	Ref.
C. mundana			U			U	U													16
C. mundana astigmata			U			U	U													16
C. mutabilis	U			U	U	U	U				U	U			U					12
C. palla	U			U	U	U	U				U	U			U					7
C. peterfi	U		U	U	U	U	U				U	U			U					12
C. radiata	U			U	U	U	U				U	U			U					12
C. reinhardii	U		?	U	U	U[i,j]	U	U			U	U			U					12, 40
C. sectilis	U		U	U	U	U	U				U	U			U					12
C. typhlos	U[g]		U	U	U	U	U[f]				U	U			U			?	?[a]	48
Chlamydomonas sp.	U	U	U	U	U		U				U	U			U					4, 5
Chlorella protothecoides	U		U	U	U		U				U	U		U	U				U[a]	4, 5, 9
C. vulgaris	U		U	U	U		U		u	u	u	U	u		U	U	u	u	U[a]	48
Chlorella sp.				U			U					U			U		U	U	ψ[h]	29
Diplostauron elegans	U		U	U	U		U				U	U			U					19
Dunaliella salina			U				U												V[a]	7, 48
D. tertiolecta			U	U	U		U	U							U					19
D. viridis			U	U	U		U													19
Furcilla stigmatophora	U	U	U	U	U	U	U	U	U	U	U	U	U	U	U	U	U	U	V[a]	11
Gloeocystis gigas	U	U	U	U	U	U	U	U	U	U	U	U	U	U	U	U	U	U		12
G. maxima	U	U	U	U	U		U	U	U	U	U	U	U	U	U	U	U	U	U[a], u[i]	12
Haematococcus (Balticola) buetschlii			U	U	U		U													15
H. droebakensis			U				U	U			U	U	U		U	U		U	U[a]	15
H. pluvialis	U	U	U	U	U		U	U	U	U	U	U	U	U	U	U	U	U	u[i]	15
Lobomonas pyriformis			U				U						U			u				34
L. sphaerica			U		U															37
Nannochloris atomus							U					U							U[j]	39
N. oculata		U	U	U	?		U	U	U	U	U	U	U		U	U	U	U	U[k]	7, 13

g Contaminated.
h Glycylglycine.
i Threonine, by one strain; not utilized by one strain.
j Cystine.
k Methionine.

Table 3 (continued)
AMINO ACIDS

Chlorophyceae (continued)

Species (synonym)	Alanine	Arginine	Asparagine	Aspartic acid	Glutamic acid	Glutamine	Glycine	Histidine	Isoleucine	Leucine	Lysine	Ornithine	Phenylalanine	Proline	Serine	Tryptophan	Tyrosine	Valine	Other	Ref.
Polytoma caudatum	u		U				U													1
P. obtusum	U	u	U				U	u												1
P. ocellatum	U		U				U													1
P. uvella	u		U				U	u		U					u	u		u		3, 6
Scenedesmus obliquus	U		U	U	U	U	U	u		U	U	U			U					12
Sphaerobotrys fluviatilis	U		U	U	U	U	U	U			U	U			U					12
Sporotetras pyriformis	U	U	U	U	U	U	U				U	U			U					15
Stephanosphaera pluvialis	U		U	U	U	U	U	U	U	U	U	U	U	U	U	U	U	U	U^a	39
Stichococcus cylindricus?		U	U				U												U^j	

Table 3 (continued)
AMINO ACIDS

REFERENCES

1. Albritton, E. C., *Standard Values in Nutrition and Metabolism,* W. B. Saunders, Philadelphia, 1954.
2. Aldrich, D. V., *Science,* 137, 988, 1962.
3. Algeus, S., *Physiol. Plant.,* 1, 66, 1948.
4. Algeus, S., *Physiol. Plant.,* 2, 266, 1949.
5. Algeus, S., *Physiol. Plant.,* 3, 370, 1950.
6. Algeus, S., *Physiol. Plant.,* 4, 459, 1951.
7. Antia, N. J., Berland, B. R., Bonin, D. J., and Maestrini, S. Y., *J. Mar. Biol. Assoc. U.K.,* 55, 519–539, 1975.
8. Antia, N. J. and Chorney, V., *J. Protozool.,* 15, 198–201, 1968.
9. Arnow, P., Oleson, J. J., and Williams, J. H., *Am. J. Bot.,* 40, 100, 1953.
10. Belcher, J. H. and Fogg, G. E., *Arch. Mikrobiol.,* 30, 17, 1958.
11. Belcher, J. H., *Arch. Mikrobiol.,* 58, 181–185, 1967.
12. Cain, J., *Can. J. Bot.,* 43, 1367, 1965.
13. Droop, M. R., *J. Mar. Biol. Assoc. U.K.,* 34, 229, 1955.
14. Droop, M. R., *J. Gen. Microbiol.,* 16, 286, 1957.
15. Droop, M. R., *Rev. Algol.,* 5(4), 247, 1961.
16. Eppley, R. W. and Maciasr, F. M., *Physiol. Plant.,* 15, 72, 1962.
17. Fries, L., *Physiol. Plant.,* 16, 695–708, 1963.
18. Fries, L. and Pettersson, H., *Br. Phycol. Bull.,* 3(3), 417–422, 1968.
19. Gibor, A., *Biol. Bull.,* 111, 223, 1956.
20. Guillard, R. R. L., in *Symposium on Marine Microbiology,* Oppenheimer, C. H., Ed., Charles C Thomas, Springfield, Ill., 1963, 93.
21. Hayward, J., *Physiol. Plant.,* 18, 201–207, 1964.
22. Hellebust, J. A. and Guillard, R. R. L., *J. Phycol.,* 3, 132–136, 1967.
23. Holz, G. G., *J. Protozool.,* 1, 114, 1954.
24. Hutner, S. H., Provasoli, L., and Filfus, J., *Ann. N.Y. Acad. Sci.,* 56, 852, 1953.
25. Iwasaki, H. and Sasada, K., *Bull. Jpn. Soc. Sci. Fish.,* 35, 943–947, 1969.
26. Kiyohara, T. et al., *J. Gen. Appl. Microbiol.,* 6, 176, 1960.
27. Kratz, W. A. and Meyers, J., *Am. J. Bot.,* 42, 282, 1955.
28. Lewin, J. and Lewin, R. A., *J. Gen. Microbiol.,* 46, 361, 1967.
29. Lynn, R. L. and Starr, R. C., *Arch Prokistenkd.,* 112, 283–302, 1970.
30. McLaughlin, J. J. A., *J. Protozool.,* 5, 75, 1958.
31. Miller, J. D. A. and Fogg, G. E., *Arch. Mikrobiol.,* 30, 1, 1958.
32. Neilson, A. H. and Doudoroff, M., *Arch. Mikrobiol.,* 89, 15–22, 1973.
33. North, B. B. and Stephens, G. C., *Biol. Bull.,* 133, 391–400, 1967.
34. Osterud, K. L., *Physiol. Zool.,* 19, 19, 1946.
35. Pintner, I. J. and Provasoli, L., *Bull. Misaki Mar. Biol. Inst. Kyoto Univ.,* 12, 25–31, 1968.
36. Pintner, I. J. and Provasoli, L., *J. Gen. Microbiol.,* 18, 190, 1958.
37. Pringsheim, E. G., *Arch. Microbiol.,* 46, 227, 1963.
38. Provasoli, L. and McLaughlin, J. J. A., in *Symposium on Marine Microbiology,* Oppenheimer, C. H., Ed., Charles C Thomas, Springfield, Ill., 1963, 105.
39. Ryther, J. H., *Biol. Bull.,* 106, 198, 1954.
40. Sager, R. and Granick, S., *Ann. N.Y. Acad. Sci.,* 56, 831–838, 1953.
41. Stanier, R. Y., Kunisawa, R., Mandel, M., and Cohen-Bazire, G., *Bacteriol. Rev.,* 35, 171–205, 1971.
42. Swift, M. J. and McLaughlin, J. J. A., *Ann. N.Y. Acad. Sci.,* 175, 577–600, 1970.
43. Thomas, W. H., *Limnol. Oceanogr.,* 11, 393, 1966.
44. Turner, M. F., *Br. Phycol. J.,* 5, 15–18, 1970.
45. Turner, M. F., in Scottish Marine Biological Association, Report of the Council for 1969/70, Oban, Scotland, 1970, 13–44.
46. Van Baalen, C., *Bot. Mar.,* 4, 129, 1962.
47. Wyatt, J. T., Martin, T. C., and Jackson, J. W., *Phycologia,* 12, 153–161, 1973.
48. Wheeler, P. A., North, B. B., and Stephens, G. C., *Limnol. Oceanogr.,* 19(2), 249–259, 1974.

Table 4
OTHER NITROGENOUS SUBSTANCES

Criterion for determining utilization of nitrogenous substances: growth of alga on a given substrate in the absence of another nitrogenous compound.

Species (synonym)	Acetamide	Adenine	Ammonium	Casein hydrolysate	Cytosine	Nitrate	Nitrite	Nitrogen	Peptone	Succinamide	Uracil	Urea	Uric acid	Miscellaneous	Ref.
Cyanophyceae															
Agmenellum quadruplicatum			U	U		U	U	∅[65] U[57]				U	U[65], U[64]	U,[a] ∅[b]	8, 57, 64, 65
Anabaena ambigua				U		U		U							73
A. azollae								U							74
A. cycadeae								U							75
A. cylindrica			U	u?		U		U				∅			4, 58, 76, 77
A. fertilissima								U							73
A. flos-aquae								U						U[c]	58, 78
A. gelatinosa								U							16
A. humicola								U							77
A. levanderi								U							79
A. naviculoides								U						U[c]	16
A. spiroides								U							33
A. torulosa[d]								∅[e,36] U							80
A. variabilis			U	?		U		∅[65] U[57] U				U			33, 36, 77, 80
Anabaenopsis circularis			U			U		∅							33, 81, 82
Anacystis marina			U	∅		U	U	∅				U	∅	∅[a,b]	8, 57, 64, 65
A. nidulans			U			U	U	∅				∅	∅		12, 36, 57, 58
Aphanizomenon flos-aquae			U			U		U							70
Aphanocapsa sp.								U							46
Aulosira fertilissima								U							73
Calothrix aeruginea[d]								U							80
C. brevissima								U							66, 81, 82
C. crustacea[d]								U							5A
C. desertica								U							33
C. elenkinii								U							83, 84

[a] Glucosamine.
[b] Hypoxanthine.
[c] Xanthine.
[d] Unialgal cultures, not bacteria free.
[e] Some strains do fix nitrogen.

Note: Blue-green algae (Cyanophyceae) characteristically show slower growth rates when growing heterotrophically in the dark than when growing photosynthetically in the light.

Table 4 (continued)
OTHER NITROGENOUS SUBSTANCES

Cyanophyceae (continued)

Species (synonym)	Acetamide	Adenine	Ammonium	Casein hydrolysate	Cytosine	Nitrate	Nitrite	Nitrogen	Peptone	Succinamide	Uracil	Urea	Uric acid	Miscellaneous	Ref.
C. parietina			U	U		U		U				∅			4, 33
C. scopulorum				U		U		U							59
Chlorogloea fritschii			U	U		U		∅,33,57 U				∅			22, 33, 57, 85
Chroococcus turgidus			U			U		U							4
Chroococcus sp.								U							72
Coccochloris elabens				U		U		∅				U	U,65 ∅64		57, 64, 65
C. peniocystis						U		∅							57, 64, 70
Cylindrospermum gorakhporense								U							73
C. licheniforme								U							77
C. maius								U							77
C. sphaerica								U							86
Cylindrospermum sp.								U							72
Fischerella major								U							87
F. musicola								U							87
Fremyella diplosiphon				U		U		∅f				u			71
Fremyella sp.				U		U		U				U			71
Gleocapsa alpicola								∅							57
G. dimidiata								U							70
G. membranina								U							70
Gleocapsa sp.								U							57, 72, 88
Hapalosiphon fontinalis			U			U		U							83, 84
Lyngbya aestuarii						U		U				∅			4, 65
L. lagerheimii						U		U					U		64, 65
Mastigocladus laminosus								U							58
Microchaete sp.[d]								∅							80
Microcoleus chthonoplastes				U		U		∅				∅	U,65 u?64		64, 65
M. tenerrimus								U							65
M. vaginatus						U		∅							33
Microcystis aeruginosa (Diplocystis aeruginosa)								U				∅			57, 70
Nodularia harveyana[d]								U							80
N. sphaerocarpa								U							33
N. spumigena[d]								U							80
Nostoc calcicola								U							77

f Not by one strain, but poorly by another.

Table 4 (continued)
OTHER NITROGENOUS SUBSTANCES

Cyanophyceae (continued)

Species (synonym)	Acetamide	Adenine	Ammonium	Casein hydrolysate	Cytosine	Nitrate	Nitrite	Nitrogen	Peptone	Succinamide	Uracil	Urea	Uric acid	Miscellaneous	Ref.
N. commune								U							72
N. cycadae								U							67
N. entophytum								U							58, 59
N. linckia[d]								U							58
N. muscorum			U	g		U		U				U			36, 58, 72
N. paludosum								U							77
N. punctiforme								U							75, 89
N. sphaericum								U							90
Nostoc sp.								U							33, 58
Oscillatoria amphibia								U[g]							65
O. brevis								U							59
O. subtilissima								U							65
O. willansii													n		64
O. woronichinii			U			U		U				U			41
Oscillatoria sp.[h,i]			U					U				U[g]			33
Phormidium foveolarum				U				U				U[g]			4, 72
P. luridum			U	U				U							4
P. persicinum			U			U		U[g]							51
P. tenue						U		U							70
Phormidium sp.[h]						U		U							33, 91
Plectonema boryanum[h]						U		U[g]							33, 91, 92
Plectonema nostocorum								U							70
P. notatum			U	U		U		U				U			4
P. terebrans				U				U				U			65
Plectonema sp.[h]								U[g]					U		33, 82
Raphidiopsis indica[h]								U							91
Scytonema arcangelii								U							79
S. hofmanni								U							79

g Allen reports utilization.
h Only under microaerophilic conditions.
i By 4 strains, not by 1.

Table 4 (continued)

OTHER NITROGENOUS SUBSTANCES

Species (synonym)	Acetamide	Adenine	Ammonium	Casein hydrolysate	Cytosine	Nitrate	Nitrite	Nitrogen	Peptone	Succinamide	Uracil	Urea	Uric acid	Miscellaneous	Ref.
Cyanophyceae (continued)															
Stigonema dendroideum			U[j]			U[j]		U							93
Synechococcus cedrorum				U		U		Ø				Ø	Ø		4, 12, 23, 57
S. elongatus						U		Ø							57
S. lividus						U	U	Ø							57
Synechococcus sp.			U	U		U		U				U		u,a,Ø[b]	8, 46, 57
Tolypothrix distorta			U[k]			U[l]		U[l]							72
T. tenuis															33, 34, 35, 58, 68, 69
Westiellopsis prolifica								U							94

j Allen reports nonutilization.
k Will also support growth in the dark.
l Will not support growth in the dark.

Table 4 (continued)
OTHER NITROGENOUS SUBSTANCES

Rhodophyceae

Species (synonym)	Acetamide	Adenine	Ammonium	Casein hydrolysate	Cytosine	Nitrate	Nitrite	Nitrogen	Peptone	Succinamide	Uracil	Urea	Uric acid	Miscellaneous	Ref.
Asterocytis ramosa		U̸	U									U	U		25
Goniotrichum elegans			U	U		U	U								24
Nemalion multifidum			U	U		U	U								24
Porphyridium cf. *cruentum*			U			U	U					U		U̸[a], U[b]	41
Porphyridium sp.		U̸	U			U						U̸			8, 69
Rhodella maculata		U?				U					U̸	U	U	U̸[c], U[d]	62
Rhodosorus marinus												U	U	U̸[e]	24

[a] Glucosamine.
[b] Hypoxanthine.
[c] Guanine, thymine, cytidylic acid.
[d] Guanidine carbonate.
[e] Guanine.

Table 4 (continued)
OTHER NITROGENOUS SUBSTANCES

Species (synonym)	Acetamide	Adenine	Ammonium	Casein hydrolysate	Cytosine	Nitrate	Nitrite	Nitrogen	Peptone	Succinamide	Uracil	Urea	Uric acid	Miscellaneous	Ref.
Cryptophyceae															
Chilomonas paramecium[a]		U												U[b]	30
Chroomonas salina			U			U	U					U		U[c]U[d]	8, 9
Hemiselmis virescens			U		U[f]	U[f]	u[e]U[f,g]				U[f]	U		U[d,e,g] U[f,g] U[g,8,9]	8
Rhodomonas lens			U			U	U					U	U	U[c,f]d	8

a Colorless.
b Various amides tested as carbon sources were not utilized.
c Glucosamine.
d Hypoxanthine.
e Alloxan, xanthine, guanine.
f Guanidine, thymine.
g Glucosamine, galactosamine.

Table 4 (continued)
OTHER NITROGENOUS SUBSTANCES

Dinophyceae

Species (synonym)	Acetamide	Adenine	Ammonium	Casein hydrolysate	Cytosine	Nitrate	Nitrite	Nitrogen	Peptone	Succinamide	Uracil	Urea	Uric acid	Miscellaneous	Ref.
Amphidinium carteri			U			U						U		U[a],[b]	8, 69
Cachonina niei						U[c]	U					U			41, 69
Exuviaella sp.												Ø			41
Gymnodinium brevis														Ø[d]	2
G. nelsoni			U			U						Ø			41
G. simplex			U				U					Ø			61
G. splendens															41
Heterocapsa kollmeriana			U			U						U			60
Heterosigma inlandica													u		32
Oxyrrhis marina												U		R[e]	18
Polykrikos schwartzi			U			U						Ø			31
Prorocentrum micans															41

a Glucosamine.
b Hypoxanthine.
c Contaminated.
d Several nitrogenous organic compounds tested as carbon sources were not utilized.
e Amino nitrogen.

Table 4 (continued)
OTHER NITROGENOUS SUBSTANCES

Bacillariophyceae

Species (synonym)	Acetamide	Adenine	Ammonium	Casein hydrolysate	Cytosine	Nitrate	Nitrite	Nitrogen	Peptone	Succinamide	Uracil	Urea	Uric acid	Miscellaneous	Ref.
Amphiprora alata			U			U						U		∅, a, U^b	9, 14
Amphora hyalina			U			U	U					U			8
Biddulphia aurita			U			U	U								39
Chaetoceros affinis						U							U		69
C. didymus			U			U	u					U			41
C. gracilis												?	u		61
C. lorenzianus												∅	∅		27
C. pelagicus												∅			27
C. simplex			U			U						U			14
Chaetoceros sp.			U			U						U	u		14, 41
Coscinodiscus asteromphalus													u		27
C. granii												u			41
C. wailesii												∅			41
Cyclotella caspia			U			U	U					U	U	u^a,b	27
C. cryptica			U			U	U					U^c	U^c		8
C. nana			U			U	U					U		u^a,U^b,d	27, 41, 42, 69
Cylindrotheca closterium			U			U						U			8
C. fusiformes (Nitzschia closterium)									U						1, 41
Ditylum brightwellii			U			U	U					u	u		27
Eucampia zoodiacus			U			U									21, 41
Fragilaria pinnata			U				U					∅		u^a,U^b	8
Guinardia flaccida												U			41
Lauderia sp.												∅			41
Lithodesmium undulatum												∅			41
Melosira juergensii			U			U						U			60
Melosira sp.						u						U			69
Navicula biskanteri			U			U	U					U		u^a,U^b	8
N. incerta						U						U			69

a Glucosamine.
b Hypoxanthine.
c By two strains; poorly utilized by one strain.
d Colorless.

Table 4 (continued)
OTHER NITROGENOUS SUBSTANCES

Bacillariophyceae (continued)

Species (synonym)	Acetamide	Adenine	Ammonium	Casein hydrolysate	Cytosine	Nitrate	Nitrite	Nitrogen	Peptone	Succinamide	Uracil	Urea	Uric acid	Miscellaneous	Ref.
N. salinarum			U			u						U		u,[a] U,[b]	69
Nitzschia acicularis[d]						U	U								8
N. alba[d]						U									37
N. frustulum						U									69
N. leucosigma[d]						U									37
N. ovalis						U									69
N. putrida[d]															37
N. seriata						u						u	u		27
Nitzschia sp.			U			U	U					U		U,[e,29]	69
Phaeodactylum tricornutum						U						U		u,[a,8,29] U,[b,8]	8, 29, 69
Rhizosolenia alata												U			41
R. calcar avis												U	U		41
R. setigera												U	U		27
R. stolterfothii												U			41
Schroderella sp.												U			41
Skeletonema costatum						U						U	U		27, 41, 42, 69
Skeletonema sp.			U			U						U			14, 41
Stephanopyxis costata			U			U[f]						U			14
S. turris												U			41, 42, 69
Stephanopyxis sp.												U			41
Thalassiosira fluviatilis						U						u,[27] U	u		27, 41, 42, 69
T. nordenskioldii												U	U		27
T. rotula												U			41

e Creatine, creatinine, hydroxylamine.
f Contaminated.

Table 4 (continued)
OTHER NITROGENOUS SUBSTANCES

Haptophyceae

Species (synonym)	Acetamide	Adenine	Ammonium	Casein hydrolysate	Cytosine	Nitrate	Nitrite	Nitrogen	Peptone	Succinamide	Uracil	Urea	Uric acid	Miscellaneous	Ref.
Chrysochromulina breviflium														U[a]	50
C. kappa						U								U[a]	50
C. strobilus						U								U[a]	50
Chrysochromulina sp.			U			U						U			14
Cricosphaera carteri						U									69
Cricosphaera sp.															41
Emiliana (*Coccolithus*) *huxleyi*			U			U	U					U[b,8] / U[e,14,27,41]	u[e]	u[c,d]	8, 14, 27, 41, 69
Hymenomonas elongata			Ø			U U						?	U		17
Isochrysis galbana			U			U	U					U	U	u,[c]u[f]	8, 41, 69
Prymnesium parvum			U[g]			U						Ø	?		17

[a] Adenylic acid.
[b] By 2 strains.
[c] Glucosamine.
[d] Hypoxanthine.
[e] By three strains.
[f] Hypoxanthine (23-day lag).
[g] Toxic under some conditions (e.g., basic pH and low salinity) and at high concentrations.[43]

Table 4 (continued)
OTHER NITROGENOUS SUBSTANCES

Species (synonym)	Acetamide	Adenine	Ammonium	Casein hydrolysate	Cytosine	Nitrate	Nitrite	Nitrogen	Peptone	Succinamide	Uracil	Urea	Uric acid	Miscellaneous	Ref.
Chrysophyceae															
Monochrysis lutherii			U			U	U					U	U	U,[a] U[b]	8, 17, 41
Ochromonas malhamensis			U			U						U			38
Pavlova gyrans														U[c]	50

[a] Glucosamine.
[b] Hypoxanthine.
[c] Guanine.

Table 4 (continued)
OTHER NITROGENOUS SUBSTANCES

Species (synonym)	Acetamide	Adenine	Ammonium	Casein hydrolysate	Cytosine	Nitrate	Nitrite	Nitrogen	Peptone	Succinamide	Uracil	Urea	Uric acid	Miscellaneous	Ref.
Xanthophyceae															
Heterothrix sp.			U			U	U					U		u,[a]U[b]	8
Monallantus salina			U			U	U					U		u,[a]U[b]	8
Monodus subterraneus	U		U			U			u	U		U			44, 45
Tribonema aequale										U		U	u		11

[a] Glucosamine.
[b] Hypoxanthine.

Table 4 (continued)
OTHER NITROGENOUS SUBSTANCES

Species (synonym)	Acetamide	Adenine	Ammonium	Casein hydrolysate	Cytosine	Nitrate	Nitrite	Nitrogen	Peptone	Succinamide	Uracil	Urea	Uric acid	Miscellaneous	Ref.
Euglenophyceae															
Astasia longa (A. chattoni, A. klebsii)			U						U						1
A. quartana															1
Euglena anabaena minor			U			U			U						1
E. deses[a]			U,[1] U[1,5]			U U			U						1
E. gracilis var. bacillaris			U			U u			U						1, 15
E. gracilis typica			U			u U			U						1
E. gracilis urophora						u U			U						1
E. klebsii[b]			u			U			U						1
E. pisciformis			U						U						1
E. stellata			U			U			U						1

a According to E. G. Pringsheim, the organism used in these studies was E. geniculata.
b According to E. G. Pringsheim, the organism used in these studies was E. mutabilis.

Table 4 (continued)
OTHER NITROGENOUS SUBSTANCES

Prasinophyceae

Species (synonym)	Acetamide	Adenine	Ammonium	Casein hydrolysate	Cytosine	Nitrate	Nitrite	Nitrogen	Peptone	Succinamide	Uracil	Urea	Uric acid	Miscellaneous	Ref.
Platymonas chui														U[a]	63
P. subcordiformis			U			U								U[a]	47, 63, 69
P. tetrathele														U[a]	63
Prasinocladus marinus			U			U	U					U		U[b,c]	8
Pyramimonas sp.												U			41
Stephanoptera gracilis			U			U						U	U	U[b,d]	26
Tetraselmis maculata			U			U	U					U			8
T. striata			U			U	U					U		U[b,d]	8
Tetraselmis sp.[e]		U									U		U	U[f]	63

[a] Thymine when grown heterotrophically in the dark.
[b] Glucosamine.
[c] Hypoxanthine (18-day lag).
[d] Hypoxanthine.
[e] For good growth in the dark, a source of reduced nitrogen in the form of a purine or pyrimidine was essential.
[f] Guanine, xanthine, orotic acid, thymine.

Table 4 (continued)
OTHER NITROGENOUS SUBSTANCES

Chlorophyceae

Species (synonym)	Acetamide	Adenine	Ammonium	Casein hydrolysate	Cytosine	Nitrate	Nitrite	Nitrogen	Peptone	Succinamide	Uracil	Urea	Uric acid	Miscellaneous	Ref.
Asterococcus superbus	U	U	U		G	U	U			G	G	U	G		13
Brachiomonas submarina	G	U	U			U					U	U	U	U[a]	19
Carteria crucifera	G	U	U		U	U	U			U	G	G	G		13
Chlamydomonas actinochloris	U	U	U			U	U		U	U	G	G	G		1
C. aeloeformis	U	U	U		U	U	U			U	U	U	U		13
C. calyptrata	G	U	U		G	U	U			G	G	U	G		13
C. carrosa[b]	U	G	U		G	U	U			G	G	U	G		13
C. chlamydogama	U	U	U[c]		U	U	U			U	U	U	U		13
C. eugametos	U	U	U[c]		G	U	U			U	U	U	U		13
C. gloeogama	U	U	U		U	U	U			U	U	U	U		13
C. gloeopara	U	U	U		U	U	U			U	U	U	U		13
C. inflexa	U	U	U		U	U	U			U	U	U	U		13
C. kakosmos	U	U	U		U	U	U			U	U	U	U		13
C. mexicana	G	U	U		G	U	U			G	G	U	G		13
C. microsphaera acuta	U	U	U		U	U	U			U	U	U	U		13
C. microsphaerella	U	U	U		U	U	U			U	U	U	G		13
C. minuta	U	U	U		U	U	G			U	U	U	U		13
C. moewusii	U	U	U		U	U	U			U	U	U	G		13
C. moewusii rotunda	G		U		G	U				G	G	U?			20
C. mundana						G						U?			20
C. mundana astigmata	G	U	U		G	G	U			G	G	U	G		13
C. mutabilis		U	U			U	U					U			13
C. palla	U	G	U		U	U	U			G	G	G	G	U, U[e]	8
C. peterfi		G				U						U			13

[a] Guanine.
[b] Four strains.
[c] By male strain; poorly utilized by female strain.
[d] Glucosamine.
[e] Hypoxanthine.

Table 4 (continued)
OTHER NITROGENOUS SUBSTANCES

Species (synonym)	Acetamide	Adenine	Ammonium	Casein hydrolysate	Cytosine	Nitrate	Nitrite	Nitrogen	Peptone	Succinamide	Uracil	Urea	Uric acid	Miscellaneous	Ref.
Chlorophyceae (continued)															
C. pulsatilla	U				U	U				U	U	U		U[f]	19
C. radiata	U	U			U	U	U			U[g]	U	U	U		13
C. reinhardii	U	U	U		U	U	U				U[g]	U[g]	U	U[g]	12, 13, 53
C. sectilis	U	U	U[h]		U	U	U			U	U	U	U[g]		13
C. typhlos	U	U	U			U	U				U	U	U		13
Chlamydomonas sp.												U			41
Chlorella autotrophica			U	U		u									55
C. candida			U	U		U									55
C. ellipsoidea			U	U		U									55
C. emersonii			U	U		U									55
C. emersonii globosa			U	U		U									55
C. fusca			U	U		U									55
C. fusca vacuolata			U	U		U									55
C. infusionum			U	U		U									55
C. infusionum auxenophila			U	U		U									55
C. miniata			U	U		U									55
C. mutabilis			U	U		U									55
C. nocturna			U	U		U									55
C. photophila			U	U		U									55

f Organic nitrogen; individual compounds not listed.

g Xanthine.

h As ammonium nitrate.

Table 4 (continued)
OTHER NITROGENOUS SUBSTANCES

Species (synonym)	Acetamide	Adenine	Ammonium	Casein hydrolysate	Cytosine	Nitrate	Nitrite	Nitrogen	Peptone	Succinamide	Uracil	Urea	Uric acid	Miscellaneous	Ref.
Chlorophyceae (continued)															
C. pringsheimii			U	U		U									55
C. protothecoides			U	U		U̶				U					3, 55
C. protothecoides communis			U	U		U̶									55
C. protothecoides galactophila			U	U		U̶									55
C. protothecoides mannophila			U	U		U						U	U	U^{g,i}	7, 12, 54
C. pyrenoidosa		U	U	U		U									55
C. regularis			U	U		U									55
C. regularis aprica			U	U		U									55
C. regularis imbricata			U	U		U									55
C. saccharophila			U	U		U									55
C. simplex			U	U		U									55
C. sorokiniana			U	U		U									55
C. vannielii			U	U		U									55
C. variabilis			U̶	U̶		U̶				U		U	U	U^g	3, 6, 12, 55
C. vulgaris			U	U		U									55
C. vulgaris luteoviridis			U			U									69
Chlorella sp.			U	U		U						U			4, 5
Cyanidium caldarium[j]			U			U?		U̶				U			40
Diplostauron elegans			U			U						U			26
Dunaliella salina			U			U						U	U	U̶^d, U^e	8, 41, 69
D. tertiolecta			U			U	U					U	U		26
D. viridis			U			U						U	U	U^k	10
Furcilla stigmatophora	U̶	U	U̶		U̶	U̶				U		U	U̶		13
Gloeocystis gigas	U	U	U		U̶	U	U			U	U̶	U	U̶		13
G. maxima		U	U			U	U			U	U̶	U	U	U^a	19
Haematococcus (Balticola) buetschlii		U	U								U̶^e	U	U	U^a	19
H. droebakensis	U	U	U								U̶	U	U	U^a	19
H. droebakensis pluvialis	U		U												48
Lobomonas pyriformis			U			U						U			52
Nannochloris atomus			U			U	U					U	U		8, 17
N. oculata			U			U	U					U		U^{d,e}	

i Hypoxanthine, xanthine.
j Genus of uncertain taxonomic position.[5,6]
k Protease, peptone.
l By one strain; not utilized by one strain.

Table 4 (continued)
OTHER NITROGENOUS SUBSTANCES

Chlorophyceae (continued)

Species (synonym)	Acetamide	Adenine	Ammonium	Casein hydrolysate	Cytosine	Nitrate	Nitrite	Nitrogen	Peptone	Succinamide	Uracil	Urea	Uric acid	Miscellaneous	Ref.
Pandorina morum			U			U						U			49
Platydorina caudata			U			U	U					U			28
Polytoma caudatum			U			U[a]			U						1
P. obtusum			U			U			U						1
P. ocellatum			U			U			U						1
P. uvella			U			U[a]			U						1
Polytomella caeca			U						U						1
Prototheca zopfii			u						U						3, 12
Scenedesmus obliquus		U	U		U[a]	U	U			U	U[a]	U	U	U[g]	13
Sphaerobotrys fluviatilis	U	U	U		U[a]	U	U			U	U[a]	U	U		13
Sporotetras pyriformis	U	U[a]	U			U				U[a]	U[a]	U	U	U[a]	19
Stephanosphaera pluvialis						U	U					U	U[a]		52
Stichococcus cylindricus?													U		

Table 4 (continued)
OTHER NITROGENOUS SUBSTANCES

REFERENCES

1. Albritton, E. C., *Standard Values in Nutrition and Metabolism*, W. B. Saunders, Philadelphia, 1954.
2. Aldrich, D. V., *Science*, 137, 988, 1962.
3. Algeus, S., *Physiol. Plant.*, 3, 370, 1950.
4. Allen, M. B., *Arch. Mikrobiol.*, 17, 34, 1952.
5. Allen, M. B., *Arch. Mikrobiol.*, 32, 270, 1959.
5A. Allen, M. B., in *Symposium on Marine Microbiology*, Oppenheimer, C. H., Ed., Charles C Thomas, Springfield, Ill., 1963, 85.
6. Arnow, P., Oleson, J. J., and Williams, J. H., *Am. J. Bot.*, 40, 100, 1953.
7. Ammann, E. C. B. and Lynch, V. H., *Biochim. Biophys. Acta*, 87, 370–379, 1964.
8. Antia, N. J., Berland, B. R., Bonin, D. J., and Maestrini, S. Y., *J. Mar. Biol. Assoc. U.K.*, 55, 519–539, 1975.
9. Antia, N. J. and Chorney, V., *J. Protozool.*, 15, 198–201, 1968.
10. Belcher, J. H., *Arch. Mikrobiol.*, 58, 181–185, 1967.
11. Belcher, J. H. and Fogg, G. E., *Arch. Mikrobiol.*, 30, 17, 1958.
12. Birdsey, E. and Lynch, V., *Science*, 137, 763, 1962.
13. Cain, J., *Can. J. Bot.*, 43, 1367, 1965.
14. Carpenter, E. J., Remsen, C. C., and Watson, S. W., *Limnol. Oceanogr.*, 17, 265–269, 1972.
15. Cramer, M. and Myers, J., *Arch. Mikrobiol.*, 17, 384–402, 1952.
16. De, P. K., *Proc. R. Soc. London Ser. B*, 127, 121, 1939.
17. Droop, M. R., *J. Mar. Biol. Assoc. U.K.*, 34, 229, 1955.
18. Droop, M. R., *J. Mar. Biol. Assoc. U.K.*, 37, 323, 1958.
19. Droop, M. R., *Rev. Algol.*, 5(4), 247, 1961.
20. Eppley, R. W. and Maciasr, F. M., *Physiol. Plant.*, 15, 72, 1962.
21. Eppley, R. W. and Rogers, J. N., *J. Phycol.*, 6, 344–351, 1970.
22. Fay, P., *J. Gen. Microbiol.*, 39, 11–20, 1965.
23. Fogg, G. E. and Wolfe, M., *Symp. Soc. Gen. Microbiol.*, 4, 99, 1954.
24. Fries, L., *Physiol. Plant.*, 16, 695–708, 1963.
25. Fries, L. and Pettersson, H., *Br. Phycol. Bull.*, 3(3), 417–422, 1968.
26. Gibor, A., *Biol. Bull.*, 111, 223, 1956.
27. Guillard, R. R. L., in *Symposium on Marine Microbiology*, Oppenheimer, C. H., Ed., Charles C Thomas, Springfield, Ill., 1963, 93.
28. Harris, D. O., *J. Phycol.*, 5, 205–210, 1969.
29. Hayward, J., *Physiol. Plant.*, 18, 201–207, 1964.
30. Holz, G. G., *J. Protozool.*, 1, 114, 1954.
31. Iwasaki, H., *Bull. Jpn. Soc. Sci. Fish.*, 37, 606–609, 1971.
32. Iwasaki, H. and Sasada, K., *Bull. Jpn. Soc. Sci. Fish.*, 35, 943–944, 1969.
33. Kenyon, C. N., Rippka, R., and Stanier, R. Y., *Arch. Mikrobiol.*, 83, 216–236, 1971.
34. Khoja, T. and Whitton, B. A., *Arch. Mikrobiol.*, 79, 280–282, 1971.
35. Kiyohara, T. et al., *J. Gen. Appl. Microbiol.*, 6, 176, 1960.
36. Kratz, W. A. and Myers, J., *Am. J. Bot.*, 42, 282, 1955.
37. Lewin, J. and Lewin, R. A., *J. Gen. Microbiol.*, 46, 361, 1967.
38. Lui, N. S. T. and Roels, O. A., *Arch. Biochem. Biophys.*, 139, 269–277, 1970.
39. Lui, N. S. T. and Roels, O. A., *J. Phycol.*, 8, 259–264, 1972.
40. Lynn, R. L. and Starr, R. C., *Arch. Protistenkd.*, 112, 283–302, 1970.
41. McCarthy, J. J., The Role of Area in Marine Phytoplankton Ecology, Ph.D. thesis, University of California, San Diego, 1971.
42. McCarthy, J. J., *J. Phycol.*, 8, 216–222, 1972.
43. McLaughlin, J. J. A., *J. Protozool.*, 5, 75, 1958.
44. Miller, J. D. A. and Fogg, G. E., *Arch. Mikrobiol.*, 28, 1, 1957.
45. Miller, J. D. A. and Fogg, G. E., *Arch. Mikrobiol.*, 30, 1, 1958.
46. Neilson, A. H. and Doudoroff, M., *Arch. Mikrobiol.*, 89, 15–22, 1973.
47. North, B. B. and Stephens, G. C., *Biol. Bull.*, 133, 391–400, 1967.
48. Osterud, K. L., *Physiol. Zool.*, 19, 19, 1946.
49. Palmer, E. G. and Starr, R. C., *J. Phycol.*, 7, 85–89, 1971.
50. Pintner, I. J. and Provasoli, L., *Bull. Misaki Mar. Biol. Inst. Kyoto Univ.*, 12, 25–31, 1968.
51. Pintner, I. J. and Provasoli, L., *J. Gen. Microbiol.*, 18, 190, 1958.
52. Ryther, J. H., *Biol. Bull.*, 106, 198, 1954.

Table 4 (continued)
OTHER NITROGENOUS SUBSTANCES

53.   Sager, R. and Granick, S., *Ann. N.Y. Acad. Sci.,* 56, 831–838, 1953.
54.   Samejima, H. and Myers, J., *J. Gen. Microbiol.,* 18, 107, 1958.
55.   Shihira, I. and Krauss, R. W., *Chlorella: Physiology and Taxonomy of 41 Isolates,* University of Maryland Press, College Park, 1965.
56.   Silva, P., in *Physiology and Biochemistry of Algae,* Lewin, R. A., Ed., Academic Press, New York, 1962, 827.
57.   Stanier, R. Y., Kunisawa, R., Mandel, M., and Cohen-Bazire, G., *Bacteriol. Rev.,* 35, 171–205, 1971.
58.   Stewart, W. D. P., Fitzgerald, F. P., and Burris, R. H., *Arch. Mikrobiol.,* 62, 336–348, 1968.
59.   Stewart, W. D. P., *Ann. Bot.* (London), 26, 439, 1962.
60.   Swift, M. J. and McLaughlin, J. J. A., *Ann. N.Y. Acad. Sci.,* 175, 577–600, 1970.
61.   Thomas, W. H., *Limnol. Oceanogr.,* 11, 393, 1966.
62.   Turner, M. F., *Br. Phycol. J.,* 5, 15–18, 1970.
63.   Turner, M. F., in Scottish Marine Biological Association, Report of the Council for 1969/1970, Oban, Scotland, 1970, 13–14.
64.   Van Baalen, C. and Marler, J. E., *J. Gen. Microbiol.,* 32, 457–463, 1963.
65.   Van Baalen, C., *Bot. Mar.,* 4, 129, 1962.
66.   Watanabe, A., *Arch. Biochem. Biophys.,* 34, 50, 1951.
67.   Watanabe, A. and Kiyohara, T., in *Microalgae and Photosynthetic Bacteria,* Japanese Society of Plant Physiology, Tokyo, 1963, 189–196.
68.   Watanabe, A., Nishigaki, S., and Konishi, C., *Nature,* 168, 748, 1951.
69.   Wheeler, P. A., North, B. B., and Stephens, G. C., *Limnol. Oceanogr.,* 19(2), 249–259, 1974.
70.   Williams, A. E. and Burris, R. H., *Am. J. Bot.,* 39, 340, 1951.
71.   Wyatt, J. T., Martin, T. C., and Jackson, J. W., *Phycologia,* 13, 153–161, 1973.
72.   Wyatt, J. T. and Silvey, J. K. G., *Science,* 165, 908–909, 1969.
73.   Singh, R. N., *Indian J. Agric. Sci.,* 12, 743–756, 1942.
74.   Venkataraman, G. S., *Indian J. Agric. Sci.,* 32, 22–24, 1962.
75.   Winter, G., *Beitr. Biol. Pflanz.,* 23, 295–335, 1935.
76.   Allen, M. B. and Arnon, D. I., *Plant Physiol.,* 30, 366–372, 1955.
77.   Bortels, H., *Arch. Mikrobiol.,* 11, 155–166, 1940.
78.   Davis, E. B., Tischer, R. G., and Brown, L. R., *Physiol. Plant.,* 19, 823–826, 1966.
79.   Cameron, R. E. and Fuller, W. H., *Soil Sci. Soc. Am. Proc.,* 24, 353–356, 1960.
80.   Stewart, W. D. P., in *Fertility of the Sea,* Costlow, J. D., Ed., Gordon & Breach, London, 1971, 537–564.
81.   Watanabe, A., *Arch. Biochem. Biophys.,* 34, 50–55, 1951.
82.   Watanabe, A., *J. Gen. Appl. Microbiol.,* 5, 21–27, 1959.
83.   Taha, M. S., *Microbiology* (USSR), 32, 421–425, 1963.
84.   Taha, M. S., *Microbiology* (USSR), 33, 352–358, 1964.
85.   Fay, P. and Fogg, G. E., *Arch. Mikrobiol.,* 42, 310–321, 1962.
86.   Ventkataraman, G. S., *Proc. Natl. Acad. Sci. India,* 31, 100–104, 1961.
87.   Pankow, H., *Naturwissenschaften,* 51, 274–275, 1964.
88.   Rippka, R., Neilson, A., Kunisawa, R., and Cohen-Bazire, G., *Arch. Mikrobiol.,* 76, 341–348, 1971.
89.   Silvester, W. B. and Smith, D. R., *Nature,* 224, 1231, 1969.
90.   Pankow, H. and Martens, B., *Arch. Mikrobiol.,* 48, 203–212, 1964.
91.   Singh, R. N., Physiology and Biochemistry of Nitrogen Fixation by Blue-green Algae, Final Tech. Rep. 1967–1972, Dep. of Botany, Banaras Hindu University, Varanasi-5, 1972.
92.   Stewart, W. D. P. and Lex, M., *Arch. Mikrobiol.,* 73, 250–260, 1970.
93.   Venkataraman, G. S., *Indian J. Agric. Sci.,* 31, 213–215, 1961.
94.   Pattnaik, H., *Ann. Bot.* (London), 30, 231–238, 1966.

QUALITATIVE REQUIREMENTS AND UTILIZATION OF NUTRIENTS: FUNGI

D. J. D. Nicholas

INTRODUCTION

The subject of nutrients for fungi can be considered under two main headings, namely mineral nutrients and vitamins. Before considering these topics in detail, however, it is important to define the concept of an "essential element" for growth. Arnon[1] has proposed the following criteria for an essential element for the growth of higher plants and these are also applicable to fungi: (1) the organism is unable to complete its life cycle without the element; (2) the element must have an effect specific to the nutrient and cannot be replaced by another; and (3) the element must be directly involved in the organism's metabolism, not merely indirectly involved through correction of unfavorable conditions in the growth medium, e.g., change in pH. However, the second criterion has been questioned because it is known that one or two metals function at the same point in the metabolism of fungi. Nicholas[12] has proposed that the term "metabolism or functional nutrient" is preferable to that of "essential nutrient". This would include any nutrient which functions in some precise way in metabolism, irrespective of whether or not it is completely specific or indispensable for growth.

Two approaches have been used to establish nutrient requirements in fungi: (1) pure culture experiments, whereby nutrients are shown to be indispensable for growth and (2) biochemical studies to determine a function for the nutrients in metabolism. These methods have been used in a complementary way to establish which nutrients are essential for microorganisms. An example of this is the observation that molybdenum deficiency is more severe when fungi are grown with nitrate rather than ammonia as a sole source of nitrogen. This led to the discovery that molybdenum is a component of the nitrate reductase enzymes in microorganisms and green plants.[6,11,13-16]

MINERAL NUTRIENTS

At the present time, the mineral nutrients found in fungi may be subdivided into two arbitrary groups: (1) essential elements (N, P, K, Mg, Ca, S, Fe, Cu, Zn, Mn, and Mo), known to be indispensable for the growth of at least one organism, and (2) other elements (Ni, Ti, Se, Co, Pb, Ag, Au, Br, Cl, I, B, etc.), often present in the ash of fungi, but not yet shown to be essential for their metabolism. Not all of the elements listed under (1) are universally required, but all of them have been found to be necessary for some type of fungus.

The essential mineral elements are sometimes subdivided into two main groups: (1) the major or macronutrients and (2) the micronutrients, trace elements or oligoelements; as the names suggest, these groups reflect a large or a small requirement for the elements. On this basis, the major elements required by fungi include N, P, K, and Mg. Contrary to the higher plant requirements, Ca is essential in micro amounts only, along with Fe, Cu, Zn, Mn, and Mo. Vanadium was claimed by Bertrand[1a] to be essential for the growth of *Aspergillus niger,* but this small response has not been confirmed by others. Equally, Steinberg[19] recorded increases in the growth of this fungus on adding gallium and even scandium (the latter when glycerol was the carbon source), but these results have not been reproduced.[20]

Fungi accumulate minerals against a concentration gradient and this necessitates the expenditure of energy, usually in the form of ATP. There is good evidence for carrier systems located in semipermeable membranes of fungal cells and involved in the transport of cations across membranes.[5]

Major Elements

Nitrogen — Nitrates and ammonia may be assimilated by fungi into cell-nitrogen compounds. Nitrates are readily utilized by many fungi, but some groups which do not assimilate this form include the Basidiomycetes, Saprolegniaceae, and Blastocladiales.[3] Nitrite is not readily assimilated by a number of fungi, as it can be toxic, especially in acid media where the free, unionized nitrous acid is inhibitory. However, some fungi (e.g., *Fusarium, Coprinus, Phymatotrichum, Scopulariopsis,* and *Rhizophlyctis*) utilize nitrite.[3] The enzymes nitrate and nitrite reductases in fungi are induced by nitrate and repressed by ammonia.[14-16] Ammonium nitrate is utilized by numerous fungi (e.g., *Neurospora crassa* and *Aspergillus niger*); the ammonium ion is utilized first and then nitrate is taken up with the concomitant production of the nitrate and nitrite reductase enzymes, reducing nitrate to ammonia. In other fungi (e.g., *Helminthosporium gramineum*) there is a different relation, wherein there is an assimilation of nitrate even in the presence of ammonium.[4,9] Brief mention should be made of the capacity of some Actinomycetes to fix N_2 gas in association with root nodules of nonlegume plants (e.g., *Alnus, Casuarina, Myrica,* etc.).[2] Although specific amino acid requirements may be produced in fungi by mutagenic agents, they are uncommon in nature. Fungi produce a range of enzymes (exopeptidases) which degrade low-molecular-weight peptides. Some fungi decompose proteins in soil. In pure cultures, it appears that saprophytic fungi and actinomycetes utilize gelatin, casein, and albumin as nitrogen sources. Keratin is a source of nitrogen for dermatophytic fungi.

Phosphorus — Orthophosphate is the usual source of the element, and a mixture of the monobasic and dibasic phosphate is often used to buffer the culture solutions against the production of organic acids (e.g., citric acid by *Aspergillus niger*). However, a number of fungi can utilize organic sources of phosphorus such as phytic acid, nucleotide or nucleoside phosphates, and even casein. A deficiency of phosphate results in metabolic disorders including a depleted utilization of glucose, as glucose is essential for the production of nucleotide phosphates which are involved in phosphorylating reactions in fungi, including the production of RNA and DNA in living systems. The absorption of phosphate is an energy-requiring process and is, therefore, dependent on respiration. Many fungi produce phosphatase enzymes, which hydrolyze organic phosphates to inorganic forms and are then absorbed by the hyphae.

Potassium — Potassium is normally required by all fungi in similar amounts to carbohydrates. It is claimed that sodium may have a sparing effect on potassium in a number of *Aspergilli,* but the degree of replacement is small. The function of potassium in metabolism is not well defined, but a few enzymes in the glycolytic pathway are dependent on this nutrient for maximal activity.

Sulfur — Fungi usually absorb and utilize sulfate, as they activate it to adenosine 5′-sulfatophosphate (APS); in addition, some phosphorylate this sulfur nucleotide to adenosine 3′-phosphate-5′-sulfatophosphate (PAPS). These sulfur nucleotides are reduced to the level of thiol, which is incorporated into cysteine via *O*-acetylserine. Organic compounds known to contain sulfur include the amino acids, cysteine, cystine and methionine, the peptide glutathione, the vitamins thiamine and biotin, and antibiotics such as penicillin and gliotoxin. There is evidence that some members of the fungal groups, such as *Phycomyces, Saprolegniales,* and *Blastocladiales,* cannot utilize sulfate, but rely on reduced forms such as sulfur, sulfide, or even cysteine and methionine as sources of the element.

Magnesium — This is indispensable for all fungi since it is required for many phosphorylating reactions involving ATP. However, Mn can substitute for Mg in some of these reactions. The absorption of the element is slower at neutral than at acid pH values, but at high pH it is precipitated as a phosphate and becomes unavailable to the fungus.

Trace Elements

A summary of the data on trace element requirements for the fungi is presented in Table 1. It is of interest that calcium, which is a macronutrient for higher plants, is only required in relatively small amounts by fungi.

VITAMINS

In addition to mineral elements, fungi require minute amounts of specific organic substances for growth. Some fungi synthesize these from basic materials, whereas others require an external supply. Growth factors such as amino acids, purines, pyrimidines, etc. are not included in this section because only vitamin requirements will be considered. These compounds usually serve as coenzymes in cellular processes and some are components of nucleic acids such as adenine.

The early work on the vitamin requirements of fungi has been reviewed by Janke,[7] Robbins and Kavanagh,[17] Schopfer,[18] Knight,[8] and Cochrane.[3] Many fungi have been used as test organisms for the bioassay of vitamins. The data in Table 2 summarize the information on vitamin requirements for the fungi.

Table 1
ESSENTIAL TRACE ELEMENTS FOR FUNGI

Element	Range of requirement (μg/ml)	Fungi known to require nutrient	Function in metabolism
Fe	0.1–0.3	All	Catalase, peroxidase, cytochromes, nonheme Fe compounds, aspergillin in *Aspergillus*, purcherrimin in *Torulopsis*
Zn	0.01–1.0	All	Component of a number of dehydrogenases, e.g., glutamic, alcohol
Cu	0.01–0.2	All	Polyphenol oxidases, cytochrome oxidase
Mn	0.01–0.1	All	Utilization of glucose enzymes of citric acid cycle
Mo	0.001–0.05	All when utilizing nitrates	Component of nitrate reductase
		Actinomycetes in association with nodules in nonlegumes	Nitrogen-fixing enzyme nitrogenase
Ca	0.5–1.0	Species in about 19 genera representing all the major groups of fungi; strontium replaces Ca in *Allomyces arbuscula*	Not known

Table 2
VITAMIN REQUIREMENTS OF FUNGI

Vitamin	Fungi known to require vitamin	Function in metabolism
Thiamine	Vitamin most generally required by fungi, e.g., *Phytophthora,* yeasts	Thiamine pyrophosphate – coenzyme in decarboxylation reactions, e.g., pyruvic acid
Biotin	Wide range of fungi, e.g., *Neurospora crassa,* yeast	Synthesis of aspartic acid, i.e., carboxylation of pyruvic acid to oxalacetate and its conversion to aspartate
Pyridoxine	*Ceratostemella* sp.	Pyridoxal-5-phosphate – coenzyme for enzymes in amino acid metabolism, e.g., decarboxylases, transaminases
Riboflavin	Rarely required by fungi in nature; mutant strains; *Dictyostelium* has an absolute requirement	Riboflavin 5'-phosphate (FMN) and flavin adenine dinucleotide (FAD); important factors in respiratory chain enzymes as well as for the assimilatory nitrate and nitrite reductase enzymes
Nicotinic acid	*Microsporum* sp., *Blastocladia* sp., *Venturia* sp., *Glomerella* sp.	Coenzymes, NAD, NADP – for dehydrogenase enzymes
Inositol	Some yeasts, *Rhizopus* sp., *Sclerotinia* sp., *Trichophyton* sp.	Inositol polyphosphates, biosynthesis of phospholipids
Pantothenic acid	Yeasts, *Polyporus* sp., *Penicillium* sp.	Constituent of coenzyme A involved in acyl transfer reactions
Para-aminobenzoic acid	*Blastocladia* sp., yeasts	Introduction of 1-carbon intermediates into other compounds

REFERENCES

1. **Arnon, D. I.,** Criteria of essentiality of inorganic micronutrients for plants, in *Trace Elements in Plant Physiology,* Wallace, T., Ed., Chronica Botanica, Waltham, Mass., 1950, 31–39.

1a. **Bertrand, D.,** Importance of the trace element vanadium for *Aspergillus niger, C. R. Acad. Sci.,* 218, 254–257, 1941.

2. **Bond, G.,** Fixation of nitrogen in non-legume root nodule plants, in *13th Symp. Soc. for Experimental Biology,* Porter, H. K., Ed., Cambridge University Press, Cambridge, 1959, 59–72.

3. **Cochrane, V. W.,** *Physiology of Fungi,* John Wiley & Sons, New York, 1958.

4. **Converse, R. H.,** The influence of nitrogenous compounds on the growth of *Helminthosporium gramineum* in culture, *Mycologia,* 45, 335–344, 1953.

5. **Foulkes, E. C.,** Cation transport in yeast, *J. Gen. Physiol.,* 39, 687–704, 1956.

6. **Hewitt, E. J.,** The metabolism of micronutrient elements in plants, *Biol. Rev. Cambridge Philos. Soc.,* 34, 333–377, 1959.

7. **Janke, A.,** Die Wuchsstoff-Frage in der Mikrobiologie, *Zentralbl. Bakteriol. Parasitenkd. Infektionskr. Hyg. Abt. 2,* 100, 409–459, 1939.

8. **Knight, B. C. J. G.,** Growth factors in microbiology. Some wider aspects of nutritional studies with microorganisms, *Vitam. Horm.* (N.Y.), 3, 108–228, 1945.

9. **Morton, A. G. and MacMillan, A.,** The assimilation of nitrogen from ammonium salts and nitrate by fungi, *J. Exp. Bot.,* 5, 232–252, 1954.

10. **Nicholas, D. J. D.,** The use of fungi for determining trace metals in biological materials, *Analyst* (London), 77, 629–642, 1952.

11. **Nicholas, D. J. D.,** Metallo-enzymes in nitrate assimilation of plants with special reference to microorganisms, in *13th Symp. Soc. for Experimental Biology,* Porter, H. K., Ed., Cambridge University Press, Cambridge, 1959, 1–23.

12. **Nicholas, D. J. D.,** Minor mineral nutrients, *Annu. Rev. Plant Physiol.,* 12, 63–90, 1961.

13. **Nicholas, D. J. D.,** The metabolism of inorganic nitrogen and its compounds in microorganisms, *Biol. Rev. Cambridge Philos. Soc.,* 38, 530–568, 1963.

14. **Nicholas, D. J. D. and Nason, A.,** Molybdenum and nitrate reductase II. Molybdenum as a constituent of nitrate reductase, *J. Biol. Chem.,* 207, 353–354, 1954.

15. **Nicholas, D. J. D. and Nason, A.,** Mechanism of action of nitrate reductase from *Neurospora, J. Biol. Chem.,* 211, 183–197, 1954.

16. **Nicholas, D. J. D., Nason, A., and McElroy, W. D.,** Molybdenum and nitrate reductase. I. Effect of molybdenum deficiency on the *Neurospora* enzyme, *J. Biol. Chem.,* 207, 341–351, 1954.

17. **Robbins, W. J. and Kavanagh, V.,** Vitamin deficiencies of the filamentous fungi, *Bot. Rev.,* 8, 411–471, 1942.

18. **Schopfer, W. H.,** *Plants and Vitamins,* Chronica Botanica, Waltham, Mass., 1943, 293.

19. **Steinberg, R. A.,** The essentiality of gallium to growth and reproduction of *Aspergillus niger, J. Agric. Res.,* 57, 569–574, 1938.

20. **Steinberg, R. A.,** Growth of fungi in synthetic nutrient solutions, *Bot. Rev.,* 15, 327–350, 1939.

QUALITATIVE REQUIREMENTS AND UTILIZATION
OF NUTRIENTS: YEAST

E. Oura and H. Suomalainen

The name "yeasts" is traditionally restricted to those fungi in which the unicellular form is predominant; it does not refer to any taxonomically uniform group. Although forming a comparatively small group, some 350 yeast species belonging to 39 genera are recognized.[1] A large volume of information has accumulated about the nutritional requirements of yeasts as a result of their extensive use as research objects and in industrial production. The scope of this review is to some extent necessarily restricted because of the relatively large number of yeast species and the multitude of possible nutrients.

We have included a representative species (referred to by Lodder[2] as the "type species of the genus") for each genus, and a representative of the genuinely new genera.[3] In addition, a few other species that are often used in investigations have been selected. If the data in this review are to be applied to species not included here, it must be stressed that even closely related species within a genus can vary in their nutritional requirements.[4,5] As for nomenclature, we have added in parentheses the name used in the reference cited if this has not been adopted as such by Lodder[2] or by Barnett and Pankhurst.[3]

It has not been possible to include data on all the compounds that have been reported in the literature, but the most common and important have been selected. In order to keep the number of references within bounds, most of the publications cited deal with more than one yeast species and more than one compound. In certain cases results have not been set out in tables; instead, a reference is given to the original publication.

Some general reviews of the nutritional requirements of yeasts have appeared.[6–11] Details that are not covered in this review may be obtained from these or other review articles referred to below.

Yeasts use minerals in the formation of structural components of the cell, to act as functional groups in enzymes, and as enzyme activators and protein stabilizers. All yeasts obviously require nitrogen, phosphorus, and sulfur. The ability of yeasts to utilize various inorganic and organic compounds as sources of sulfur and nitrogen is presented in Tables 1, 6, and 7. Yeasts usually satisfy their requirements for phosphorus with *ortho*-phosphate, but in a phosphate-free medium many yeasts can take up the necessary phosphate from phosphate-esters.[12] Yeasts require a source of the following minerals in order to grow: potassium and magnesium,[13,14] iron,[15,16] copper,[15–17] and zinc.[16,17]

Sodium does not appear to be necessary for growth,[9] but it and other alkali metals can partly satisfy the requirement for potassium.[13]

Chlorine is another element that appears to be essential for growth.[9] Although apparently not necessary for the growth of yeast cells, calcium does stimulate it.[18,19] The same has been shown to be true for low concentrations of manganese,[17] although the opposite has also been reported for both manganese and calcium.[16] According to Greaves et al.,[20] iodine, either as the element or as iodide, stimulates growth, but Olson and Johnson[16] were unable to reproduce this effect. The latter also reported that the addition of boron, cobalt, tin, or thallium to the medium had no effect on the yield of *Saccharomyces cerevisiae*. On the other hand, Richards[21] has shown that the presence of thallium stimulated growth, as did Hébert[22] with chromium. Various trace elements, selected as indicated by the ash composition of the yeast, are often included in the synthetic media on which yeasts are cultivated. Accordingly, traces of aluminium, boron, cobalt, molybdenum, nickel, and strontium are also frequently included in recipes. The literature is often vague and contradictory about which of these is necessary for growth and which can stimulate growth. The uncertainties arise from the difficulties in ensuring that the basal medium is entirely free of the elements under investigation and that some apparently essential elements can be partly or completely replaced by others.

It is to be noted that increasing the concentration of the trace elements above a certain, usually very low, level can strongly inhibit or even completely prevent growth (for example, see Reference 23). The mineral requirements of yeasts have been dealt with in a number of review articles (for example, References 9-11, 24, and 25).

Table 1
MINERALS: UTILIZATION OF SULFUR COMPOUNDS[a]

Yeast species	Sodium sulfate (Na_2SO_4)	Sodium bisulfate ($NaHSO_3$)	Sodium thiosulfate ($Na_2S_2O_3 \cdot 5 H_2O$)	Sodium sulfide ($Na_2S \cdot 9 H_2O$)	Cysteine	Cystine	Methionine	Glutathione	Ref.
1 Candida pseudotropicalis (Torula cremoris)	U	U	U	U		Ψ	U	U	26
2 Candida utilis (Torula utilis)	U	U	U	U	U	U	U	U	26
3 Pichia membranaefaciens (Willia belciga)	Ψ				U	U	u	U	26
4 Saccharomyces cerevisiae	U	U	U	U	Ψ	Ψ	u	Ψ	26
5 Saccharomyces cerevisiae[b]	U		U	Ψ	U	U	U	U	27
6 (Saccharomyces cerevisiae, str. anamensis)	U	U	U	U	Ψ	Ψ	u	U	26
7 Saccharomyces uvarum (Saccharomyces carlsbergensis var. mandshuricus I)	U	U	U	U	u	u	U	U	26

[a] U = utilized; u = poorly utilized; Ψ = not utilized.
[b] Sodium sulfite (Na_2SO_3): U; ammonium thiocyanate: Ψ; thiourea: u; homocysteine: U; ethionine: Ψ; biotin: Ψ; thiamine: Ψ.

Table 2
VITAMIN REQUIREMENTS[a]

Besides the vitamins presented in the table, in anaerobic conditions some yeasts also require ergosterol and unsaturated fatty acids,[28,29] which can, however, be replaced by slight aeration. Nicotinic acid is an essential growth factor for Saccharomyces cerevisiae in anaerobic conditions.[30,31] Besides the references mentioned in the table see also the review of Koser.[32]

	Yeast species	Number of strains	Thiamine	Riboflavin	Nicotinic acid	B_6	Biotin	Pantothenic acid	Folic acid	p-Aminobenzoic acid	Inositol	Ref.
1	Brettanomyces bruxellensis	1	E		E	E	E	E			E	33
2	Candida albicans	1	E		E	E	E	E			E	34
3		50	E		E	E	E	E			E	35
4		40	S		E	E	E	E			E	36
5	Candida guilliermondii	5	S		E	E	E	E			E	37
6		5	E		E	E	E	E			E	35
7		1	E	E	E	E	E	E			E	38
8		1	E	E	E	E	E	E			E	39
9		1	E		E	E	E	E			E	34
10	(Candida melibiosi)	7	S(3)		E(7)		E(1)				E(7)	40
11	Candida krusei	2	E(4)				E(6)	E				36
12		30	E(10)	E	E(2)	E(6)	E(26)	E			E	35
13	Candida lipolytica	1	S		E	E	E	E			E	38
14	Candida membranaefaciens	1	E		E	E	E	E			E	40
15	(Candida melibiosi var membranaefaciens)		E		S	E	E				S	40
16	Candida stellatoidea	4	S		S(1)	S(1)	E	S(1)			S(1)	37
17	Candida tropicalis	10	E		E(3)	E(3)	E	E(3)			E(3)	35
18		2	E	E	E	E	E	E		E	E	38
19		12[b]	S(1)		E	E	E	E			E	41
20	Candida utilis (Torulopsis utilis)		E		E	E	E	E			E	33
21	(Torula utilis)	1	E		E	E	E	E			E	34
22		1	E		E	E	E	E			E	36
23		1	S		E	E	E	E			E	40
24	Chryptococcus albidus	1	S(1)		S	S	S					42
25		6[b]	E(5)		E	E	E	E			E	41

[a] E = essential; Ẹ = not essential; S = stimulatory, supportive growth; I = inhibitory.

Table 2 (continued)
VITAMIN REQUIREMENTS[a]

	Yeast species	Number of strains	Thiamine	Riboflavin	Nicotinic acid	B_6	Biotin	Pantothenic acid	Folic acid	p-Aminobenzoic acid	Inositol	Ref.
26	Debaryomyces dekkeri		E		E	E	E	E			E	33
27	Hansenula anomala		E		E	E	E	E			E	33
28	Hansenula lambica	13[b]	E		E	E	E	E			E	41
29	Hansenula wingei	24	E		E	E	E	E			E	33
30	Kloeckera apiculata	2	S		E	E	S(1)	E			E	42
31	Kluyveromyces africanus	1	S	I	E	S	S	I	I	I	E	42
32	Kluyveromyces fragilis	1	E	I	E	S	E	E	I	I	E	42
33	(Saccharomyces fragilis)	1	E		E	E	E	E			E	34
34	Pachysolen tannophilus	2	E		E	E	S	E			E	33
35	Pichia belgica	1	E		E	E	E	E			E	42
36	Pichia kluyveri	1	E		E	E	E	E			E	33
37	Pichia membranaefaciens	1	E		E	E	E	E			E	33
38	Pichia pseudopolymorpha	1					S					42
39	Pichia ohmeri (Candida guilliermondii var. membranaefaciens)	1	E		E		E				E	40
41	Rhodotorula glutinis	78[b]	E		E	E	E	E	E	E	E	43
42	Rhodotorula glutinis	1[c]	S(9)		E	E	E	E	E	E	E	43
43	Rhodotorula rubra	131[b]	E(122)		E	E	E	E	E	E	E	43
44	Rhodotorula rubra	1[c]	E		E	E	E	E	E	E	E	43
45	Saccharomyces capensis	1	E(1)		E	S	S	S		E	E	44
46	Saccharomyces cerevisiae	10	E(9)			E	E	S(1) E(9)			E(4)	45
47	(Saccharomyces cerevisiae var. ellipsoideus)	2	E		E	E	E	E			E(6)	33
48	Saccharomyces chevalieri	1	E		E	E	E	E			E	34
49	Saccharomyces italicus (Saccharomyces chodati)	3	E		E	E	E	E			E	33
50	Saccharomyces uvarum	1	E	E	E	E	S	E			E	38
51		1	E		E	E	E	E			E	33
52	(Saccharomyces carlsbergensis)	21[d]	S(4) E(1) E(16)	S(1) E(20)	S(3) E(4) E(14)	S(6) E(1) E(14)	E	S(1) E(20)		S(2) E(19)	S(4) E(7) E(10)	46

[c]Terrestrial strains.
[d]Uracil and guanine were essential for 11 strains.

Table 2 (continued)
VITAMIN REQUIREMENTS[a]

	Yeast species	Number of strains	Thiamine	Riboflavin	Nicotinic acid	B_6	Biotin	Pantothenic acid	Folic acid	p-Aminobenzoic acid	Inositol	Ref.
53	Schizoblastosporion starkeyi-henricii		E		E	E	E	E			E	33
54	Schizosaccharomyces pombe	8	S(2) E(6)	S(2) E(6)	E(2) S(6)	S(1) E(7)	E(2) E(6)	E(7) E(1)			E	47
55		1	E		E	E	E	E			E	37
56	Torulopsis candida	1	E	E	E	E	E	E			E	38
57	Torulopsis colliculosa	1	E		E	E	E	E			E	48

Table 3
FERMENTATION OF CARBOHYDRATES[a]

The data on the ability of yeasts to produce carbon dioxide from different sugars are restricted to the information appearing in Lodder[2] and Barnett and Pankhurst.[3]

Yeast species	Cellobiose	Galactose	Glucose	Lactose	Maltose	Melezitose	Melibiose	Raffinose[b]	Sucrose	Trehalose	Inulin	α-Methyl-D-glucoside	Soluble starch	Maltotriose	Ref.
1 Ambrosiozyma cicatricosa	F̸	F̸	F	F̸	F̸	F̸	F̸	F̸	F̸	F̸	F̸		F̸		3
2 Brettanomyces bruxellensis	F̸	F̸	F	F̸	F̸	F̸	F̸	F̸	F̸	F̸	F̸	F	F̸		2
3 Brettanomyces claussenii	F̸	F̸	F	F̸	F̸	F̸	F̸	f	F̸	F̸	F̸	F			2
4 Bullera alba	F̸	F̸	F̸	F̸	F̸	F̸	F̸	F̸	F̸	F̸	F̸	F̸	F̸		3
5 Candida albicans	F̸	F̸	F	F̸	F̸	F̸	F̸	F̸	F̸	f	F̸				2
6 Candida boidinii	F̸	F̸	F	F̸	F̸	F̸	F̸	F̸	F̸	F̸	F̸	F̸	F̸		2
7 Candida guilliermondii var. guilliermondii	F̸	F̸	F	F̸	F̸	F̸	f	F̸	F̸	F̸	F̸	F̸			2
8 Candida krusei	F̸	F̸	F	F̸	F̸	F̸	F̸	F̸	F̸	F̸	F̸	F̸	F̸		2
9 Candida lipolytica var. lipolytica	F̸	F̸	F̸	F̸	F̸	f	F̸	F̸	F̸	F̸	F̸	F̸	F̸		2
10 Candida tropicalis	F̸	F̸	F	F̸	F̸	f	F̸	F̸	F̸	F̸					2
11 Candida utilis	F̸	F̸	F	F̸	F̸	F̸	F̸	F̸	F̸	F̸	f				2
12 Citeromyces matritensis	F̸	F̸	F	F̸	F̸	F̸	F̸	F̸	F̸	F̸					2
13 Chryptococcus albidus var. albidus	F̸	F̸	F	F̸	F̸	F̸	F̸	F̸	F̸	F̸	F̸	F̸	F̸		2
14 Chryptococcus neoformans	F̸	F̸	f	F̸	f	F̸	F̸	F̸	f	F̸	F̸		F̸		2
15 Debaryomyces hansenii		f	f	F̸	f			f	f	f					2
16 Dekkera bruxellensis	F̸	F̸	F	F̸	F̸	F	F̸	F̸	F̸	F̸	F̸	F	F̸		2
17 Endomycopsis capsularis	F̸	F̸	F	F̸	F̸	F̸	F̸	F̸	F̸	F̸	F̸				2
18 Endomycopsis fibuligera	F̸	F̸	F	F̸	F̸	F̸	F̸	f	F̸	F̸	F̸				2
19 Filobasidium floriforme	F̸	F̸	F̸	F̸	F̸	f	F̸	F̸	F̸	F̸	F̸		f		3
20 Hanseniaspora valbyensis	f	f	F	F̸	f			f	F̸	F̸	F̸	F̸	F̸		2
21 Hansenula anomala var. anomala	F̸	f	F̸	F̸	f			F̸	F̸	F̸					2
22 Hansenula bimundalis var. bimundalis	F̸	F̸	F̸	F̸	F̸			F̸	F̸						2
23 Hansenula capsulata	F̸	F̸	F̸	F̸	F̸			F̸	F̸						2
24 Hansenula fabianii	F̸	F̸	F̸	f	F̸			F̸	F̸						2
25 Hansenula saturnus var. saturnus	F̸	F̸	F̸	F̸	F̸			F̸	F̸						2
26 Kloeckera apiculata	F̸	F̸	F	F̸	F̸	F̸	F̸	F̸	F̸	F̸	F̸	F̸	F̸		2
27 Kluyveromyces bulgaricus	F̸	F̸	F	f	F̸	f	F̸	F̸	F̸	F̸	F	F̸	F̸		2
28 Kluyveromyces lactis	F̸	F̸	F	F	f	f	F̸	f	f	f	F̸	F̸	F̸		2
29 Kluyveromyces polysporus	F̸	F̸	F	F̸	F̸	F̸	F̸	F̸	F̸	F̸	F̸	F̸	F̸		2
30 Leucosporidium scottii	F̸	F̸	F	F̸	F̸	F̸	F̸	F̸	F̸	F̸	F̸	F̸	F̸		2
31 Lipomyces starkeyi	F̸	f	F	F̸	f			F̸	f	F̸	F̸		F̸		2
32 Lodderomyces elongisporus	F̸	F̸	F	f	F̸			F̸	f	F̸	F̸	F̸	F̸		2
33 Metschnikowia bicuspidata var. bicuspidata	F̸	F̸	F̸	F̸	F̸			F̸	F̸	F̸	F̸	F̸	F̸		2
34 Nadsonia fulvescens	F	F	F	F	F			F	F						2

[a] F = fermented; F̸ = not fermented; f = reaction varies.

[b] Raffinose is marked with F even when only one third fermented.

Table 3 (continued)
FERMENTATION OF CARBOHYDRATES[a]

Yeast species	Cellobiose	Galactose	Glucose	Lactose	Maltose	Melezitose	Melibiose	Raffinose[b]	Sucrose	Trehalose	Inulin	α-Methyl-D-glucoside	Soluble starch	Maltotriose	Ref.
35 Nematospora coryli		F	F	F	F	F		F	F	F	F				2
36 Oosporidium margaritiferum	f	f	F	F	f	f	f	F	F	F	F	F	F		2
37 Pachysolen tannophilus		f	F	F	f		f		f	f					2
38 Pichia farinosa		f	F	F	f				f						2
39 Pichia membranaefaciens	f	F	f	F	F	F	F	F	F	F	f	F	F		2
40 Pityrosporum pachydermatis	F	F	F	F	F	f	F	F	F	F	f	F	F		2
41 Rhodosporium toruloides	F	F	F	F	F		F	F	F	F	f	f	F		2
42 Rhodotorula glutinis var. glutinis	F	F	F	F	F	f	F	F	F	F	f	f	F		2
43 Rhodotorula rubra	F	F	F	F	F	f	F	F	F	F	f	f	F		2
44 Saccharomyces cerevisiae	F	F	F	F	F	f	F	F	F	F	f	f	F		2
45 Saccharomyces chevalieri	F	F	F	F	f	f	F	F	F	f	f	f	F		2
46 Saccharomyces kluyveri	F	F	F	F	f	f	F	F	f	f	f	f	F		2
47 Saccharomyces rosei	F	F	F	F	F	f	F	F	F	F	f	f	F		2
48 Saccharomyces rouxii	F	F	F	F	F	f	F	F	f	f	f	f	F		2
49 Saccharomyces uvarum	F	F	F	F	F	f	F	F	F	F	f	F	F		2
50 Schizoblastosporion starkeyi-henricii	F	F	F	F	F		F	F	f				F	F	2
51 Schizosaccharomyces octosporus	F	F	F	F	F	F	F	F	F		f		F	F	3
52 Schizosaccharomyces pombe		F	F	F	F		F	F	F		F				2
53 Schwanniomyces occidentalis		F	F	F	F		F	F	F	F	F		F		2
54 Selenotila intestinalis	F	F	F	F	F	F	F	F	F	F	F	F	F		2
55 Sporidiobolus johnsonii	F	F	F	F	F		F	F	F	F	F	F	F		2
56 Sporobolomyces roseus	F	F	F	F	F		F	F	F	F	F	F	F		2
57 Sterigmatomyces halophilus	F	F	F	F	F		F	F	F	F	F	F	F		3
58 Sympodiomyces parvus	F	F	F	F	F		F	F	F	F	F	F	F		2
59 Torulopsis colliculosa	F	F	F	F	F		F	F	F	F	F		F		2
60 Trichosporon cutaneum	F	F	F	F	F	f	F	F	F	F	F	F	F		2
61 Trigonopsis variabilis	F	F	F	F	F		F	F	F	F	F		f		2
62 Wickerhamia fluorescens		F	F	F	F				F		F	F	F		2
63 Wingea robertsii	f	f	F	F	F		F	F	F	F	F	F	F		2

Table 4
OXIDATIVE UTILIZATION OF CARBOHYDRATES[a,b]

The table is based on Lodder[2] and Barnett and Pankhurst.[3]

Yeast species	D-Arabinose	L-Arabinose	Cellobiose	Galactose	Glucose	Lactose	Maltose	Melezitose	Melibiose	Ref.
1 Ambrosiozyma cicatricosa	A̸	A	A	A̸	A	A̸	A	A	A	3
2 Brettanomyces bruxellensis	A̸	A̸	A̸	A̸	A	A̸	A	A	A̸	2
3 Brettanomyces claussenii	A̸	A̸	A	A	A	A	A	A	A̸	2
4 Bullera alba	A	A	A	a	A	A	A	A	a	2
5 Candida albicans	A̸	a	a	A	A	A̸	A	a	A	2
6 Candida boidinii	A̸	a	A̸	A̸	A	A̸	A̸	A̸	A̸	2
7 Candida guilliermondii var. guilliermondii	A	A	A	A	A	A̸	A	A	A	2
8 Candida krusei	A̸	A̸	A̸	A̸	A	A̸	A̸	A̸	A̸	2
9 Candida lipolytica var. lipolytica	A̸	A̸	A̸	A̸	A	A̸	a	A̸	A̸	2
10 Candida tropicalis	A̸	a	a	A	A	A̸	A	A	A̸	2
11 Candida utilis	A̸	A̸	A	A̸	A	A̸	A	A	A̸	2
12 Citeromyces matritensis	A̸	A̸	A̸	A̸	A	A̸	A	A̸	A̸	2
13 Chryptococcus albidus var. albidus	a	A	A	a	A	A	A	A	a	2
14 Chryptococcus neoformans	A	a	A	A	A	A̸	A	A	A̸	2
15 Debaryomyces hansenii	a	A	A	A	A	a	A	A	a	2
16 Dekkera bruxellensis	A̸	A̸	A̸	A̸	A	A̸	A	A	A̸	2
17 Endomycopsis capsularis	a	A̸	A	A̸	A	A̸	A	A̸	A̸	2
18 Endomycopsis fibuligera	A̸	A̸	A	A̸	A	A̸	A	A	A̸	2
19 Filobasidium floriforme	A	A	A	A	A	A̸	A	A	A̸	3
20 Hanseniaspora valbyensis	A̸	A̸	A̸	A̸	A	A̸	A̸	A̸	A̸	2
21 Hansenula anomala var. anomala	A̸	a	A̸	a	A	A̸	A	A	A̸	2
22 Hansenula bimundalis var. bimundalis	A	a	A	A̸	A	A̸	A	a	A̸	2
23 Hansenula capsulata	a	a	A	A̸	A	a	A	A	A̸	2
24 Hansenula fabianii	A̸	A̸	A	A̸	A	A̸	A	A	A̸	2
25 Hansenula saturnus var. saturnus	A̸	A̸	A	A̸	A	A̸	A̸	A	A̸	2
26 Kloeckera apiculata	A̸	A̸	A̸	A̸	A	A̸	A̸	A	A̸	2
27 Kluyveromyces bulgaricus	A̸	a	A̸	A	A	A	A̸	A̸	A̸	2
28 Kluyveromyces lactis	A̸	A̸	A	A	A	A	a	A	A̸	2
29 Kluyveromyces polysporus	A̸	A̸	A̸	A̸	A	A̸	A̸	A̸	A̸	2
30 Leucosporidium scottii	a	a	A	a	A	a	A	A	A̸	2

[a] Based on Lodder[2] and Barnett and Pankhurst.[3]

[b] A = assimilated; A̸ - not assimilated; a = reaction varies.

Table 4 (continued)
OXIDATIVE UTILIZATION OF CARBOHYDRATES[a,b]

Yeast species	D-Arabinose	L-Arabinose	Cellobiose	Galactose	Glucose	Lactose	Maltose	Melezitose	Melibiose	Ref.
31 Lipomyces starkeyi	a	A	a	A	A	a	A	A	A	2
32 Lodderomyces elongisporus	A	a	A	A	A	A	A	A	A	2
33 Metschnikowia bicuspidata var. bicuspidata	A	A	A	A	A	A	A	A	A	2
34 Nadsonia fulvescens	A	A	A	A	A	A	A	A	A	2
35 Nematospora coryli	A	A	A	a	A	A	A	A	A	2
36 Oosporidium margaritiferum	A	A	A	A	A	A	A	A	A	2
37 Pachysolen tannophilus	A	A	A	A	A	A	A	A	A	2
38 Pichia farinosa	A	a	a	A	A	a	A	A	A	2
39 Pichia membranaefaciens	A	A	A	A	A	A	A	A	A	2
40 Pityrosporum pachydermatis	A	A	A	A	A	A	A	A	A	2
41 Rhodosporium toruloides	A	A	a	A	A	A	A	A	A	2
42 Rhodotorula glutinis var. glutinis	a	a	a	a	A	A	A	A	A	2
43 Rhodotorula rubra	a	a	a	a	A	A	A	A	A	2
44 Saccharomyces cerevisiae	a	a	a	A	A	A	A	a	A	2
45 Saccharomyces chevalieri	A	A	A	A	A	A	A	A	A	2
46 Saccharomyces kluyveri	A	A	a	A	A	A	A	a	A	2
47 Saccharomyces rosei	A	A	A	A	A	A	a	A	A	2
48 Saccharomyces rouxii	A	A	A	a	A	A	A	A	A	2
49 Saccharomyces uvarum	A	A	A	A	A	A	A	a	A	2
50 Schizoblastosporion starkeyi-henricii	A	A	A	A	A	A	A	A	A	2
51 Schizosaccharomyces octosporus[c]	A	A	A	A	A	A	A	A	A	2
52 Schizosaccharomyces pombe[c]	A	A	A	A	A	A	A	A	A	2
53 Schwanniomyces occidentalis	A	A	A	A	A	A	A	A	A	3
54 Selenotila intestinalis	A	A	A	A	A	A	A	A	A	2
55 Sporidiobolus johnsonii	A	A	A	a	A	A	A	A	A	2
56 Sporobolomyces roseus	A	a	a	a	A	A	A	A	A	2
57 Sterigmatomyces halophilus	a	A	a	A	A	a	A	A	A	3
58 Sympodiomyces parvus	A	A	A	A	A	A	A	A	A	2
59 Torulopsis colliculosa	A	A	A	A	A	A		A	A	2
60 Trichosporon cutaneum	a	a	a	a	A	A	a	a	a	2
61 Trigonopsis variabilis	A	a	a	a	A	A	A	a	A	2
62 Wickerhamia fluorescens	A	A	A	A	A	A	A	A	A	2
63 Wingea robertsii	A	A	A	A	A	A	A	A	A	2

c Vitamins added until fivefold to the N-base.

Table 4 (continued)
OXIDATIVE UTILIZATION OF CARBOHYDRATES[a,b]

Yeast species	Raffinose[d]	L-Rhamnose	D-Ribose	L-Sorbose	Sucrose	Trehalose	D-Xylose	Inulin	Soluble starch	Ref.
1 Ambrosiozyma cicatricosa	A	A	A	A	A	A	A	A	A	3
2 Brettanomyces bruxellensis	A	A	a	A	A	A	A	A	A	2
3 Brettanomyces claussenii	A	A	a	A	A	A	A	a	A	2
4 Bullera alba	A	A	a	a	A	A	A	A	a	2
5 Candida albicans	A	A	A	a	A	A	A	A	A	2
6 Candida boidinii	A	A	A	A	A	A	A	A	A	2
7 Candida guilliermondii var. guilliermondii	A	a	a	A	A	A	A	A	a	2
8 Candida krusei	A	A	A	A	A	A	A	A	A	2
9 Candida lipolytica var. lipolytica	A	A	A	a	A	A	A	A	A	2
10 Candida tropicalis	A	A	A	a	A	A	A	A	A	2
11 Candida utilis	A	A	A	A	A	A	A	A	A	2
12 Citeromyces matritensis	A	A	A	A	A	A	A	A	A	2
13 Chryptococcus albidus var. albidus	A	a	a	a	A	A	A	A	a	2
14 Chryptococcus neoformans	A	A	A	a	A	A	A	A	a	2
15 Debaryomyces hansenii	A	a	a	a	A	A	A	a	A	2
16 Dekkera bruxellensis	A	A	a	A	A	A	A	A	A	2
17 Endomycopsis capsularis	A	A	a	A	A	A	A	A	A	2
18 Endomycopsis fibuligera	a	A	A	A	A	a	A	A	A	2
19 Filobasidium floriforme	A	A	A	A	A	A	A	A	A	3
20 Hanseniaspora valbyensis	A	A	A	A	A	A	A	A	A	2
21 Hansenula anomala var. anomala	A	A	a	A	A	A	a	A	A	2
22 Hansenula bimundalis var. bimundalis	A	A	A	A	A	A	A	A	A	2
23 Hansenula capsulata	A	a	A	A	A	A	A	A	A	2
24 Hansenula fabianii	A	A	A	A	A	A	A	A	A	2
25 Hansenula saturnus var. saturnus	A	A	A	A	A	A	A	a	A	2
26 Kloeckera apiculata	A	A	A	A	A	A	A	A	A	2
27 Kluyveromyces bulgaricus	A	A	A	A	A	a	A	A	A	2
28 Kluyveromyces lactis	a	A	A	a	A	A	a	a	A	2
29 Kluyveromyces polysporus	A	A	A	A	A	A	A	a	A	2
30 Leucosporidium scottii	A	A	a	A	A	A	A	A	A	2
31 Lipomyces starkeyi	A	a	a	A	A	A	A	A	A	2
32 Lodderomyces elongisporus	A	A	A	A	A	A	A	A	A	2
33 Metschnikowia bicuspidata var. bicuspidata	A	A	a	A	A	A	A	A	A	2

[d] Raffinose is marked with A even when only one third assimilated.

Table 4 (continued)
OXIDATIVE UTILIZATION OF CARBOHYDRATES[a,b]

Yeast species	Raffinose[d]	L-Rhamnose	D-Ribose	L-Sorbose	Sucrose	Trehalose	D-Xylose	Inulin	Soluble starch	Ref.
34 Nadsonia fulvescens	A	A	A	A	A	A	A	A	A	2
35 Nematospora coryli	a	A	A	A	A	A	A	A	A	2
36 Oosporidium margaritiferum	A	A	A	A	A	A	A	A	A	2
37 Pachysolen tannophilus	A	A	A	A	A	A	A	A	A	2
38 Pichia farinosa	A	A	A	a	A	a	a	A	A	2
39 Pichia membranaefaciens	A	A	A	a	A	A	a	A	A	2
40 Pityrosporum pachydermatis	A	A	A	A	A	A	A	A	A	2
41 Rhodosporium toruloides	A	A	A	a	A	A	A	A	A	2
42 Rhodotorula glutinis var. glutinis	a	a	a	a	A	A	A	a	A	2
43 Rhodotorula rubra	A	a	a	a	A	A	A	a	A	2
44 Saccharomyces cerevisiae	A	A	a	a	A	a	A	a	A	2
45 Saccharomyces chevalieri	A	A	A	A	A	A	A	a	A	2
46 Saccharomyces kluyveri	A	A	A	a	A	A	A	a	A	2
47 Saccharomyces rosei	A	A	A	a	A	A	A	A	A	2
48 Saccharomyces rouxi	A	A	A	A	a	a	A	a	A	2
49 Saccharomyces uvarum	A	A	A	A	A	a	A	a	A	2
50 Schizoblastosporion starkeyi-henricii	A	A	A	a	A	a	A	a	A	2
51 Schizosaccharomyces octosporus[c]	A	A	A	A	A	A	A	A	A	2
52 Schizosaccharomyces pombe[c]	A	A	A	A	a	A	A	a	A	2
53 Schwanniomyces occidentalis	A	A	A	A	A	A	A	A	A	3
54 Selenotila intestinalis	A	A	A	A	A	A	A	A	A	2
55 Sporidiobolus johnsonii	A	A	a	A	A	A	A	a	A	2
56 Sporobolomyces roseus	A	A	a	a	a	A	a	A	A	2
57 Sterigmatomyces halophilus	A	A	A	A	A	A	A	A	A	3
58 Sympodiomyces parvus	A	A	A	A	A	A	A	A	A	2
59 Torulopsis colliculosa	A	A	A	A	A	A	A	A	a	2
60 Trichosporon cutaneum	a	a	a	a	a	a	A	A	A	2
61 Trigonopsis variabilis	A	A	A	a	A	a	A	A	A	2
62 Wickerhamia fluorescens	A	A	A	A	A	A	A	A	A	2
63 Wingea robertsii	A	A	a	A	A	A	A	A		2

Table 5
OXIDATIVE UTILIZATION OF OTHER CARBON COMPOUNDS[a]

The table presents the capacity of some yeasts to assimilate various sugar derivatives, alcohols, and organic acids. See Abadie[49,50] for the ability of amino acids and some other nitrogen compounds to act simultaneously as sources of carbon and nitrogen. Shennan and Levi[51] have listed the ability of 150 yeasts to grow on hydrocarbons, and data for the assimilation of various hydrocarbons have been given by, for example, Munk et al.[52] Yeasts assimilate carbon dioxide,[53,54] and the presence of carbon dioxide has been shown to be essential for growth.[55]

Yeast species	Erythritol	Galactitol	D-Glucitol	Glycerol	D-Mannitol	α-Methyl-D-glucoside	Ribitol	Salicin	Ethanol	Methanol	Inositol	Citric acid	Lactic acid	Succinic acid	2-Keto-gluconic acid	5-Keto-gluconic acid	Glucono-δ-lactone	Ref.
1 Ambrosiozyma cicatricosa	A	Ⱥ	A	A	A	A	A	A	A		Ⱥ	A	A	A				3
2 Brettanomyces bruxellensis	Ⱥ	Ⱥ	a	A	Ⱥ	A	a	Ⱥ	A		Ⱥ	Ⱥ	a	a				2
3 Brettanomyces claussenii	Ⱥ	Ⱥ	a	A	Ⱥ	A	a	Ⱥ	A		Ⱥ	A	A	a				2
4 Bullera alba	a	a	A	a	a	A	a	A	A		A	A	A	A	A	A		2
5 Candida albicans	Ⱥ	Ⱥ	A	a	A	a	A	a	A		Ⱥ	a	A	a	A	A	A	2
6 Candida boidinii	Ⱥ	Ⱥ	A	A	A	Ⱥ	A	A	A	A[b]	Ⱥ	Ⱥ	a	a				2
7 Candida guilliermondii var. guilliermondii	Ⱥ	A	A	A	A	A	A	A	a		Ⱥ	A	a	A				2
8 Candida krusei	Ⱥ	Ⱥ	Ⱥ	a	Ⱥ	Ⱥ	Ⱥ	Ⱥ	A		Ⱥ	Ⱥ	A	A				2
9 Candida lipolytica var. lipolytica	A	Ⱥ	Ⱥ	A	A	Ⱥ	A	Ⱥ	A		Ⱥ	A	A	A				2
10 Candida tropicalis	Ⱥ	Ⱥ	A	a	A	A	A	a	A		Ⱥ	a	a	A				2
11 Candida utilis	Ⱥ	Ⱥ	Ⱥ	A	a	A	Ⱥ	A	A		Ⱥ	A	A	A				2
12 Citeromyces matritensis	Ⱥ	Ⱥ	A	A	A	A	Ⱥ	Ⱥ	A		Ⱥ	A	a	Ⱥ				2
13 Cryptococcus albidus var. albidus	a	a	a	a	a	A	a	A	A		Ⱥ	A	a	A	A	A	a	2
14 Chryptococcus neoformans	a	A	A	a	A	A	A	A	A		A	a	a	a	A	A	A	2
15 Debaryomyces hansenii	a	a	A	A	A	A	A	A	A		Ⱥ	a	a	A				2
16 Dekkera bruxellensis	Ⱥ	Ⱥ	A	A	Ⱥ	A	a	Ⱥ	A		Ⱥ	a	a	a				2
17 Endomycopsis capsularis	A	Ⱥ	A	A	Ⱥ	A	a	A	A		A	a	A	A				2
18 Endomycopsis fibuligera	a	Ⱥ	A	A	Ⱥ	A	a	A	A		a	a	a	a				2
19 Filobasidium floriforme	Ⱥ		A	A	A	A	A	A			A	A	Ⱥ	Ⱥ				3
20 Hanseniaspora valbyensis	Ⱥ	Ⱥ	A	A	A	A	Ⱥ	A	Ⱥ		Ⱥ	A	Ⱥ	Ⱥ				2
21 Hansenula anomala var. anomala	A	Ⱥ	A	A	A	A	a	A	A		Ⱥ	A	A	Ⱥ	a	Ⱥ	A	2
22 Hansenula bimundalis var. bimundalis	Ⱥ	Ⱥ	A	A	A	a	Ⱥ	A	A		Ⱥ	A	A	A				2
23 Hansenula capsulata	A	Ⱥ	A	A	A	a	A	A	A		Ⱥ	Ⱥ	A	a				2
24 Hansenula fabianii	A	Ⱥ	A	A	A	A	Ⱥ	A	A		Ⱥ	A	A	A				2
25 Hansenula saturnus var. saturnus	Ⱥ	Ⱥ	A	A	A	Ⱥ	A	A	A		Ⱥ	a	A	A				2
26 Kloeckera apiculata	Ⱥ	Ⱥ	Ⱥ	Ⱥ	Ⱥ	Ⱥ	Ⱥ	Ⱥ	Ⱥ		Ⱥ	Ⱥ	Ⱥ	Ⱥ	A	Ⱥ	A	2
27 Kluyveromyces bulgaricus	Ⱥ	Ⱥ	A	A	a	A	a	a	A		Ⱥ	A	A	A				2
28 Kluyveromyces lactis	Ⱥ	Ⱥ	A	a	Ⱥ	a	A	Ⱥ	A		Ⱥ	A	A	A				2
29 Kluyveromyces polysporus	Ⱥ	Ⱥ	Ⱥ	A	Ⱥ	Ⱥ	Ⱥ	Ⱥ	a		Ⱥ	A	Ⱥ	A				2
30 Leucosporidium scottii	A	Ⱥ	a	a	A	A	a	a	a		Ⱥ	A	A	A	A	A		2
31 Lipomyces starkeyi	Ⱥ	A	A	a	A	A	A	a	a		a	a	a	a		A		2
32 Lodderomyces elongisporus	Ⱥ	Ⱥ	A	a	A	A	A	Ⱥ	A		Ⱥ	A	A	A	A		A	2
33 Metschnikowia bicuspidata var. bicuspidata	Ⱥ	A	Ⱥ	a	A	A	A	Ⱥ			Ⱥ	A	A	a	A	Ⱥ	A	2
34 Nadsonia fulvescens	Ⱥ	A	A	A	A	A	A	Ⱥ	A		Ⱥ	A	Ⱥ	A	Ⱥ	Ⱥ	A	2

[a] A = assimilated; Ⱥ = not assimilated; a = reaction varies.
[b] Reference 56.

Table 5 (continued)
OXIDATIVE UTILIZATION OF OTHER CARBON COMPOUNDS[a]

Yeast species	Erythritol	Galactitol	D-Glucitol	Glycerol	D-Mannitol	α-Methyl-D-glucoside	Ribitol	Salicin	Ethanol	Methanol	Inositol	Citric acid	Lactic acid	Succinic acid	2-Keto-gluconic acid	5-Keto-gluconic acid	Glucono-δ-lactone	Ref.
35 Nematospora coryli	A	A	A	A	A	A	A	A	a		A	A	A	A				2
36 Oosporidium margaritiferum	A	A	a	A	A	A	A	A	A		A	A	A	A				2
37 Pachysolen tannophilus	A	A	A	A	A	A	A	a	A		A	A	A	A				2
38 Pichia farinosa	A	A	A	A	A	A	A	a	A		A	A	A	a				2
39 Pichia membranaefaciens	A	A	A	a	A	A	A	A	A		A	a	a	a				2
40 Pityrosporum pachydermatis	A	A	A	a	A	A	A	A	A		A	A	A	a				2
41 Rhodosporium toruloides	A	A	A	A	A	A	A	A	A		A	A	A	A	A	A		2
42 Rhodotorula glutinis var. glutinis	A	a	a	a	a	a	a	A	a		A	a	a	A	a	A	A	2
43 Rhodotorula rubra	A	a	a	a	a	a	a	a	a		A	A	a	A	A	A	A	2
44 Saccharomyces cerevisiae	A	A	a	a	a	a	A	A	a		A	A	A	a				2
45 Saccharomyces chevalieri	A	A	a	a	a	a	A	A	a		A	A	A	A				2
46 Saccharomyces kluyveri	A	A	A	a	A	A	A	a	A		A	A	A	A				2
47 Saccharomyces rosei	A	A	A	a	A	a	A	A	A		A	A	A	A				2
48 Saccharomyces rouxii	A	A	A	a	A	a	a	A	A		A	A	A	A				2
49 Saccharomyces uvarum	A	A	a	a	a	a	A	A	A		A	A	A	a				2
50 Schizoblastosporion starkeyi-henricii	A	A	a	A	a	A	A	A	a		A	a	A	A				2
51 Schizosaccharomyces octosporus	A	A	A	A	A	A	A	A	A		A	A	A	A			A	2
52 Schizosaccharomyces pombe	A	A	A	A	A	a	A	A	A		A	A	A	A	A	A		2
53 Schwanniomyces occidentalis	A	A	A	A	A	A	A	A	A		A	A	A	A		A	A	3
54 Selenotila intestinalis	A	A	A	A	A		A	A	A		A		A	A	A	A	A	2
55 Sporidiobolus johnsonii	A	A	A	A	A	A	a	A	A		A	A	a	A	A	A	A	2
56 Sporobolomyces roseus	A	A	A	A	A	a	a	A	A		A	A	A	A	A	A		2
57 Stergmatomyces halophilus	A	A	A	A	A	A	A	A	A		A	A	A	A				3
58 Sympodiomyces parvus	A	A	A	A		A	A	A	A			A	A	A				3
59 Torulopsis colliculosa	A	A	A	A		A	A	A	A			A	A	A				2
60 Trichosporon cutaneum	a	a	a	a	a	a	a	a	A		a	a	a	a				2
61 Trigonopsis variabilis	A	A	a	A	A	A	A	A	A		A	a	A	A	A	A		2
62 Wickerhamia fluorescens	A	A	a	A	A	A	A	A	A		A	A	A	A		A	A	2
63 Wingea robertsii	A	A	A	A	A	A	A	A	A		A	A	a	A				2

Table 6

UTILIZATION OF INORGANIC AND ORGANIC COMPOUNDS (EXCLUDING AMINO ACIDS) AS NITROGEN SOURCES[a]

All species listed assimilate ammonium sulfate. Values for the utilization of nitrate as nitrogen source are from Lodder.[2] For the capacity of some yeasts to assimilate molecular nitrogen, see, for example, Mandel and Bieth.[57] In many cases the utilization of the different inorganic and organic nitrogen compounds originally appeared as quantitative data, but the information is presented here in a qualitative form.

Yeast species	Nitrate	Nitrite	Urea	Hydroxyl-amine	Methyl-amine	Dimethyl-amine	Ethyl-amine	Diethyl-amine	Ethanol-amine	Ethylene-diamine	Benzyl-amine	Cyclohexyl-amine	D-Glucos-amine	Acetamide	Adenine	Guanine	Ref.
1 Brettanomyces bruxellensis	U[b]		U												U	U	58
2 Brettanomyces claussenii	U		U												U	U	58
3 Bullera alba	U̸	u	U												U	U	58
4 Candida humicola	U̸		U	U		U	U	U	U				U		U	U	50
5 Candida krusei	U̸		U	U			U	U	U								58
6 Candida lipolytica	U̸		U	U̸		U	U	U	U				U				59
7 Candida melinii	U		U														50
8 Candida tropicalis	U̸		U												U	U	58
9 Candida utilis	U̸		U												U	U	58
10 (Torulopsis utilis)																	60
11 Citeromyces matritensis	U		U												U	U	58
12 Chryptococcus albidus	U	U	U	U̸		U	U	u	U				u		u	u	50
13																	58
14 Debaryomyces hansenii (Debaryomyces kloeckeri)	U̸	u	U	U̸		u	U	u	U				U		U	U	50
15 (Debaryomyces subglobosus)	U̸		U												u	U	58
16 Endomycopsis capsularis	U̸		U												U	U	58
17 Hanseniaspora osmophila (Hanseniaspora vinea)																	58
18 Hansenula anomala	U		U	U̸		U	U	U	U				u		U	U	50
19			U										U		u	u	58
20 Kloeckera apiculata	U̸	U	U	U̸		u	U	u	U						U	U	50
21		u	U												U	U	58
22 Kluyveromyces fragilis	U̸				U												58
23 (Saccharomyces fragilis)					u							U		U			5
24 Kluyveromyces polysporus	U̸	U	U̸	U̸			U		U̸	U̸	U̸	U̸		U̸	U̸	U̸	61
25 Lipomyces starkeyi	U̸	U	U̸				U̸		U̸						U̸	U̸	58
26 Metschnikowia bicuspidata var. bicuspidata (Metschnikowia kamienski)	U̸																58
27 Nematospora coryli	U		U			U		U	U			U	U		u	U̸	58
28 Pachysolen tannophilus	U	U	U	U̸		U		U	U						U	U	50
29			U												U	U	58
30 Pichia farinosa	U̸	U	U	U̸		U		U	U				U		U	U	58
31 Pichia membranaefaciens	U̸		U												U	U	58
32 Pichia pseudopolymorpha	U̸	U	U	U̸		U		U	U				U		U	U	50
33 Pichia rhodanensis	U̸		U	U̸		U		U	U				u		U	U	58
34		u	U			U		U	U				U		U	U	50
35 Rhodotorula glutinis	U	u	U	U̸		u		u	U				U		U	U	58
36 Rhodotorula rubra	U̸	U̸	U	U̸					U						U̸	U	50
37 (Rhodotorula mucilaginosa)			U												u	u	58
38 Saccharomyces cerevisiae	U̸	u	U	U̸	U̸	u	U̸		U̸	U̸	U̸	U̸	u	U̸	U		50
39	U̸	U̸	U		U̸		U̸		U̸								58
40																	61
41																	5
42 Saccharomyces kluyveri	U̸		U												U		58

[a] U = utilized; u = poorly utilized; U̸ = not utilized.
[b] Reaction varies.

Table 6 (continued)
UTILIZATION OF INORGANIC AND ORGANIC COMPOUNDS (EXCLUDING AMINO ACIDS) AS NITROGEN SOURCES[a]

Yeast species	Nitrate	Nitrite	Urea	Hydroxyl-amine	Methyl-amine	Dimethyl-amine	Ethyl-amine	Diethyl-amine	Ethanol-amine	Benzyl-amine	Cyclohexyl-amine	D-Glucos-amine	Acetamide	Adenine	Guanine	Ref.
43																5
44 Saccharomyces rosei	ψ				U		U		ψ	ψ	U		ψ			61
45					U		U									5
46 Saccharomyces rouxii[c]	ψ		U		U		U		U	U(2) ψ(1)	U		U	U	U	58, 61
47																5
48 Saccharomyces uvarum	ψ				ψ		U		ψ	ψ	ψ		ψ	ψ	ψ	58, 61
49 Schizosaccharomyces octosporus	ψ		U		ψ		ψ		ψ					ψ	ψ	58
50 Schizosaccharomyces pombe	ψ		U	ψ			ψ	ψ	ψ			ψ		U	ψ	50
51			U													58
52 Schwanniomyces occidentalis	ψ		U				U							U	U	58
53 Sporobolomyces roseus	U	ψ	U			U	U	U						U	U	58
54 Sporobolomyces salmonicolor	ψ	U	U	U			U		U			U		ψ	ψ	50
55 Torulopsis colliculosa	ψ						U							U	U	58
56 Trichosporon cutaneum	ψ						U							ψ	ψ	58
57 Trigonopsis variabilis	ψ															58

Yeast species	Hypo-xanthine	Xanthine	Cytocine	Thymine	Uracil	Uric acid	Allantoin	Allantoic acid	Betaine	Choline	Creatine	Creatinine	Dimethyl-glycine	Guanidine	Sarcosine	Taurine	Ref.
1 Brettanomyces bruxellensis	U	U	U	ψ	U	U	U	U	U				U	U	U		58
2 Brettanomyces claussenii	U	U	U	ψ	ψ	U	U	U	U				U	U	U		58
3 Bullera alba	U ψ	U	ψ	ψ	U	U	U	U	U				U	U	U		58
4 Candida humicola		U	U	U	ψ	U	U	U		U			U	U			50
5 Candida krusei	U	U			U	U	U										59
6 Candida lipolytica					ψ		U										50
7 Candida melinii	U	U	U	ψ	U	U	U	U	U	U			U	U			58
8 Candida tropicalis	U	U	U	ψ	ψ	U	U	U	U	U			U	U			58
9 Candida utilis	L				U	U	U										60
10 (Torulopsis utilis)	U ψ	U	U	U	U	U	U	U	U	U			ψ	U	U		58
11 Citeromyces matritensis	U L	U	U	U	ψ	U	U	U	U	U			U	U			50
12 Cryptococcus albidus	U	U	U	U	U	U	U	U		U			ψ	U	U		58
13	■	U		ψ	U	U	U		■	U			ψ	U	U		50
14 Debaryomyces hansenii	U				U	U	U										
15 (Debaryomyces kloeckeri)	L ψ ■	U	U	ψ	ψ	U	U	U					U	U			58
16 Endomycopsis capsularis	■ ■	U	U	ψ	ψ	U	U	U					U	U			58
17 Hanseniaspora osmophila	U	U	U	ψ	ψ	U	U						U	U			58
18 (Hanseniaspora vinea)																	
19 Hansenula anomala	■ ■	U	U	ψ	U	U	U		U	U			U	U	U		50
20 Kloeckera apiculata	U ■	U			U ■ U	U	■ U		U	U			U	U	U		58
21	U ■	U	U	ψ	U ψ	U	U										58
22 Kluyveromyces fragilis	U	U	U	ψ		U				U			ψ	ψ			58
23 (Saccharomyces fragilis)														■			5
24 Kluyveromyces polysporus	U	U			U	U								■		ψ	61
25 Lipomyces starkeyi	U	U	U	U	U	U	U	U	U								58

[c] Three strains tested.

Table 6 (continued)

UTILIZATION OF INORGANIC AND ORGANIC COMPOUNDS (EXCLUDING AMINO ACIDS) AS NITROGEN SOURCES[a]

Yeast species	Hypo-xanthine	Xanthine	Cytocine	Thymine	Uracil	Uric acid	Allantoin	Allantoic acid	Betaine	Choline	Creatine	Creatinine	Dimethyl-glycine	Guanidine	Sarcosine	Taurine	Ref.
26 *Metschnikowia bicuspidata* var. *bicuspidata* (*Metschnikowia kamienski*)	v	v	v	v	v	v	v	v									58
27 *Nematospora coryli*	v	U	v	v	U	v	U	U									58
28 *Pachysolen tannophilus*	U	U	U	U	U	U	U	U	U	U	v	U	v	U	U		50
29	U	U	U	U	U	U	U	U									58
30 *Pichia farinosa*		U	U	v	U	U	U	U									58
31 *Pichia membranaefaciens*	U	U	U	v	v	U	U	U	U	U	U	U	U	U	U		58
32 *Pichia pseudopolymorpha*		v	u		u	U	U	U									50
33 *Pichia rhodanensis*	U	U	U	v	U	U	U	U	U	U	v	U	U	U	U		58
34		U			U	U	U	U									58
35 *Rhodotorula glutinis*	U	U	U	U	U	U	U	U	U	U	v	U	U	v	U		50
36 *Rhodotorula rubra*		U		v	U	U	U	U									58
37 (*Rhodotorula mucilaginosa*)	U	U	U	v	v	U	U	U	U	u	v	U	v	v	U		50
38 *Saccharomyces cerevisiae*	v	v	u	v	v	v	U	U	U	u	u	U	u	U	U		58
39					U	u	U	U									58
40							U										50
41							U									v	61
42 *Saccharomyces kluyveri*	U	U	U	U	U	U	U	U									5
43							U										58
44 *Saccharomyces rosei*							U									U	5
45																	61
46 *Saccharomyces rouxii*[c]	v	U	U	v	U	U	U	U								U	5
47																v	58, 61
48 *Saccharomyces uvarum*	U	U	u	v	u	U	U	U	U	u	v	v	u	U	u	v	5
49 *Schizosaccharomyces octosporus*	U	U	v	v	U	U	U	U									58, 61
50 *Schizosaccharomyces pombe*	U	U	v	v	U	U	U	U	U		U						58
51	U	U	v	v	v	U	U	U									50
52 *Schwanniomyces occidentalis*	U	U	u	v	u	U	U	U	U	u	u	u	u	U	u		58
53 *Sporobolomyces roseus*	v	U	U	v	u	U	U	U									58
54 *Sporobolomyces salmonicolor*		U	U	v	U	U	U	U	U	U	U	U	U	u	u		58
55 *Torulopsis colliculosa*	v	U	U	v	U	U	U	U									50
56 *Trichosporon cutaneum*	u	u	U	v	U	U	U	U									58
57 *Trigonopsis variabilis*	U	U	U	v	v	U	U	U									58

Table 7
UTILIZATION OF AMINO ACIDS AS NITROGEN SOURCES[a]

The values given in original publications for the ability of amino acids to serve as nitrogen sources for yeasts are often given quantitatively but are handled here in a qualitative form.

Yeast species	Alanine	2-Aminobutyric acid	Arginine	Asparagine	Aspartic acid	Citrulline	Cysteine	Glutamic acid	Glutamine	Glycine	Histidine	Homoserine	Ref.
1 Candida guilliermondii	U		U		U		U	U		U	U		62
2 Candida humicola	U	U	U		U			U		U	U	U	49
3 Candida melinii	U	U	U		U	U		U		U	U	U	49
4 Candida utilis (Torula utilis)	U		U	U	U		u	U		U	u		62
5 (Torulopsis utilis)			U	U		U							60
6 Chryptococcus albidus	U	U	U		U	U		U		U	U	U	49
7 Debaryomyces hansenii (Debaryomyces kloeckerii)	U	U	U		U	U		U		U	U	U	49
8 Hansenula anomala	U	U	U		U	U		U		U	U	U	49
9 Hansenula lambica	U		U	U	U			U		U	u		62
10 Kloeckera apiculata	U	U	U		U	U	u	U		U	U	U	49
11 Kloeckera brevis	U		U	U	U		u	U		U	U		62
12 Kluyveromyces lactis (Zygosaccharomyces lactis)	U		U	U	U		u	U		U	U		62
13 Kluyveromyces fragilis (Saccharomyces fragilis)	U[b]												61
14 Pachysolen tannophilus	U	U	U		U	U		U		U	U	U	49
15 Pichia belgica	U	U	U	U	U		u	U		U	U		62
16 Pichia kluyveri	U		U	U	U		u	U		U	U		62
17 Pichia pseudopolymorpha	U	U	U		U	U		U		U	U	U	49
18 Pichia rhodanensis	U	U	U		U	U		U		U	U	U	49
19 Rhodotorula rubra (Rhodotorula mucilaginosa)	U	U	U		U	U		U		U	U	U	49
20 Saccharomyces sp.[c]	U		U	U		U	u	U		U(2) u(38)	U(1) u(39)	U	62
21 Saccharomyces cerevisiae	U	U	U	U	U		Ʋ	U		U	u	U	49
22	U[d]	U	U		U			U	U	U	U		63

[a] U = utilized; u = poorly utilized; Ʋ = not utilized.
[b] D-Amino acid.
[c] 40 strains tested.
[d] β-Alanine Ʋ.

Table 7 (continued)
UTILIZATION OF AMINO ACIDS AS NITROGEN SOURCES[a]

Yeast species	Alanine	2-Aminobutyric acid	Arginine	Asparagine	Aspartic acid	Citrulline	Cysteine	Glutamic acid	Glutamine	Glycine	Histidine	Homoserine	Ref.
23	Ʊ[b]												61
24 /Saccharomyces cerevisiae str. anamensis)	U		U	U	U		Ʊ	U		Ʊ	Ʊ		62
25 Schizosaccharomyces pombe	U	U	u		U	U		U		U	u	u	49
26 Sporobolomyces salmonicolor	U	U	U		U	U		U		U	U	U	49
27 Torulopsis stellata	U		U	U	U		u	U		U	U		62

Yeast species	Isoleucine	Leucine	Lysine	Methionine	Ornithine	Phenyl-alanine	Proline	Serine	Threonine	Tryptophan	Tyrosine	Valine	Ref.
1 Candida guilliermondii	U	U	U	U		U	U	U	U	U	U	U	62
2 Candida humicola		U	U	U	U	U	U	U	U	U	U	U	49
3 Candida melinii		U	U	U	U	U	U	U	U	U	U	U	49
4 Candida utilis (Torula utilis)	U	U	U	U		U	U	U		U	U	U	62
5 (Torulopsis utilis)					U								60
6 Chryptococcus albidus		U	U	U		U	u	u	U	U	U	U	49
7 Debaryomyces hansenii (Debaryomyces kloeckerii)		U	U	U		U	U	U	U	U	U	U	49
8 Hansenula anomala		U	U	U	U	U	U	U	u	U	U	U	49
9 Hansenula lambica	U	U	u	U		U	U	U	U	U	U	u	62
10 Kloeckera apiculata		U	U	U	U	U	Ʊ	U	u	U	U	U	49
11 Kloeckera brevis	U	U	U	U		U	U	u	U		U	U	62
12 Kluyveromyces lactis (Zygosaccharomyces lactis)	U	U	U	U		U	U	U	U	U	u	U	62
13 Kluyveromyces fragilis (Saccharomyces fragilis)			U[e]				U[b]						61
14 Pachysolen tannophilus		U	U	U	U	U	Ʊ	U	U	U	U	U	49
15 Pichia belgica	U	U	U	U		U	U	U	U	U	U	U	62
16 Pichia kluyveri	U	U	U	U	U	U	U	U	u	U	u	U	62
17 Pichia pseudopolymorpha		U	U	U	U	U	U	U		U	U	U	49
18 Pichia rhodanensis		U	U	U	U	U	U	U		U	U	U	49
19 Rhodotorula rubra (Rhodotorula mucilaginosa)		U	U			U	U	U		U	U	U	49
20 Saccharomyces sp.[c]	U(32) u(8)	U	U(1) u(39)	U(36) u(4)		U(37) u(3)	U(33) u(7)	U(38) u(2)	u	U(36) u(4)	U(33) u(7)	U	62

[e] D- and L-Amino acids.

Table 7 (continued)
UTILIZATION OF AMINO ACIDS AS NITROGEN SOURCES[a]

Yeast species	Isoleucine	Leucine	Lysine	Methionine	Ornithine	Phenyl-alanine	Proline	Serine	Threonine	Tryptophan	Tyrosine	Valine	Ref.
21 *Saccharomyces cerevisiae*		U[f]	U	U	U	U	U	U	U	U	U	U	49
22	U	U	U	U	U	U	U	U	U	U	U	U	63
23			U[e]				U[b]	U					61
24 (*Saccharomyces cerevisiae* str. *anamensis*)	U	U	U	U		U	U	U	U	U	U	U	62
25 *Schizosaccharomyces pombe*		U	u	U	U	U	U	U	U	U	u	U	49
26 *Sporobolomyces salmonicolor*		U	u	U	U	U	u	U	U	u	U	U	49
27 *Torulopsis stellata*	U	U	U	U		U	U	U	u	U	u	U	62

f   D-Leucine U.

REFERENCES

1. **Kreger-van Rij, N. J. W.,** in *The Yeasts,* Vol. 1., Rose, A. H. and Harrison, J. S., Eds., Academic Press, London, 1969, 5–78.
2. **Lodder, J.,** Ed., *The Yeasts, A Taxonomic Study,* North Holland, Amsterdam, 1970.
3. **Barnett, J. A. and Pankhurst, R. J.,** *A New Key to the Yeasts,* North Holland, Amsterdam; American Elsevier, New York, 1974.
4. **Takahashi, M.,** *Nippon Nôgei Kagaku Kaishi,* 28, 395–404, 1954; *C.A.,* 52, 18640i, 1958.
5. **Brady, B. L.,** *Antonie van Leeuwenhoek; J. Microbiol. Serol.,* 31, 95–102, 1965.
6. **Haehn, H.,** *Biochemie der Gärungen,* Walter de Gruyter & Co., Berlin, 1952.
7. **Ingraham, M. A.,** *An Introduction to the Biology of Yeasts,* Sir Isaac Pitman & Sons, London, 1955.
8. **Pyke, M.,** in *Yeasts,* Roman, W., Ed., Dr. W. Junk, Publishers, The Hague, 1957.
9. **Morris, E. O.,** in *The Chemistry and Biology of Yeasts,* Cook, A. H., Ed., Academic Press, New York, 1958, 251–321.
10. **Uhl, A.,** in *Die Hefen, Band 1, Die Hefen in der Wissenschaft,* Reiff, F., Kautzmann, R., Lüers, H., and Lindemann, M., Eds., Verlag Hans Carl, Nürnberg, 1960, 209–320.
11. **Suomalainen, H. and Oura, E.,** in *The Yeasts,* Vol. 2., Rose, A. H. and Harrison, J. S., Eds., Academic Press, London, 1971, 3–74.
12. **Günther, Th. and Kattner, W.,** *Z. Naturforsch. Teil B,* 23, 77–80, 1968.
13. **Lasnitzki, A. and Szörényi, E.,** *Biochem. J.,* 28, 1678–1683, 1934.
14. **Fulmer, E. I., Underkofler, L. A., and Lesh, J. B.,** *J. Am. Chem. Soc.,* 58, 1356–1358, 1936.
15. **Elvehjem, C. A.,** *J. Biol. Chem.,* 90, 111–132, 1931.
16. **Olson, B. H. and Johnson, M. J.,** *J. Bacteriol.,* 57, 235–246, 1949.
17. **McHargue, J. S. and Calfee, R. K.,** *Plant Physiol.,* 6, 559–566, 1931.
18. **Fulmer, E. I., Nelson, V. E., and Sherwood, F. F.,** *J. Am. Chem. Soc.,* 43, 191–199, 1921.
19. **Richards, O. W.,** *J. Am. Chem. Soc.,* 47, 1671–1676, 1925.
20. **Greaves, J. E., Zobell, C. E., and Greaves, J. D.,** *J. Bacteriol.,* 16, 409–430, 1928.
21. **Richards, O. W.,** *J. Biol. Chem.,* 96, 405–418, 1932.
22. **Hébert, A.,** *Bull. Soc. Chim. Fr. Ser. 4,* 1, 1026–1032, 1907.
23. **White, J.,** *Yeast Technology,* Chapman & Hall, London, 1954, 286–293.
24. **Joslyn, M. A.,** *Wallerstein Lab. Commun.,* 4, 49–65, 1941.
25. **Kautzmann, R.,** *Branntweinwirtschaft,* 109, 193–200, 1969.
26. **Schultz, A. S. and McManus, D. K.,** *Arch. Biochem.,* 25, 401–409, 1950.
27. **Maw, G. A.,** *J. Inst. Brew. London,* 66, 162–167, 1960.
28. **Andreasen, A. A. and Stier, T. J. B.,** *J. Cell. Comp. Physiol.,* 41, 23–36, 1953.
29. **Jollow, D., Kellerman, G. M., and Linnane, A. W.,** *J. Cell Biol.,* 37, 221–230, 1968.
30. **Suomalainen, H., Nurminen, T., Vihervaara, K., and Oura, E.,** *J. Inst. Brew. London,* 71, 227–231, 1965.
31. **Vavra, J. J. and Johnson, M. J.,** *J. Biol. Chem.,* 220, 33–43, 1956.
32. **Koser, S. A.,** *Vitamin Requirements of Bacteria and Yeasts,* Charles C Thomas, Springfield, Ill., 1968.
33. **Burkholder, P. R., McVeigh, I., and Moyer, D.,** *J. Bacteriol.,* 48, 385–391, 1944.
34. **Schultz, A. S. and Atkin, L.,** *Arch. Biochem.,* 14, 369–380, 1947.
35. **Drouhet, E. and Vieu, M.,** *Ann. Inst. Pasteur Paris,* 92, 825–831, 1957.
36. **Miyashita, S., Miwatani, T., and Fujino, T.,** *Biken J.,* 1, 45–49, 1958.
37. **Burkholder, P. R. and Moyer, D.,** *Bull. Torrey Bot. Club,* 70, 372–377, 1943.
38. **Burkholder, P. R.,** *Am. J. Bot.,* 30, 206–211, 1943.
39. **Emery, W. B., McLeod, N., and Robinson, F. A.,** *Biochem. J.,* 40, 426–432, 1946.
40. **Shavlovsky, G. M. and Ksheminskaya, G. P.,** *Mikrobiologiya,* 34, 53–60, 1965.
41. **Ahearn, D. G. and Roth, F. J.,** *Dev. Ind. Microbiol.,* 3, 163–173, 1962.
42. **Abadie, F.,** *Mycopathol. Mycol. Appl.,* 36, 81-93, 1968.
43. **Ahearn, D. G., Roth, F. J., and Meyers, S. P.,** *Can. J. Microbiol.,* 8, 121–132, 1962.
44. **Wilkinson, R. A. and Gray, W. D.,** *Mycopathol. Mycol. Appl.,* 42, 281–288, 1970.
45. **Leonian, L. H. and Lilly, V. G.,** *Am. J. Bot.,* 29, 459–464, 1942.
46. **Weinfurtner, F., Eschenbecher, F., and Borges, W.-D.,** *Zentralbl. Bakteriol. Parasitenk. Infektionskr. Hyg. Abt. 2,* 113, 134–162, 1959.
47. **Ahmad, M., Chaudhury, A. R., and Ahmad, K. U.,** *Mycologia,* 46, 708–720, 1954.
48. **Volkova, L. P.,** *Mikrobiologiya,* 32, 778–782, 1963.
49. **Abadie, F.,** *Ann. Inst. Pasteur Paris,* 113, 81–95, 1967.
50. **Abadie, F.,** *Ann. Inst. Pasteur Paris,* 115, 197–211, 1968.
51. **Shennan, J. L. and Levi, J. D.,** *Prog. Ind. Microbiol.,* 13, 1–57, 1974.

52.   Munk, V., Volfová, O., Dostálek, M., Mostecky, J., and Pecka, K., *Folia Microbiol. Prague,* 14, 334–344, 1969.

53.   Liener, I. F. and Buchanan, D. L., *J. Bacteriol.,* 61, 527–534, 1951.

54.   Oura, E., see Katinger, H. W. D., *Proc. 4th Int. Symp. Yeasts,* Part II, Vienna, 1974, 58–59.

55.   Rockwell, G. E. and Highberger, J. H., *J. Infect. Dis.,* 40, 438–446, 1957.

56.   Volfová, O., *Folia Microbiol. Prague,* 20, 307–319, 1975.

57.   Mandel, P. and Bieth, R., in *Die Hefen, Band 1, Die Hefen in der Wissenschaft,* Reiff, F., Kautzmann, R., Lüers, H., and Lindemann, M., Eds., Verlag Hans Carl, Nürnberg, 1960, 722–778.

58.   LaRue T. A. and Spencer, J. F. T., *Can. J. Microbiol.,* 14, 79–86, 1968.

59.   Norkrans, B., *Acta Chem. Scand.,* 23, 1457–1459, 1969.

60.   Steiner, M., *Symp. Soc. Exp. Biol.,* 13, 177–192, 1959.

61.   LaRue, T. A. and Spencer, J. F. T., *Antonie van Leeuwenhoek; J. Microbiol. Serol.,* 34, 153–158, 1968.

62.   Schultz, A. S. and Pomper, S., *Arch. Biochem.,* 19, 184–192, 1948.

63.   Nielsen, N., *Ergeb. Biol.,* 19, 375–408, 1943.

QUALITATIVE REQUIREMENTS AND UTILIZATION OF NUTRIENTS: MYXOMYCETES

W. D. Gray

According to Lister,[1] the first recorded reference to a member of this small but unique group of organisms was Pankow's description of *Lycogala epidendrum* which appeared in 1654. Thus, it may be said that the study of myxomycetes is now over three centuries old. Seventy-five years after Pankow's work appeared, Micheli[2] described several species and made reference to the plasmodium — the assimilative structure with which we must be primarily concerned in any discussion of the nutrition of myxomycetes, although in a complete discussion, the nutritive requirements of the gametes (which in this group may assimilate nutrients and reproduce themselves by simple division) as well as the nutritive requirements for sclerotization and reproduction* must also be considered.

For the early development of the area of myxomycete studies, much credit must be given to DeBary,[3-8] who first studied myxomycetes in detail and elucidated the general nature of the developmental cycle, and to Cienkowski,[9] who produced an excellent early study of the plasmodium. Thus, over a century ago these two earlier workers provided the groundwork for studies of the biology of myxomycetes.

It was noted quite early by students of myxomycetes that a plasmodium has the capability of engulfing small particles of organic matter and completely or partially digesting them (depending upon the nature of the particulates). It was also noted that the gametes (myxamoebae $\rightleftharpoons$ swarm cells) could engulf and digest very small particulates such as bacteria. Hence, we have known for many years how myxomycetes feed, but unfortunately we still know very little about their nutritive requirements. In addition to their known ability to engulf small particulates, it has been assumed for many years that these organisms can also obtain nutrients in solution by diffusion through the membranes surrounding the plasmodium — a very reasonable assumption in view of the known behavior of other types of living organisms. The probability is great that plasmodia can obtain dissolved nutrients in this manner, but the discovery of pinocytosis in the plasmodium of *Physarum polycephalum* by Guttes and Guttes[10] in 1960 now makes it possible to ask whether or not myxomycetes do in fact obtain nutrients in solution by simple diffusion. This question was raised by Gray and Alexpoulos[11] and discussed in greater detail by Gray.[12]

In order to speak with any degree of authority regarding the nutritional requirements of such organisms as fungi, bacteria, protozoans, myxomycetes, etc., it is necessary that they be obtained in pure culture and studied under rigidly controlled conditions, and herein lies the major obstacle that exists with respect to our present understanding of the nutritional requirements of myxomycetes. Of the approximately 425 species of myxomycetes, "spore-to-spore" cultivation in the laboratory has been reported for less than 15% of the species, and of these the plasmodia of only three species have been obtained in pure culture on chemically defined media. The undoubted "spore-to-spore" cultivation of any species of myxomycete under pure culture conditions in chemically defined medium has yet to be reported. Thus, as Gray[12] has pointed out, the problem of bringing a greater number of species into culture in the laboratory is probably the major problem facing myxomycete students today, and represents the first step in development

---

* Since it has become more and more obvious that myxomycetes are not fungi and certainly not plants, it is suggested that it is now time to discard some of the terminology that has been used in connection with these organisms. As a starter, the terms spore, sporangium, sporulation, fructification, and vegetation will for the most part be eliminated from the present discussion.

of pure culture techniques for a greater number of species which can then be used to present a broader picture of the nutritional requirements of the group as a whole.

NUTRITION OF PLASMODIA

Since the plasmodium is the principal assimilative structure, a discussion of the nutritional requirements of myxomycetes must be concerned largely with the nutrition of this structure. However, since the plasmodia of only three species have been cultured axenically in chemically defined media, the discussion must of necessity be somewhat limited, and it should be emphasized that the nutritional requirements of these three species may not necessarily be those of myxomycetes in general. This note of caution is introduced for four very sound reasons: (1) all three species are members of the genus *Physarum*, (2) *Physarum polycephalum*, the most studied species, can scarcely be considered a "typical" myxomycete, (3) all three species have phaneroplasmodia, so we know nothing about the nutritive requirements of protoplasmodia or aphanoplasmodia, and (4) all are pigmented.

Although there had been some slight successes in the laboratory cultivation of several species of myxomycetes prior to 1931, the development of a reliable and easy method for the cultivation of *P. polycephalum* by Howard[13] in that year marked the first instance in which myxomycete students could provide themselves with large quantities of plasmodia in the laboratory at all seasons of the year. The events following Howard's discovery have been recorded by Gray[12] and need not be repeated here. However, in view of the great preponderance of effort that was devoted to studies of *P. polycephalum*, it was almost inevitable that the plasmodium of this species would be the first to be brought into pure culture in chemically defined medium. Thus, Daniel and Rusch[14] isolated a wild strain of the plasmodium of this species in pure culture and first cultivated it in medium containing tryptone, glucose, yeast extract, $CaCO_3$, inorganic salts, and a small amount of chick embryo extract. Single large plasmodia could be produced in surface cultures or suspensions of tiny plasmodia could be produced in submerged culture. Components of the complete growth medium employed in these initial studies are listed in Table 1.

In a later publication, Daniel, Kelley, and Rusch[15] reported their successful cultivation of *P. polycephalum* in chemically defined media, having found that the chick embryo extract could be replaced by hematin or certain hemoproteins and the tryptone and yeast extract by a combination of various organic compounds. The media employed

Table 1
COMPLETE CHEMICALLY UNDEFINED MEDIUM FOR GROWTH OF PLASMODIA OF *PHYSARUM POLYCEPHALUM*

Component	Concentration (g/100 ml medium)	Component	Concentration (g/100 ml medium)
Tryptone (Difco)®	1.0	$MnCl_2 \cdot 4H_2O$	0.0084
Yeast extract (Difco)	0.15	$ZnSO_4 \cdot 7H_2O$	0.0034
Glucose, anhydrous	1.0	Citric acid $\cdot H_2O$	0.048
KH_2PO_4	0.20	HCl, concentrated	0.006 ml
$CaCl_2 \cdot 2H_2O$	0.06	Distilled water	To 100 ml
$MgSO_4 \cdot 7H_2O$	0.06	$CaCO_3$	0.30
$FeCl_2 \cdot 4H_2O$	0.006	Chick embryo extract[a]	1.5 ml

[a] Difco ampoule containing 2 ml of a lyophilized 50% extract reconstituted with 8.3 ml distilled H_2O.

From Daniel, J. W. and Rusch, H. P., *J. Gen. Microbiol.*, 25, 47, 1961. With permission.

in these later studies contained citric acid, glucose, 11 inorganic salts, HCl, inositol, choline HCl, 10 vitamins, 13 amino acids, and hematin. These studies were further expanded and Daniel et al[16] then reported that this species has absolute requirements for D- or L-methionine, biotin, thiamin, and hematin. Optimal concentrations of these absolute organic requirements were established. Components of the media used by Daniel et al. (with the exception of hematin) are presented in Tables 2 and 3. Hematin was added to media by adding 0.1 ml of a 0.05% hematin in 1% NaOH solution for each 10 ml of medium.

Ross[17] reported the pure culture of *Physarella oblonga* and *Physarum flavicomum.* He found that the first medium developed by Daniel and Rusch[14] for the culture of *P. polycephalum* did not support growth of the two species he was studying, but by modifying the Daniel and Rusch medium he successfully cultured them. Composition of the chemically undefined medium used by Ross is described in Table 4.

Ross established no nutritional requirements (other than for hematin) for the species he was studying, making note of the fact that the requirement for hematin is common to *Physarum polycephalum, P. flavicomum,* and *Physarella oblonga.* He did not succeed in obtaining the plasmodium of *Badhamia curtisii* in pure culture, but as a result of his observations he concluded that this latter species does not have a hematin requirement. This conclusion, based on somewhat dubious grounds, must be verified or denied when *B. curtisii* is obtained in pure culture on chemically defined medium. However, if it is verified, further weight will be added to the present writer's contention that what may be true of *P. polycephalum* may not necessarily be true for myxomycetes in general.

After the report by Ross of the pure culture of plasmodia of *P. flavicomum* in 1964, it was noted by Ross and Sunshine[18] that the rate of growth was gradually declining. Twenty-one days were now required to attain the growth formerly attained in 3 to 4 days. Various compounds were added to the medium, and it was found that 10^{-2} *M* chlorogenic acid or quinic acid enhanced the growth rate. Concentrations lower than 10^{-2} *M* had no effect, while higher concentrations were lethal. The growth-enhancing effect was duplicated to a lesser extent by gallic acid, but caffeic acid had no effect.

Henney and Henney[19] also obtained *P. flavicomum* in pure culture as well as another species, *P. rigidum.* While the medium employed by these investigators also was not chemically defined, it was much simpler than that used by Ross[17] and Ross and Sunshine.[18] In a later paper, Henney and Lynch[20] reported the growth of *P. rigidum* and *P. flavicomum* as well as *P. polycephalum* in a chemically defined medium. They found that while *P. rigidum* and *P. flavicomum* would not grow in the medium developed by Daniel et al.[16] for *P. polycephalum,* this latter species would grow in the medium they had developed for *P. rigidum* and *P. flavicomum* — evidence that the nutritional requirements of all myxomycetes may not necessarily be similar and conceivably could be quite different. As had been shown by Daniel et al.[16] for *P. polycephalum,* both *P. flavicomum* and *P. rigidum* required biotin, thiamin, and hematin. Minimal medium containing L-methionine, glycine, and L-arginine supported growth of *P. flavicomum* and *P. polycephalum,* but *P. rigidum* also required valine. Some curious results were obtained when Henney and Lynch[20] employed seven different combinations of amino acids as well as homocysteine thiolactone in media. These results serve well to demonstrate the need for expanded investigations of the nutrition of myxomycetes. The composition of the defined minimal medium developed by Henney and Lynch is presented in Table 5.

NUTRITION OF SWARM CELLS AND MYXAMOEBAE*

Although the role of the haploid swarm cells and myxamoebae as gametes of myxomycetes has been known for many years, it has only been in recent years that

---

* Since swarm cells and myxamoebae readily convert one to the other, they will here be referred to simply as the haploid phase.

Table 2
COMPOSITION OF SYNTHETIC MEDIA A, AV-40, AND OV-40[a]

I		II		III	
Component	Concentration (mg/l)	Component	Concentration (mg/l)	Component	Concentration (mg/l)
Citric acid·H_2O	2850	Inositol	11.9	DL-*Methionine*	252
Glucose	10,250	Choline hydrochloride	8.57	*Glycine*	454
NH_4Cl	2020	*Biotin*	0.158[b]	L-*Arginine hydrochloride*	605
KH_2PO_4	656	*Thiamine hydrochloride*	0.424[b]	L-Cysteine hydrochloride	502
K_2HPO_4·$3H_2O$	875	Pyridoxal hydrochloride	60.9	·H_2O	502
$FeCl_2$·$4H_2O$	46.5	Pyridoxine hydrochloride	8.72	L-Histidine hydrochloride	268
$MnCl$·$4H_2O$	65	Niacin	4.22	·H_2O	268
$ZnSO_4$·$7H_2O$	33.6	Calcium pantothenate	4.5	L-Leucine	524
Na_2SO_4	300	*p-Aminobenzoic acid*	0.816	L-Lysine hydrochloride	630
$CuCl_2$·$2H_2O$	2.56	Folic acid	0.407	DL-Isoleucine	348
$CoCl_2$·$6H_2O$	0.36	Vitamin B_{12}	0.0049	DL-Phenylalanine	434
$MgSO_4$·$7H_2O$	232	Riboflavine	4.36	DL-Tryptophan	177
$CaCl_2$·$2H_2O$	933			DL-Serine	470
				DL-Threonine	376
				DL-Valine	432
				(DL-Alanine)[c]	2437

[a] Columns I and II comprise the constituents for the 4× Basal for medium A. Column I comprises the constituents of the 4× Basal for media AV-40 and CV-40. Only the italicized vitamins (column II) are used in the AV-40 and OV-40 media. Column III gives the amino acid composition of the A and AV-40 media. Only the italicized amino acids (column III) appear in the OV-40 medium.

[b] For AV-40 and OV-40 media; × 100 for A medium.

[c] DL-Alanine appears only in the OV-40 medium.

From Daniel, J. W., Babcock, K. L., Sievert, A. H., and Rusch, H. P., *J. Bacteriol.*, 86, 324, 1963. With permission.

Table 3
VOLUMES OF STOCK SOLUTIONS USED TO PREPARE MEDIA A, AV-40, AND OV-40

Medium	4 × Basal, with vitamins (ml)	4 × Basal, no vitamins (ml)	Thiamine (ml)	Biotin (ml)	Amino acids (ml)	Water (ml)
A	252	–	–	–	260	581
AV-40	–	252	20	20	260	541
OV-40	–	252	20	20	120	680

From Daniel, J. W., Babcock, K. L., Sievert, A. H., and Rusch, H. P., *J. Bacteriol.,* 86, 324, 1963. With permission.

Table 4
CONSTITUENTS OF THE CHEMICALLY UNDEFINED MEDIUM DEVELOPED FOR THE PURE CULTURE OF PLASMODIA OF *PHYSARUM FLAVICOMUM* AND *PHYSARELLA OBLONGA*

	Agar (g/l)	Liquid (g/l)
BBL corn meal agar	15.0	b
Tryptone	5.0	2.5
Dextrose	5.0	–
Yeast extract	0.75	0.375
KH_2PO_4 [d]	0.8	0.8
Na_2HPO_4 [d]	0.2	0.2
$CaCl_2$ [d]	0.1	0.1
$MgSO_4$ [d]	0.1	0.1
$MnCl_2$ [d]	0.014	0.014
$ZnSO_4$ [d]	0.006	0.006
$FeCl_2$	0.01	0.01
Citric acid	0.2	0.2
Chick embryo extract	8.0 ml[a,c]	8.0 ml[a,c]
Distilled water	1000.0	See Footnote b
Final pH	5.6	5.6

[a] One ampoule Difco® chick embryo extract reconstituted with 8 ml sterile distilled water and added aseptically to the sterile medium.

[b] Corn meal extract: 30 g BBL corn meal agar suspended in one liter distilled water, stirred for 5 min, filtered, and filtrate made to one liter and used in place of water of liquid medium.

[c] Hematin substituted for chick embryo extract in later work: 0.05 g hemin dissolved in 100 ml of 1% NaOH; 0.1 ml added per 10 ml medium before sterilization.

[d] The salt solution referred to in the text contained these salts in the amounts shown per liter.

From Ross, I. K., *Bull. Torrey Bot. Club,* 91, 23, 1964. With permission.

Table 5

DEFINED MINIMAL MEDIA FOR GROWTH OF
PHYSARUM FLAVICOMUM,
P. POLYCEPHALUM, AND *P. RIGIDUM*[a]

Component	Amount
Basal salts mixture solution (pH 4.0)[b]	100.00 ml
Trace elements solution[c]	0.10 ml
Glucose	5.00 g
Biotin	0.10 mg
Thiamine	0.20 mg
L-Methionine	0.16 g
Glycine	1.80 g
L-Arginine	1.00 g
L-Valine[d]	1.00 g
Glass-distilled water	To 1000.00 ml

[a] Autoclave media (121°C, 15 min) and cool; add sterile hematin solution (2.5 μg/ml, final concentration). Final pH is about 4.2.

[b] Add successively to about 700 ml of glass-distilled water (final volume to 1000 ml) with stirring (grams): citric acid, 29.78; K_2HPO_4, 33.10; NaCl, 2.50; $MgSO_4 \cdot 7H_2O$, 1.00; $CaCl_2 \cdot 2H_2O$, 0.50. Final pH is 4.0.

[c] See Henney and Henney.[19]

[d] Required only by *P. rigidum*; omit for *P. flavicomum* and *P. polycephalum*.

From Henney, H. R., Jr. and Lynch, T., *J. Bacteriol.,* 99, 531, 1969. With permission.

attempts have been made to obtain these protozoanlike structures in pure culture. Because of their ability to engulf and digest small contaminant organisms such as bacteria, it was possible to maintain fairly clean and usable cultures of the haploid phase — sufficiently clean at least for use in a variety of very creditable genetic studies.

In his 1964 report of the successful pure culture of the plasmodia of three species, Ross[17] also noted his attempt to culture the plasmodium of a fourth species (*Badhamia curtisii*), in which he was unsuccessful. However, he was able to obtain the haploid phase of this latter species in pure culture but not on chemically defined medium. The fact that the haploid phase of *B. curtisii* could be cultured but the diploid phase could not indicate that in some species at least the nutritional requirements of the two phases may be different.

Henney[21] maintained clones of *Physarum flavicomum* haploid phase in monoxenic culture with *Aerobacter aerogenes* and using such material was able to investigate the mating type system of *P. flavicomum*. This investigation was later expanded by Henney and Henney[22] and pure cultures were obtained but formalin-killed bacteria were supplied as food. Later, Goodman[23] reported the pure culture of the haploid phase of *Physarum polycephalum*. However, since the medium that he used was not chemically defined, little is known of the nutritional requirements of the haploid phase of this species.

In 1974, Henney, Asgari, and Henney[24] reported the successful growth of both the haploid phase and plasmodia of *P. flavicomum* in pure culture in partially defined media, with casein hydrolysate being the only chemically undefined component. In aerobic shake cultures they obtained cell yields of 2×10^7 to 4×10^7 cells per milliliter. The following year, Henney and Asgari[25] reported the successful growth of the haploid phase in chemically defined minimal media. It was established that biotin, thiamin, and hematin were required for growth and that a minimal medium must also contain glucose and three amino acids, one of which (arginine) was required for growth. Highest growth rate and cell yield occurred when the three-amino-acid combination consisted of arginine, glycine, and methionine. This successful growth of both the haploid and diploid phases of the same species of myxomycete in the same chemically defined medium now makes possible much more extensive and meaningful physiological and biochemical studies. It also makes possible the isolation of nutritional mutants and thus could prove to be quite valuable in genetic studies.

SPECIAL NUTRITIONAL REQUIREMENTS FOR REPRODUCTION

The most striking and easily observed morphogenetic change that occurs in myxomycetes is that which occurs when a plasmodium becomes transformed into the reproductive structures which contain the bodies that will later give rise to gametes. In 1938, Gray[26] offered experimental evidence that visible radiation was the factor that triggered the initiation of the reproductive phase in yellow-pigmented plasmodia, a finding that has since been verified by a considerable number of other investigators with a variety of different species. However, Gray was unable to discover what factor or factors triggered reproduction in nonpigmented plasmodia. These factors still remain unknown, primarily because a thorough investigation of the problem has yet to be conducted.

Daniel and Rusch,[27,28] in their studies of *Physarum polycephalum*, agreed that light was necessary for initiation of reproduction in this species, but in addition found that reproduction would not occur even in illuminated cultures unless the plasmodium (prior to light exposure) was incubated for 4 days in the dark on a salts medium containing niacin, niacinamide, or tryptophane. Thus, in order for reproduction to occur in this species there is an absolute requirement for niacin or some related compound. Curiously enough, niacin is not required for growth of the plasmodium. Unfortunately, there are no experimental data to indicate whether or not other species may also have specific nutritive requirements for reproduction.

NUTRITIVE REQUIREMENTS FOR SCLEROTIZATION

Another less obvious and certainly less complicated (at least morphologically) type of morphogenesis also may be observed in myxomycetes. This happens when adverse conditions (for growth of plasmodium) occur, and the plasmodium becomes transformed into a resistant structure (the sclerotium) consisting of many roughly spherical, walled, multinucleate structures formerly called scleriotospores or spherules but for which Jump[29] proposed the name "macrocyst." The sclerotium does not occupy an essential position in the life cycle, and Jump has pointed out that the earlier workers probably paid little attention to it for this reason.

Jump[29] found that sclerotization of plasmodia of *Physarum polycephalum* may occur as a result of the application of several factors, one of which is starvation. However, Lonert,[30] working at Turtox Laboratories, developed a practical method for producing sclerotia of this species which involved overfeeding with rolled oats — a method that certainly does not involve starvation and supplies some evidence that nutritional factors may be involved. Lynch and Henney[31] have recently reported on some investigations of carbohydrate metabolism during sclerotization of the plasmodium of *Physarum flavicomum*, and Henney and Maxey[32] have studied the nutritional control of sclerotization of this same species. The latter work yielded results that indicate that the process of sclerotization is triggered by nutritional imbalance, and Henney and Maxey stated that the unavailability of an adequate spectrum of amino acids initiates sclerotization.

SUMMATION

From the above brief account it seems obvious that we have much yet to learn concerning the nutritive requirements of myxomycetes as a group of living organisms. Since less than 1% of the total number of species have been grown in pure culture in chemically defined media, and unquestioned "spore-to-spore" culture under such conditions has yet to be reported for any species, it would be foolhardy to speak with certainty of the nutritional requirements of the group at this time. This is especially true in view of the fact that the small amount of evidence accumulated to date is sufficient to hint strongly that the nutritional requirements of the plasmodium of one species may not be the same as that of another. Nothing at all is known about the nutritional requirements of species with nonpigmented plasmodia, just as nothing is known about the nutritional requirements of either protoplasmodia or aphanoplasmodia.

The matter is not a simple one because learning about the complete nutrition of a single species requires investigation of the nutritional requirements of four different phases of that organism. The first step, of course, is to bring a much greater number of species into culture. Then, as more is learned about each species, attempts can be made to bring each into pure culture on chemically defined media. At that point we will be in a position to study their nutritional requirements. It is to be hoped that developments will occur much more rapidly than they have during the entire 321 years that myxomycetes have been known to science, for if the rate of progress remains constant it will be slightly over 45 millenia before we can speak authoritatively concerning the nutrition of myxomycetes as a group! Even if knowledge of the pure culture of myxomycetes accumulates at the accelerated rate exhibited during the years 1961 through 1975, it will be over 2000 years before our information is complete.

REFERENCES

1. Lister, A., *A Monograph of the Mycetozoa,* 3rd. ed. (revised by G. Lister), British Museum of Natural History, London, 1925.
2. Micheli, P. A., *Nova Plantarum Genera,* Florence, 1729.
3. DeBary, A., *Bot. Z.,* 16, 357–358, 361–369, 1858.
4. DeBary, A., *Z. Wiss. Zool.,* 10, 88–175, 1859.
5. DeBary, A., *Die Mycetozoen (Schleimpilze) Ein Beitrag zur Kenntnis der niedersten Organismen,* 2nd ed., W. Engleman, Leipzig, 1864.
6. DeBary, A., *Morphologie und Physiologie der Pilze, Flechten und Myxomyceten,* Leipzig, 1866.
7. DeBary, A., *Vergleichende Morphologie und Biologie der Pilze, Mycetozoen, und Bacterien,* Leipzig, 1884.
8. DeBary, A., *Comparative Morphology and Biology of the Fungi, Mycetozoa and Bacteria* (English translation), Clarendon Press, London, 1887.
9. Cienkowski, L., *Jahrb. Wiss. Bot.,* 3, 400–441, 1863.
10. Guttes, E. and Guttes, S., *Exp. Cell Res.,* 20, 239–241, 1960.
11. Gray, W. D. and Alexopoulos, C. J., *Biology of the Myxomycetes,* Ronald Press, New York, 1968.
12. Gray, W. D., *CRC Crit. Rev. Microbiol.,* 4, 225–248, 1976.
13. Howard, F. L., *Am. J. Bot.,* 18, 624–628, 1931.
14. Daniel, J. W. and Rusch, H. P., *J. Gen. Microbiol.,* 25, 47–59, 1961.
15. Daniel, J. W., Kelley, J., and Rusch, H. P., *J. Bacteriol.,* 84, 1104–1110, 1962.
16. Daniel, J. W., Babcock, K. L., Sievert, A. H., and Rusch, H. P., *J. Bacteriol.,* 86, 324–331, 1963.
17. Ross, I. K., *Bull. Torrey Bot. Club,* 91, 23–31, 1964.
18. Ross, I. K. and Sunshine, L. D., *Mycologia,* 57, 360–367, 1965.
19. Henney, H. R., Jr. and Henney, M., *J. Gen. Microbiol.,* 53, 333–339, 1968.
20. Henney, H. R., Jr. and Lynch, T., *J. Bacteriol.,* 99, 531–534, 1969.
21. Henney, M. R., *Mycologia,* 59, 637–652, 1967.
22. Henney, M. R. and Henney, H. R., Jr., *J. Gen. Microbiol.,* 53, 321–332, 1968.
23. Goodman, E. M., *J. Bacteriol.,* 111, 242–247, 1972.
24. Henney, H. R., Jr., Asgari, M., and Henney, M. R., *Can. J. Microbiol.,* 20, 967–970, 1974.
25. Henney, H. R., Jr. and Asgari, M., *Arch. Microbiol.,* 102, 175–178, 1975.
26. Gray, W. D., *Am. J. Bot.,* 25, 511–522, 1938.
27. Daniel, J. W. and Rusch, H. P., *J. Bacteriol.,* 83, 234–240, 1962.
28. Daniel, J. W. and Rusch, H. P., *J. Bacteriol.,* 83, 1244–1250, 1962.
29. Jump, J. A., *Am. J. Bot.,* 41, 561–567, 1954.
30. Lonert, A. C., *Turtox News,* 43, 98–102, 1965.
31. Lynch, T. and Henney, H. R., Jr., *Arch. Microbiol.,* 90, 189–198, 1973.
32. Henney, H. R., Jr. and Maxey, G., *Can. J. Biochem.,* 53, 810–818, 1975.

QUALITATIVE REQUIREMENTS AND UTILIZATION OF NUTRIENTS: LICHENS

V. Ahmadjian

Table 1
NUTRIENT REQUIREMENTS AND UTILIZATION: LICHEN FUNGI — VITAMINS AND CARBON SOURCES[a,b]

There is very little information on the nutritional relationships between naturally occurring lichens and their substrates. It has been assumed that lichens derive nutrients from their substrates but there is no direct evidence to prove this is the case, let alone determine the types of nutrients obtained from the substrate. The information that does exist on the nutritional preferences of lichens comes from studies of the isolated symbionts in axenic cultures. Whether such information can be extended to the natural behavior of the lichenized symbionts is debatable. We know that the physiology of the algal symbiont *Trebouxia* changes considerably when the alga is removed from its fungal association and grown separately. The same may be true for lichen fungi. Thus, one must be cautious in relating the nutritional preferences of the symbionts in culture to their natural preferences. The criterion used to determine if lichen fungi require vitamins consisted of comparing growth in vitamin-free media with growth in media to which vitamins were added. In studies determining use of different carbon sources, growth was compared between that in carbon-free media and that in media to which specific carbon sources were added. With the exception of two studies[4,12] where growth was determined by colony size or by direct measurements of hyphal length, comparisons were made between dry weights of mycelia that had grown in liquid media.

QUALITATIVE REQUIREMENTS AND UTILIZATION OF NUTRIENTS: LICHENS

Table 1
NUTRIENT REQUIREMENTS AND UTILIZATION: LICHEN FUNGI – VITAMINS AND CARBON SOURCES[a,b]

Lichen (synonym)	Vitamins			Carbon sources																								Ref.
	Thiamine	Biotin	Other	Arabinose	Cellobiose	Cellulose	Dextrin	Erythritol	Fructose	Galactose	Glucose	Glycerol	Lactose	Maltose	Mannitol	Mannose	Raffinose	Rhamnose	Ribose	Sorbose	Starch	Sucrose	Trehalose	Xylose	Other	
Acarospora fuscata	R	R	Ɍ[c]		U					U	U		u	U						u	u	U				1, 2, 3, 7
Anthracothecium sp.						U					U		U	U			U			u	ψ	u				2, 3
Arthonia sp.						U					u	u		u	U						u		u				2, 3
Buellia stillingiana	R	R	Ɍ[c]	ψ	U		U	ψ	U	u	U		ψ	U	U					U		U			ψ[d]	6, 7
Caloplaca aurantiaca	Ɍ		Ɍ[e]				U		U					U	U		U									12
Cladonia cristatella	R	R	Ɍ[c]		U					U	U		u	U						u	ψ	U				2, 3
Cladonia decipiens	R	Ɍ	Ɍ[f]		ψ					U	U		u	U						u		u				2, 3
Collema tenax	R	R	Ɍ[c]						U		U					U	U	U	u					u	U[g]	8
Glyphis lepida											u		U	U		U		U		U		U				2, 3
Graphina bipartita	R	R	Ɍ[c]			U					u		U	U		U				U	ψ	u				2, 3
Graphina virginalis	Ɍ	R	Ɍ[e]			U					u		U	U		U				u	U	U				2, 3
Graphis scripta	Ɍ										U		U	U							U					12
Hypogymnia physodes	R	R																								11
Lecanora muralis (*Placodium saxicola*)	Ɍ		Ɍ[e]														U									12
Lecidea elaeochroma v. *flavicans* (*L. parasema*)	Ɍ		Ɍ[e]																							12
Physconia pulverulenta (*Physcia pulverulenta*)	Ɍ		Ɍ[e]																							12

[a] Taxonomically, only the fungal symbiont and the algal symbiont of a lichen have names. The association itself does not have a separate name. Names earlier given to the association are now considered to belong only to the fungal partner.

[b] R = required; Ɍ = not required; U = utilized; ψ = not utilized; u = poorly utilized.

[c] Inositol and pyridoxine.

[d] Citrate.

[e] Culture medium contained the following vitamins: B_1, B_2, B_6, C, D_2, E, biotin, beta-carotene, m-inositol, 2-methyl naphthoquinone, nicotinamide, pantothenic acid, and tocopherolacetate.

[f] Mixture of biotin, calcium d pantothenate, choline, folic acid, inositol, riboflavin, nicotinic acid, para-aminobenzoic acid, and pyridoxin.

Table 1 (continued)

NUTRIENT REQUIREMENTS AND UTILIZATION: LICHEN FUNGI – VITAMINS AND CARBON SOURCES[a,b]

Lichen (synonym)	Vitamins			Carbon sources																							Ref.
	Thiamine	Biotin	Other	Arabinose	Cellobiose	Cellulose	Dextrin	Erythritol	Fructose	Galactose	Glucose	Glycerol	Lactose	Maltose	Mannitol	Mannose	Raffinose	Rhamnose	Ribose	Sorbose	Starch	Sucrose	Trehalose	Xylose	Other	
Ramalina ecklonii	R	R	R[c]						U							U									U[h]	5
Sarcographa rechingeri	R	R	R[c]							U	u		U	U						U		U				2, 3
Sarcographina sandwicensis	R	R	R[c]		U		U			U	U		U	U						U		U				2, 3
Sarcogyne similis	R	R	R[c]	u	U		U			U	U		U	U	U	U				u		U				6, 7
Stereocaulon saxatile v. *evolutoides*	R	R	R[c]	U		u	u	u	u	ψ	u	ψ	u	u	U		u		ψ	U	u	U	u	ψ	u[i]	9
Stereocaulon tomentosum	R	R	R[c]																							9
Stereocaulon vulcani	R	R	R[c]								u	U	U	U						u		U				2, 3
Xanthoria parietina								U			U	U	U	U											U[j], ψ[k]	4, 10

h   Inositol.
i   Inulin, glucosamine, glycogen, pectin.
j   U = amygdalin, citric acid, glycogen, malic acid, pectin, peptone, salicin, succinic acid.
k   ψ = inositol, salicylic acid.

REFERENCES

1.   **Ahmadjian, V.**, *Bryologist*, 64, 168–179, 1961.
2.   **Ahmadjian, V.**, *Bryologist*, 67, 87–98, 1964.
3.   **Ahmadjian, V.**, unpublished results.
4.   **Am Ende, I.**, *Arch. Mikrobiol.*, 15, 185–202, 1950.
5.   **Fox, C. H.**, *Physiol. Plant.*, 19, 830–839, 1966.
6.   **Hale, M. E.**, *Bull. Torrey Bot. Club*, 85, 182–187, 1958.
7.   **Hale, M. E.**, *Lichen Handbook*, Publ. 4434, Smithsonian Institution, Washington, D.C., 1961.
8.   **Henriksson, E.**, *Sven. Bot. Tidskr.*, 58, 361–370, 1964.
9.   **Hutchinson, W. A.**, *Stereocaulon: Ecology and Physiology of Some Boreal Species and their Isolated Components,* Ph.D. dissertation, University of Massachusetts, Amherst, 1969.
10. **Quispel, A.**, *Recl. Trav. Bot. Neerl.*, 40, 413–541, 1943-1945.
11. **Scott, G. D.**, Recent studies of lichens, *Adv. Sci.*, 31, 244–248, 1964.
12. **Zehnder, A.**, *Ber. Schweiz. Bot. Ges.*, 59, 201–267, 1949.

Table 2
NUTRIENT REQUIREMENTS AND UTILIZATION: LICHEN FUNGI – AMINO ACIDS[a,b]

Most of the studies that determined utilization of amino acids as a nitrogen source compared growth in a N-free medium (control) with growth in media to which single amino acids were added. In two studies[4],[8] the control series contained KNO_3 as a nitrogen source, and growth of other series of amino acids was compared to that control.

Lichen (synonym)	Alanine	Arginine	Asparagine	Aspartic acid	Cysteine	Cystine	Glutamine	Glutamic acid	Glycine	Histidine	Isoleucine	Leucine	Lysine	Methionine	Phenylalanine	Proline	Serine	Threonine	Thymine	Tyrosine	Tryptophane	Valine	Ref.
Acarospora fuscata	U	U	U[c]	U		u		U	u	u	Ψ	u	u[c]	Ψ	u	U	U	Ψ		Ψ	Ψ	u	1, 4, 5
Cladonia cristatella	U	U	U	U		Ψ		U	U	u	u	u	U	u	u	U	u	u	Ψ	u	Ψ	u	1, 5
Lecanora tephroeceta	U	u	U	Ψ	Ψ		U	u		Ψ		U	Ψ	u	Ψ=	Ψ	u	Ψ	Ψ	Ψ	Ψ		8
Lecidea macrocarpa (*L. steriza*)	Ψ	U	U	Ψ				U		U	U		Ψ		Ψ	U				U			9
Ramalina ecklonii		U																					3
Sarcogyne similis	U	U	U	u	Ψ			U	Ψ	U	U		U[c]	u	u	U		U		U	u	U	4, 9
Stereocaulon saxatile v. *evolutoides*	U	U	U	u	Ψ	u		U	Ψ	Ψ	U		U	Ψ	u	U		U		U	Ψ	U	6
Xanthoria parietina		U							Ψ											U			2, 7

[a] Taxonomically, only the fungal symbiont and the algal symbiont of a lichen have names. The association itself does not have a separate name. Names earlier given to the association are now considered to belong only to the fungal partner.

[b] U = utilized; Ψ = not utilized; u = poorly utilized.

[c] By one strain; not utilized by one strain.

REFERENCES

1. **Ahmadjian, V.,** *Bryologist*, 67, 87–98, 1964.
2. **Am Ende, I.,** *Arch. Mikrobiol.*, 15, 185–202, 1950.
3. **Fox, C. H.,** *Physiol. Plant.*, 19, 830–839, 1966.
4. **Furnari, F. and Luciani, F.,** *Boll. Ist. Bot. Univ. Catania*, 3, 39–47, 1962.
5. **Gross, M. and Ahmadjian, V.,** *Sven. Bot. Tidskr.*, 60, 74–80, 1966.
6. **Hutchinson, W. A.,** *Stereocaulon: Ecology and Physiology of Some Boreal Species and their Isolated Components*, Ph.D. dissertation, University of Massachusetts, Amherst, 1969.
7. **Quispel, A.,** *Rec. Trav. Bot. Neerl.*, 40, 413–541, 1943-1945.
8. **Schofield, E. and Ahmadjian, V.,** in *Antarctic Terrestrial Biology*, Antarctic Res. Series, Vol. 20, Llano, G. A., Ed., American Geophysical Union, Washington, D.C., 1972, 97, 142.
9. **Tomaselli, R.,** *Atti Ist. Bot. Univ. Lab. Crittogam. Pavia*, 16, 180–191, 1959.

Table 3
NUTRIENT REQUIREMENTS AND UTILIZATION: LICHEN FUNGI — OTHER NITROGENOUS SUBSTANCES[a,b]

Most of the studies to determine utilization of nitrogenous substances compared growth in N-free medium (control) with growth in media to which specific nitrogenous substances were added. In two studies[4,8] the control series contained KNO_3 as a nitrogen source, and growth of other series of nitrogenous compounds was compared to that control.

Lichen (synonym)	Adenine	Adenosine	Allantoin	Ammonium	Casein hydrolysate	Cytidine	Cytosine	Gelatin	Nitrate	Nitrite	Peptone	Thymidine	Uracil	Urea	Uric acid	Uridine	Xanthine	Ref.
Acarospora fuscata				U	u				U	u				U				1, 4
Cladonia cristatella				U	U				U	U				U				1
Lecanora tephroeceta[c]	Ø	Ø	Ø	U	u	Ø	Ø		u		Ø	Ø	Ø	U	U	Ø	U	7
Lecidea macrocarpa (*L. steriza*)									u									8
Ramalina ecklonii				U					u									3
Sarcogyne similis									U[d]									4, 8
Stereocaulon saxatile v. *evolutoides*				U	U				U	Ø	U			Ø				5
Xanthoria parietina				U				U	U		U			Ø[d]				2, 6

[a] Taxonomically, only the fungal symbiont and the algal symbiont of a lichen have names. The association itself does not have a separate name. Names earlier given to the association are now considered to belong only to the fungal partner.

[b] U = utilized; Ø = not utilized; u = poorly utilized.

[c] Lichen collected in a snow petrel rookery in west Antarctica.

[d] By one strain; utilized poorly by one strain.

REFERENCES

1. **Ahmadjian, V.,** *Bryologist,* 67, 87–98, 1964.
2. **Am Ende, I.,** *Arch. Mikrobiol.,* 15, 185–202, 1950.
3. **Fox, C. H.,** *Physiol. Plant.,* 19, 830–839, 1966.
4. **Furnari, F. and Luciani, F.,** *Boll. Ist. Bot. Univ. Catania,* 3, 39–47, 1962.
5. **Hutchinson, W. A.,** *Stereocaulon: Ecology and Physiology of Some Boreal Species and their Isolated Components,* Ph.D. dissertation, University of Massachusetts, Amherst, 1969.
6. **Quispel, A.,** *Recl. Trav. Bot. Neerl.,* 40, 413–541, 1943–1945.
7. **Schofield, E. and Ahmadjian, V.,** in *Antarctic Terrestrial Biology,* Antarctic Res. Series, Vol. 20, Llano, G. A., Ed., American Geophysical Union, Washington, D.C., 1972, 97–142.
8. **Tomaselli, R.,** *Atti Ist. Bot. Univ. Lab. Crittogam. Pavia,* 16, 180–191, 1959.

Table 4
NUTRIENT REQUIREMENTS AND UTILIZATION: LICHEN ALGAE —
VITAMINS, SUGARS, AND ALCOHOLS[a,b]

The criterion used to determine whether or not lichen algae require vitamins compared growth in vitamin-plus media with growth in vitamin-free media. Early studies simply added different vitamins to a mineral-glucose-agar medium and compared the different sizes of algal colonies with those in control flasks. A later study[3] used dry weight measurements of growth and was meticulous in its technique to prevent contamination of the media and glassware with trace vitamins. The criterion for determining utilization of sugars and alcohols as a carbon source compared growth in media to which specific carbon sources were added to a carbon-free medium. In most of the studies the cultures were grown under low intensities of light. Light has been shown to increase growth of *Trebouxia* in a sugar medium, especially if the sugar is at a high concentration.

Table 4
NUTRIENT REQUIREMENTS AND UTILIZATION: LICHEN ALGAE — VITAMINS, SUGARS, AND ALCOHOLS[a,b]

Trebouxia Symbiont

Lichen (synonym)	Thiamine	Biotin	Other (Vit.)	Arabinose	Cellobiose	Dextrin	Fructose	Galactose	Glucose	Glycogen	Inulin	Lactose	Maltose	Mannose	Raffinose	Rhamnose	Ribose	Saccharose	Sorbose	Starch	Sucrose	Trehalose	Xylose	Dulcitol	Erythritol	Glycerol	Mannitol	Salicin	Other (Alc.)	Ref.
Acarospora fuscata	R	R																												1
Caloplaca aurantiaca	R	R	Ɍ[c]																											6
Cladonia bacillaris	R	R	Ɍ[d]																											3
Cladonia cristatella				Ψ	Ψ		U	U	U		Ψ	Ψ	Ψ	Ψ	Ψ	Ψ			Ψ		Ψ			Ψ	Ψ	Ψ	U	Ψ		2
Cladonia furcata								U	U				u					u											u[e]	4
Cladonia pyxidata								U	U				u					u											u[e]	4
Cladonia squamosa	R	R	Ɍ[d]																											3
Lecanora muralis (*Placodium saxicola*)	R	R	Ɍ[c]																											6
Lecanora radiosa (*Placodium circinatum*)	R		Ɍ[c]																											6
Parmelia acetabulum	R	R	Ɍ[f]	Ψ	Ψ		U	U	U			Ψ	Ψ								U				Ψ	Ψ	U		u[i]	5
Physconia pulverulenta (*Physcia pulverulenta*)	R	R	Ɍ[f,g] S[h]	U			U	U	U			Ψ	Ψ								U				Ψ	Ψ	U		u[i]	5
Stereocaulon dactylophyllum v. occidentale	R	R	Ɍ[d]	u	u	U	U	U	U	u	u	u	U		u		u		u	u	u	U	U			u			3
Stereocaulon pileatum	R	R	Ɍ[d]																											3
Xanthoria parietina	R	R	Ɍ[c,f]	Ψ			U	u	U			Ψ	Ψ								U				Ψ	Ψ	U		u[i]	5, 6

Table 4 (continued)

NUTRIENT REQUIREMENTS AND UTILIZATION: LICHEN ALGAE — VITAMINS, SUGARS, AND ALCOHOLS[a,b]

Lichen (synonym)	Thiamine	Biotin	Other	Arabinose	Cellobiose	Dextrin	Fructose	Galactose	Glucose	Glycogen	Inulin	Lactose	Maltose	Mannose	Raffinose	Rhamnose	Ribose	Saccharose	Sorbose	Starch	Sucrose	Trehalose	Xylose	Dulcitol	Erythritol	Glycerol	Mannitol	Salicin	Other	Ref.
	Vitamins			Carbon sources — Sugars																				Alcohols						
Chlorella Symbiont																														
Lecidea elaeochroma v. flavicans (L. parasema)	R		R[c]																											6
Coccomyxa Symbiont																														
Icmadophila ericetorum	R		R[c]																											6
Peltigera aphthosa	R		R[c]																											6
Solorina saccata	R		R[c]																											6

REFERENCES

1. Ahmadjian, V., Am. J. Bot., 49, 277–283, 1962.
2. Fox, C. H., Physiol. Plant., 20, 251–262, 1967.
3. Hutchinson, W. A., Stereocaulon: Ecology and Physiology of Some Boreal Species and their Isolated Components, Ph.D. Dissertation, University of Massachusetts, Amherst, 1965.
4. Korniloff, M., Bull. Soc. Bot. Genève, 5, 114–132, 1913.
5. Quispel, A., Recl. Trav. Bot. Neerl., 40, 413–541, 1943-1945.
6. Zehnder, A., Ber. Schweiz. Bot. Ges., 59, 201–267, 1949.

Table 5
NUTRIENT REQUIREMENTS AND UTILIZATION: LICHEN ALGAE – AMINO ACIDS[a,b]

The criterion used by most studies to determine utilization of amino acids as a nitrogen source compared growth in a N-free medium (control) with growth in media to which single amino acids were added. In one study[2] growth was expressed as a percentage of that obtained in a control-medium that contained 0.01 M nitrate.

Lichen (synonym)	Alanine	Arginine	Asparagine	Aspartic acid	Cysteine	Cystine	Glutamine	Glycine	Histidine	Isoleucine	Leucine	Lysine	Methionine	Phenylalanine	Proline	Serine	Threonine	Thymine	Tyrosine	Tryptophane	Valine	Ref.
Trebouxia Symbiont																						
Alectoria implexa	U		U					U			u											9
Anaptychia ciliaris (*Physcia ciliaris*)	U		U					U			u											9
Buellia pernigra		ψ	ψ																			8
Chaenotheca chrysocephala	U		U					U														7
Cladonia bacilliformis v. irregularis (*C. irregularis*)	U		U					U														4
Cladonia cristatella	U			U	U	u	U	U	u	u		u	u	u	U	u	u	u	u	u	u	2
Cladonia endiviaefolia	U		U	U				U														4
Cladonia furcata	U		U					U														4, 5
Cladonia pyxidata	U		U					U														4, 5
Lecanora rubina	u	U		U	U	u	U	u	u	u		u	ψ	u	U	ψ	ψ	u	u	u	u	2
Parmelia acetabulum	u		U[c]					u														4, 6
Parmelia caperata	U		U					U														4
Parmelia saxatilis	u		u					u														4
Parmelia scortea	u		u					u														4
Parmelia sulcata	u		u					u			u											4
Physconia pulverulenta (*Physcia pulverulenta*)	U		U					U														9

a   Many of the algal symbionts that have been isolated from different lichens have not been classified into acceptable species. To avoid confusion, the algal symbionts in this table are identified according to the lichen from which they were isolated.

b   U = utilized; ψ = poorly utilized; u = not utilized.

c   By one isolate; utilized poorly by one isolate.

Table 5 (continued)
NUTRIENT REQUIREMENTS AND UTILIZATION: LICHEN ALGAE — AMINO ACIDS[a,b]

Lichen (synonym)	Alanine	Arginine	Asparagine	Aspartic acid	Cysteine	Cystine	Glutamine	Glycine	Histidine	Isoleucine	Leucine	Lysine	Methionine	Phenylalanine	Proline	Serine	Threonine	Thymine	Tyrosine	Tryptophane	Valine	Ref.
Trebouxia Symbiont (continued)																						
Polycauliona citrina		U	U																			8
Pseudevernia furfuracea (*Parmelia furfuracea*)	U		U					U			U											9
Ramalina fraxinea	U	U	U		U			U		U	U											9
Stereocaulon dactylophyllum v. occidentale	U	U	U	U	U	U		U	U	U	U	U	Ū	U	Ū	U	U		U	U	U	3
Umbilicaria deusta (*Gyrophora flocculosa*)	U		U					U		U	U	U										9
Xanthoria parietina	U	U	U	U	U	U	U	U U[c]	U	U	U	U	U	U	U	U	U	U	U	U	U	2, 9
Chlorella Symbiont																						
Calcicium chlorinum v. exsertum	U		U				U	U														7
Coccomyxa Symbiont																						
Peltigera aphthosa	U		U					U			U											9
Solorina sp.	U		U																			1
Stichococcus Symbiont																						
Chaenotheca stemonea	U		U					U														7
Coniocybe furfuracea	U		U					U														7
Dermatocarpon miniatum[d]	U		U								U											9

d   The phycobiont of this lichen was described by Warén[10] as a species of *Hyalococcus*. My isolates of this algal symbiont, however, indicate that it is a species of *Stichococcus*.

Table 5 (continued)

NUTRIENT REQUIREMENTS AND UTILIZATION: LICHEN ALGAE – AMINO ACIDS[a,b]

REFERENCES

1. Chodat, R., *Monographie d'Algues en Culture Pure, Matér. Flore Cryptogam. Suisse*, 4, 1–266, 1913.
2. Fox, C. H., *Physiol. Plant.*, 20, 251–262, 1967.
3. Hutchinson, W. A., *Stereocaulon: Ecology and Physiology of Some Boreal Species and their Isolated Components*, Ph.D. dissertation, University of Massachusetts, Amherst, 1969.
4. Jaag, O., *Bull. Soc. Bot. Génève*, 21, 1–119, 1929.
5. Korniloff, M., *Bull. Soc. Bot. Génève*, 5, 114–132, 1913.
6. Quispel, A., *Recl. Trav. Bot. Neerl.*, 40, 413–541, 1943-1945.
7. Raths, H., *Ber. Schweiz. Bot. Ges.*, 48, 329–416, 1938.
8. Schofield, E. and Ahmadjian, V., in *Antarctic Terrestrial Biology*, Antarctic Res. Series, Vol. 20, Llano, G. A., Ed., American Geophysical Union, Washington, D.C., 1972, 97–142.
9. Warén, H., *Overs. Finsk. Vetenskaps.-Soc. Forh.*, 61, 1–79, 1918-1919.

Table 6
NUTRIENT REQUIREMENTS AND UTILIZATION: LICHEN ALGAE – OTHER NITROGENOUS SUBSTANCES[a,b]

The criterion used by most studies to determine utilization of nitrogenous substances compared growth in a N-free medium (control) with growth in media to which specific nitrogenous substances were added. In one study[2] growth was expressed as a percentage of that obtained in a control medium that contained 0.01 M nitrate.

Lichen (synonym)	Acetamide	Adenine	Ammonium	Casamino acids	Casein hydrolysate	Cytidine	Glutathione	Guanosine	Guanine	Nitrate	Nitrite	Peptone	Urea	Other	Ref.
					Trebouxia Symbiont										
Alectoria implexa	Ψ		U							u		U	Ψ		11
Anaptychia ciliaris (*Physcia ciliaris*)	u		U							u		u	Ψ		11
Buellia pernigra[c]			u							u		U	Ψ		9
Caloplaca cerina										U		U			10
Caloplaca murorum										U		U			10
Chaenotheca chrysocephala			U							U	Ψ				8
Cladonia bacilliformis v. *irregularis* (*C. irregularis*)			U								Ψ				4
Cladonia cristatella		u	U	U		u	u	U	U	U		U		u[d]	2
Cladonia digitata										U		U			10
Cladonia endiviaefolia			U								Ψ				4
Cladonia fimbriata										U		U			10
Cladonia furcata			U							u	Ψ[e]	U			4, 5
Cladonia pyxidata			U							U[d]	Ψ[e]	U			4, 5, 10
Cladonia squamosa										U		U			10
Lecanora rubina		u	U	U		u		U	U					u[d]	2
Parmelia acetabulum			U[e]							U	u	U			4, 7
Parmelia caperata			U							U	U	U			4, 6
Parmelia saxatilis			U								u				4
Parmelia scortea			u								u				4
Parmelia sulcata			u								u				4
Physconia pulverulenta (*Physcia pulverulenta*)	U		U							u		u	Ψ		11
Polycauliona citrina			u							u		U	u		9
Pseudevernia furfuracea (*Parmelia furfuracea*)	u		u							u		u	Ψ		11
Ramalina fraxinea	u		U							u		u	Ψ		11

[a] Many of the algal symbionts that have been isolated from different lichens have not been classified into acceptable species. To avoid confusion, the algal symbionts in this table are identified according to the lichen from which they were isolated.

[b] U = utilized; Ψ = not utilized; u = poorly utilized.

[c] Lichen collected from the Cape Hallett region of Antarctica near penguin and skua breeding areas. Peptone was by far the preferred nitrogen source for the algal symbiont, even better than ammonium compounds and urea.

[d] Uracil.

[e] By one isolate; utilized poorly by one isolate.

Table 6 (continued)
NUTRIENT REQUIREMENTS AND UTILIZATION: LICHEN ALGAE – OTHER NITROGENOUS SUBSTANCES[a,b]

Lichen (synonym)	Acetamide	Adenine	Ammonium	Casamino acids	Casein hydrolysate	Cytidine	Glutathione	Guanosine	Guanine	Nitrate	Nitrite	Peptone	Urea	Other	Ref.
Trebouxia Symbiont (continued)															
Stereocaulon dactylophyllum v. *occidentale*			U		u					u	u	U			3
Umbilicaria deusta (*Gyrophora flocculosa*)	u		u							u		u	∅		11
Xanthoria parietina	u	u	U	U		u	∅	∅		U[f]	u	U[f]	∅	u[d]	2, 6, 10, 11
Chlorella Symbiont															
Calicium chlorinum v. *exsertum*			U							U	U				8
Coccomyxa Symbiont															
Peltigera aphthosa	u		U							U		U	∅		11
Solorina sp.	U		U							U		U			1
Stichococcus Symbiont															
Chaenotheca stemonea			U							U	U				8
Coniocybe furfuracea			U							U	U				8
Dermatocarpon miniatum	u		U							U		U	∅		11

[f]   By two isolates; utilized poorly by one isolate.

REFERENCES

1. **Chodat, R.,** *Monographie d'Algues en Culture Pure, Matér. Flore Cryptogam. Suisse,* 4, 1–266, 1913.
2. **Fox, C. H.,** *Physiol. Plant.,* 20, 251–262, 1967.
3. **Hutchinson, W. A.,** *Stereocaulon: Ecology and Physiology of Some Boreal Species and their Isolated Components,* Ph.D. dissertation, University of Massachusetts, Amherst, 1969.
4. **Jaag, O.,** *Bull. Soc. Bot. Génève,* 21, 1–119, 1929.
5. **Korniloff, M.,** *Bull. Soc. Bot. Génève,* 5, 114–132, 1913.
6. **Manco, P. A.,** *A Study of Two Lichen Phycobionts of the Genus Trebouxia in Culture,* M.S. thesis, University of Tennessee, Knoxville, 1962.
7. **Quispel, A.,** *Recl. Trav. Bot. Neerl.,* 40, 413–541, 1943–1945.
8. **Raths, H.,** *Ber. Schweiz. Bot. Ges.,* 48, 329–416, 1938.
9. **Schofield, E. and Ahmadjian, V.,** in *Antarctic Terrestrial Biology,* Antarctic Res. Series, Vol. 20, Llano, G. A., Ed., American Geophysical Union, Washington, D.C., 1972, 97–142.
10. **Thomas, E. A.,** *Beitr. Kryptogamenflora Schweiz,* 9, 1–206, 1939.
11. **Warén, H.,** *Overs. Finsk. Vetenskaps.-Soc. Forh.,* 61, 1–79, 1918–1919.

Plants

QUALITATIVE REQUIREMENTS AND UTILIZATION OF NUTRIENTS: PTERIDOPHYTES

A. E. DeMaggio and M. Krasnoff

In examining the nutritional requirements of ferns, a distinction has been made between gametophytes and sporophytes. These plants represent two independent and separate phases in the fern life cycle. Gametophytes are produced from the germination of spores. They are haploid, free-living, photosynthetic plants. Ordinarily, they grow in shaded, woodland habitats and are small and inconspicuous. They are essentially two-dimensional, chordate or filament shaped, and only one to several cell layers thick. The plants grow prostrate in contact with the soil, and sex organs containing sperms and eggs are produced on them. In the presence of water, sperms are released and the eggs fertilized. The sporophyte is the diploid plant that develops from the fertilized egg and produces shoots, roots, and leaves. It differentiates into the adult fern plant which is easily recognized.

Like that of other vascular plants, the nutrition of fern plants is relatively simple. Fern gametophytes and sporophytes are photosynthetically functional and produce a variety of organic constituents, some of which are utilized during the growth and development of the plants. Since gametophytes and sporophytes grow as individual plants, they are able to obtain most of the mineral elements necessary for their functioning directly from the soil. Systematic studies have not been performed to determine exactly which elements can be extracted from the soil by ferns and which elements must be provided for them. It is generally believed that fern plants require the same major and minor elements that are known to be essential for most other vascular plants. The elements required in rather large amounts, which make up the major elements, are calcium, magnesium, nitrogen, phosphorus, potassium, and sulfur. Other elements such as boron, chlorine, copper, iron, manganese, molybdenum, and zinc are generally required in very small amounts for plant growth, and are the minor elements. Table 1* indicates the essential major and minor elements and the form in which they are available to plants. Additional elements, including cobalt, sodium, and selenium, are sometimes beneficial and improve growth of specific plants, but they do not appear to play an important role in the nutrition of ferns.

The development of techniques for growing ferns under sterile conditions has contributed only slightly to our understanding of their nutritional requirements.[28] For most ferns studied, solutions containing major elements alone or a combination of major and minor elements will satisfactorily support spore germination, gametophyte development, and growth of the sporophyte. Whether or not each element is required or utilized by a particular fern species remains to be determined.

It is clear that the concentration of individual major or minor elements can have a profound influence on the growth of various ferns. Our survey revealed that the effects of most compounds varied with the concentration supplied to the plant. Even sugars, which generally promote growth of fern plants, were found to inhibit growth at elevated concentrations. In addition, the effect of a particular concentration of a compound is influenced by the stage of development or age of the plant. These important aspects of nutrition in fern plants are not included here, but information on these topics is provided by Miller.[28]

Many compounds exert a particular effect when supplied individually to fern plants. However, when supplied in combination with other compounds the effect may be enhanced or inhibited. For example, some data are available indicating that certain amino

* All tables appear at end of text.

acids are not well tolerated by fern gametophytes when supplied individually.[38] When the same amino acids are combined, plant growth is promoted. Testing of additional compounds on many more species of ferns is needed before general trends concerning the synergistic and antagonistic action of nutrients are evident.

In Tables 2 through 7, we have compiled information concerning the influence of carbohydrates, nucleic acids, amino acids, growth regulators, and selected miscellaneous compounds on spore germination and the normal process of growth in fern gametophytes and sporophytes. Information on the precise nutritional requirements for ferns is scanty; only small numbers of ferns have been studied, and no complete or systematic investigations of their nutritional needs have been conducted. Miller[28] has reviewed much of the existing literature on nutrition of fern gametophytes. This review should be consulted for a more detailed discussion of the general nutritional requirements during various stages of gametophyte development. No similar treatment for fern sporophytes exists, but White[47] provides some information on nutrition in his consideration of experimental studies of fern sporophytes.

Certain nutrients and growth regulators are capable of causing profound alterations in the fern life cycle. In some ferns, sporophytes can be induced to grow directly from vegetative cells of the gametophyte, called apogamous sporophytes. Since there is no union of gametes in the formation of these plants, they bear the same chromosome complement as the gametophyte. The production of gametophytes from vegetative cells of the sporophyte is also possible, and in this case the gametophytes are termed aposporous. In their formation the meiotic process or chromosome reduction is by-passed, and the plants are chromosomally equivalent to the sporophyte from which they arose. Apogamy and apospory thus represent variations in the life cycle of ferns, but little is known about their nutritional regulation. There is some evidence that simple nutritional substances such as sugars or growth regulators such as ethylene which influence normal development can promote life cycle variations. It is possible that to achieve their effect these substances alter the usual nutritional pattern in ferns. Whittier[48] provides an interesting review of these processes, and should be consulted for additional information.

Table 1
ELEMENTS ESSENTIAL FOR
MOST HIGHER PLANTS

Element	Chemical symbol	Form available to plants
	Major Elements	
Calcium	Ca	Ca^{++}
Carbon	C	CO_2
Hydrogen	H	H_2O
Magnesium	Mg	Mg^{++}
Nitrogen	N	NO_3^-, NH_4^+
Oxygen	O	O_2, H_2O
Phosphorus	P	$H_2PO_4^-$, $HPO_4=$
Potassium	K	K^+
Sulfur	S	$SO_4=$
	Minor Elements	
Boron	B	$BO_3 \equiv$, $B_4O_7=$
Chlorine	Cl	Cl^-
Copper	Cu	Cu^+, Cu^{++}
Iron	Fe	Fe^{+++}, Fe^{++}
Manganese	Mn	Mn^{++}
Molybdenum	Mo	$MoO_4=$
Zinc	Zn	Zn^{++}

Modified from Salisbury, F. B. and Ross, C., *Plant Physiology*, Wadsworth, Belmont, Calif., 1969.

Table 2
NUTRITIONAL INFLUENCES ON SPORE GERMINATION

S = stimulatory; s = moderately supportive; $ = not supportive; ψ = not utilized.

	Glucose	Sucrose	Ethylene	GA (gibberellic acid)	IAA (indole-acetic acid)	NAA (naphthalene-acetic acid)	IAN (indole-aceto nitrile)	TIBA (triiodobenzoic acid)	Cycloheximide	Ammonium-3, 4-dichloro-phenoxyacetic acid	Phenylboric acid	Fatty acid	Coumarin	Ref.
Adiantum caeneatum	$													19
Alsophila australis		ψ			ψ	ψ	ψ							20, 39
Anemia phyllitidis				S										46
Gymnogramme calomelanos				S	ψ	ψ	ψ						$	39, 40
Lygodium japonicum												$a		46
Onoclea sensibilis			$									$a		32
Pteridium aquilinum	$								$		s			19
Pteris cretica					S									43
Pteris longifolia						$		$		ψ				45
Pteris sp.														

a Addition of antheridogen overcomes the inhibitory effect.

Table 3
UTILIZATION OF CARBOHYDRATES

S = stimulatory; s = moderately supportive; $ = not supportive; Ψ = not utilized.

	Glucose	Xylose	Sucrose	Soluble starch	Fructose	Lactose	Mannose	Maltose	Dextrose	Galactose	Galactiol	Inositol	Sorbitol	Ref.
Gametophyte														
Adiantum cuneatum	$													19
Alsophila australis	S		S		s			S						20
Asplenium sp.	S													18
Pteris calomelanos	S[a]		S[a]		S[a]									5
Pteris cretica				S										21
Pteris vittata		$							S					18
Todea barbara			S[b]							$				18
Osmunda japonica	S		S											22
Sporophyte														
Marsilea drummondii	S		S		S	$	$							4
Pteris longifolia	S											Ψ		10
Salvinia rotundifolia	$		S		$									15
Todea barbara			S									S	S	44
Marsilea drummondii (leaf)	S		S		S									4
Marsilea vestita (leaf)	$		S								Ψ	Ψ	Ψ	14
Marsilea vestita (roots)											Ψ			26
Osmunda cinnamomea (leaf)	S		S		s									7, 17
Pteris longifolia (root)	S													10
Pteris vittata (root)	S		S[c]											25
Todea barbara (leaf)			S											44

[a] Sucrose, glucose, and fructose effect differentiation of callus into sporophyte.
[b] Induces tracheary element differentiation.
[c] Induces sporophyte callus.

Table 4

UTILIZATION OF NUCLEIC ACID BASES AND ANALOGUES

S = stimulatory; $ = not supportive; Ψ = not utilized.

	Adenine	Guanine	Uracil	8-Azaadenine	8-Azaguanine	5-Fluorouracil	2-Thiocytosine	2-Thiouracil	Ref.
Gametophyte									
Asplenium nidus						$			34
Onoclea sensibilis					$				29
Phymatodes nigrescens[a]				$	$		$	$	49
Sporophyte									
Ceratopteris thalictroides	S								16
Marsilea drummondii (leaf)	Ψ	Ψ	Ψ						1

[a] Addition of riboflavin to these four inhibitors completely reverses their effect.[5]

Table 5

UTILIZATIONS OF AMINO ACIDS AND AMIDES

S = stimulatory; s = moderately supportive.

	Alanine	Asparagine	Aspartic acid	Glutamic acid	Glutamine	Glycine	Leucine	Phenylalanine	Serine	Valine	Alanine and phenylalanine	Ref.
Gametophyte												
Gymnogramme calomelanos	s	S	S	S	S	s	s	S	s	s	S	35—38
Sporophyte												
Marsilia drummondii (leaf)	S		S									4

Table 6
UTILIZATION OF GROWTH REGULATORS BY GAMETOPHYTES

S = stimulatory; s = moderately supportive; $ = not supportive.

	Ethylene	Phenylboric acid	IAA (indole-acetic acid)	NAA (napthalene-acetic acid)	TIBA (triiodobenzoic acid)	Kinetin	Ref.
Asplenium sp.			S				28
Onoclea sensibilis	$		$				13, 29, 42
Pteris aquilinum		s					6
Pteris longifolia			S		$	S	12
Pteris sp.			S	$			45
Pteris vittata			$				22, 23

Table 7
UTILIZATION OF GROWTH REGULATORS BY SPOROPHYTES

S = stimulatory; s = moderately supportive; $ = not supportive; Ψ = not utilized.

	GA (gibberellic acid)	IAA (indole-acetic acid)	NAA (napthalene-acetic acid)	IAN (indole-aceto nitrile)	Kinetin	2,4 D (2,4-dichloro-phenoxy-acetic acid)	Coconut water	Ref.
Marsilia drummondii				$				2
Pteris longifolia		S	S			$	S	11
Salvinia rotundifolia		s						15
Todea barbara							S	8
Ceratopteris thalictroides (leaf)	S	S[a]						16, 41
Marsilea drummondii (leaf)	S	Ψ			$			3
Marsilea drummondii (root)		$						1
Marsilea vestita (root)					$			26
Pteris aquilinum (root)		s			S		S	30

[a] Normal plantlet developed from excised meristem.

Table 8
UTILIZATION OF MISCELLANEOUS COMPOUNDS

S = stimulatory; $ = not supportive; Ψ = not utilized.

	Biotin	Casein acid hydrolysate	Colchicine	Cycloheximide	IAA (indole-acetic acid) + casein	Nicotinic acid	Pyridoxine HCl	Thiamine	Yeast extract	Ref.
Asplenium sp.			$							
Goniopteris multilineata			$							27
Marsilea vestita (leaf)		Ψ							Ψ	14
Onoclea sensibilis			S							42
Pteris aquilinum		Ψ		$		$	$	$		31, 33
Pteris longifolia	Ψ					S		Ψ		10
Pteris aquilinum (root)		S			S					30
Pteris vittata (root)		$							$	24

REFERENCES

1. Allsopp, A., *Ann. Bot.* (London), 16, 165–184, 1952.
2. Allsopp, A., *J. Exp. Bot.,* 7, 1–13, 1956.
3. Allsopp, A. and Szweykowska, A., *Nature,* 186, 813–814, 1960.
4. Allsopp, A., *J. Linn. Soc. London Bot.,* 58, 417, 1963.
5. Bristow, M. J., *Dev. Biol.,* 4, 361–375, 1962.
6. Caruso, J. L., *Can. J. Bot.,* 51, 1998–2001, 1973.
7. Clutter, M. E. and Sussex, I. M., *Bot. Gaz.,* 126, 72–78, 1965.
8. DeMaggio, A. E., *Am. J. Bot.,* 48, 551–565, 1961.
9. DeMaggio, A. E., *Bot. Gaz.,* 133, 311–317, 1972.
10. Durand-Rivieres, R., *C. R. Acad. Sci.* (Paris), 250, 2442–2444, 1960.
11. Durand-Rivieres, R. and Fillon, F., *C. R. Acad. Sci.* (Paris), 266, 471–473, 1968.
12. Durand-Rivieres, R., *Ann. Sci. Nat. Bot.,* 5, 1–86, 1964.
13. Edwards, M. E. and Miller, J. H., *Am. J. Bot.,* 59, 450–457, 1972.
14. Gaudet, J. J., *Can. J. Bot.,* 45, 1127–1134, 1967.
15. Gaudet, J. J. and Koh, D. V., *Bull. Torrey Bot. Club,* 95, 92–102, 1968.
16. Gottlieb, J. E., *Bot. Gaz.,* 133, 299–304, 1972.
17. Harvey, W. H. and Caponetti, J. D., *Can. J. Bot.,* 51, 341, 1973.
18. Hurel-Py, G., *C. R. Acad. Sci.,* 214, 571–573, 1942.
19. Hurel-Py, G., *C. R. Acad. Sci.,* 224, 950–952, 1947.
20. Hurel-Py, G., *C. R. Acad. Sci.,* 240, 1119–1121, 1955.
21. Kato, Y., *Bot. Gaz.,* 125, 33–37, 1964.
22. Kato, Y., *Cytologia,* 30, 67–74, 1965.
23. Kato, Y., *Planta,* 77, 127–134, 1967.
24. Kato, Y., *Bot. Mag.,* 85, 307–311, 1972.
25. Kato, Y., *Cytologia,* 38, 117–124, 1973.
26. Laetsch, W. H. and Briggs, W. R., *Am. J. Bot.,* 48, 369–377, 1961.
27. Mehra, P. N. and Loyal, D. S., *Ann. Bot.* (London), 20, 544–552, 1956.
28. Miller, J. H., *Bot. Rev.,* 34, 361–440, 1968.
29. Miller, J. H., *Physiol. Plant.,* 21, 699–710, 1968.
30. Munroe, M. H. and Sussex, I. M., *Can. J. Bot.,* 47, 617–621, 1969.
31. Partanen, J. N. and Partanen, C. R., *Can. J. Bot.,* 41, 1657–1661, 1963.
32. Pringle, R. B., *Plant Physiol.,* 45, 315–317, 1970.
33. Raghavan, V., *Exp. Cell. Res.,* 65, 401–407, 1971.
34. Raghavan, V., *Physiol. Plant.,* 30, 137–142, 1974.
35. Sossountzov, I., *Physiol. Plant.,* 4, 726–742, 1954.
36. Sossountzov, I., *Physiol. Plant.,* 7, 1–15, 1954.
37. Sossountzov, I., *Physiol. Plant.,* 7, 383–396, 1954.
38. Sossountzov, I., *C. R. Soc. Biol.,* 149, 1374–1377, 1955.
39. Sossountzov, I., *C. R. Soc. Biol.,* 151, 531–536, 1957.
40. Sossountzov, I., *C. R. Soc. Biol.,* 155, 1006–1010, 1961.
41. Stein, D., *Plant Physiol.,* 48, 416, 1971.
42. Stockwell, C. R. and Miller, J. H., *Am. J. Bot.,* 61, 375–378, 1974.
43. Strickler, B., *Bot. Gaz.,* 108, 101–114, 1946.
44. Sussex, I. M. and Steeves, T. A., *Bot. Gaz.,* 119, 203–208, 1958.
45. Van Onsem, J. G., *Meded. Landbouwhogesch. Opzoekingsstn. Staat Gent,* 17, 174, 1952.
46. Weinberg, E. S. and Voeller, B. R., *Am. Fern J.,* 59, 153–167, 1969.
47. White, R., *Bot. Rev.,* 37, 509–540, 1971.
48. Whittier, D., *BioScience,* 21, 225–227, 1971.
49. Yeoh, O. C. and Raghavan, V., *Plant Physiol.,* 41, 1739–1742, 1966.

QUALITATIVE REQUIREMENTS AND UTILIZATION OF NUTRIENTS: ANGIOSPERMS

J. F. Loneragan

QUALITATIVE REQUIREMENTS FOR NUTRIENTS

The chemical elements needed by plants are usually considered in three groups according to the quantities required and the source of supply. They are

1. *The structural elements* C, H, and O, which generally account for 99% of the weight of living plants and for 90 to 95% of their dry weight. In autotrophic angiosperms, these elements are supplied by CO_2 and H_2O.

2. *The macronutrient elements* Ca, Mg, K, P, S, and N, which account for most of the remaining 5 to 10% of the plant dry weight. In general, angiosperms obtain these elements by absorption of inorganic ions from solution through their roots. Some angiosperms, acting in symbiotic association with other organisms, may also obtain N from gaseous N_2 in the atmosphere.

3. *The micronutrient elements or trace elements* B, Cl, Co, Cu, Fe, Mn, Mo, Na, and Zn, which are generally absorbed as inorganic ions from solution. Current evidence suggests that, with the exception of Co and Na, all angiosperms require all of these micronutrient elements. A limited number of species appear to need Co and Na. There have also been claims that some angiosperms need Se and Si.

Cobalt appears to be necessary for nitrogen fixation in all symbiotic associations which have been investigated, including those between *Rhizobium* and legumes[1,2,27] and between fungi and *Alnus, Casuarina,* and *Myrica.*[5,16] However, in most investigations, addition of inorganic ions of N has eliminated the need for Co. Some investigators have reported responses of some species to Co in the presence of inorganic N (e.g., rubber and tomato,[4] subterranean clover and wheat[34]), but suggestions that any angiosperm requires Co for activities other than N fixation cannot yet be accepted.

Sodium is absolutely necessary for the growth of some angiosperms and promotes the growth of others. For some other species, Na only promotes growth when K is low (Table 1), while for some species Na does not promote growth under any conditions. Brownell and Crossland[7] have postulated that all plants with the C_4 dicarboxylic photosynthetic pathway require Na, while those with the C_3 pathway do not.

Silicon is essential for growth of diatoms and of the horsetail *Equisetum arvense* but so far has not been shown to be essential for the growth of higher plants. However, many higher plants respond to the addition of Si in particular situations. Jones and Handreck[19] attribute the reported responses of higher plants to Si to a variety of indirect effects of Si or SiO_2, including alleviation of Mn toxicity, increasing P uptake from soils, increasing resistance to fungal and insect attack, causing cereal leaves to assume a more erect disposition, preventing lodging in cereal crops, decreasing loss of seed from shattering of the inflorescence, and decreasing transpiration.

Selenium has been claimed to be essential for the growth of some species, but recent critical work has failed to support this claim.[8,29]

UTILIZATION OF NUTRIENTS

The order of concentration of various elements that angiosperms require for their

Table 1

RESPONSE OF SOME HIGHER PLANTS TO SODIUM, AS RELATED TO POTASSIUM SUPPLY[7,26]

Sodium essential	Response with ample potassium supply		Response with insufficient potassium supply	
	Large	Slight to moderate	Moderate to large	None to moderate
Echinochloa utilis	Beet	Cabbage	African violet	Bean, white
L. Ohwi et Yabuno	Chard	Celeriac	Alfalfa	Buckwheat
(Japanese millet)	Fodder	Coconut	Asparagus	Clover, red
Cynodon dactylon L.	Mangel	Kale	Barley	Corn
(Bermuda grass)	Sugar	Kale, marrowstem	Broccoli	Cucumber
Kyllinga brevifolia	Table	Kohlrabi	Brussels sprouts	Grass
Amaranthus tricolor	Celery	Lupine	Carrot	Bahia
L. cv. Early	Spinach	Oats	Chicory	Bermuda
Splendour	Turnip	Pea	Clover, ladino	Carpet
Atriplex nummularia		Radish	Cotton	Kentucky blue
Lindl. (oldman, giant		Rape	Flax	Pensacola Bahia
saltbush)		Rubber	Grass	Weeping love
Atriplex paludosa R.Br.		Rutabaga	Pangola	*Lespedeza sericea*
(marsh saltbush)		*Salicornia herbacea*	Sudan	Lettuce
Atriplex quinii Fv.M.			Horseradish	Onion
Atriplex semibaccata			Millet	Parsley
R.Br. (berry saltbush)			Mustard	Parsnip
Atriplex inflata Fv.M.			Rape	Peppermint
Atriplex leptocarpa			Salsify	Potato
Fv.M.			Stock	Rutabaga
Atriplex spongiosa			Tomato	Rye
Fv. M. (pop salt-			Vetch	Soybean
bush) .			Wheat	Spearmint
Atriplex semilunalaris				Squash
Aellen				Strawberry
Atriplex lindleyi Moq.				Sunflower
Atriplex vesicaria				*Chenopodium capitatum* L.
Howard ex Benth				Aschers
(bladder saltbush)				*Atriplex hortensis* L. var.
Kochia childsii Hort.				Atrosanguineae (Garden
Halogeton glomeratus				orache)
(Bieb) Meyer				*Atriplex angustifolia* Sm.
Portulacca grandi-				*Atriplex glabriuscula*
flora Hook (rose				Edmonton
moss)				*Atriplex albicans* Ait.
				Kochia pyramidata Benth.
				Aster tripolium L
				Exomis axyrioides Fenzl ex
				Moq.

growth is presented in Table 2.[15] While they are useful in broadly indicating plant requirements, the values presented in Table 2 cannot be used for accurately defining deficiency or adequacy of nutrients for plants. Problems arise from the fact that plants vary in their ability to utilize nutrients efficiently, so that the "critical concentration" at which a nutrient is just sufficient for maximum growth may vary. Nutrient utilization at the critical concentration is not a unique value because it depends upon at least two distinct physiological activities of the plant, each of which may vary with the nutrient, plant species, rate and stage of plant growth, and environmental conditions. Nutrient utilization depends upon:

1. The movement of nutrient to sites of function within the plant ("nutrient mobility")

Table 2

CONCENTRATIONS OF NUTRIENT ELEMENTS IN PLANT MATERIAL AT LEVELS CONSIDERED ADEQUATE

Element	Chemical symbol	Atomic weight	Concentrations in dry matter		Relative number of atoms with respect to molybdenum
			μmol/g	ppm or %	
				ppm	
Molybdenum	Mo	95.95	0.001	0.1	1
Copper	Cu	63.54	0.10	6	100
Zinc	Zn	65.38	0.30	20	300
Manganese	Mn	54.94	1.0	50	1,000
Iron	Fe	55.85	2.0	100	2,000
Boron	B	10.82	2.0	20	2,000
Chlorine	Cl	35.46	3.0	100	3,000
				%	
Sulfur	S	32.07	30	0.1	30,000
Phosphorus	P	30.98	60	0.2	60,000
Magnesium	Mg	24.32	80	0.2	80,000
Calcium	Ca	40.08	125	0.5	125,000
Potassium	K	39.10	250	1.0	250,000
Nitrogen	N	14.01	1,000	1.5	1,000,000
Oxygen	O	16.00	30,000	45	30,000,000
Carbon	C	12.01	40,000	45	40,000,000
Hydrogen	H	1.01	60,000	6	60,000,000

From Epstein, E., in *Plant Biochemistry,* Bonner, J. and Varner, J. E., Eds., Academic Press, New York, 1965, 438—466. With permission.

2. The amount of nutrient required at the sites of function within the plant ("functional nutrient requirement")

Nutrient mobility may have an overriding importance in nutrient utilization, as has been clearly shown for Ca. After absorption, Ca moves in the xylem with the transpiration stream to leaves and other plant organs, but little if any Ca moves in plant phloem. As a result, Ca does not move out of leaves in which it has been deposited. In addition, for their Ca supply, roots and shoots of growing plants depend upon a continuous supply of Ca in the growth medium. As soon as the external supply of Ca becomes inadequate, Ca deficiency develops, regardless of the level of Ca in old leaves.

This behavior of Ca in the plant causes the values obtained for nutrient utilization at the critical concentration to vary widely, depending upon the conditions of Ca supply under which Ca deficiency develops. Thus, subterranean clover and corn plants given a low but continuous supply of Ca and containing only 0.2 and 0.04% Ca, respectively, in their tops grew well, with no symptoms of Ca deficiency. In contrast, plants transferred from an initially high to a low concentration of Ca in solution developed severe Ca deficiency with five to ten times these concentrations of Ca in their tops.[25]

These studies indicate that nutrient mobility in the phloem is an important aspect in nutrient utilization. Nutrients which, like Ca, tend to stay in leaves in which they have been deposited will give variable values, depending upon the conditions of nutrient supply under which deficiency develops. Nutrients such as N, P, and K that appear to move rapidly from regions of excess to regions of deficiency within the plant should give values largely independent of the conditions of nutrient supply. The behavior of some other nutrients such as S and Cu which varies with the nutrient status of the plant is more complex and limits the value of nutrient concentrations of whole plants as an indication of nutrient status of the plant.[21-24]

Functional nutrient requirement may also vary in plants, resulting in large variations in nutrient utilization. Cobalt provides an extreme example, being required for the growth of N-fixing plants only when they have no supply of fixed N, as has already been mentioned. In some species, the functional K requirement also varies widely, largely as a result of the supply of Na, which is apparently able to partially replace K in its function. In these species, the critical concentration of K for diagnosing K deficiency may change dramatically with Na supply; for example, the concentration of K in the tops of Rhodes grass growing at 95% of maximum yield varied from 2.7% in plants not given Na to 0.5% in plants receiving Na.[30]

These examples show the importance of both nutrient mobility and functional nutrient requirement to nutrient utilization, but too little is known of the way these physiological activities vary with environmental conditions and stage and rate of plant growth to make further useful generalizations. However, it is clear that the variations in nutrient utilization are sufficiently large to restrict the usefulness of values of critical concentrations for diagnosing nutrient deficiencies to specific crops and situations. Under these circumstances, it may be useful to report as examples of the order of nutrient utilization encountered the concentrations of nutrients that have been measured in plants that appear healthy or that had definite symptoms of either deficiency or toxicity. A very thorough tabulation of this information has recently been made by Chapman.[9] The data are far too extensive to report in full here, but from them I have chosen a few species which encompass the common experience (Tables 3 and 4). Readers interested in more specific information on other species or in understanding more fully the implications of the data presented should refer to the original publication.

Table 3
CONCENTRATIONS OF MACRONUTRIENT ELEMENTS IN PLANTS REPORTED AS USEFUL IN INDICATING NUTRIENT STATUS

Element	Plant	Tissue sampled Type	Condition	Deficient	Low	Mid	High	Toxic	Ref.
					Range in % dry matter				
Ca	Alfalfa (*Medicago sativa*)	Tops		0.58	0.55–2.0	0.86–3.88			10
	Corn (*Zea mays*)	Tops	25 days old	0.30	—	0.70–0.80			10
		Entire plant		0.18–0.32	—	0.38–0.43			10
		Tops	Tasseling	—	—	0.45–0.61			10
	Orange (*Citrus sinensis*)	Leaves	Spring cycle from fruiting terminals	1.5–2.5	<3.0	3.0–5.5	5.0–7.0		10
		Leaves	Bloom cycle from nonfruiting terminals	1.50(?)	1.60–2.90	3.00–5.50	5.60–6.90	>7.00	10
	Tomato (*Lycopersicon esculentum*)	Leaves	65 days from seeding	0.58		5.70			10
		Tops		0.79–0.96		0.82–1.78			10
		Fruit	Green and ripe		0.042 (blossom-end rot)	0.066 (normal)			10
		Tops	5–6 weeks old		0.36–0.60	0.83–1.32			10
		Lower leaves	From plants 26 days old			3.20–4.40			10
	Wheat (*Triticum aestivum*)	Tops	Young plants	0.14		1.38			10
K	Alfalfa (*Medicago sativa*)	Plant	August	0.44–1.02		1.26–2.30			31
		Stems	One-tenth in bloom	0.48–0.94		1.50			31
		Leaves		0.14		0.83			31
		Stems		0.83		0.91			31
	Orange (*Citrus sinensis*)	Leaves	Spring cycle, from nonfruiting terminals	<0.40–0.60	0.70–1.10	1.20–1.70	1.80–2.30	>2.40	31
		Leaves	Spring cycle, from fruiting terminals	<0.25–0.35	0.40	0.40–1.50	1.50	>1.50–2.00	31

Table 3 (continued)

CONCENTRATIONS OF MACRONUTRIENT ELEMENTS IN PLANTS REPORTED AS USEFUL IN INDICATING NUTRIENT STATUS

Element	Plant	Tissue sampled Type	Condition	Range in % dry matter Deficient	Low	Mid	High	Toxic	Ref.
	Corn (Zea mays)	Leaves		0.58–0.78		0.74–1.49			31
		Leaves	Third from base, at tasseling	0.39–1.30		1.46–5.80			31
		Lower stems	August	0.029–0.10 (% in sap)		0.33–0.59 (% in sap)			31
		Plant	Early	0.88–1.37		1.00–4.54			31
	Tomato (Lycopersicon esculentum)	Leaves	April–May	0.96–1.23		1.55–3.76			31
		Stems	April–May	0.86–1.06		1.55–5.30			31
		Lower leaves	April–May	0.30–1.00		1.40–2.40			31
		Upper stems	April–May	0.20–3.20		1.90–6.00			31
		Leaf blades	From middle third (February–June)	0.48–0.95 (plus Na)					31
		Leaf blades	—	0.30–0.70 (minus Na)					31
Mg	Orange (Citrus sinensis)	Leaves	3–10 months, spring cycle, from fruiting terminals	0.05–0.15	0.16–0.20	0.20–0.60	>0.60–1.00(?)		14
	Corn (Zea mays)	Blades		0.07		0.20			14
		Leaf sheaths		0.07		0.24			14
		Stem nodes		0.06		0.21			14
		Stem inter-nodes		0.04		0.10			14
	Tobacco (Nicotiana tobacum)	Leaves		0.13		0.23–0.35			14
		Leaves		0.08–0.20		0.18–0.65			14
		Stems		0.15–0.24		0.11–0.31			14
		Top leaves		0.41–0.47		0.48–0.98			14
		Bottom leaves		0.57–0.60		0.60–1.22			14

Table 3 (continued)

CONCENTRATIONS OF MACRONUTRIENT ELEMENTS IN PLANTS REPORTED AS USEFUL IN INDICATING NUTRIENT STATUS

Element	Plant	Tissue sampled		Range in % dry matter					Ref.
		Type	Condition	Deficient	Low	Mid	High	Toxic	
N	Orange (*Citrus sinensis*)	Leaves	4–7 months (July–Oct.) from behind fruit	<2.00		2.00–3.16		>3.50	18
		Leaves	4–7 months, spring cycle from nonfruiting terminals	<2.00	2.10–2.30	2.40–2.90	3.00–3.50	>3.60	18
		Leaves	4–7 months, bloom cycle, from nonfruiting terminals	<1.90	2.00–2.30	2.40–2.70	2.80–3.20	>3.30	18
		Leaves	8–12 months old	<2.00		2.50			18
		Leaves	3–4 months, spring cycle from nonfruiting terminals	<2.20	2.20–2.40	2.40–2.60	>2.60	>2.70	18
P	Alfalfa (*Medicago sativa*)	Shoots	Flowering	0.11–0.20	0.14	0.25–0.51			3
		Leaves	Flowering		0.19	0.35–0.53			3
		Stems	Flowering		0.11	0.25–0.47			3
	Orange (*Citrus sinensis*)	Leaves	Recently matured	0.06–0.12	0.07–0.12	0.08–0.30	0.17–0.44	>0.30	3
	Corn (*Zea mays*)	Leaves	Tasseling	0.11	0.17	0.20–0.25			3
		Ear leaves	Tasseling		0.22	0.25–0.33			3
		Leaves	Silking		0.19	0.23–0.46			3
		Second below ear at silking			0.17–0.18	0.18–0.23			3
		Stover	Mature		0.04	0.30–0.37			3
		Grain	Mature		0.23	0.43–0.80			3
		Leaves	Mature		0.05	0.12–0.23			3
	Tomato (*Lycopersicon esculentum*)	Leaves	Recently matured	0.010–0.180	0.18–0.38	0.44–0.90	>0.90		3
		Leaves	Early fruiting stage	0.24–0.35		0.42–0.72			3
		Stems	Early fruiting stage	0.20–0.25		0.30–0.55			3
		Fruit	Harvest		0.29	0.55	0.85		3

Table 3 (continued)
CONCENTRATIONS OF MACRONUTRIENT ELEMENTS IN PLANTS REPORTED AS USEFUL IN INDICATING NUTRIENT STATUS

		Tissue sampled		Range in % dry matter					
Element	Plant	Type	Condition	Deficient	Low	Mid	High	Toxic	Ref.
S	Alfalfa (*Medicago sativa*)	Leaves	Mature	0.12—0.26		0.34—0.62			13
		Hay	Mature	0.10—0.16		0.20—0.33	2.24	0.30—0.75	13
		Hay	Mature	0.01—0.13 (S as SO$_4$)	0.05—0.12 (S as SO$_4$)	0.14—0.22 (S as SO$_4$)			13
		Hay	Mature	0.05—0.15 (S as org.)	0.15—0.21 (S as org.)	0.14—0.20 (S as org.)			13
	Orange (*Citrus sinensis; Citrus aurantium*)	Leaves	Young	0.08—0.10		0.20—0.26			13
		Leaves	Old	0.12—0.13		0.22—0.26			13
		Leaves	3—7 months, from fruiting terminals	0.05—0.13	0.14—0.19	0.20—0.30	0.40—0.49		13
	Corn (*Zea mays*)	Tops		0.04		0.08			13
	Tobacco (*Nicotiana tabacum*)	Leaves		0.11		0.15			13
		Plant	June	0.18		0.23—0.26			13
		Plant	July	0.13		0.15—0.18			13

Table 4
CONCENTRATIONS OF MICRONUTRIENT ELEMENTS IN PLANTS REPORTED AS USEFUL IN INDICATING NUTRIENT STATUS

Element	Plant	Tissue sampled		Range in ppm or, for Cl only, % dry matter					Ref.
		Type	Condition	Deficient	Low	Mid	High	Toxic	
B	Alfalfa (*Medicago sativa*)	Leaves		15.00	13.00–17.00	28.00–654.00		516.00–996.00	6
	Orange (*Citrus sinensis*)	Leaves	4–6 months old	16.00–25.00		78–144		262–386	6
		Leaves	6–12 months old			140–265		267–1111	6
		Leaves	Mature	5–10		21–90		460–1700	6
		Leaves	3–7 months, spring cycle, from fruiting terminals					>200	6
	Corn (*Zea mays*)	Leaves	October	1–2		27–72		179	6
		Tops	25 days old			5–8		25	6
	Tomato (*Lycopersicon esculentum*)	Leaves	July-August			34–150		253–1416	6
		Plant		14–32		34–96		91–415	6
Cl	Bean, green pod (*Phaseolus* spp.)	Blades	8 days after flower buds appeared		0.63			4.35–5.35	12
		Blades	46 days after flower buds appeared		0.88			5.02–6.17	12
	Orange (*Citrus sinensis*)	Leaves	4–7 months, spring cycle, from nonfruiting terminals			<0.30	0.40–0.60	>0.70	12
	Corn (*Zea mays*)	Leaves	4–10 months, from fruiting terminals			<0.15	0.20–0.30	>0.40	12
		Leaves	At ear, silking stage		0.20–0.32	0.34–0.53			12
		Leaves	Second from base (June)		0.97		2.11		12
		Leaves	Opposite ear (August)		0.59		2.07		12
	Tobacco	Leaves			0.07–1.13	0.85	2.9–4.2		12
Co	Alfalfa (*Medicago sativa*)	Tops	Mature			0.02–0.24			32
		Tops	Early bloom			0.04–0.29			32
	Corn (*Zea mays*)	Grain	Mature			0.01			32
		Ears	Edible part			0.01			32
		Tops	Silage			0.04			32
	Tomato (*Lycopersicon esculentum*)	Fruit	Mature			0.06–0.25			32

Table 4 (continued)

CONCENTRATIONS OF MICRONUTRIENT ELEMENTS IN PLANTS REPORTED AS USEFUL IN INDICATING NUTRIENT STATUS

Element	Plant	Tissue sampled		Range in ppm or, for Cl only, % dry matter					Ref.
		Type	Condition	Deficient	Low	Mid	High	Toxic	
Cu	Orange (Citrus sinensis)	Leaves	3–7 months, spring cycle, from fruiting terminals			1–4	4–10	15	28
		Leaves	Bloom cycle, from non-fruiting terminals	<3.5–4.0	3.6–5.9	5.0–16.0	17–22	>23	28
	Clover, subterranean (Trifolium subterraneum)	Leaves	Blooming	<3.00		7.00–12.00			28
		Tops	Blooming	3.00		3.00–32.00			28
	Oats (Avena sativa)	Tops	Harvest	1.00–2.50		1–17			28
		Leaves	6–9 weeks old	<3.00		7.00–12.00			28
		Grain	Harvest			2–12			28
	Tomato (Lycopersicon esculentum)	Leaves	Harvest			3.10–12.30			28
		Fruit	Harvest			13.00–37.00			28
		Plant	Harvest			15.00–25.00			28
Fe	Orange (Citrus sinensis)	Leaves	Recently matured	11–68		40–137			33
	Corn (Zea mays)	Golden Bantam leaves	Recently matured	24.00–56.00		56.00–178.00			33
	Soybean (Glycine soja)	Shoots	34 days old	28.00–38.00		44.00–60.00			33
	Tobacco (Nicotiana tabacum)	Leaves	Recently matured	63.00–70.00		68.00–140.00			33
		Tops	45 days old	33.00–36.00		91.00–130.00			33
Mn	Bean (Phaseolus spp.)	Leaves		32–38		68.00		1000–3000	20
		Tops		3–13		200–1000			20
	Orange (Citrus sinensis)	Leaves		15		14–60		>200	20
		Leaves	3–7 months, spring cycle, from fruiting terminals		16–24	20–80			20
	Oats (Avena sativa)	Leaves	4–7 months old	15		25–200	300–500	1000	20
		Tops	Flowering	5–14		7–82			20
		Tops	After flowering	6–13		11–111			20
		Plants	11 weeks old	10		64–140			20

Table 4 (continued)
CONCENTRATIONS OF MICRONUTRIENT ELEMENTS IN PLANTS REPORTED AS USEFUL IN INDICATING NUTRIENT STATUS

Element	Plant	Tissue sampled		Range in ppm or, for Cl only, % dry matter					Ref.
		Type	Condition	Deficient	Low	Mid	High	Toxic	
	Tomato (*Lycopersicon esculentum*)	Leaves	October	5–7		46–398			20
		Fruit	Harvest	0.20		2.00			20
Mo	Alfalfa (*Medicago sativa*)	Leaves	10% in bloom	0.28		0.34			17
		Leaves	Pre-bloom (?)	<0.10		0.30–0.60			17
	Barley (*Hordeum vulgare*)	Blades	8 weeks old			<0.03			17
	Orange (*Citrus sinensis*)	Leaves		0.031–0.083		0.059–0.088			17
	Tomato (*Lycopersicon esculentum*)	Leaves	8 weeks old	0.13		0.68			17
Zn	Alfalfa (*Medicago sativa*)	Top half of shoots	In bloom	8.0		13.80			11
		Tops	12 weeks old	13.0		39.00–48.00			11
	Corn (*Zea mays*)	Lower leaves	Tasseling	9.0–9.30		31.10–36.60			11
		Leaves	Sixth node from base, at silking	14.60–15.10		>15.00			11
	Orange (*Citrus sinensis*)	Leaves	4–10 months, from fruiting terminals	4.00–15.00	15.00–24.00	25.00–100.00	110.00–200.00	>200.00(?)	11
	Tomato (*Lycopersicon esculentum*)	Leaves	Midseason	6.00–8.70		13.00			11
		Leaves	From middle third of plant	14.40		26.90			11
		Leaves	Basal and median, at fruit-setting stage	9.00–15.00		65.00–198.00		526.00–1489.00	11

REFERENCES

1. **Ahmed, S. and Evans, H. J.,** Effect of cobalt on the growth of soybeans in the absence of supplied nitrogen, *Biochem. Biophys. Res. Commun.,* 1, 271—275, 1959.
2. **Ahmed, S. and Evans, H. J.,** Cobalt: a micronutrient element for the growth of soybean plants under symbiotic conditions, *Soil Sci.,* 90, 205—210, 1960.
3. **Bingham, F. T.,** Phosphorus, in *Diagnostic Criteria for Plants and Soils,* Chapman, H. D., Ed., Division of Agricultural Sciences, University of California, 1966, 325—361.
4. **Bolle-Jones, W. W. and Mallikarjuneswara, V. R.,** Beneficial effect of cobalt on the growth of rubber plant (*Hevea braziliensis*), *Nature,* 179, 738—739, 1957.
5. **Bond, G. and Hewitt, E. J.,** Cobalt and the fixation of nitrogen by root nodules of *Alnus* and *Casuarina, Nature,* 195, 94—95, 1962.
6. **Bradford, G. R.,** Boron, in *Diagnostic Criteria for Plants and Soils,* Chapman, H. D., Ed., Division of Agricultural Sciences, University of California, 1966, 43—61.
7. **Brownell, P. F. and Crossland, C. J.,** The requirement for sodium as a micronutrient by species having the C_4 dicarboxylic photosynthetic pathway, *Plant Physiol.,* 49, 794—797, 1972.
8. **Broyer, T. C., Lee, D. C., and Asher, C. J.,** Selenium nutrition of green plants. Effect of selenite supply on growth and selenium content of alfalfa and subterranean clover, *Plant Physiol.,* 41, 1425—1428, 1966.
9. **Chapman, H. D., Ed.,** *Diagnostic Criteria for Plants and Soils,* Division of Agricultural Sciences, University of California, 1966.
10. **Chapman, H. D.,** Calcium, in *Diagnostic Criteria for Plants and Soils,* Chapman, H. D., Ed., Division of Agricultural Sciences, University of California, 1966, 65—91.
11. **Chapman, H. D.,** Zinc, in *Diagnostic Criteria for Plants and Soils,* Chapman, H. D., Ed., Division of Agricultural Sciences, University of California, 1966, 485—499.
12. **Eaton, F. M.,** Chlorine, in *Diagnostic Criteria for Plants and Soils,* Chapman, H. D., Ed., Division of Agricultural Sciences, University of California, 1966, 99—135.
13. **Eaton, F. M.,** Sulfur, in *Diagnostic Criteria for Plants and Soils,* Chapman, H. D., Ed., Division of Agricultural Sciences, University of California, 1966, 445—475.
14. **Embleton, T. W.,** Magnesium, in *Diagnostic Criteria for Plants and Soils,* Chapman, H. D., Ed., Division of Agricultural Sciences, University of California, 1966, 225—263.
15. **Epstein, E.,** Mineral metabolism, in *Plant Biochemistry,* Bonner, J. and Varner, J. E., Eds., Academic Press, New York, 1965, 438—466.
16. **Hewitt, E. J. and Bond, G.,** The cobalt requirement of nonlegume root nodule plants, *J. Exp. Bot.,* 17, 480—491, 1966.
17. **Johnson, C. M.,** Molybdenum, in *Diagnostic Criteria for Plants and Soils,* Chapman, H. D., Ed., Division of Agricultural Sciences, University of California, 1966, 286—301.
18. **Jones, W. W.,** Nitrogen, in *Diagnostic Criteria for Plants and Soils,* Chapman, H. D., Ed., Division of Agricultural Sciences, University of California, 1966, 310—323.
19. **Jones, L. H. P. and Handreck, K. A.,** Silica in soils, plants and animals, *Adv. Agron.,* 19, 107—149, 1967.
20. **Labanauskas, C. K.,** Manganese, in *Diagnostic Criteria for Plants and Soils,* Chapman, H. D., Ed., Division of Agricultural Sciences, University of California, 1966, 264—285.
21. **Loneragan, J. F.,** Nutrient concentration, nutrient flux, and plant growth, *Trans. 9th Int. Congr. Soil Sci.* (Adelaide), 2, 173—182, 1968.
22. **Loneragan, J. F.,** Nutrient requirements of plants, *Nature,* 188, 1307—1308, 1968.
23. **Loneragan, J. F.,** The availability and absorption of trace elements in soil-plant systems and their relation to movement and concentrations of trace elements in plants, in *Trace Elements in Soil-Plant-Animal systems,* Nicholas, D. J. D. and Egan, A., Eds., Academic Press, New York, 1975, 109—134.
24. **Loneragan, J. F., Robson, A. D., and Snowball, K.,** Nutrient mobilisation and its significance in plant nutrition, in *Transport and Transfer Processes in Plants,* Wardlaw, I., Ed., Academic Press, London, 1976, 463—469.
25. **Loneragan, J. F. and Snowball, K.,** Calcium requirements of plants, *Aust. J. Agric. Res.,* 20, 465—478, 1969.
26. **Lunt, O. R.,** Sodium, in *Diagnostic Criteria for Plants and Soils,* Chapman, H. D., Ed., Division of Agricultural Sciences, University of California, 1966, 409—432.
27. **Reisenauer, H. M.,** Cobalt in nitrogen fixation by a legume, *Nature,* 186, 375—376, 1960.
28. **Reuther, W. and Labanauskas, C. K.,** Copper, in *Diagnostic Criteria for Plants and Soils,* Chapman, H. D., Ed., Division of Agricultural Sciences, University of California, 1966, 157—179.

29. Shrift, A., Aspects of selenium metabolism in higher plants, *Ann. Rev. Plant Physiol.*, 20, 475–494, 1969.
30. Smith, F. W., The effect of sodium on potassium nutrition and ionic relations in Rhodes grass, *Aust. J. Agric. Res.*, 25, 407–414, 1974.
31. Ulrich, A. and Ohki, K., Potassium, in *Diagnostic Criteria for Plants and Soils*, Chapman, H. D., Ed., Division of Agricultural Sciences, University of California, 1966, 362–393.
32. Vanselow, A. P., Cobalt, in *Diagnostic Criteria for Plants and Soils*, Chapman, H. D., Ed., Division of Agricultural Sciences, University of California, 1966, 157–179.
33. Wallihan, E. F., Iron, in *Diagnostic Criteria for Plants and Soils*, Chapman, H. D., Ed., Division of Agricultural Sciences, University of California, 1966, 203–211.
34. Wilson, S. B. and Nicholas, D. J. D., A cobalt requirement for non-nodulated legumes and for wheat, *Phytochemistry*, 6, 1057–1066, 1967.

Invertebrates

QUALITATIVE REQUIREMENTS AND UTILIZATION OF NUTRIENTS: PORIFERA

H. M. Reiswig

Our knowledge of the details of nutrition in members of the phylum Porifera remains at a primitive state. The natural diets of a few marine sponges have only recently been determined. Attempts to bring sponges into laboratory culture still invariably fail despite advances. Because of our lack of dietary control of culture specimens, the obvious absence of mouth, anus, and digestive system in sponges, significant resolution of the details of sponge nutrition are not likely to be exposed easily or soon. The following compilation may thus appear crude when compared with data from other metazoan groups, but it does reflect the special difficulties encountered in unravelling the biology of Porifera. Recent reviews by Jørgensen,[1] Rasmont,[2] and Frost[3] (see Table 1 references) reflect the rate of advance and the direction of recent research efforts. Abbreviations in the following tables are U = utilized, u = poorly utilized, Ø = not utilized, R = required.

Table 1
MINERALS

Species (synonym)	Ca	CO$_3$	Mg	Sr	Na	SO$_4$	SiO$_2$	Other	Ref.
Amphiute paulini	R	R	u	u	u	u	–	–	4
Clathrina coriacea	R	R	u	u	u	u	–	–	4
Ephydatia fluviatilis (Meyenia)	–	–	–	–	–	–	R[a]	–	5–8
Ephydatia mulleri (Meyenia)	–	–	–	–	–	–	R	–	8–10
Grantia compressa	R	R	u	u	u	u	–	–	4
Heteromeyenia baileyi	–	–	–	–	–	–	R	–	10
Heteromeyenia repens	–	–	–	–	–	–	R	–	11
Haliclona rosea	–	–	–	–	–	–	R	–	12
Leucandra aspera	R	R	u	u	u	u	–	u[b]	13
Leuconia nivea	R	R	u	u	u	u	–	–	4
Leuconia pumila	R	R	u	u	u	u	–	–	4
Leucosolenia complicata	R	R	u	u	u	u	–	–	4
Spongilla fragilis	–	–	–	–	–	–	R[c]	–	8
Spongilla lacustris	–	R[c]	–	–	–	–	R[a]	–	11, 14
Sycandra sp.	–	R[d]	–	–	–	–	–	–	15
Sycon ciliatum	R	R	u	u	u	u	–	–	4

[a] Required as dissolved silicic acid.
[b] Traces of Li, Ba, and Mn.
[c] SiO$_2$ and CO$_3$ concentrations limit growth and distribution.
[d] Must be available as carbonate ion.

REFERENCES

1. **Jørgensen, C. B.,** *Biology of Suspension Feeding,* Vol. 27, Int. Ser. Monogr. Pure Appl. Biol., Zool. Div., Pergamon, London, 1966.
2. **Rasmont, R.,** in *Chemical Zoology,* Vol. 2, Florkin, M. and Scheer, B. T., Eds., Academic Press, New York, 1968, 43–50.
3. **Frost, T. M.,** in *Aspects of Sponge Biology,* Harrison, F. W. and Cowden, R. R., Eds., Academic Press, New York, 1976, 283–298.
4. **Jones, W. C. and Jenkins, D. A.,** *Calcif. Tissue Res.,* 4, 314–329, 1970.
5. **Pé, J.,** *Arch. Biol.,* 84, 147–173, 1973.
6. **Kilian, E. F.,** *Zool. Beitr.,* 10, 85–159, 1964.
7. **Sarà, M. and Vacelet, J.,** in *Traité de Zoologie,* Grassé, P. P., Ed., Masson et Cie, Paris, 1973, 462–576.
8. **Jewell, M. E.,** *Ecol. Monogr.,* 5, 461–504, 1935.
9. **Elvin, D. W.,** *Trans. Am. Microsc. Soc.,* 90, 219–224, 1971.
10. **Elvin, D. W.,** *Exp. Cell Res.,* 72, 551–553, 1972.
11. **Jewell, M. E.,** *Ecology,* 20, 11–28, 1939.
12. **Garrone, R.,** *J. Microsc. Paris,* 8, 581–598, 1969.
13. **Jones, W. C.,** *Symp. Zool. Soc. London,* 25, 91–123, 1970.
14. **Simpson, R. L. and Vaccaro, C. A.,** *J. Ultrastruct. Res.,* 47, 296–309, 1974.
15. **Maas, O.,** *Arch. Entwicklungsmech. Org.,* 22, 581–602, 1906.

Table 2
ORGANIC MATERIALS

Species (synonyms)	Dissolved	Colloidal	Bacteria	Algae	Fungi	Detritus	Products of symbiont algae	Other	Ref.
Corvomeyenia carolinensis	u[b,c]	—	U[a]	—	—	—	—	—	1
Ephydatia fluviatilis (Meyenia)	—	U	U[d]	U[e]	Ụ	—	U	u[f]	2–10
Ephydatia mulleri (Meyenia)	u[c]	—	u[a,d]	U[e]	Ụ	—	—	u[f]	2–4, 9, 11
Halichondria panicea	—	—	U	—	—	—	—	—	12
Haliclona permollis	—	—	U	—	—	—	—	—	13
Heteromeyenia (?) sp.	—	—	—	—	—	—	U[g]	—	14
Ircinia variabilis (Hircinia)	—	—	—	U	—	—	U	—	15
Leuconia aspera	—	—	—	—	—	U	—	—	16
Microciona prolifera	Ụ[b]	—	U[a,h]	U[a,h]	U	U[a]	—	—	17–23
Microciona spinosa	—	—	U[a]	U[a]	—	U[a]	—	—	21
Mycale sp.	Ụ	U	U	U	u	Ụ	—	—	24, 25
Ophlitaspongia seriata	—	—	U[i]	U[i]	—	—	—	—	26
Reniera simulans	—	—	U	—	—	—	—	—	27
Spongilla lacustris	U[f]	—	U	U[e]	Ụ	—	U[g]	u[f]	6, 8, 9, 14, 28, 29
Spongilla proliferans	—	u	—	U	—	u	U	U[j]	30
Spongilla sp.	—	u	—	—	—	—	u[k]	U[l]	31, 32
Suberites ficus (Ficulina)	—	U	U	U	U	—	—	—	13
Suberites massa	u[m]	—	U	U	—	U	—	—	33
Sycon lingua	—	—	—	—	U	—	—	—	34
Sycon raphanus	—	—	U[a]	U[a]	—	U	—	—	35
Tedania ignis	Ụ	—	U	U[a]	u	U[a]	—	—	21
Tethya crypta (Cryptotethya)	Ụ	U	U	U[a]	—	Ụ	—	—	24, 25
Thalysias juniperina	—	—	U[a]	U[a]	—	U[a]	—	—	21
Thalysias schoenus	—	—	U[a]	U[a]	—	U[a]	—	—	21
Verongia aerophoba (Luffaria)	—	—	U[n]	U[n]	—	—	u	—	36
Verongia cavernicola (Luffaria)	—	—	U[n]	—	—	—	—	—	37, 38
Verongia gigantea (Luffaria)	U	U	U[h,n]	U	U	Ụ	—	—	24, 25

[a] Ingestion of particles by epithelial cells.

[b] References cite conflicting data.

[c] Radio-labeled phenylalanine, fluorescent-labeled casein, and rabbit serum proteins taken up from solution.

[d] Grown on killed Escherichia coli and Staphylococcus aureus suspensions.

[e] Algae added to medium from cultures: Chlamydomonas sp., Chlorella sp., Hematococcus sp.

[f] Various kinds of organic materials added to medium, ranging from bean broth to apple juice to individual vitamin species; almost all attempts to supplement natural foods showed short-term growth followed by bacterial infection and death of the sponges.

[g] Demonstrated ^{14}C fixation by zoochlorellae and transfer to host sponge.

[h] Ingestion following particle capture by choanocytes.

Table 2 (continued)
ORGANIC MATERIALS

i Algae and bacteria added to supplement diet: *Pseudomonas, Arthrobacter, Micromonas, Isochrysis, Monochrysis,* and *Tetraselmis* spp.
j Non-natural food materials offered and digested: milk, toad sperm, blood cells, etc.
k Glucose excreted from algae in culture; not shown to be used by host sponge.
l Fed trout fry started food with success.
m Glycine.
n Symbionts are cropped (digested).

REFERENCES

1. Harrison, F. W., *Hydrobiologia,* 39, 495–508, 1972.
2. Rasmont, R., *Ann. Soc. R. Zool. Belg.,* 91, 147–156, 1961.
3. Rasmont, R., *Dev. Biol.,* 8, 243–271, 1963.
4. Schmidt, I., *Z. Vgl. Physiol.,* 66, 398–420, 1970.
5. Kilian, E. F., *Z. Vgl. Physiol.,* 34, 407–447, 1952.
6. Kilian, E. F., *Zool. Beitr.,* 10, 85–159, 1964.
7. Efremova, S. M., *Symp. Zool. Soc. London,* 25, 399–413, 1970.
8. Trigt, H. Van, *Tijdschr. Ned. Dierkd. Ver.,* Ser. 2(17), 1–220, 1919.
9. Simon, L., *Zool. Jahrb. Abt. Allg. Zool. Physiol. Tiere,* 64, 206–234, 1953.
10. Pourbaix, N., *Ann. Soc. R. Sci. Med. Nat. Bruxelles,* 2, 1–15, 1934.
11. Atamanova, M. V. and Zolotarev, O. B., *Vestn. Leningr. Univ. Biol.,* 4, 5–11, 1974.
12. Jørgensen, C. B., *Nature,* 163, 912, 1949.
13. Reiswig, H. M., *Can. J. Zool.,* 53, 582–589, 1975.
14. Gilbert, J. J. and Allen, H. L., *Int. Rev. Gesamten Hydrobiol.,* 58, 633–658, 1973.
15. Sarà, M., *Mar. Biol.,* 11, 214–221, 1971.
16. Bidder, G. P., *Proc. Linn. Soc. London,* 149, 119–145, 1937.
17. Bagby, R. M., *J. Exp. Zool.,* 180, 217–243, 1972.
18. Kunen, S., Claus, G., Madri, P., and Peyer, L., *Hydrobiologia,* 38, 565–576, 1971.
19. Simpson, T. L., *J. Exp. Zool.,* 154, 135–152, 1963.
20. Spiegel, M. and Metcalf, C., *Biol. Bull.,* 113, 356, 1957.
21. Simpson, T. L., *Peabody Mus. Nat. Hist. Yale Univ. Bull.,* 25, 1–141, 1968.
22. Simpson, R. L., *J. Exp. Mar. Biol. Ecol.,* 2, 252–277, 1968.
23. Gordon, E., Spiegel, M., and Villee, C. A., *J. Cell. Comp. Physiol.,* 45, 479–483, 1955.
24. Reiswig, H. M., *Biol. Bull.,* 141, 568–591, 1971.
25. Reiswig, H. M., *Limnol. Oceanogr.,* 17, 341–348, 1972.
26. Fry, W. G., *Symp. Eur. Mar. Biol.,* 4, 155–178, 1971.
27. Pourbaix, N., *Ann. Soc. R. Zool. Belg.,* 64, 11–20, 1933.
28. Frost, T. M., in *Aspects of Sponge Biology,* Harrison, F. W. and Cowden, R. R., Eds., Academic Press, New York, 1976, 283–298.
29. Castro-Rodriguez, G., *Ann. Soc. R. Zool. Belg.,* 61, 113–122, 1931.
30. van Weel, P. B., *Physiol. Comp. Oecol.,* 1, 110–126, 1949.
31. Imlay, M. J. and Paige, M. L., *Prog. Fish Cult.,* 34, 210–215, 1972.
32. Muscatine, L., Karakashian, S. J., and Karakashian, M. W., *Comp. Biochem. Physiol.,* 20, 1–12, 1967.
33. Pütter, A., *Z. Allg. Physiol.,* 16, 65–114, 1914.
34. Efremova, S. M., *Vestn. Leningr. Univ. Biol.,* 20, 17–23, 1965.
35. Tuzet, O., in *Traité de Zoologie,* Grassé, P. P., Ed., Masson et Cie, Paris, 1973, 27–132.
36. Vacelet, J., *J. Microsc. Paris,* 12, 363–380, 1971.
37. Vacelet, J., *J. Microsc. Paris,* 9, 333–346, 1970.
38. Bertrand, J. C. and Vacelet, J., *C. R. Acad. Sci. Paris Nat. Hist.,* 273, 638–641, 1971.

NUTRITIONAL RESEARCH IN THE CNIDARIA

H. M. Lenhoff and W. Heagy

Students of animal nutrition continue to be intrigued by cnidarians because they represent transitional stages in the evolution from intracellular to extracellular digestion. Yet, virtually none of the factors that facilitate nutritional research in the other aquatic metazoans apply to the cnidarians i.e., (a) axenic conditions, (b) simple and reliable methods of quantifying growth, (c) permeability to nutrients, and (c) availability of nutritional mutants. No cnidarian has been grown axenically; most have considerable bacteria, fungi, and protozoans on their surfaces and/or in their gastric cavities. The growth of only a few (hydras, some hydroids, and a few sea anemones) has been quantified. The fresh-water hydras, which are the cnidarians most amenable to investigation, are generally impermeable to ions. And, lastly, nutritional mutants in the cnidarians have never been reported.

As a result, current research on feeding, digestion, and nutrition among the cnidarians is progressing along four avenues, none of which has yielded any information about specific nutritional requirements per se and only one of which has yielded concrete information expected to eventually lead to significant nutritional studies. These research areas are (a) ionic requirements for growth,[5] (b) supplying of nutrients by endosymbiotic algae,[16] (c) uptake of organic small molecules from the environment,[21] and (d) activation of feeding behaviors by specific small molecules. This last research area has been very active in recent years, and currently appears to be the only one that could lead to significant nutritional studies, for knowledge of the compound that activates feeding will allow investigators to control ingestion in the cnidarians and then to feed known nutritional mixtures. Thus, the remainder of this article and the accompanying survey (Table 1) will deal entirely with molecules that activate feeding in the Cnidaria.

CHEMICAL ACTIVATION OF FEEDING BEHAVIOR BY SPECIFIC MOLECULES EMITTED FROM CAPTURED PREY

A common mode of feeding in the cnidarians consists of the animals detecting a chemical(s) emitted from their captured, wounded prey. The wound results from a puncture by the harpoon-like action of the nematocysts that line the cnidarian's tentacles. Since Loomis[12] first discovered that the tripeptide-reduced glutathione specifically activated the feeding behavior of *Hydra littoralis,* many compounds have been shown to act in similar ways in a wide number of cnidarians (Table 1).

The data listed in Table 1 show that cnidarians tested from every class and most families elicit a feeding response to either one or a few small molecules. The molecules most commonly found to initiate a feeding response were reduced glutathione and the imino acid proline. All other compounds triggering a response were amino acids, with no carbohydrates or large molecules being effective. In every member of the Hydrozoa investigated, the feeding response was induced by a single specific molecule, whereas among the Scyphozoa and Anthozoa either one or more compounds were reported to activate feeding behaviors.

Considering first the Hydrozoa, proline seems to be especially prevalent as an activator among the families Clavidae and Williidae of the suborder Gymnoblastea. For example, all members of these groups thus far tested, i.e., *Cordylophora, Pennaria,* and *Proboscidactyla,* respond only to proline. All members of the Hydridae (suborder Gymnoblastea) tested responded only to GSH. Other hydrozoans tested were one calyptoblast and two siphonophores; all responded to GSH only.

Turning to the Anthozoa, we see three general characteristics emerging which correlate somewhat with three orders of that class. Among the order Actinaria, the results show that although a number of different compounds trigger feeding behaviors in different anemones, there is a fair degree of specificity by each species for their respective feeding activator. For example, *Boloceroides* responds primarily to valine, *Anthopleura* to GSH, *Haliplanella* to leucine, *Actinia* to glutamic acid, and *Calliactis* to GSH. The specificity broadens when we consider the order Zoanthidia. Whereas *Zoanthus* responds primarily to GSH, *Palythoa* responds to relatively high concentrations of either GSH or proline, or to low amounts of these two activators acting synergistically. Lastly, corals (order Madreporaria) seem to respond best to proline alone or to GSH alone, as well as to a varying number of other amino acids at relatively higher concentrations. (See the discussion at the end of this chapter for difficulties in working with anthozoans.)

There are not enough examples of the Scyphozoa to make a generalization as yet. The one tested, *Crysaora*, seems to respond to GSH and to a wide number of amino acids.

QUALIFICATIONS

We are not necessarily convinced that all the compounds tested are true activators. Likewise, we also feel that other feeding activators for those organisms listed may yet be detected. We pose three criteria that should be met before a compound is accepted as a natural activator of a feeding behavior: (a) The compound should be active in low concentrations; i.e., 10^{-4} M or less, (b) sufficient analogs of the presumed activator should be tested to show that the receptor has some degree of specificity, and (c) an analog that is a nontoxic competitive inhibitor should be able to reversibly inhibit the response activated by natural tissue extracts.

Thus far, in only one instance, that of hydra, have all these criteria been met.[5] The next most thoroughly studied case is Fulton's work[2] showing that the proline activation of feeding in *Cordylophora lacustris* could take place in less than 10^{-6} M and that the specificity resided in the imino region of a heterocyclic α-imino acid which is neither substituted nor unsaturated in such a way as to affect the imino acid group.

A third case, which shows interesting experiments with analogs, is that dealing with the swimming sea anenome *Boloceroides* sp.[8] The feeding behavior of this animal was controlled by the branched amino acid valine. Isoleucine, which is valine with an ethyl group replacing one of the branched methyls, is an effective competitive inhibitor. On the other hand, leucine, identical to valine in all respects except that the branch point is separated from the α-carbon by an additional methylene group, is not effective either as an activator or as an inhibitor.

These criteria are more difficult to establish with members of the Anthozoa because of the nature of the animals themselves. For example, it is difficult to handle corals without damaging the tissues; as a result, either an activator or possibly even a feeding inhibitor may be released and contaminate the test solution. Similarly, sea anemones, when agitated, may release mucus or soluble materials. Another difficulty in working with anthozoans is the complex nature of many of their feeding behaviors, e.g., the feeding activators in many cases do not cause the mouth to open widely, but rather aid the animal to swallow objects via ciliary action. Thus, it becomes extremely difficult to measure exact concentrations of the potential feeding activator in such a system.

ACKNOWLEDGMENTS

This survey was collected while under sponsorship of Grant NS 11171 from the National Institutes of Health.

Table 1

CHEMICAL ACTIVATORS OF FEEDING AMONG THE CNIDARIA

Classification	Genus and species	Compound	Concentration range (M)	Ref.
Class Hydrozoa				
Order Hydroida				
Suborder Gymnoblastea (=Athecata)	*Hydra littoralis*	GSH	10^{-8} – 10^{-7}	6, 12
		Tyr[a]	$<10^{-4}$	1
	Hydra attenuata	GSH	10^{-7} – 10^{-6}	26
	Hydra pirardi	GSH	10^{-6}	24
	Hydra pseudoligactis	GSH	10^{-6}	24
	Hydra viridis	GSH	10^{-6}	25
	Cordolophora lacustris	Pro	10^{-7} – 10^{-6}	2
	Pennaria tiarella	Pro	10^{-6}	17
	Proboscidactyla flavicirrata	Pro	10^{-5}	22
Suborder Calyptoblastea (=Thecata)	*Campanularia flexuosa*	GSH	$<10^{-5}$	4
Order Siphonophora	*Physalia physalis*	GSH	$<10^{-5}$	4
	Nanomia cara	GSH	(?)	13
Class Anthozoa				
Order Actiniaria	*Anthopleura elegantissima*	GSH[b]	$<10^{-4}$	9
		Asn[b]	10^{-5}	
	Boloceroides sp.	Val	10^{-4}	8
		Isoleu(?)[c]		
	Actinia equina	Glu	(?)	23
	Haliplanella luciae	Leu	(?)	10
	Calliactis polypus	GSH	10^{-6}	20
		Pro	10^{-4}	
		Leu	10^{-4} – 10^{-3}	
		Eight amino acids	$<10^{-1}$	
Order Madreporaria	*Fungia scutaria*	Pro	10^{-7}	14
		GSH	10^{-7}	
	Cyphastrea ocellina	Pro	10^{-7}	14
		GSH	10^{-6} – 10^{-5}	
	Pocillopora damicornis	Pro	10^{-7}	14
		GSH	$<10^{-5}$	15
		Phe[d]	$<10^{-1}$	15

a Tyrosine activates an enteroreceptor and acts in conjunction with GSH to stimulate another behavioral response.
b Asparagine stimulates tentacle bending and GSH stimulates mouth opening.
c Possibly a weak activator, but also effective as a competitive inhibitor.
d Caused a slight mouth opening, but no contraction of tentacles.

Table 1 (continued)
CHEMICAL ACTIVATORS OF FEEDING AMONG THE CNIDARIA

Classification	Genus and species	Compound	Concentration range (M)	Ref.
Order Madreporaria (continued)	Pocillopora meandria	Pro(+++)	$<10^{-1}$	15
		GSH(+++)	$<10^{-1}$	
	Porites compressa	Pro(++)	$<10^{-1}$	15
		GSH(++)	$<10^{-1}$	
	Leptastrea bottae	Pro(++)	$<10^{-1}$	15
		GSH(+++)	$<10^{-1}$	
	Montastrea cavernosa	GLU	$<10^{-3}$	3
		Asp	$<10^{-1}$	
		Pro	$<10^{-1}$	
		Arg	$<10^{-1}$	
	Tubastrea manni	Pro(+++)	$<10^{-1}$	15
		GSH(++)	$<10^{-1}$	
Order Zoanthidea	Palythoa psammophilia	Pro	$<10^{-4}$	18
		GSH	$<10^{-4}$	
		Pro and GSH[e]	5×10^{-5} and 10^{-6}	
		Ala	$<10^{-1}$	
		Ser	$<10^{-1}$	
		Lys	$<10^{-1}$	
	Zoanthus pacificus	GSH	10^{-6}	19
		Gly	10^{-3}—10^{-4}	
Class Scyphozoa	Crysaora quinquecirrha	20 amino acids	10^{-8}—10^{-4}	11
		GSH	10^{-12}	
		Glycylglycine	10^{-8}	

[e] Act synergistically.

Note: The work of the noted coral biologist, Thomas Goreau, published posthumously,[27] stated that "very low concentrations of . . . glycine, alanine, phenylalanine and leucine could trigger . . . feeding responses . . . "in a number of Jamaican reef corals. One was reported to respond to alanine and glycine (10^{-9} M).

REFERENCES

1. **Blanquet, R., and Lenhoff, H. M.,** Tyrosine enteroreceptor of hydra: its function in eliciting a behavior modification, *Science,* 159, 633–634, 1968.

2. **Fulton, C.,** Proline control of the feeding reaction of *Cordylorphora, J. Gen. Physiol.,* 46, 823–837, 1963.

3. **Lehman, J. T. and Porter, J. W.,** Chemical activation of feeding in the Caribbean reef building coral *Montastrea cavernosa, Biol. Bull.,* 145, 140–149, 1973.

4. **Lenhoff, H. M. and Schneiderman, H. A.,** The chemical control of feeding in the Portuguese man-of-war, *Physalia physalis* L., and its bearing on the evolution of the Cnidaria, *Biol. Bull.,* 116, 452–460, 1959.

5. **Lenhoff, H. M.,** Chemical perspectives on the feeding response, digestion and nutrition of selected coelenterates, in *Chemical Zoology,* Vol. 1, Florkin, M. and Scheer, B., Eds., Academic Press, New York, 1968, 157–221.

6. **Lenhoff, H. M.,** pH profile of a peptide receptor, *Comp. Biochem. Physiol.,* 28, 571–586, 1969.

7. **Lenhoff, H. M.,** On the mechanism of action and evolution of receptors associated with feeding and digestion, in *Coelenterate Biology: Review and New Perspectives,* Muscatine, L. and Lenhoff, H., Eds., Academic Press, New York, 1974, 359–384.

8. **Lindstedt, K. J., Muscatine, L., and Lenhoff, H. M.,** Valine activation of feeding in the sea anemone *Boloceroides, Comp. Biochem. Physiol.,* 26, 567–572, 1968.

9. **Lindstedt, K. J.,** Chemical control of feeding behavior, *Comp. Biochem. Physiol.,* 39, 553–581, 1971.

10. **Lindstedt, K. J.,** Biphasic feeding response in a sea anemone: control by asparagine and glutathione, *Science,* 173, 333–334, 1971.

11. **Loeb, M. and Blanquet, R.,** Feeding behavior in polyps of the Chesapeake Bay sea nettle *Chysaora quinquecirrha* (Desor, 1848), *Biol. Bull.,* 145, 150–158, 1973.

12. **Loomis, W. F.,** Glutathione control of the specific feeding reactions of hydra, *Ann. N.Y. Acad. Sci.,* 62, 209–228, 1955.

13. **Mackie, G. O. and Boag, D. A.,** Fishing, feeding and digestion in siphonophores, *Pubbl. St. Zool. Napoli* 33, 178–196, 1963.

14. **Mariscal, R. N. and Lenhoff, H. M.,** The chemical control of feeding behavior in *Cyphastrea ocellina* and some other Hawaiian corals, *J. Exp. Biol.,* 49, 689–699, 1968.

15. **Mariscal, R. N.,** The chemical control of the feeding behavior in some Hawaiian corals, in *Experimental Coelenterate Biology,* Lenhoff, H. M., Muscatine, L., and Davis, V., Eds., University of Hawaii Press, Honolulu, 1971, 100–118.

16. **Muscatine, L.,** Endosymbiosis of cnidarians and algae, in *Coelenterate Biology: Reviews and New Perspectives,* Muscatine, L. and Lenhoff, H. M., Eds., Academic Press, New York, 1974, 359–389.

17. **Pardy, R. L. and Lenhoff, H. M.,** The feeding biology of the gymnoblastic hydroid, *Pennaria tiarella, J. Exp. Zool.,* 168, 197–202, 1968.

18. **Reimer, A. A.,** Feeding behavior in the Hawaiian zoanthids *Palythoa* and *Zoanthus, Pac. Sci.,* 25, 512–520, 1971.

19. **Reimer, A. A.,** Chemical control of feeding behavior in *Palythoa* (Zoanthidea, Coelenterata), *Comp. Gen. Pharmacol.,* 2, 383–396, 1971.

20. **Reimer, A. A.,** Feeding behavior in the sea anemone *Calliactis polypus* (Forsköl, 1775), *Comp. Biochem. Physiol.,* 44, 1289–1301, 1973.

21. **Schlicter, D.,** Der Einfluss physikalischer und chemischer Faktoren auf die Aufnahme in Meerwasser geloester Aminosaeuren durch Aktinien, *Mar. Biol.,* 25, 279–290, 1974.

22. **Spencer A. N.,** Behavior and electrical activity in the hydrozoan *Proboscidactyla flavicirrata* (Brandt), *Biol. Bull.,* 146, 100–115, 1974.

23. **Steiner, G.,** Über die chemische Nahrungswahl von *Actinia equina* (L.), *Naturwissenschaften,* 44, 70–71, 1957.

24. **Lenhoff, H. M.,** unpublished.

25. **Mariscal, R. N.,** unpublished.

26. **Heagy, W. et al.,** unpublished.

27. **Goreau, T. F., Goreau, N. I., and Yonge, C. M.,** Reef corals: autotrophs or heterotrophs?, *Biol. Bull.,* 141, 247–260, 1971.

QUALITATIVE REQUIREMENTS AND UTILIZATION OF NUTRIENTS: PLATYHELMINTHES

D. F. Mettrick and D. J. Jackson

The Platyhelminthes are found in almost all environments, but the majority of studies completed on nutrient requirements and utilization pertain to the parasitic Trematoda and Cestoda. These pose unique problems in that the life cycles of individual members of the group are bizarre and complex, involving one or more larval stages; secondly, the adults, and some larval forms, are always parasitic upon another animal. The very nature of a parasite makes it impossible to study it in isolation in its natural environment because its whole being, including its nutritional physiology, is completely integrated with its host.

As a result, much of the in vivo information on platyhelminth nutrient requirements is secondhand, and follows this course: Changes in the quality or quantity of exogenous nutrient supply to host species "A" result in observed increased/decreased growth rate of helminth species "X" parasitic on or in host species "A." The link(s) between the supposed cause and effect is (are), in most cases, entirely speculative.

The alternative approach has been to study helminths in vitro, which, of course, depends on the development of appropriate culture methods. Considerable progress has been made in maintaining certain stages of the life cycles of various platyhelminths under in vitro conditions. Attempts to cultivate particular helminth species through all stages of their life cycles under in vitro conditions have been less successful.

The reasons for this lack of success center around the fact that each stage of development may require specific physicochemical environmental conditions and specific, but different, nutrient requirements. In most cases the in vivo physicochemical conditions are unknown, and work done on this aspect of the host-parasite association has shown that a helminth may completely change the physicochemical conditions of its environmental niche from those normally found in uninfected host animals.[1] In addition, platyhelminths are extremely catholic in their nutrient requirements, feeding on material such as bile, blood, tissue, exudates, intestinal contents, etc. These are all complex food substrates, and it is both time consuming and difficult to determine precisely the component(s) of the food material upon which the helminths are actually feeding.

A final difficulty in the use of in vitro techniques is a lack of uniformity in published data on the determination of the biological integrity of the helminth being cultured. Mere survival *per se* of a cultured worm is of little value in determining its nutrient requirements and utilization. Maturation and the production of viable eggs are more useful standards, but even then differences in the rate of maturation and in the quantitative production of eggs raise questions regarding the similarity of the in vitro conditions to the normal in vivo environment. A recent review of the problems of in vitro culture of *Echinococcus granulosus* illustrates this point more fully.[2]

For example, it has been common practice to place helminths such as tapeworms in an in vitro isotonic medium and determine the rate of uptake of a particular metabolite. Active transport of the metabolite was deemed to indicate that the substance in question was "required" by the worms. Such conclusions may indeed be true, but the experimental methodology previously and currently used for such studies certainly does not allow unequivocal acceptance of the fact. Changes in the metabolism of the worms during any kind of in vitro preparation must be considered an evaluating nutrient requirements and utilization in particular and other metabolic studies in general.

As an introduction to the problems and techniques of in vitro culture, the reader is referred to Taylor and Baker[3] and Silverman and Hansen.[4] It should be noted that to date no adult or larval platyhelminth has been successfully cultured in a fully defined chemical medium.

Without question, the most favored site for adult platyhelminths, represented by the digenetic trematodes and the cestodes, is the alimentary canal of various vertebrate hosts. There are three principal nutrient sources for such worms. The first, exogenous nutrients ingested by the host animal, and the second, secretions and exfoliated material of endogenous origin, contribute to the luminal contents of the gastrointestinal tract. The third nutrient source is the cells of the tissue layers surrounding the intestinal lumen; some helminths migrate through these tissues while others browse on the intestinal mucosa while their actual body remains in the intestinal luminal cavity.

A recent review of mammalian gastrointestinal physiology[5] illustrates the considerable variation in basic nutrients such as carbohydrates, amino acids, fats, bile acids, etc. to which an intestinal helminth will be exposed. In addition, changing luminal pH, pCO_2, Eh, and other physicochemical characteristics significantly affect the degree to which the helminths can utilize the available nutrients. The complexity of the host-parasite interaction[6] and its involvement in the host's neural-hormonal gastrointestinal control system[7-9] illustrate why it is virtually impossible, at present, to subclassify nutrient requirement and utilization for platyhelminths.

The following tables summarize the information currently available on nutrient utilization and requirements by the Trematoda, Cestoda, and Turbellaria. The fact that only some 40 species have been studied eloquently demonstrates the amount of work still to be done with the Platyhelminthes.

Table 1

AMINO ACIDS UTILIZED BY LARVAL AND ADULT TREMATODA[a]

Species	Alanine	Arginine	Asparagine	Aspartic acid	Cystine	Glutamic acid	Glutamine	Glycine	Histidine	Isoleucine	Leucine	Lysine	Methionine	Phenylalanine	Proline	Threonine	Tryptophan	Tyrosine	Valine
Cercaria emasculans	U[10]			U[10]		U[10]													
Cryptocotyle lingua	U[10]			U[10]		U[10]													
Fasciola hepatica (adult)	U[11]	U[11] R[12]		U[11]					U[13]	U[14]	U[14]		U[11]	U[11]	U[11]			U[11]	U[14]
Fasciolopsis buski								U[15]											
Himasthla quissetensis						U[16]									U[16]				
Philophthalmus megalurus											U[15,17]							U[17]	
Schistosoma mansoni (adult)	U[18]	U[18,19]		U[18,19]		U[19]	U[20]	U[18]	U[19]				U[21]		U[22]		U[19]		
S. mansoni (larval)													U[21]						

[a] U = utilized.

Table 2

CARBOHYDRATES UTILIZED BY LARVAL AND ADULT TREMATODA[a]

Species	Dextrin	Fructose	Galactose	Glucose	Lactose	Maltose	Mannose	Saccharose	Sucrose	Trehalose
Diplostomum phoxini				U[23]						
Entobdella bumpusi		Ʉ[24]		Ʉ[24]						
Fasciola hepatica (adult)				U[25-27]						
Gorgoderina sp.				U[28]						
Haematolechus medioplexus				U[28]						
Himasthla quissetensis				U[16,29]			U[16]			
Macraspis cristata			Ʉ[24]	U[24]		Ʉ[24]				
Philophthalmus megalurus				U[17]						
Schistosoma japonicum		U[30]		U[30]		U[30]				
Schistosoma mansoni (adult)		U[20,30]		U[30]		U[30]				
Schistosoma mansoni (larva)		U[31]	Ʉ[31]	U[31-33]	Ʉ[31]	Ʉ[31]	U[31]	U[31]		

U = utilized; Ʉ = not utilized.

Table 3

LIPIDS AND FATTY ACIDS UTILIZED BY LARVAL AND ADULT TREMATODA[a]

Species	Linoleic acid	Monoolein	Octanoic acid	Oleic acid	Palmitic acid	Stearic acid	Cholesterol	Glycerol
Dicrocoelium dendriticum								U[34]
Schistosoma mansoni (adult)			U[35]	U[35]		U[35]	R[36]	

U = utilized; R = required.

Table 4

NUCLEIC ACIDS UTILIZED BY LARVAL AND ADULT TREMATODA[a]

Species	Adenine	Cytidine	Thymidine	Thymine
Philophthalmus megalurus			U[17]	

[a] U = utilized.

Table 5

ORGANIC ACIDS, VITAMINS, AND MINERALS UTILIZED BY LARVAL AND ADULT TREMATODA[a]

Species	Organic acids						Vitamins and minerals		
	Acetate	Fumarate	Hydroxymethyl glutarate	Malate	Mevalonate	Pyruvate	Haemin	Nicotinic acid	Pyridoxine
Dicrocoelium dendriticum									R[37]
Schistosoma mansoni (adult)				U[38]		U[38]			
Schistosoma mansoni (larva)						U[38]			

[a] R = required; U = utilized.

Table 6
AMINO ACIDS UTILIZED BY LARVAL AND ADULT CESTODA[a]

Species	Alanine	Arginine	Asparagine	Aspartic acid	Cysteine	Glutamic acid	Glycine	Isoleucine	Leucine	Lysine	Methionine	Phenylalanine	Proline	Serine	Tyrosine	Valine
Calliobothrium verticillatum	U[41]								U[39]							U[39,40]
Hymenolepis citelli (adult)			U[41]	U[41]		U[41]										
Hymenolepis diminuta (adult)	U[43]	U[43,44]	U[43]	U[43]	U[45]	U[43,44]			U[46]	U[44,46]	U[45,46]	U[44]	U[47]	U[44,47]	U[48]	U[46]
Hymenolepis diminuta (cysticercoids)					U[49]											
Hymenolepis nana			U[41]	U[41]		U[41]										
Mesocestoides corti							U[50]	U[50]								

[a] U = utilized.

Table 7
CARBOHYDRATES UTILIZED BY LARVAL AND ADULT CESTODA[a]

Species	Dextrin	Fructose	Galactose	Glucose	Lactose	Maltose	Mannose	Saccharose	Sucrose	Trehalose
Anthrobothrium variable				U[51]						
Calliobothrium verticillatum		u[51]	U[51]	U[51]		Ψ[51]	u[51]	Ψ[51]		Ψ[51]
Chandlerella hawkingi				U[52]						
Cittotaenia sp.		Ψ[53]	U[53]	U[53]	Ψ[53]	U[53]	Ψ[53]	U[53]		Ψ[53]
Disculiceps pileatum				U[51]						
Hydatigera taeniaeformis		U[54]								
Hymenolepis citelli (adult)		U[53]	U[53]	U[53,55]	Ψ[53]	Ψ[53]	Ψ[53]	Ψ[53]		Ψ[53]
Hymenolepis diminuta (adult)	S[56]	Ψ[57]	U;[48,57,58] $[56]	U;[57,59-62] S[56]	U[57]	S[56]	U[57]	Ψ[57]	S[56]	Ψ[57]
(Hymenolepis nana)		Ψ[53]	U[53]	U[53]	Ψ[53]	Ψ[53]	Ψ[53]	Ψ[53]		Ψ[53]
Inermiphyllidium pulvinatum				U[51]						
Lacistorhynchus tenuis		U[63]	U[63]	U[63]		Ψ[63]	Ψ[63]	Ψ[63]		Ψ[63]
Multiceps serialis (adult)		U[64]	Ψ[64]	U[64]				Ψ[64]		
Multiceps serialis (larva)		Ψ[64]	Ψ[64]	U[64]				Ψ[64]		
Onchobothrium pseudo-uncinatum				U[51]						
Orygmatobothrium dohrni				U[51]						
Phyllobothrium foliatum				U[51]		U[51]				
Taenia taeniaeformis (adult)		Ψ[54]	U[54]	U[54,65]				u[54]		
Taenia taeniaformis (larva)		u[54]	U[54]	U[54,65]				u[54]		

[a]U = utilized; u = poorly utilized, Ψ = not utilized; S = stimulatory; $ = non-stimulatory.

Table 8

LIPIDS AND FATTY ACIDS UTILIZED BY LARVAL AND ADULT CESTODA[a]

Species	Linoleic acid	Monoolein	Oleic acid	Palmitic acid	Stearic acid	Cholesterol	Glycerol
Hymenolepis diminuta (adult)	U[66-68]	U[66]	U[66]	U[69]	U[69]	U[66,71]	R[70]

[a] U = utilized; R = required.

Table 9

NUCLEIC ACIDS UTILIZED BY LARVAL AND ADULT CESTODA[a]

Species	Adenine	Cytidine	Thymidine	Thymine
Hymenolepis diminuta (adult)	u[72]			
Hymenolepis microstoma		U[73]		U[73]
Raillientina cesticillus		U[74]		U[74]

[a] U = utilized; u = poorly utilized.

Table 10

ORGANIC ACIDS, VITAMINS, AND MINERALS UTILIZED BY LARVAL
AND ADULT CESTODA[a]

| Species | Organic acids | | | | | | Vitamins and minerals | | |
	Acetate	Fumarate	Hydroxymethyl glutarate	Malate	Mevalonate	Pyruvate	Haemin	Nicotinic acid	Pyridoxine
Echinococcus granulosus	U[75]	U[76]							
Hymenolepis diminuta (adult)	U[77]		U[77]		U[77]	U[78]		S[79]	R[80]
Hymenolepis microstoma							R[81]		
Mesocestoides corti							R[82]		
Taenia hydratigenia	U[76]								

[a] U = utilized; S = stimulatory; R = required.

Table 11

NUTRITIONAL COMPOUNDS UTILIZED BY TURBELLARIA[a]

| Species | Carbohydrates | Lipids and fatty acids | | | Organic acids and vitamins | |
	Glucose	Sterate	Oleate	Octanoate	Acetate	Antistiffness factor[86,87]
Dugesia dorotocephala	U[35]	U[35]	U[35]	U[35]	U[35]	
Dugesia tigrina						R[83-85]

[a] U = utilized; R = required.

REFERENCES

1. Podesta, R. B. and Mettrick, D. F., *Int. J. Parasitol.*, 4, 277, 1974.
2. Smyth, J. D. and Davies, Z., *Int. J. Parasitol.*, 4, 631, 1974.
3. Taylor, A. E. R. and Baker, J. R., *The Cultivation of Parasites in Vivo*, Blackwell, Oxford, Engl., 1968, 159.
4. Silverman, P. and Hansen, E. L., *Adv. Parasitol.*, 9, 227, 1971.
5. Mettrick, D. F. and Podesta, R. B., *Adv. Parasitol.*, 12, 183, 1974.
6. Mettrick, D. F., *Can. J. Public Health*, 64, 70, 1973.
7. Anderson, S., *Annu. Rev. Physiol.*, 35, 431, 1973.
8. Rogers, Q. R. and Leung, P. M. B., *Fed. Proc. Fed. Am. Soc. Exp. Biol.*, 32, 709, 1973.
9. Lepkovsky, S., *Fed. Proc. Fed. Am. Soc. Ex. Biol.*, 32, 1705, 1973.
10. Watts, S. D. M., *Parasitology*, 61, 499, 1970.
11. Daugherty, J. W., *Exp. Parasitol.*, 1, 331, 1952.
12. Kurelec, B., *Acta Parasitologica Jugosl.*, 5, 33, 1974.
13. Kurelec, B., Rijavec, M., and Klepac, R., *Comp. Biochem. Physiol.*, 29, 885, 1969.
14. Lahoud, H., Prichard, R. K., McManus, W. R., and Schofield, P. J., *Int. J. Parasitol.*, 1, 223, 1971.
15. Cain, G. D., *J. Parasitol.*, 55, 307, 1969.
16. Vernberg, W. B. and Hunter, W. S., *Exp. Parasitol.*, 14, 311, 1963.
17. Nollen, P. M., *J. Parasitol.*, 54, 295, 1968.
18. Garson, S. and Williams, J. S., *J. Parasitol.*, 43, Suppl. 27, 1957.
19. Senft, A. W., *Ann. N.Y. Acad. Sci.*, 113, 272, 1963.
20. Bruce, J. I., Weiss, E., Stirewalt, M. A., and Lincicome, D. R., *Exp. Parasitol.*, 26, 29, 1969.
21. Chappell, L. H., *Int. J. Parasitol.*, 4, 361, 1974.
22. Senft, A. W., *Comp. Biochem. Physiol.*, 27, 251, 1968.
23. Bibby, M. C. and Rees, G., *Z. Parasitenkd.*, 37, 187, 1971.
24. Laurie, J. S., *Comp. Biochem. Physiol.*, 4, 63, 1961.
25. Mansour, T. E., *Biochim. Biophys. Acta*, 34, 456, 1959.
26. Thorsell, W., Appelgren, L. E., and Kippar, M., *Z. Parasitenkd.*, 31, 113, 1968.
27. Bryant, C. and Williams, J. P. G., *Exp. Parasitol.*, 12, 372, 1962.
28. Parkening, T. A. and Johnson, A. D., *Exp. Parasitol.*, 25, 358, 1969.
29. Chappell, L. H., *Exp. Parasitol.*, 35, 61, 1974.
30. Bruce, J. I., Ruff, M. D., Davidson, D. E., and Crum, J. W., *Comp. Biochem. Physiol.*, 49B, 157, 1974.
31. Bueding, E., in *Chemical Physiology of Endoparasitic Animals*, Von Brand, T., Ed., Academic Press, New York, 1952.
32. Coles, G. C., *Int. J. Parasitol.*, 3, 783, 1973.
33. Bueding, E., Peters, L., and Waite, J. F., *Proc. Soc. Exp. Biol. Med.*, 64, 111, 1947.
34. Eckert, J. and Lehner, B., *Z. Parasitenkd.*, 37, 288, 1971.
35. Meyer, F., Meyer, H., and Bueding, E., *Biochim. Biophys. Acta*, 210, 257, 1970.
36. Smith, T. M., Brooks, T. J., Jr., and Lockard, V. G., *Lipids*, 5, 854, 1970.
37. Williams, M. O., Hopkins, C. A., and Wyllie, M. R., *Exp. Parasitol.*, 11, 121, 1961.
38. Oya, H., Hayashi, H., and Aoki, T., *Jpn. J. Parasitol.*, 20, 269, 1971.
39. Read, C. P., Simmons, J. E., and Rothman, A. H., *J. Parasitol.*, 46, 33, 1960.
40. Fisher, F. M., Jr., and Starling, J. A., *J. Parasitol.*, 56, 103, 1970.
41. Wertheim, G., Zeledon, R., and Read, C. P., *J. Parasitol.*, 46, 497, 1960.
42. Voge, M. and Green, J., *J. Parasitol.*, 61, 291, 1975.
43. Aldrich, D. V., Chandler, A. C., and Daugherty, J. W., *Exp. Parasitol.*, 3, 173, 1954.
44. Read, C. P., Rothman, A. H., and Simmons, J. E., *Ann. N.Y. Acad. Sci.*, 113, 154, 1963.
45. Daugherty, J. W., *Exp. Parasitol.*, 6, 60, 1957.
46. Harris, B. G. and Read, C. P., *Comp. Biochem. Physiol.*, 28, 645, 1969.
47. Parker, R. D., *J. Parasitol.*, 56, 259, 1970.
48. Webb, R. A. and Mettrick, D. F., *Int. J. Parasitol.*, 3, 47, 1973.
49. Voge, M., *J. Parasitol.*, 61, 563, 1975.
50. Heath, R. L., *J. Parasitol.*, 56, 98, 1970.
51. Laurie, J. S., *Comp. Biochem. Physiol.*, 4, 63, 1961.
52. Srivastava, V. M. L., Ghatak, S., and Krishna Murti, C. R., *Exp. Parasitol.*, 23, 339, 1968.
53. Read, C. P. and Rothman, A. H., *Exp. Parasitol.*, 7, 217, 1958.
54. Von Brand, T., McMahon, P., Gibbs, E., and Higgins, H., *Exp. Parasitol.*, 15, 410, 1964.
55. Tkachuck, R. D. and MacInnis, A. J., *Comp. Biochem. Physiol.*, 40B, 993, 1971.
56. Dunkley, L. C. and Mettrick, D. F., *Exp. Parasitol.*, 25, 146, 1969.

57. Laurie, J. S., *Exp. Parasitol.,* 6, 245, 1957.
58. Read, C. P. and Rothman, A. H., *Exp. Parasitol.,* 7, 217, 1958.
59. Fairbairn, D., Wertheim, G., Harpur, R. P., and Schiller, E. L., *Exp. Parasitol.,* 11, 248, 1961.
60. Colucci, A. V., Orrell, S. A., Saz, H. J., and Bueding, E., *J. Biol. Chem.,* 241, 464, 1966.
61. Ginger, C. D. and Fairbairn, D., *J. Parasitol.,* 52, 1086, 1966.
62. Webb, R. A. and Mettrick, D. F., *Int. J. Parasitol.,* 5, 107, 1975.
63. Read, C. P., *Exp. Parasitol.,* 6, 288, 1957.
64. Esch, G. W., *J. Parasitol.,* 50, 72, 1964.
65. Von Brand, T., Churchwell, F., and Eckert, J., *Exp. Parasitol.,* 23, 309, 1968.
66. Bailey, H. H. and Fairbairn, D., *Comp. Biochem. Physiol.,* 26, 819, 1968.
67. Lumsden, R. D. and Harrington, G. W., *J. Parasitol.,* 52, 695, 1966.
68. King, J. W. and Lumsden, R. D., *J. Parasitol.,* 55, 250, 1969.
69. Jacobsen, N. S. and Fairbairn, D., *J. Parasitol.,* 53, 355, 1967.
70. Frayha, G. J. and Fairbairn, D., *J. Parasitol.,* 54, 1144, 1968.
71. Buteau, G. H. and Fairbairn, D., *Exp. Parasitol.,* 25, 265, 1969.
72. Prescott, D. M. and Voge, M., *J. Parasitol.,* 45, 587, 1959.
73. Dvorak, J. A. and Jones, A. W., *Exp. Parasitol.,* 14, 316, 1963.
74. Daugherty, J. W. and Foster, W. B., *Exp. Parasitol.,* 7, 99, 1958.
75. Frayha, G. J., *Comp. Biochem. Physiol.,* 49, 93, 1974.
76. Bryant, C. and Morseth, D. J., *Comp. Biochem. Physiol.,* 25, 541, 1968.
77. Frayha, G. J. and Fairbairn, D., *Comp. Biochem. Physiol.,* 28, 1115, 1969.
78. Daugherty, J. W., *J. Parasitol.,* 42, 17, 1956.
79. Read, C. P. and Simmons, J. E., *Physiol. Rev.,* 43, 263, 1963.
80. Platzer, E. G. and Roberts, L. S., *Comp. Biochem. Physiol.,* 35, 535, 1970.
81. Seidel, J. S., *J. Parasitol.,* 57, 566, 1971.
82. Voge, M. and Seidel, J., *J. Parasitol.,* 54, 269, 1968.
83. Wulzen, R. and Bahrs, A. M., *Physiol. Zool.,* 8, 457, 1935.
84. Bahrs, A. M. and Wulzen, R., *Proc. Soc. Exp. Biol. Med.,* 33, 528, 1936.
85. Wulzen, R. and Bahrs, A. M., *Physiol. Zool.,* 9, 508, 1936.
86. Wagtendonk, W. J. van and Wulzen, R., *Vitam. Horm. N.Y.,* 8, 69, 1950.
87. Cheldelin, V. H., in *The Vitamins,* Sebrell, W. H. and Harris, R. S., Eds., Academic Press, New York, 1972, 395.

QUALITATIVE REQUIREMENTS AND UTILIZATION OF NUTRIENTS: NEMATODA*

W. F. Hieb

INTRODUCTION

A majority of the experiments cited in these tables were carried out under axenic culture conditions, i.e., germ free; see Reference 1 for nomenclature. The exceptions are noted. Culture media usually contain a basal component which is chemically defined, i.e., carbohydrates, amino acids, vitamins, nucleotides, salts, etc. Additional supplements of varying complexity, e.g., tissue extracts, serum, killed bacteria, are often necessary for growth or reproduction. Some basal media are composed completely of undefined materials, e.g., yeast extract, peptones. Most experiments involve organisms cultivated in continuous axenic stock cultures, while others involve animals grown initially in the presence of other living organisms (xenic) and then axenized just before use in the experiment (e.g., *Chiloplacus*). Experimental systems generally fall into two categories: (a) the observation of growth (e.g., *Nippostrongylus*), reproduction, or survival of individual organisms (or eggs) throughout the experiment, and (b) the estimation of changes in large populations (mass cultures). Results based on reproduction may extend for only one generation or for several successive generations (subcultures). The latter procedure is a desirable means of determining a nutritional requirement, as the organism eventually becomes depleted in any vital component that may be carried over from the original medium. More detailed information pertaining to the above can be found in a review of the culture of free-living and parasitic nematodes.[2]

REFERENCES

1. Dougherty, E. C., *Parasitology,* 42, 259–261, 1953.
2. Rothstein, M. and Nicholas, W. L., in *Chemical Zoology,* Vol. III, Florkin, M. and Scheer, B. T., Eds., Academic Press, New York, 1969, 289–328.

*Preparation was supported in part by grant AI-07145 from the National Institute of Allergy and Infectious Diseases, U.S.P.H.S.

Table 1
PROTEINS AND PEPTIDES

The nutritional activity of proteins and peptides listed below, unless otherwise indicated, is related to their ability to provide unidentified growth factors. This does not preclude their role as sources of amino acids and nitrogen. A majority of nematode species that have been studied require an unidentified growth factor. In addition to a basal defined medium (including amino acids) containing a source of heme (see Table 5) and sterols (see Table 6), a source of growth factor must be provided in order to obtain more than a minimal rate of reproduction. Although the nature of the unknown nutrient(s) has not been determined, the growth factor requirement can be satisfied by a variety of proteins or peptides, and to some extent by polysaccharides (see Table 3) or phospholipids (see Table 7). Evidence suggests that there are at least two growth factor requirements (see Footnote m). Larval individuals of *Caenorhabditis briggsae* also appear to have a physical requirement for particulates in the culture medium (see Table 7, Footnote 1). This adds a complication to a bioassay based on the individual larval method[1] as compared with a mass culture method. The latter is particularly suitable for showing a quantitative nutritional response to growth-factor-containing substances,[2] and the need for particles is not evident. *Abbreviations*: S = moderately or highly stimulatory for growth or reproduction; s = slightly stimulatory; $ = not stimulatory.

Table 1 (continued)

Species	Albumin	Casein[a]	Cytochrome c	Fetuin[b]	α-Globulin	β-Globulin	γ-Globulin[c]	Hemoglobin	Insulin[d]
Aphelenchoides rutgersi[h]	—	—	—	—	—	—	S	—	—
Aphelenchus avenae[i]	—	—	—	S	S	—	S	S	—
Caenorhabditis briggsae[j,k]	s	S[m]	S[n]	S	s	s	S	S[n]	s
Turbatrix aceti	—	S[r]	S	—	—	—	s[s]	S	S[s]

Species (continued)	β-Lacto-globulin	Leucyl-phenylalanine[e]	β-Lipoprotein[f]	Myoglobin	Ribonuclease	Soy peptone[g]	Other	Ref.
Aphelenchoides rutgersi[h]	—	—	S	—	—	—	—	3
Aphelenchus avenae[i]	—	—	S	—	—	—	—	4
Caenorhabditis briggsae[j,k]	s	S[o]	S	S[n]	S	S[p]	S[q]	2, 4—14
Turbatrix aceti	—	—	—	S	—	S[t]	S[u]	4, 5, 6, 12

a   Insoluble casein stimulated the reproduction of mass cultures of *Caenohabditis briggsae* and *Turbatrix aceti* in an undefined medium.[5] Sodium caseinate (almost completely soluble) was stimulatory to *Caenorhabditis briggsae*, *Caenorhabditis elegans*, *Neoaplectana carpocapsae*, *Pangrellus redivivus*, and *Turbatrix aceti* (see Footnote m).[12] Enzymatic or acid hydrolysates of casein were beneficial for the cultivation of *Rhabditis marina*.[15]

b   Fetuin is a glycoprotein found in fetal calf serum.

c   Also supported reproduction of *Caenorhabditis elegans*, *Panagrellus redivivus*, and *Neoaplectana carpocapsae*.[4]

d   Low-zinc porcine insulin.

e   Constituent amino acids were L-isomers.

f   Cohn fraction IV-4 of blood plasma.

g   An enzymatic digest of soy bean meal.

h   The medium did not contain an added heme source.

i   The response to these proteins was not increased by the addition of hemin to the defined basal medium. The nutritional activity of proteins may depend on their natural heme content. For example, it was found that α-globulin and β-lipoprotein, but not γ-globulin and albumins, had a spectrophotometric absorbance peak at 407 nm (the maximum absorption wavelength for hemeproteins).

j   All compounds were assayed in a defined basal medium in the presence of hemin unless otherwise specified. Hemin, being free of protein, is chemically more defined than hemeproteins and less likely to have growth factors associated with it (see Footnote m).

Table 1 (continued)

k   Albumin, fetuin, α-globulin, and β-globulin were assayed with individual larval cultures without the use of methods that increase the formation of insoluble particles in the medium.[4] This procedure seems to be necessary, or at least beneficial, for the assay of growth factor substances with *C. briggsae* larvae in the absence of natural particulates (see Table 7, Footnote 1).

l   Bovine serum albumin, egg albumin (primarily ovalbumin),[4] or in addition conalbumin.[7] It should be noted that serum albumins often contain bound fatty acids. Egg albumin stimulated reproduction of *C. briggsae* mass cultures in an undefined medium.[5]

m   Potassium caseinate is an effective growth factor supplement for mass cultures grown in a defined basal medium containing cytochrome *c*;[2,8] however, it gave a poor response with hemin as a heme source.[9] This suggests that an unknown nutrient may be associated with cytochrome *c* (and possibly other hemeproteins) which is not present in casein. Enzymatic digests and acid hydrolysates of casein also supported reproduction, but could not be replaced by a synthetic mixture of the individual component amino acids.[8]

n   Cytochrome *c* will not serve as a source of both heme and growth factor for mass cultures after repeated subculturing.[13] This is also true for hemoglobin and myoglobin.[11]

o   Histidine-containing peptides were less active than leucyl-phenylalanine in supporting reproduction of mass cultures in the presence of cytochrome *c* as a heme source, and glycine-containing peptides were inactive.[8] Leucyl-phenylalanine was less active than either casein or soy peptone on a dry weight basis.

p   It was found that soy peptone is a good source of growth factor for mass cultures with either cytochrome *c*[11] or hemin[14] as a heme source. Upon chromatographic fractionation by ion-exchange and gel filtration, the active component appeared to have a molecular weight of less than 1000.[11,14] It should be noted that soy peptone contains as much as 30% carbohydrate on a dry weight basis.

q   Soy peptone was found to be the most effective of several protein hydrolysates (including enzymatic digests of casein or lactalbumin or acid-hydrolyzed casein) which were tested as substitutes for the amino acid component of the defined basal medium (not as sources of growth factor).[10] Bovine mucin (a glycoprotein) supported reproduction of mass cultures in defined medium.[14] Tobacco mosaic virus protein,[6] ferritin, or transferrin[7] had growth factor activity in the presence of hemin as a heme source.

r   See Footnote a.

s   Assayed with hemin as a heme source.[4,6]

t   Slightly inhibitory to reproduction.[11]

u   Tobacco mosaic virus protein with hemin as a heme source.[6]

Table 1 (continued)

REFERENCES

1.   Lower, W. R., Hansen, E., and Yarwood, E. A., *J. Exp. Zool.,* 161, 29–35, 1966.
2.   Pinnock, C. B., Hieb, W. F., and Stokstad, E. L. R., *Nematologica,* 21, 1-4, 1975.
3.   Thirugnanam, M., *J. Nematol.,* 6, 153, 1974.
4.   Buecher, E. J., Hansen, E. L., and Yarwood, E. A., *Nematologica,* 16, 403–409, 1970.
5.   Rothstein, M. and Cook, E., *Comp. Biochem. Physiol.,* 17, 683–692, 1966.
6.   Buecher, E. J., Hansen, E. L., and Yarwood, E. A., *J. Nematol.,* 3, 89–90, 1971.
7.   Vanfleteren, J. R., *Nature,* 248, 255–257, 1974.
8.   Pinnock, C., Shane, B., and Stokstad, E. L. R., *Proc. Soc. Exp. Biol. Med.,* 148, 710–713, 1975.
9.   Pinnock, C. B. and Stokstad, E. L. R., *Nematologica,* 21, 258–261, 1975.
10.   Tomlinson, G. A. and Rothstein, M., *Biochim. Biophys. Acta,* 63, 465–470, 1962.
11.   Rothstein, M., *Comp. Biochem. Physiol.,* 49B, 669–678, 1974.
12.   Buecher, E. J. and Hansen, E. L., *J. Nematol.,* 2, 199–200, 1971.
13.   Rothstein, M., private communication.
14.   Huang, L. S., M. A. thesis, State University of New York, Buffalo, 1970.
15.   Tietjen, J. H. and Lee, J. J., *Cah. Biol. Mar.,* 16, 685–693, 1975.

Table 2
AMINO ACIDS

The dietary amino acid requirements of nematodes are similar to those of most multicellular animals. However, biochemical findings suggest that some species are capable of synthesizing "essential" amino acids. For example, radioactivity from ^{14}C-labeled acetate or glucose was incorporated into tryptophan which was isolated from a mixture of two *Meloidogyne* species.[1] *Aphelenchoides rutgersi* and *Caenorhabditis briggsae* were reported to synthesize a variety of essential amino acids from radioactively labeled compounds (see Footnotes b and j).[2,3] Such observations may explain the partial nutritional requirement of *A. rutgersi* for arginine and valine; on the other hand, the relatively poor incorporation of radioactivity into several of the nonessential amino acids in *A. rutgersi* (see Footnote b) may be related to their stimulatory effect on reproduction. In connection with nematode amino acid requirements, it should be noted that there may be a relationship between the nutritional availability of amino acids and the biological activity of the protein growth factor or growth-promoting peptides (see Table 1). *Abbreviations:* R = absolute requirements; S = moderately or highly stimulatory for growth or reproduction; S = unspecified degree of stimulation; $ = not stimulatory.

Table 2 (continued)

Essential amino acids[a]

Species	Arginine	Histidine	Isoleucine	Leucine	Lysine	Methionine	Phenylalanine	Threonine	Tryptophan	Valine	Other	Ref.
Aphelenchoides rutgersi[b]	S	R[c]	R	R	R	R	R	R	R[c]	S	S[h]	3,4
Caenorhabditis briggsae[i,j]	R	R	R	R	R	R[k]	R	R	R	R	–	5
Chiloplacus lentus	–	–	–	–	–	–	–	–	–	–	S[l]	6
Neoplectana glaseri	R	R	R	R	R	R	R	R	R	R	S[m]	7

Nonessential amino acids[a]

Species (continued)	Alanine	Asparagine	Aspartic acid	Cysteine	Glutamic acid	Glutamine	Glycine	Proline	Serine	Tyrosine
Aphelenchoides rutgersi[b]	S	S[d]	S[d]	S[e]	S[d]	S[d]	S	S[f]	S	S[g]
Caenorhabditis briggsae[i,j]	S	S	–	S	S	S	S	S	S	S
Chiloplacus lentus	–	–	–	–	–	–	–	–	–	–
Neoplectana glaseri	S	–	S	–	S	–	S	S	S	S

a   All amino acids were L-isomers or a DL-mixture, unless not applicable.

b   *Aphelenchoides* sp. was shown to incorporate significant amounts of radioactivity from ^{14}C-labeled acetate into arginine, isoleucine-leucine, lysine, phenylalanine, threonine, and valine. Less radioactivity was incorporated into alanine, glycine, proline, and serine than would be expected as compared with the labeling of glutamic and aspartic acids.[3]

c   Inhibitory at high concentration.[4]

d   Omission of these amino acids from the medium in various combinations resulted in a substantial decline in nematode population.[3]

e   Cystine and cysteic acid were also negative.[4] Earlier results showed that the deletion of all three amino acids (cysteine, cystine, and cysteic acid) from the medium simultaneously caused a decline in population.[3]

f   Plus hydroxyproline.[3]

g   Spared phenylalanine.[4]

h   No effect was observed when the following amino acids were omitted from the medium: α-aminoadipic acid, α-amino-*n*-butyric acid, α-ε-diaminopimelic acid, homoserine, or ornithine;[3] β-alanine, or γ-aminobutyric acid.[4]

i   Amino acids, when omitted from the medium, were replaced by a mixture of equimolar amounts of acetate, citrate, pyruvate, glucose, and glycine (the mixture contained no glycine when this amino acid was tested). Glutathione was omitted from the medium when glutamic acid, glutamine, cysteine, or glycine were tested. It was shown that deletion of nonessential amino acids as a group from the medium did not affect reproduction.

j   *C. briggsae* was reported to synthesize the following amino acids from ^{14}C-labeled acetate, glycine or glucose: arginine, histidine, isoleucine, leucine, lysine, threonine, and tyrosine (not phenylalanine).[2]

k   Homocysteine (as homocysteine thiolactone) can be utilized in place of methionine in the medium.[8]

l   Addition of amino acids in the following two groups resulted in a nutritional response: (a) alanine, glycine, isoleucine, leucine, serine, threonine, and valine, and (b) arginine, asparagine, citrulline, glutamine, histidine, hydroxylysine, and lysine.

m   Taurine.

Table 2 (continued)

REFERENCES

1. Myers, R. F. and Krusberg, L. R., *Phytopathology,* 55, 429–437, 1965.
2. Rothstein, M. and Tomlinson, G., *Biochim. Biophys. Acta,* 63, 471–480, 1962.
3. Balasubramanian, M. and Myers, R. F., *Exp. Parasitol.,* 29, 330–336, 1971.
4. Myers, R. F. and Balasubramanian, M., *Exp. Parasitol.,* 34, 123–131, 1973.
5. Vanfleteren, J. R., *Nematologica,* 19, 93–99, 1973.
6. Roy, T. K., *Nematologica,* 21, 12–18, 1975.
7. Jackson, G. J., *Exp. Parasitol.,* 34, 111–114, 1973.
8. Lu, N., Hieb, W. F., and Stokstad, E. L. R., *Proc. Soc. Exp. Biol. Med.,* 151, 701–706, 1976.

Table 3
CARBOHYDRATES

Abbreviations: S = moderately or highly stimulatory to growth or reproduction; s = slightly stimulatory; $ = not stimulatory.

Polysaccharides[a]

Species	Glycogen[b]	Guar gum (anionic)[c]	Guar gum (neutral)[c]	Guar gum (cationic)[c]	Lipopolysaccharide[d]	Methylcellulose	Other	Ref.
Aphelenchoides rutgersi	–	–	–	–	–	–	–	1
Caenorhabditis briggsae[f]	S[g]	s	s	s[h]	S[i]	s	$[j]	2, 3, 4
Caenorhabditis elegans	S[k]	–	–	–	–	–	–	2, 5
Chiloplacus lentus	–	–	–	–	–	–	–	6
Panagrellus redivivus[m]	–	–	–	–	–	–	–	2
Trichinella spiralis[n]	–	–	–	–	–	–	–	7

Sugars

Species (continued)	Erythrose	Fructose	Galactose	Glucose	Maltose	Mannose	Raffinose	Ribose	Sucrose	Trehalose	Xylose	Other
Aphelenchoides rutgersi	–	$	$	S	$	S	S	S	s	S	$	$[e]
Caenorhabditis briggsae[f]	–	–	–	S	–	–	–	$	$	S	–	s
Caenorhabditis elegans	–	–	s	S	–	–	–	$	$	S	s	–[l]
Chiloplacus lentus	S	S	–	S	s	S	S	S	–	S	s	s[l]
Panagrellus redivivus[m]	–	–	–	$	–	–	–	$	$	$	–	–
Trichinella spiralis[n]	–	–	–	S	S	–	–	–	$	S[o]	–	$[p]

a   The nutritional activity of these substances is primarily related to their ability to provide unidentified growth factors, and not necessarily to their role as carbohydrate sources (see introduction to Table 1). A carbohydrate such as glucose is normally present in the medium.

b   *Turbatrix aceti* reproduced initially in a basal defined medium containing hemin and ethanol in addition to glycogen, but failed to reproduce when subcultured; *Neoaplectana carpocapsae* and *Aphelenchus avenae* did not reproduce.[3]

c   A polymer of mannose and galactose.

d   A purified lipopolysaccharide fraction from *Escherichia coli* cell walls.

e   Arabinose.

f   *Rhabditis anomala* gave a similar response to the sugars listed for *Caenorhabditis briggsae.*[2]

g   Larval cultures were maintained for ten serial subcultures in a defined basal medium containing hemin as the source of heme.[3] Substitution of glycogen with cornstarch, rice starch, rice flour, trehalose, and other carbohydrates individually in the medium failed to support reproduction.[3]

h   Mass cultures with either hemin or cytochrome c as the source of heme in the medium gave no response.[4] Substitution of cationic guar gum with a proportional synthetic mixture of the component sugars in the medium gave no response.[4]

i   Individual components of typical lipopolysaccharide complexes, including galactosamine, glucosamine, and mannosamine, and their acetyl derivatives, acetyl neuraminic acid, fucose, galactose, mannose, rhamnose, ribose, and xylose, were ineffective as growth factors.[8,9]

Table 3 (continued)

j   Agar and ficoll (a high-molecular-weight polymer of sucrose).[4]
k   Mass cultures were grown in a medium containing soy peptone, yeast extract, dextrose, vitamins, and hemin; reproduction was superior to an identical medium containing liver extract instead of glycogen.[5]
l   Rhamose, sorbose, glucose, or a mixture of sucrose and lactose.
m   The reproduction of *Pauagrellus redivivus* was unchanged by the presence of carbohydrate in the medium.[2]
n   As determined by survival of juveniles.
o   Survival of adults in the presence of trehalose was poor.
p   Lactose.

REFERENCES

1. Petriello, R. P. and Myers, R., *Exp. Parasitol.*, 29, 423—432, 1971.
2. Hansen, E. L. and Buecher, E. J., *J. Nematol.*, 2, 1-6, 1970.
3. Hansen, E. L., Perez-Mendez, G., and Buecher, E. J., *Proc. Soc. Exp. Biol. Med.*, 137, 1352-1354, 1971.
4. Pinnock, C. B. and Stokstad, E. L. R., *Nematologica*, 21, 258-261, 1975.
5. Buecher, E. J. and Hansen, E. L., *J. Nematol.*, 3, 199-200, 1971.
6. Roy, T. K., *Nematologica*, 21, 12-18, 1975.
7. Castro, G. A. and Roy, S. A., *J. Parasitol.*, 60, 887-889, 1974.
8. Rothstein, M. and Hieb, W. F., *J. Parasitol.*, 56 (No. 4, Sec. II, Part 1), 292-293, 1970.
9. Huang, L. S., M.A. thesis, State University of New York, Buffalo, 1970.

Table 4
VITAMINS

B-vitamin requirements for nematodes appear to be similar to those of higher animals. Determination of these requirements has often been burdened by the use of culture media containing undefined materials such as tissue extracts (e.g., chick embryo extract, liver extract). Media supplements that are specifically deficient in one or more nutrients (e.g., *Escherichia coli* cells, soy-peptone, vitamin-free casein) have made it less difficult to determine absolute B-vitamin requirements. It is also possible to arrive at conclusions regarding the lack of requirement for certain nutrients. For instance, it is known that *E. coli* does not contain derivatives of the vitamin A, C, D, or E groups unless these were present in the medium in which the cells were grown originally. Therefore, it is unlikely that either vitamin C or the fat-soluble vitamins (except vitamin K) are required for *Caenohabditis briggsae* or for other species which are cultivated with *E. coli* as a primary source of nutrients (see also Table 7). This observation does not exclude the possibility that these compounds may still be beneficial when added to the medium. Though not covered in the table below, there are reports which indicate that the addition of vitamins as a group has a beneficial effect for several species of parasitic nematodes.[1-4] *Abbreviations:* R = absolute requirement; S = moderately or highly stimulatory for growth or reproduction; S = unspecified degree of stimulation; $ = not stimulatory.

Table 4 (continued)

Species	Biotin	Choline	Folacin	Niacin	Pantothenic acid	Riboflavin	Thiamin	Vitamin B-6[a]	Vitamin B-12[b]	Other	Ref.
Aphelenchoides rutgersi	–	–	S[c]	–	–	–	–	–	–	–	5
Caenorhabditis briggsae	R[d]	S[e]	R[f]	S	S	S	S	S	R[g]	–	6–9
Neoaplectana glaseri	S	–	R[h]	–	–	–	–	–	–	S[i]	10–12
Nippostrongylus spp.[j]	–	–	S[k]	S	–	S	S	S	S	S[l]	13, 14

a   Pyridoxine.

b   Vitamin B-12 was concentrated in the coelomocytes of various strongylid nematodes when they were cultured in media containing the vitamin.[14,15] After absorption into the intestine of *Ascaris suum*,[16] Co-labeled vitamin B-12 was bound principally to mucoprotein[16] and converted to metabolically active coenzyme in the body wall.[17]

c   Determined in the presence of the folic acid antagonist aminopterin (4-aminopteroylglutamic acid) and folinic acid (N^5-formylpteroylglutamic acid).

d   The deficiency was produced by the addition of avidin to the medium.[7]

e   Organisms did not show a response after addition of choline to a basal medium containing autoclaved *Escherichia coli* cell residue (e.g., washed free of soluble contents).[8] *E. coli* does not contain phosphatidyl choline;[18] therefore, the residue should not contain significant amounts of choline.

f   Either folic acid (pteroylglutamic acid) or folinic acid (N^5-formylpteroylglutamic acid) were nutritionally active.[19] Biopterin (an unconjugated pteridine), one form of "*Crithidia* factor," had no biological activity when substituted for folic acid.[20]

g   The requirement for vitamin B-12 was shown to be relatively high when *C. briggsae* was cultured in a basal medium containing homocysteine in place of methionine.[8] Both *E. coli* cell residue (see Footnote e) and soy peptone (see Table 1) were used as supplements since the former is deficient in vitamin B-12 (not synthesized), folic acid, methionine, and choline. The latter, being a plant source, is deficient in vitamin B-12 and methionine. The vitamin B-12 requirement was less in the presence of methionine, making it necessary to subculture the organisms many times in order to deplete their stores of the vitamin.[9]

h   Pteroylglutamic acid.[11]

i   Vitamins of the B group were required by *N. glaseri* at concentrations present in the basal medium.[10] This mixture included *p*-aminobenzoic acid and inositol. It was later reported that cultures were not inhibited by the addition of antagonists of *p*-aminobenzoic acid.[11]

j   No details were given. It was noted that certain other vitamins could be omitted without affecting the growth response.[13]

k   Pteroylglutamic acid.[13]

l   *p*-Aminobenzoic acid.[13]

Table 4 (continued)

REFERENCES

1. Douvres, F. W., *J. Parasitol.*, 48, 314–320, 1962.
2. Leland, S. E., Jr., *J. Parasitol.*, 49, 600–611, 1963.
3. Meerovitch, E., *Can. J. Zool.*, 43, 69–79, 1965.
4. Silverman, P. H., Alger, N. E., and Hansen, E. L., *Ann. N.Y. Acad. Sci.*, 139, 124–142, 1966.
5. Thirugnanam, M. and Myers, R. F., *Exp. Parasitol.*, 36, 202–209, 1974.
6. Nicholas, W. L., Hansen, E. L., and Dougherty, E. C., *Nematologica*, 8, 129–135, 1962.
7. Nicholas, W. L. and Jantunen, R., *Nematologica*, 9, 332–336, 1963.
8. Lu, N., Hieb, W. F., and Stokstad, E. L. R., *Proc. Soc. Exp. Biol. Med.*, 151, 701–706, 1976.
9. Lu, N., Hieb, W. F., and Stokstad, E. L. R., unpublished results.
10. Jackson, G. J., *Exp. Parasitol.*, 12, 25–32, 1962.
11. Jackson, G. J. and Siddiqui, W. A., *J. Parasitol.*, 51, 727–739, 1965.
12. Jackson, G. J. and Platzer, E. G., *J. Parasitol.*, 60, 453–457, 1974.
13. Weinstein, P. P. and Jones, M. F., *Am. J. Trop. Med. Hyg.*, 6, 480–484, 1957.
14. Weinstein, P. P., *J. Parasitol.*, Suppl. 51, 55, 1965.
15. Weinstein, P. P., *J. Parasitol.*, Suppl. 47, 23, 1961.
16. Zam, S. G. and Martin, W. E., *J. Parasitol.*, 55, 480–485, 1969.
17. Oya, H. and Weinstein, P. P., *Comp. Biochem. Physiol.* 50B, 435–442, 1975.
18. Kates, M., in *Advances in Lipid Research*, Paoletti, R. and Kritchevsky, D., Eds., Academic Press, New York, 1964, 17–90.
19. Hansen, E. L. and Dougherty, E. C., *Fed. Proc.*, 16(No. 1, Part 1) 304–305, 1957.
20. Dougherty, E. C. and Hansen, E. L., *Anat. Rec.*, 128, 541–542, 1957.

Table 5
PORPHYRINS AND RELATED COMPOUNDS

It appears that nematodes have a general nutritional requirement for heme (iron-containing porphyrin compounds). Although heme requirements are widespread throughout the unicellular kingdom (e.g., bacteria, fungi, protozoa), very few examples of a nutritional need for heme by multicellular animals other than nematodes have been reported. The requirement in nematodes could be attributed to metabolic deficiencies such as the following: (a) an inability to synthesize the porphyrin ring structure, (b) a defect in the mechanism by which iron is introduced into the porphyrin ring, or (c) a problem in the transport of iron through the intestinal cells. Both nutritional and biochemical evidence are too limited to point clearly to any of these explanations. However, some related experimental observations should be noted. The hemoglobin content of perienteric fluid in *Ascaris lumbricoides* was reported to increase when porphyrins (protoporphyrin, hemin, and chlorophyll) and hemoproteins were added to the incubation medium.[1] Also, it was found that ^{14}C-labeled hematin was incorporated into coproporphyrin I and protoporphyrin IX by *Nippostrongylus brasiliensis* in such a manner as to suggest that this organism is capable of converting hemin or protoporphyrin to coproporphyrin, but not to uroporphyrin.[2] *Abbreviations:* S = moderately or highly stimulatory to growth or reproduction; S̲ = unspecified degree of stimulation; s = slightly stimulatory; $ = not stimulatory.

Table 5 (continued)

Species	δ-Aminolevulinic acid	Bilirubin	Biliverdin	Chlorophyll	Coproporphyrin[a]	Cytochrome c	Ferrichrome[b]	Ferridoxin	Other	Ref.
Aphelenchoides rutgersi	—	—	—	—	—	—	—	S	—	3
Caenorhabditis briggsae[f]	S	S	S	S	—	S	S[g]	—	S[i]	4–8
Nippostrongylus brasiliensis[j]	S	S	S	—	S[k]	S	—	—	—	2
Turbatrix aceti	—	—	—	—	—	—	—	—	—	7, 9
Rhabditis marina	—	—	—	—	—	—	—	—	S[n]	10

Species (continued)	Hemin[c]	Hemoglobin	Myoglobin	Porphobilinogen	Protoporphyrin[d]	Uroporphyrin
Aphelenchoides rutgersi	S[e]	—	—	—	—	—
Caenorhabditis briggsae[f]	S[h]	S	S	S	S	—
Nippostrongylus brasiliensis[j]	S	S	S	S	S[k]	—
Turbatrix aceti	S[l]	S[m]	—	—	—	—
Rhabditis marina	—	—	—	—	—	—

a   Coproporphyrin I and coproporphyrin III.

b   One of many iron-containing growth factors for heme-requiring microorganisms that belong to the family of such compounds known as siderochromes.[1,2] These compounds are naturally occurring complex organic iron-chelating agents which are derivatives of hydroxamic acid.

c   Aphelenchus avenae reproduction was unchanged by the presence of hemin (see Table 1, Footnote i).[4]

d   Protoporphyrin IX.

e   The results were obtained using a defined basal medium supplemented with 20% chick embryo extract. The latter was previously treated with activated charcoal to remove the heme-containing substances.

f   Culture medium contained a defined basal component and autoclaved Escherichia coli cells.[5] Although C. briggsae will reproduce in a medium containing viable E. coli cells (see Table 5, Footnote e), it will not reproduce if the medium contains autoclaved E. coli unless a heme-containing compound is added. It was concluded that the nutritionally active heme compounds of E. coli are destroyed by heat.

g   Although ferrichrome was found to be nutritionally inadequate as a substitute for heme,[8] it was able to spare the requirement for a liver supplement to some extent.[11]

h   It has been reported that hemin, if precipitated under special conditions, will support continued and substantial reproduction of individual larvae in the absence of an additional growth factor source (see Table 7, Footnote l).[13]

i   Ferric chloride.[4]

j   Organisms were cultured in a liquid medium containing either viable or formalin-killed E. coli.[2] All experiments were based on the development of eggs as far as third-stage filariform larvae, at which time the size of the larvae were measured.[15]

k   Tetramethyl esters (Waldenstrom type).[2]

l   Also reported in another study.[13] No data given.

m   Sterilized by filtration. Hemoglobin was nutritionally inactive when autoclaved.

n   Ferritin, hematin (hydroxy derivative of hemin).

Table 5 (continued)

REFERENCES

1. Smith, M. H. and Lee, D. L., *Proc. R. Soc. London Ser. B,* 157, 234–257, 1963.
2. Bolla, R. I., Weinstein, P. P., and Lou, C., *Comp. Biochem. Physiol.,* 48B, 147–157, 1974.
3. Thirugnanam, M., *J. Nematol.,* 6, 153, 1974.
4. Buecher, E. J., Hansen, E. L., and Yarwood, E. A., *Nematologica,* 16, 403–409, 1970.
5. Hieb, W. F., Stokstad, E. L. R., and Rothstein, M., *Science,* 168, 143–144, 1970.
6. Rothstein, M. and Hieb, W. F., *J. Parasitol.* (No. 4, Sec. II, Part 1), 292–293, 1970.
7. Rothstein, M., *Comp. Biochem. Physiol.,* 49B, 669–678, 1974.
8. Hieb, W. F. and Stokstad, E. L. R., unpublished results.
9. Buecher, E. J., Hansen, E. L., and Yarwood, E. A., *J. Nematol.,* 3, 89–90, 1971.
10. Tietjen, J. H. and Lee, J. J., *Cah. Biol. Mar.,* 16, 685–693, 1975.
11. Nicholas, W. L., Dougherty, E. C., and Hansen, E. L., *Ann. N.Y. Acad. Sci.,* 77, 218–236, 1959.
12. Nielands, J. B., *Science,* 156, 1443–1447, 1967.
13. Vanfleteren, J. R., *Nature,* 248, 255–257, 1974.
14. Pinnock, C., Shane, B., and Stokstad, E. L. R., *Proc. Soc. Exp. Biol. Med.,* 148, 710–713, 1975.
15. Bolla, R. I., Weinstein, P. P., and Lou, C., *Comp. Biochem. Physiol.* 43B, 487–501, 1972.

Table 6
STEROLS AND STEROL PRECURSORS

All of those nematodes studied have a dietary requirement for sterols. Biochemical observations indicate that several free-living species lack the ability to biosynthesize cholesterol from ^{14}C-labeled acetate or mevalonate.[1,2] However, it has recently been reported that ^{3}H-labeled squalene oxide is converted to lanosterol in *Panagrellus redivivus.*[3] This result suggests that at least one nematode species has the enzymatic machinery to synthesize sterols from squalene. Nutritional information supports these findings since squalene possesses biological activity for at least three species. However, in some instances the response was suboptimal, indicating perhaps only a partial ability to cyclize squalene. The remainder of the pathway seems to be intact, as lanosterol and desmosterol are utilized. Mevalonic acid, the precursor of squalene, is inactive. Therefore, a metabolic block in sterol biosynthesis could occur in the conversion of mevalonate to squalene. Previous steps in the pathway may also be lacking. The capacity for sterol synthesis in nematodes appears to consist of a combination of both a lack of early steps in synthesis and limited conversion of pre-sterol compounds. Available evidence suggests that cholesterol is not the ultimately required sterol. A wide range of sterols serve satisfactorily in meeting the requirement. Biochemical observations also indicate that nematodes have a considerable ability to interconvert various sterols.[2] Nevertheless, a pattern of nutritional specificity is emerging. Whereas those sterols required for the growth of insects must possess a closed planar ring system, a β-hydroxyl group at C3, and a hydrocarbon side chain at C17 similar in length to that of cholesterol,[4] nematodes seem to be capable of utilizing such compounds as cholestenone (keto group at C3), 25-norcholesterol (the last three carbons of the side chain absent), and cholestane. Compounds belonging to the coprostane series, e.g., coprostanol and Δ^7-coprosten-3-ol, consistently failed to support reproduction. *Abbreviations:* S = stimulatory to growth or reproduction; s = slightly stimulatory; S̲ = unspecified stimulation; $\cancel{S}$ = not stimulatory.

Table 6 (continued)

	Sterol precursors[a]				Sterol and related compounds[a]							
Species	Acetic acid	Farnesol	Mevalonic acid	Squalene	Cholestane[b]	Δ^4-Cholesten-3-one	Cholesterol	Coprostanol[c]	Δ^7-Coprosten-3-ol[c]	Other	7-Dehydrocholesterol	Ref.
Caenorhabditis briggsae[d]	S	S	S[e]	S	S	—	S	—	—	S[f]	S	5—8
Neoaplectana carpocapsae[g]	—	—	—	—	—	S	S	—	—	S[h]	S	9
Nippostrongylus brasiliensis[i]	S	S	S	s	S	—	S	S[j]	—	S[k]	S	10
Panagrellus redivivus[l]	—	—	—	—	—	—	S	—	—	—	—	11
Turbatrix aceti[l]	S	S	S[e]	s	—	S	S	S	S	S[m]	S	6—8, 11

	Sterol and related compounds[a]							
Species (continued)	Desmosterol	Ergosterol	Fucosterol	Lanosterol	Lathosterol	25-Norcholesterol	β-Sitosterol	Stigmasterol
Caenorhabditis briggsae[e]	—	S	—	S	—	—	S	S
Neoaplectana carpocapsae[g]	—	S	S	—	S	—	S	S
Nippostrongylus brasiliensis[i]	—	S	S	s	—	—	S	S
Panagrellus redivivus[l]	—	—	—	—	—	—	—	—
Turbatrix aceti[l]	S	S	S	S	S	S	S	S

a   Much of the data in these tables is based on cultures that contained bacteria (viable or killed). This type of medium (either xenic or axenic) provides a convenient system for determining sterol requirements since bacteria do not synthesize sterols or precursors such as farnesol and squalene,[12] nor do they contain these compounds unless present in the original growth medium. However, the use of cultures containing live bacteria brings about the possibility that sterols will be metabolized in such a way as to change their nutritional activity for the nematodes.

b   Cholestane is a steroid hydrocarbon which would not normally be expected to have biological activity. It may serve a "sparing" action for some portion of the sterol requirement. In the presence of live bacteria, e.g., *C. briggsae* cultures, it is possible that cholestane is converted to nutritionally active compounds.

c   Steroids belonging to the coprostane series differ from those related to cholestane principally by the *cis* (normal) conformation of the A and B rings of the steroid nucleus; the hydrogen at C5 is in a β configuration.

d   Nutritional effects of cholestane, 7-dehydrocholesterol, ergosterol, and stigmasterol were determined using *C. briggsae* cultivated in a liquid, monoxenic medium containing viable *Escherichia coli* cells (originally grown on a minimal medium) and salts.[5]

e   As DL-mevalonic acid lactone.[10]

f   Vitamin A, a mixture of vitamin E, linoleic and arachadonic acids, or Tween 80 (used to disperse lipids in medium) had no effect on reproduction when substituted in place of sterols in the medium.[5] Vitamins D_2 (ergocalciferol) and D_3 (cholecalciferol) were also ineffective.[13]

g   Cultivated on monoxenic peptone-glucose agar medium with either viable or "dead" bacterial cells.

h   22-Dihydrobrassicasterol and stigmasterol.

Table 6 (continued)

i See Table 5, Footnote j.

j Slightly inhibitory at higher concentrations.

k Coprostanone.

l All sterols, except cholesterol, β-sitosterol, and stigmasterol for *T. aceti*, were tested with nematodes cultured on a monoxenic peptone-glucose agar medium containing viable *Bacillus subtilis* endospores.[11]

m Campesterol, cholestanol, cholestenone, 24-methylenecholesterol.[11]

REFERENCES

1. Rothstein, M., *Comp. Biochem. Physiol.*, 27, 309–317, 1968.
2. Cole, R. J. and Krusberg, L. R., *Life Sci.*, 7, 713–724, 1968.
3. Willett, J. D. and Downey, W. L., *Biochem. J.*, 138, 233–237, 1974.
4. Clayton, R. B., *J. Lipid Res.*, 5, 3–19, 1964.
5. Hieb, W. F. and Rothstein, M., *Science*, 160, 778–780, 1968.
6. Buecher, E. J., Hansen, E. L., and Yarwood, E. A., *J. Nematol.*, 3, 89–90, 1971.
7. Rothstein, M., *Comp. Biochem. Physiol.*, 49B, 669–678, 1974.
8. Lu, N., Newton, C., and Stokstad, E. L. R., private communication.
9. Dutky, S. R., Robbins, W. E., and Thompson, J. V., *Nematologica*, 13, 140, 1967.
10. Bolla, R. I., Weinstein, P. P., and Lou, C., *Comp. Biochem. Physiol.* 43B, 487–501, 1972.
11. Cole, R. J. and Dutky, S. R., *J. Nematol.*, 1, 72–75, 1969.
12. Bloch, K., in *Taxonomic Biochemistry and Serology*, Leone, C. A., Ed., Ronald Press, New York, 1964, 377–390.
13. Hieb, W. F. and Rothstein, M., unpublished results.

Table 7
MISCELLANEOUS

The ability of nematodes to utilize two-carbon compounds such as acetic acid and ethanol for growth may be an indication of the importance of the glyoxylate cycle enzymes[1,2] in the metabolism of these organisms. This anaplerotic pathway makes possible the net conversion of two-carbon compounds (including glycine and acetyl CoA from fatty acid catabolism) into carbohydrate and amino acids. Nematodes are unique among multicellular animals in this potential. Another unusual characteristic of nematodes is the *de novo* biosynthesis of polyunsaturated fatty acids. *Turbatrix aceti, Caenorhabditis briggsae,* and *Panagrellus redivivus* were shown to incorporate ^{14}C-labeled acetate throughout the entire carbon chains of polyunsaturated "essential" fatty acids (e.g., linoleic, linolenic, and arachidonic acids) and not by the addition of two-carbon units to existing precursors.[3,4] These findings lend support to the observation that some nematodes such as *C. briggsae* can reproduce quite well in media containing *Escherichia coli* as a major component (see introduction to Table 4), although *E. coli* does not synthesize polyunsaturated fatty acids.[5] However, as noted below, fatty acids (or lecithin) had a beneficial effect on the reproduction of several species (including *C. briggsae*). Biochemical information relating to the nutritional effect of nucleic acid substituents is limited. It has been shown that the incorporation of ^{3}H-labeled orotic acid into pyrimidines by *Aphelenchoides rutgersi* was many times greater than the incorporation of ^{3}H-labeled glycine into purines.[6] This tends to agree with the reported nutritional stimulation by purine compounds in this species. *Abbreviations:* S = moderately or highly stimulatory for growth or reproduction; S̲ = unspecified degree of stimulation; $ = not stimulatory.

Table 7 (continued)

Species	Acetic acid	Ethanol	Fatty acids	Glutathione[a]	Lecithin	Nucleic acid substituents	Particulates	Purine compounds	Pyrimidine compounds	Ribonucleic acid	Other[b]	Ref.
Aphelenchoides rutgersi	—	—	—	—	—	S[c]	—	S[d]	§[e]	—	§[f]	6,7
Aphelenchus avenae	—	S[g]	—	—	—	—	§[h]	—	—	—	—	8,9
Caenorhabditis briggsae	—	§[i]	—	§	S[j]	S[k]	S[l]	—	—	§[m]	—	10—15
Neoaplectana glaseri	§[p]	—	—	—	—	§[n]	—	—	—	—	§[o]	16,17
Panagrellus redivivus	§[p]	—	S[q]	—	—	—	—	—	—	—	—	18,19
Rhabditis marina	—	—	S[s]	—	S[t]	—	—	—	—	S[r]	S[u]	20
Turbatrix aceti	S	S	—	—	—	S[k]	—	—	—	—	—	15,18

a   Although glutathione is a peptide (γ-glutamyl-cysteinyl-glycine), it is included here mainly for its function as a physiological oxidation-reduction compound.

b   Intrinsic factor (a mucoprotein which facilitates vitamin B-12 absorption in higher animals) stimulated the initial development of third-stage *Hemonchus contortus* to young adults.[21]

c   A mixture of all the nucleic acid substituents listed in Footnotes d and e.

d   A mixture of adenosine-2'(3')-phosphate, guanosine-2'(3')-phosphate, deoxyadenosine-5'-phosphate, and deoxyguanosine.[22]

e   A mixture of cytosine-2'(3')-phosphate, uridine-2'(3')-phosphate, deoxycytidine, thymidine, and thymine.[22] Data suggest that pyrimidines may also be stimulatory to reproduction.[23]

f   A mixture of cholesterol, linoleic acid, linolenic acid, oleic acid, palmitoleic acid, palmitic acid, and Tween 80.

g   Reproduction was also enhanced by *n*-propanol.[8]

h   The reproduction of this organism was unaffected by particulates.[9] This is probably due to its stylet mode of feeding.

i   Although ethanol appears to be toxic to *C. briggsae* when present in culture media,[15] certain amounts of ethanol are both utilized and produced by an unnamed species of *Caenorhabditis*.[8]

j   Tested as a complex with hemin as a heme source;[13] thus, egg lecithin appears to have growth factor activity (see Table 1). The nutritional activity can be attributed to fatty acids in this compound since choline was present in the medium (see also Table 4 for lack of choline requirement in *C. briggsae*).

k   A mixture of adenosine-2'(3')-phosphate, guanosine-2'(3')-phosphate, cytidine-2'(3')-phosphate, uridine-2'(3')-phosphate, and thymine.[14],[15]

l   The biological activity of various supplements for *C. briggsae* larvae was increased by precipitation into finely divided (colloidal) particles or by the addition of particles to the medium.[10] Other such supplements include proteins (see Table 1), glycogen (see Table 3), and lecithin (listed in this table). The extent of the contribution of particles to the biological activity of these substances has not yet been determined. There does not seem to be any direct evidence for a stimulatory effect of particulates on the reproduction of mass cultures (see introduction to Table 1).

m   Removal of ribonucleic acid (RNA) from a yeast preparation by protamine extraction did not affect the nutritional activity of yeast as a supplement to a basal medium containing ribonucleotides.[11]

n   A mixture of adenosine, cytidylic acid, guanine, xanthine, and uracil.[16]

o   Urea.[16]

p   Addition to medium caused a lag in reproduction when compared with a control at a similar pH adjusted with hydrochloric acid.[18]

q   A mixture of linoleic, linolenic, and ethyl arachidonic acids.[19]

r   Basal medium contained ribonucleotides.[19]

s   A mixture of linoleic, linolenic, oleic, palmitic, and stearic acids.

t   Egg lecithin and lecithin of animal origin were stimulatory to reproduction; however, soy lecithin was ineffective.

u   Tween 80, caphalin, or lactalbumin (a source of fatty acids).

Table 7 (continued)

REFERENCES

1. Rothstein, M. and Mayoh, H., *Comp. Biochem. Physiol.,* 16, 361–365, 1965.
2. Rothstein, M. and Mayoh, H., *Comp. Biochem. Physiol.,* 17, 1181–1188, 1966.
3. Rothstein, M. and Götz, P., *Arch. Biochem. Biophys.,* 126, 131–140, 1968.
4. Rothstein, M., *Int. J. Biochem.,* 1, 422–428, 1970.
5. Bloch, K., in *Taxonomic Biochemistry and Serology,* Leone, C. A., Ed., Ronald Press, New York, 1964, 377–390.
6. Petriello, R. P. and Myers, R. F., *Exp. Parasitol.,* 29, 423–432, 1971.
7. Thirugnanam, M., *J. Nematol.,* 6, 153, 1974.
8. Cooper, A. F., Jr. and Van Gundy, S. D., *J. Nematol.,* 3, 205–214, 1971.
9. Hansen, E., Buecher, E. J., and Evans, A. A. F., *Nematologica,* 16, 328, 1970.
10. Buecher, E. J., Perez-Mendez, G., and Hansen, E. L., *Proc. Soc. Exp. Biol. Med.,* 132, 724–728, 1969.
11. Buecher, E. J., Hansen, E. L., and Gottfried, T., *J. Nematol.,* 2, 93–98, 1970.
12. Vanfleteren, J. R., *Nematologica,* 19, 93–99, 1973.
13. Vanfleteren, J. R., *Nature,* 248, 255–257, 1974.
14. Lu, N., Hieb, W. F., and Stokstad, E. L. R., unpublished results.
15. Rothstein, M., private communication.
16. Jackson, G. J., *Exp. Parasitol.,* 34, 111–114, 1973.
17. Jackson, G. J. and Platzer, E. G., *J. Parasitol.,* 60, 453–457, 1974.
18. Rothstein, M. and Cook, E., *Comp. Biochem. Physiol.,* 17, 683–692, 1966.
19. Deubert, K. H. and Zuckerman, B. M., *Exp. Parasitol.,* 21, 209–214, 1967.
20. Tietjen, J. H. and Lee, J. J., *Cah. Biol. Mar.,* 16, 685–693, 1975.
21. Silverman, P. H., Alger, N. E., and Hansen, E. L., *Ann. N.Y. Acad. Sci.,* 139, 124–142, 1966.
22. Myers, R. F. and Balasubramanian, M., *Exp. Parasitol.,* 34, 123–131, 1973.
23. Thirugnanam, M. and Myers, R. F., *Exp. Parasitol.,* 36, 202–209, 1974.

QUALITATIVE REQUIREMENTS AND UTILIZATION
OF NUTRIENTS: ACANTHOCEPHALA

D. W. T. Crompton

INTRODUCTION

About a thousand species of acanthocephalan worm have been described. All are endoparasites which attain sexual maturity in the alimentary tract of vertebrates.[1,2] Acanthocephalans may be recognized by the possession of a retractile, hook-bearing proboscis by which they become attached to the host's intestinal wall. The proboscis and its associated organs are known as the praesoma and the remainder of the body, which is usually exposed to the host's intestinal lumen, is called the metasoma. There is a body cavity, but no alimentary tract has been observed at any stage of development.[2,3] The acanthocephalan life cycle, where known, has been found to involve an arthropod intermediate host, and infection of the vertebrate host occurs when intermediate hosts containing the cystacanth stage are swallowed.[3-5] A summary of the life cycles of the acanthocephalan species discussed in this article and a guide to some of the literature on acanthocephalan nutrition are presented in Table 1. The nutrients required by both partners of a host-parasite relationship are obtained by the host, and it is likely that an acanthocephalan parasite's requirements will be similar to those of its host. This prediction cannot be investigated until acanthocephalans have been grown in vitro in chemically defined media from which sera and other crude extracts have been omitted.[6] The manner in which acanthocephalans assimilate and metabolize a variety of substances has been investigated.[2,3,7-11] The knowledge that has been gained depends on evidence obtained from morphological and histochemical techniques, from the results of short-term experiments in vitro, and from studies on the course of acanthocephalan infection in laboratory hosts fed diets of known composition.

Table 1

SOME HOST-PARASITE RELATIONSHIPS INVOLVING ACANTHOCEPHALAN WORMS

The references are concerned with studies on the feeding, metabolism, and nutrition of the species mentioned. Review articles are cited in the text.[2,3,6-11,55,60]

Acanthocephalan species[a]	Host[b]	Ref.
(P) *Acanthocephalus ranae*	Crustaceans and toads	13, 14, 25
(P) *Echinorhynchus gadi*	Amphipods and flounders (m)	15, 43
(P) *E. truttae*	Amphipods and salmonid fish (fw)	14
(P) *Filicollis anatis*	Isopods and ducks	16
(A) *Macracanthorhynchus hirudinaceus*	Beetles and pigs	16, 27, 52, 57, 61
(A) *Moniliformis dubius*	Cockroaches and rats	12, 17–22, 27–31, 35–40, 44–48, 51, 53, 54, 56, 58, 59, 61–63
(E) *Neoechinorhynchus emydis* and *N. pseudemydis*	Ostracods, snails, and turtles (fw)	41, 42
(E) *Paulisentis actus*	Copepods and creek chub (fw)	26
(P) *Polymorphus minutus*	Amphipods and ducks	24, 32–34, 49, 50
(P) *Pomphorhynchus laevis*	Amphipods and fish (fw)	14, 23

[a] The phylum Acanthocephala is considered to be composed of three orders: Palaeacanthocephala (P), Archiacanthocephala (A), and Eoacanthocephala (E).[4]

[b] The intermediate hosts are invertebrates and the definitive hosts are vertebrates; fw indicates fresh water and m indicates marine.

ASSIMILATION OF NUTRIENTS

Adult acanthocephalans have been found to feed by absorbing substances from the lumen of that part of the host's intestine in which they live. When rats infected with *Moniliformis dubius* were given either oral or intraperitoneal doses of radioactively labeled compounds, much more radioactivity was found in worms from the rats given the oral doses than in those from the injected rats.*[12] Ultrastructural studies of the body surfaces of *Acanthocephalus ranae, Echinorhynchus gadi, E. truttae, Filicollis anatis, Macracanthorhynchus hirudinaceus, Moniliformis dubius, Polymorphus minutus,* and *Pomphorhynchus laevis* have revealed the presence of many pores leading into membrane-lined channels which extend into the body wall.[13-24] The pores and channels may be interpreted as a device to increase the absorptive surface area of the parasites. The absorptive surface area per unit area seems to be greater for the metasoma than for the praesoma, which is usually in contact with the host's tissues and probably does not participate to any great extent in nutrient absorption.**[24-26]

A variety of substances have been shown to be absorbed by adult acanthocephalans, and both active transport and diffusion contribute to the uptake of certain amino acids.[27-29] The feeding activities of adult acanthocephalans may involve more than the direct uptake of small molecules that have been released by the host's digestive processes. Some evidence indicates that proteolytic activity of parasite origin and amylytic activity of host origin occur at the surface of *M. dubius*.[30,31] A possible function of these enzymes is that they may facilitate the uptake of nutrients by the parasite in competition with the host by liberating small molecules in the immediate vicinity of the parasite's uptake mechanisms. The filamentous deposit of mucopolysaccharide, which has been detected on the surface of some species, might provide a suitable surface for trapping large molecules.[20,22,32]

Direct and circumstantial evidence indicate that the L-isomers of alanine, glutamic acid, isoleucine, leucine, methionine, and serine and the DL-isomers of methionine, serine, and valine are taken up by *M. dubius* from its surroundings.[12,27-30] Similarly, L-methionine is taken up by *A. ranae*, L-tyrosine by *Paulisentis fractus*, and the L-isomers of alanine, isoleucine, leucine, methionine, and serine by *Macracanthorhynchus hirudinaceus*.[25-27] Glucose has been found to be absorbed by all the species of Acanthocephala studied to date.[25,26,33-43] Circumstantial evidence suggests that fructose, galactose, mannose, maltose, trehalose, and possibly sucrose may be absorbed by *Moniliformis dubius*.[28,35,36,41] The disaccharides may have been hydrolyzed at the surface of the worm rather than absorbed intact.[31] The detection of radioactivity in the tissues of *A. ranae* and in those of *P. fractus* after incubation with tritiated glyceryl trioleate and tritiated thymidine, respectively, suggests that these molecules are absorbed by acanthocephalan worms.[25,26]

NUTRITIONAL REQUIREMENTS

The survival, growth, and reproduction of *M. dubius* in rats have been found to be dependent on the ingestion of starch by the hosts.[44,45] The establishment of *M. dubius* and its emigratory behavior in the small intestine during the first part of the course of the infection are not dependent on the presence of starch in the diet of the host.[45-47] Many worms from established populations of *M. dubius* and *Polymorphus minutus* are lost from

---

*   [32]P disodium hydrogen phosphate and [14]C L-leucine.
** Autoradiographic examination of tissue sections of worms following incubation in vitro with glyceryl trioleate-9,10-[3]H (*A. ranae*) and with ^{3}H glucose, ^{3}H tyrosine, and ^{3}H thymidine (*Paulisentis fractus*).

a host whose diet of a standard commercial ration is either withheld for a short time or replaced by a starch-free ration.[45,46,48-50] In contrast to their requirements for a source of glucose, the survival and growth of established *M. dubius* in rats deprived of essential amino acids for 14 days did not appear to be affected.[45]

METABOLISM OF NUTRIENTS

Glycolysis is considered to be the major pathway for energy production in the Acanthocephala.[7,8,11,30] Fermentation reactions, resulting in the excretion of incomplete oxidation products, occur whether oxygen is present or not.[11,33-39,41-43,51-53] The adult stages of most species of Acanthocephala live in regions of low oxygen tension. Therefore, it is possible that the complete tricarboxylic acid cycle or electron transport system will not be found to function in adult acanthocephalans, but some stages in these pathways may be of significance.[8,11,53-55] Glycogen and trehalose have been shown to be synthesized from glucose.[26,34,36-38,40,56] Homogenates of *Macracanthorhynchus hirudinaceus* are able to act on a variety of sugars, some of which may be important in the nutrition and metabolism of the parasite.[8,57] Evidence for the fixation of CO_2 into organic acids, which are then metabolized during energy production, has been obtained from *E. gadi* and *Moniliformis dubius*.[11,43,58-60] Radioactivity from labeled glucose has been detected in amino acids extracted from *M. dubius* and *Polymorphus minutus*.[34,38] Little information is available about the metabolism of proteins and lipids by acanthocephalan worms.[7,8,11,61]

NUTRITION OF DEVELOPING STAGES

Most of the development of acanthocephalan parasites occurs in the body cavity of arthropods, where a wide variety of nutrient molecules will be available for assimilation and metabolism.[3] From the few studies that have been made to date, the feeding and nutrition of the developmental stages seem to resemble those of the adults.[7,8,11] For example, glucose is absorbed by larval *P. minutus* and glycolysis and CO_2 fixation appear to function in larval *M. dubius*.[3,51,59,62,63]

REFERENCES

1. **Schmidt, G. D.,** Acanthocephala as agents of disease in wild mammals, *Wildl. Dis.*, 53, 10 pages (microfiche), 1969.
2. **Crompton, D. W. T.,** Relationships between Acanthocephala and their hosts, in *Symbiosis,* Jennings, D. H. and Lee, D. L., Eds., *Symp. Soc. Exp. Biol.*, 29, 467–504, 1975.
3. **Crompton, D. W. T.,** *An Ecological Approach to Acanthocephalan Physiology,* The University Press, Cambridge, Engl., 1970.
4. **Bullock, W. L.,** Morphological features as tools and pitfalls in acanthocephalan systematics, in *Problems in Systematics of Parasites,* Schmidt, G. D., Ed., University Park Press, Baltimore, 1969, 9–45.
5. **Lackie, A. M.,** The activation of infective stages of endoparasites of vertebrates, *Biol. Rev. Cambridge Philos. Soc.,* 50, 285–323, 1975.
6. **Lackie, A. M.,** Acanthocephala, in *Methods of Cultivating Parasites In Vitro,* Taylor, A. E. R. and Baker, J. R., Eds., Academic Press, New York, in press, 1977.
7. **Nicholas, W. L.,** The biology of the Acanthocephala, in *Advances in Parasitology,* Vol. 5, Dawes, B., Ed., Academic Press, New York, 1967, 205–246.
8. **Nicholas, W. L.,** The biology of the Acanthocephala, in *Advances in Parasitology,* Vol. 11, Dawes, B., Ed., Academic Press, New York, 1973, 671–706.
9. **Crompton, D. W. T.,** The sites occupied by some parasitic helminths in the alimentary tracts of vertebrates, *Biol. Rev. Cambridge Philos. Soc.,* 48, 27–83, 1973.
10. **Mettrick, D. F. and Podesta, R. B.,** Ecological and physiological aspects of helminth-host interactions in the mammalian gastrointestinal canal, in *Advances in Parasitology,* Vol. 12, Dawes, B., Ed., Academic Press, New York, 1974, 183–278.

11. Brand, T. von., *Biochemistry of Parasites,* 2nd ed., Academic Press, New York, 1973.

12. Edmonds, S. J., Some experiments on the nutrition of *Moniliformis dubius* Meyer (Acanthocephala), *Parasitology,* 55, 337–347, 1965.

13. Hammond, R. A., The fine structure of the trunk and praesoma wall of *Acanthocephalus ranae* (Schrank, 1788), Lühe, 1911, *Parasitology,* 57, 475–486, 1967.

14. Hammond, R. A., Observations on the body surface of some acanthocephalans, *Nature,* 218, 872–873, 1968.

15. Lange, H., Über Struktur und Histochemie des Integuments von *Echinorhynchus gadi* Müller (Acanthocephala), *Z. Zellforsch. Mikrosk. Anat.,* 104, 149–164, 1970.

16. Barabashova, V. N., Fine structure of the epidermis of two species of Acanthocephala — *Filicollis anatis* and *Macracanthorhynchus hirudinaceus, Mater. Nauch. Konf. Vses. Obshch. Gel'mint.,* Part III, 25–29, 1965 (in Russian).

17. Nicholas, W. L. and Mercer, E. H., The ultrastructure of the tegument of *Moniliformis dubius* (Acanthocephala), *Q. J. Microsc. Sci.,* 106, 137–146, 1965.

18. Edmonds, S. J. and Dixon, B. R., Uptake of small particles by *Moniliformis dubius* (Acanthocephala), *Nature,* 209, 99, 1966.

19. Rothman, A. H., Ultrastructural enzyme localization in the surface of *Moniliformis dubius* (Acanthocephala), *Exp. Parasitol.,* 21, 42–46, 1967.

20. Wright, R. D. and Lumsden, R. D., Ultrastructural and histochemical properties of the acanthocephalan epicuticle, *J. Parasitol.,* 54, 1111–1123, 1968.

21. Wright, R. D. and Lumsden, R. D., Ultrastructure of the tegumentary pore-canal system of the acanthocephalan *Moniliformis dubius, J. Parasitol.,* 55, 993–1003, 1969.

22. Rothman, A. H. and Elder, J. E., Histochemical nature of an acanthocephalan, a cestode and trematode absorbing surface, *Comp. Biochem. Physiol.,* 33, 745–762, 1970.

23. Stranack, F. R., Woodhouse, M. A., and Griffin, R. L., Preliminary observations on the ultrastructure of the body wall of *Pomphorhynchus laevis* (Acanthocephala), *J. Helminthol.,* 40, 395–402, 1966.

24. Crompton, D. W. T. and Lee, D. L., The fine structure of the body wall of *Polymorphus minutus* (Goeze, 1782) (Acanthocephala), *Parasitology,* 55, 357–364, 1965.

25. Hammond, R. A., Some observations on the role of the body wall of *Acanthocephalus ranae* in lipid uptake, *J. Exp. Biol.,* 48, 217–225, 1968.

26. Hibbard, K. M. and Cable, R. M., The uptake and metabolism of tritiated glucose, tyrosine, and thymidine by adult *Paulisentis fractus* Van Cleave and Bangham, 1949 (Acanthocephala: Neoechinorhynchidae), *J. Parasitol.,* 54, 517–523, 1968.

27. Rothman, A. H. and Fisher, F. M., Permeation of amino acids in *Moniliformis* and *Macracanthorhynchus* (Acanthocephala), *J. Parasitol.,* 50, 410–414, 1964.

28. Branch, S. I., Accumulation of amino acids by *Moniliformis dubius* (Acanthocephala), *Exp. Parasitol.,* 25, 95–99, 1970.

29. Uglem, G. L. and Read, C. P., *Moniliformis dubius:* uptake of leucine and alanine by adults, *Exp. Parasitol.,* 34, 148–153, 1973.

30. Uglem, G. L., Pappas, P. W., and Read, C. P., Surface aminopeptidase in *Moniliformis dubius* and its relation to amino acid uptake, *Parasitology,* 67, 185–195, 1973.

31. Ruff, M. D., Uglem, G. L., and Read, C. P., Interactions of *Moniliformis dubius* with pancreatic enzymes, *J. Parasitol.,* 59, 839–843, 1973.

32. Crompton, D. W. T., Morphological and histochemical observations on *Polymorphus minutus* (Goeze, 1782) with special reference to the body wall, *Parasitology,* 53, 662–685, 1963.

33. Crompton, D. W. T. and Ward, P. F. V., Lactic and succinic acids as excretory products of *Polymorphus minutus* (Acanthocephala) *in vitro, J. Exp. Biol.,* 46, 423–430, 1967.

34. Crompton, D. W. T. and Lockwood, A. P. M., Studies on the absorption and metabolism of D-(u-^{14}C) glucose by *Polymorphus minutus* (Acanthocephala) *in vitro, J. Exp. Biol.,* 48, 411–425, 1968.

35. Laurie, J. S., The *in vitro* fermentation of carbohydrates by two species of cestodes and one species of Acanthocephala, *Exp. Parasitol.,* 6, 245–260, 1957.

36. Laurie, J. S., Aerobic metabolism of *Moniliformis dubius* (Acanthocephala), *Exp. Parasitol.,* 8, 188–197, 1959.

37. Kilejian, A., The effect of carbon dioxide on glycogenesis in *Moniliformis dubius* (Acanthocephala), *J. Parasitol.,* 49, 862–863, 1963.

38. Graff, D. J., Metabolism of C^{14}-glucose by *Moniliformis dubius* (Acanthocephala), *J. Parasitol.,* 50, 230–234, 1964.

39. Ward, P. F. V. and Crompton, D. W. T., The alcoholic fermentation of glucose by *Moniliformis dubius* (Acanthocephala) *in vitro, Proc. R. Soc. London Ser. B,* 172, 65–88, 1969.

40. **McAlister, R. O. and Fisher, F. M.,** The biosynthesis of trehalose in *Moniliformis dubius* (Acanthocephala), *J. Parasitol.,* 58, 51—62, 1972.

41. **Dunagan, T. T.,** Studies on *in vitro* survival of Acanthocephala, *Proc. Helminthol. Soc. Wash.,* 29, 131—135, 1962.

42. **Dunagan, T. T.,** Studies on the carbohydrate metabolism of *Neoechinorhynchus* spp. (Acanthocephala), *Proc. Helminthol. Soc. Wash.,* 31, 166—172, 1964.

43. **Beitinger, T. L. and Hammen, C. S.,** Utilization of oxygen, glucose, and carbon dioxide by *Echinorhynchus gadi, Exp. Parasitol.,* 30, 224—232, 1971.

44. **Crompton, D. W. T. and Nesheim, M. C.,** Relationships between *Moniliformis dubius* (Acanthocephala) and the carbohydrate intake of rats, *Parasitology,* 67, ii, 1973.

45. **Nesheim, M. C., Crompton, D. W. T., Arnold, S., and Barnard, D.,** Dietary relations between *Moniliformis* (Acanthocephala) and laboratory rats, *Proceedings of the Royal Society of London,* Series B, 1977.

46. **Burlingame, P. L. and Chandler, A. C.,** Host-parasite relations of *Moniliformis dubius* (Acanthocephala) in albino rats, and the environmental nature of resistance to single and superimposed infections with this parasite, *Am. J. Hyg.,* 33, 1—21, 1941.

47. **Holmes, J. C.,** Effects of concurrent infections on *Hymenolepis diminuta* (Cestoda) and *Moniliformis dubius* (Acanthocephala). I. General effects and comparison with crowding, *J. Parasitol.,* 47, 209—216, 1961.

48. **Read, C. P. and Rothman, A. H.,** The carbohydrate requirement of *Moniliformis* (Acanthocephala), *Exp. Parasitol.,* 7, 191—197, 1958.

49. **Nicholas, W. L. and Hynes, H. B. N.,** Studies on *Polymorphus minutus* (Goeze, 1782) (Acanthocephala) as a parasite of the domestic duck, *Ann. Trop. Med. Parasitol.,* 52, 36—47, 1958.

50. **Hynes, H. B. N. and Nicholas, W. L.,** The importance of the acanthocephalan *Polymorphus minutus* as a parasite of domestic ducks in the United Kingdom, *J. Helminthol.,* 37, 185—198, 1963.

51. **Körting, W. and Fairbairn, D.,** Anaerobic energy metabolism in *Moniliformis dubius* (Acanthocephala), *J. Parasitol.,* 58, 45—50, 1972.

52. **Dunagan, T. T. and Scheifinger, C. C.,** Studies on glycolytic enzymes from *Macracathorhynchus hirudinaceus* (Acanthocephala), *J. Parasitol.,* 52, 730—734, 1966.

53. **Bryant, C. and Nicholas, W. L.,** Intermediary metabolism in *Moniliformis dubius* (Acanthocephala), *Comp. Biochem. Physiol.,* 15, 103—112, 1965.

54. **Bryant, C. and Nicholas, W. L.,** Studies on the oxidative metabolism of *Moniliformis dubius* (Acanthocephala), *Comp. Biochem. Physiol.,* 17, 825—840, 1966.

55. **Bryant, C.,** Electron transport in parasitic helminths and protozoa, in *Advances in Parasitology,* Vol. 8, Dawes, B., Ed., Academic Press, New York, 1970, 139—172.

56. **Fisher, F. M.,** Synthesis of trehalose in Acanthocephala, *J. Parasitol.,* 50, 803—804, 1964.

57. **Dunagan, T. T. and Yau, T. M.,** Oligosaccharidases from *Macracanthorhynchus hirudinaceus* (Acanthocephala) from swine, *Comp. Biochem. Physiol.,* 26, 281—289, 1968.

58. **Graff, D. J.,** The utilization of $C^{14}O_2$ in the production of acid metabolites by *Moniliformis dubius* (Acanthocephala), *J. Parasitol.,* 51, 72—75, 1965.

59. **Horvath, K. and Fisher, F. M.,** Enzymes of CO_2 fixation in larval and adult *Moniliformis dubius, J. Parasitol.,* 57, 440—442, 1971.

60. **Bryant, C.,** Carbon dioxide utilization and the regulation of respiratory metabolic pathways in parasitic helminths, in *Advances in Parasitology,* Vol. 13, Dawes, B., Ed., Academic Press, New York, 1975, 36—69.

61. **Barrett, J., Cain, G. D., and Fairbairn, D.,** Sterols in *Ascaris lumbricoides* (Nematoda), *Macracanthorhynchus hirudinaceus* and *Moniliformis dubius* (Acanthocephala) and *Echinostoma revolution* (Trematoda), *J. Parasitol.,* 56, 1004—1008, 1970.

62. **Horvath, K.,** Glycogen metabolism in larval *Moniliformis dubius* (Acanthocephala), *J. Parasitol.,* 57, 132—136, 1971.

63. **Horvath, K.,** Glycolytic enzymes in larval *Moniliformis dubius, J. Parasitol.,* 58, 1219—1220, 1972.

QUALITATIVE REQUIREMENTS AND UTILIZATION OF NUTRIENTS: ANNELIDA, ECHIURA, AND SIPUNCULA*

D. Dean

Since the early 1960s, there have been a number of studies on the uptake and/or utilization of substances, primarily radioactively labeled dissolved organic compounds. Even though these studies have demonstrated uptake, some authors have questioned whether there is a net uptake or influx of the substances.[1] More recently, using newer techniques, direct evidence has been accumulated for a rapid net influx of naturally occurring organic material and amino acids in two genera of annelid infauna.[2] The relative importance of the uptake of dissolved organic substances in the metabolism of marine organisms still remains obscure.

The majority of studies concerning the qualitative requirements and nutrient utilization in these three phyla has been made on members of the annelid class Polychaeta, and is summarized in Table 1. Studies on the taxa of other members of these phyla are summarized in Table 2.

---

* Contribution number 98 from the Ira C. Darling Center for Research, Teaching and Service, University of Maine at Orono, Walpole, Maine, 04573.

Table 1

QUALITATIVE REQUIREMENTS AND UTILIZATION OF NUTRIENTS BY POLYCHAETOUS ANNELIDS

A = adult; L = larvae; U = utilized or taken up; ψ = not utilized or taken up; R = required.

	Stage	Alanine	Arginine	Aspartic acid	Ca^{++}	Citric acid	Cysteine	Glucose	Glutamic acid	Glycine	Glycolic acid	Histidine	Isoleucine	Leucine	Lysine	Ref.
Family Hesionidae																
Podarke pugettensis	A															3
Family Nereidae																
Eunereis longissima	A													U^a		4
Nereis sp.	L, A															5
Nereis arenacecdentata	L, A															6
Nereis diversicolor	A			U^c	R				ψ^d	U^b						2, 7, 8, 9
Nereis limnicola	A									U^e						3, 10
Nereis succinea	A									U						10
Nereis virens	A					U		U	U	U	U					11—13
Family Nephtyidae																
Nephtys sp.	A															4
Family Glyceridae																
Glycera dibranchiata	A	U^f	U^f	U^f					U^f	U^f						3, 14, 15

a V_{max} was 3.82 X 10^{-6}, which represented 0.05% of the wet body weight/hour; larval uptake was inversely proportional to body weight.
b After 30 min exposure, 5 to 102 times as much radioactivity per unit volume was present in the animal as in the medium, depending upon size and feeding stage.
c Reported as aspartate.
d Reported as glutamate.
e K_t and V_{max} values were affected by the chlorinity of the medium.
f K_t values range from 2 X 10^{-4} to 2 X 10^{-3} and V_{max} from 9.62 X 10^{-7} to 1.06 X 10^{-6}.

Table 1 (continued)
QUALITATIVE REQUIREMENTS AND UTILIZATION OF NUTRIENTS BY POLYCHAETOUS ANNELIDS

	Stage	Alanine	Arginine	Aspartic acid	Ca++	Citric acid	Cysteine	Glucose	Glutamic acid	Glycine	Glycolic acid	Histidine	Isoleucine	Leucine	Lysine	Ref.
Family Goniadidae																
Goniada sp.	A															4
Family Lumbrineridae																
Lumbrinereis sp.	A															3
Family Dorvilleidae																
Stauronereis sp.	A	U		U[c]					U[d]	U						16
Stauronereis rudolphi	A	U	U	U												3
(*=Dorvillea articulata*)	A			U			U		U	U		U	U	U	U	17
Family Orbiniidae																
Nainereis dendritica	A															3
Family Chaetopteridae																
Chaetopterus variopedatus	A									U						13
Family Cirratulidae																
Cirriformia spirabranchia	A															3
Family Capitellidae																
Capitella capitata	A			U[c]					U[d]	U						2

Table 1 (continued)
QUALITATIVE REQUIREMENTS AND UTILIZATION OF NUTRIENTS BY POLYCHAETOUS ANNELIDS

	Stage	Alanine	Arginine	Aspartic acid	Ca++	Citric acid	Cysteine	Glucose	Glutamic acid	Glycine	Glycolic acid	Histidine	Isoleucine	Leucine	Lycine	Ref.
Family Maldanidae																
Maldanid	A															4
Clymenella torquata	A									U^g					U	1, 18, 19
Family Terebellidae																
Terebellid	A									U						4
Amphitrite ornata	A									U						13
Lanice conchilega	A							U								20

g Reference 1 confirmed uptake reported by Reference 19, but found a net efflux of dissolved free amino acids. Reference 2, using different techniques, provided "direct evidence for a rapid net influx of naturally occurring organic material and amino acids in two genera of annelid infauna."

Table 1 (continued)
QUALITATIVE REQUIREMENTS AND UTILIZATION OF NUTRIENTS BY POLYCHAETOUS ANNELIDS

	Stage	Methionine	Oleic acid	Palmitic acid	Phenylalanine	Proline	Serine	Sodium butyrate	Succinic acid	Threonine	Tryptophan	Tyrosine	Valine	Other	Ref.
Family Hesionidae															
Podarke pugettensis	A			U											3
Family Nereidae															
Eunereis longissima	A				U										4
Nereis sp.	L, A							U							5
Nereis arenaceodentata	L, A														6
Nereis diversicolor	A		U											U^h	2, 7, 8, 9
Nereis limnicola	A						U								3, 10
Nereis succinea	A														11–13
Nereis virens	A								U						
Family Nephtyidae															
Nephtys sp.	A			U											4
Family Glyceridae															
Glycera dibranchiata	A			U	U^f	U^f							U^f	U^i	3, 14, 15
Family Goniadidae															
Goniada sp.	A				U			U							4
Family Lumbrineridae															
Lumbrinereis sp.	A			U											3

h Mixed amino acids.
i Took up 12 unnamed amino acids.[15]

Table 1 (continued)

QUALITATIVE REQUIREMENTS AND UTILIZATION OF NUTRIENTS BY POLYCHAETOUS ANNELIDS

	Stage	Methionine	Oleic acid	Palmitic acid	Phenylalanine	Proline	Serine	Sodium butyrate	Succinic acid	Threonine	Tryptophan	Tyrosine	Valine	Other	Ref.
Family Dorvilleidae															
Stauronereis sp.	A				U								U		16
Stauronereis rudolphi	A	U[j]	U												3
(=*Dorvillea articulata*)	A				U		U			U	U[j]	U	U[j]		17
Family Orbiniidae															
Nainereis dendritica	A		U	U											3
Family Chaetopteridae															
Chaetopterus variopedatus	A														13
Family Cirratulidae															
Cirriformia spirabranchia	A			U											3
Family Capitellidae															
Capitella capitata	A						U								2
Family Maldanidae															
Maldanid	A													U[h]	4
Clymenella torquata	A				U								U		1, 18, 19
Family Terebellidae															
Terebellid	A				U										4
Amphitrite ornata	A														13
Lanice conchelega	A														20

Table 2
QUALITATIVE REQUIREMENTS AND UTILIZATION OF NUTRIENTS BY NONPOLYCHAETOUS ANNELIDS, ECHIURA, AND SIPUNCULA

A = adult; U = utilized or taken up; Ψ = not utilized or taken up.

	Stage	B-complex vitamins	Biotin	Ca pantothenate	Caproic acid	Cyanocobalamine	Glycine	Ref.
Class Oligochaeta								
Family Naididae								
Dero sp.	A	U[a]						21
Dero limnosa	A	U[a]						21
Family Tubificidae								
Limnodrilus hoffmeisteri	A						Ψ	22
Peloscolex multisetosa	A						U	22
Tubifex tubifex	A				U		Ψ	3, 22
Family Enchytraeidae								
Enchytraeus fragmentosus	A		Ψ	U		U		23
Phylum Echiura								
Urechis caupo	A							3
Phylum Sipuncula								
Golfingia gouldii	A						U	13, 24

[a] Thiamine hydrochloride, riboflavin, pyridoxine (4 γ each); calcium panthothenate (20 γ); nicotinic acid, *p*-aminobenzoic acid, inositol (50 γ each); biotin methyl ester (0.025 γ); and ethanol (0.0005 cc) were added to 30 cc of filtered spring water. Vitamin-enriched cultures significantly enhanced growth and fission rates over controls.

Table 2
QUALITATIVE REQUIREMENTS AND UTILIZATION OF NUTRIENTS BY NONPOLYCHAETOUS ANNELIDS, ECHIURA, AND SIPUNCULA

	Nicotinamide	Oleic acid	Pteroylglutamic acid	Pyridoxine HCl	Riboflavin	Thiamine HCl	Ref.
Class Oligochaeta							
Family Naididae							
Dero sp.							21
Dero limnosis							21
Family Tubificidae							
Limnodrilus hoffmeisteri							22
Peloscolex multisetosa							22
Tubifex tubifex		U					3, 22
Family Enchytraeidae							
Enchytraeus fragmentosus	U		U	U	U	U	23
Phylum Echiura							
Urechis caupo		U					3
Phylum Sipuncula							
Golfingia gouldii							13, 24

REFERENCES

1. Johannes, R. E. and Webb, K. L., in *Symposium on Organic Matter in Natural Waters,* Hood, D. W., Ed., Institute of Marine Science, University of Alaska, Occasional Publications, 1970, 257–273.
2. Stephens, G. C., *Biol. Bull.,* 149, 397–407, 1975.
3. Testerman, J. K., *Biol. Bull.,* 142, 160–177, 1972.
4. Southward, A. J. and Southward, E. C., *Sarsia,* 48, 61–68, 1972.
5. Bass, N., Chapman, G., and Chapman, J. H., *Nature,* 221, 476–477, 1969.
6. Reish, D. J. and Stephens, G. C., *Mar. Biol.,* 3, 352–355, 1969.
7. Beadle, L. C., *J. Exp. Biol.,* 14, 56–70, 1937.
8. Ellis, W. G., *Nature,* 132, 748, 1933.
9. Ellis, W. G., *J. Exp. Biol.,* 14, 340–350, 1937.
10. Stephens, G. C., *Biol. Bull.,* 126, 150–162, 1964.
11. Chapman, G. and Taylor, A. G., *Nature,* 217, 763–764, 1968.
12. Taylor, A. G., *Comp. Biochem. Physiol.,* 29, 243–250, 1969.
13. Stephens, G. C. and Schinske, R. A., *Limnol. Oceanogr.,* 6, 175–181, 1961.
14. Chien, P. K., Stephens, G. C., and Healey, P. L., *Biol. Bull.,* 142, 219–235, 1972.
15. Preston, R. L. and Stephens, G. C., *Am. Zool.,* 9, 1116, 1969.
16. Stephens, G. C., in *Nitrogen Metabolism and the Environment,* Campbell, J. W. and Goldstein, L., Eds., Academic Press, London, 1972, 155–184.
17. Stephens, G. C., *Am. Zool.,* 8, 95–106, 1968.
18. Stephens, G. C., *Biol. Bull.,* 123, 512, 1962.
19. Stephens, G. C., *Comp. Biochem. Physiol.,* 10, 191–202, 1963.
20. Ernst, W. and Goerke, H., *Veröff, Inst. Meeresforsch. Bremerhaven,* 11, 313–326, 1969.
21. Hauschka, T., *Growth,* 8, 51–58, 1944.
22. Brinkhurst, R. O. and Chua, K. E., *J. Fish. Res. Board Canada,* 26, 2659–2668, 1969.
23. Dougherty, E. C. and Gotthold, M. L., *Trans. Am. Microsc. Soc.,* 84, 166–167, 1965.
24. Virkar, R. A., *Biol. Bull.,* 125, 396–397, 1963.

QUALITATIVE REQUIREMENTS AND UTILIZATION OF NUTRIENTS: INSECTS

R. H. Dadd

Knowledge of insect nutritional requirements has come mainly from three experimental approaches which give different kinds of information: (a) the classic deletion method, whereby alterations in growth and development are studied when defined components are omitted from otherwise satisfactory basal diets, (b) utilizability experiments, which examine effects on survival, growth, and reproduction of substitutions within a *class* of nutrient known to be required, and (c) metabolic studies, to determine whether physiologically essential substances can or cannot be derived endogenously from other precursor materials, the presumption being that if not synthesized they must be supplied exogenously as dietary constituents. Brief consideration of these methodologies and their attendant pitfalls follows, in order to emphasize the limits within which the experimental conclusions summarized in the tables are valid. Because of the simplification entailed in a tabular summary of a large body of information, it is important to bear in mind the several possibilities for uncertainty even where experimental data seem clear. Common sources of uncertainty are listed following discussion of the methodology to which they attach.

Deletion Methods

Given that a chemically defined diet for a particular species allows good growth, development, and reproduction, if these are impaired by deletion of a single component for which there is no substituent of equivalent or lesser complexity, then most likely the component is required nutritionally. A required nutrient may be defined as essential if lack of it results in, or clearly would eventually result in, cessation of growth and reproduction. In this definition, *required* nutritents are not necessarily *essential*, since their absence may merely reduce growth or fecundity rather than negate them. A distinction between required and essential is often a mattter of subjective experience, since experiments frequently are terminated before total cessation of growth (including continued generations) has been demonstrated with respect to a deleted nutrient which background considerations suggest should be essential. Therefore, no distinction is made between essential and required in the following tables. Sources of uncertainty inherent in the deletion method are as follows:

1. Growth may be impaired because of underfeeding if a phagostimulant substance which is not required nutritionally is omitted, possibly leading to erroneous designation of the substance as nutritionally required. Among phytophagous insects particularly, phagostimulants of trivial nutritional importance are commonly needed to induce normal feeding and must therefore be taken into account in nutritional studies of such species. However, for both these insects and other species not requiring specific token feeding stimulants, many nutrients function also as phagostimulants;[1,2,4,8,10,11] it is then difficult to demonstrate unequivocally that they are required specifically as nutrients. In general, proper feeding necessitates complex olfactory and gustatory stimuli provided by combinations of substances, usually including members of several nutrient classes; yet few nutritional studies have systematically checked whether dietary ingredients whose omission retards growth can significantly affect the rate of ingestion of food.

2. A nonrequired ingredient may erroneously be designated required if it is contaminated by a required nutrient not otherwise supplied in the basal diet. This

possible error is most likely to occur with meridic diets containing refined but ill-defined complex natural materials such as RNA, vegetable oils, or plant extracts, which usually carry over appreciable quantities of vitamins, lipids, or trace minerals from the original crude source material. The converse error is also most probable with meridic diets: a required nutrient may test out as "not required" if it is cryptically present as a contaminant of some other dietary component.

3. A substance may be designated as required when it is only one of several possible alternatives within a required nutrient class, other members of which have not been tested. In such cases the range of interchangeable nutrients may form the object of utilizability studies, considered below, but until such studies are made it cannot be concluded that the particular class representative tested is necessarily required, but only that the class as a whole is required. In connection with this type of uncertainty it is important to realize that a nutrient may be necessary in a diet of particular composition but not in another of different composition or balance. In general, if two or more nutrients can interchangeably fulfill some need, each may appear to be uniquely required if tested in a basal diet lacking adequate amounts of the other interchangeable nutrients.

4. Required nutrients may appear unnecessary if supplied cryptically by micro-organisms. In work with nonsterile dry diets, such organisms may develop in diet or digestive tract; in sterile culture, intracellular symbiotes, normally present in many insects, are difficult or impossible to eliminate.

5. Because insect eggs are large relative to the mature adult, reserves of micronutrients may be sufficient to support at least one cycle of normal development without dietary supplementation; if a micronutrient deletion experiment has only this span, a designation of nonrequirement is highly uncertain. Several studies have shown that two or more consecutive generations are necessary to reveal certain deficiencies, whereas the vast majority of studies terminate after one cycle of development; it can therefore be argued that the designation "not required" with respect to micronutrients known to be required by most species is open to question until put to the rigorous test of rearing consecutive generations on deficient diet. In the following tables that deal with micronutrients such as vitamins, the designation "not required" should generally be understood to carry this caveat.

Utilizability Studies

If a particular nutrient is required, it follows that it is utilizable. Nutrients that are not required are doubtless often utilized to some extent if available in the diet; excepting certain complex nutrients which may be shown to be indigestible, it is difficult to exclude *some* utilization of nonrequired dietary components without detailed metabolic studies. However, for certain required classes of nutrients, interest centers on which members or subclasses can or cannot be utilized for the class requirement. In such cases, where use of typical class members has established a class requirement (e.g., glucose, sucrose, or starch for a carbohydrate source), substitution of related compounds allows their designation as fully or partially utilizable, or not utilizable, *with respect to the particular class function of interest.* With the accumulation of enough such data it becomes possible to more precisely define the structural requirements for utilizability within the class. Since interest in this kind of information mainly concerns sterol and carbohydrate utilization, consideration of the uncertainties involved is deferred for discussion under those headings.

Metabolic Studies

A physiologically necessary tissue component or metabolite is dietarily essential if it cannot be formed endogenously in adequate amounts from other dietary precursors. Sometimes comparative analysis of levels of certain materials in food and in the growing animal can indicate which are inessential and which essential; the isotope method for determining essentiality is more widely used, and has been reviewed with respect to insects.

If labeling from appropriate radiotagged precursors which are fed or injected into an insect fails to appear in particular members of a class of nutrient separated from the whole animal after a suitable interval for metabolic cycling, then the unlabeled members are probably dietarily essential, and heavily labeled members are perhaps not essential. In spite of the ambiguities of interpretation listed below, this methodology provides valuable confirmation of deletion studies, and supplies the only information on qualitative requirements when a lack of satisfactory synthetic diets precludes the use of deletion methods. Sources of uncertainty with the isotope method are as follows:

1. Though substantial labeling of a metabolite indicates synthesis from the precursor provided, synthesis may be too slow to keep pace with physiological requirements unless augmented by dietary intake of preformed metabolite. Further, there may be exchange of isotopic label without *de novo* synthesis. Thus, a substantially labeled material may on occasion be nutritionally required.

2. Incomplete chromatographic separation may give false labeling to a substance not synthesized from the precursor.

3. An inappropriate precursor may result in lack of label and the designation of essentiality; for example, tyrosine is not essential since it is formed from (essential) phenylalanine, but if the tagged precursor is glucose, whose label will not appear in phenylalanine, then neither will it appear in tyrosine.

4. Insufficient time for transformations of the labeled precursor could result in ambiguously low levels of label in nonessential nutrients.

5. If the insect harbors symbiotes, results will reflect the joint metabolism of insect and symbiotes.

The foregoing discussion is given to emphasize general uncertainties of interpretation underlying many of the studies here summarized; in contrast to work with large animals, for example, many of the insect species have been studied by single workers only, so that confirmatory checks tend to be few. Where the experimental results or interpretation are obviously questionable and out of line with the generality of information, or the results of different workers are in disagreement, this will be indicated by a question mark and footnote in the tables. For fuller discussion, the more recent of the numerous reviews on insect nutrition should be consulted.[2,3,5,6]

REFERENCES

1. Beck, S. D., *Annu. Rev. Entomol.,* 10, 207–232, 1965.
2. Dadd, R. H., Arthropod nutrition, in *Chemical Zoology,* Vol. 5, Florkin, M. and Scheer, B. T., Eds., Academic Press, New York, 1970, 117–145.
3. Dadd, R. H., *Annu. Rev. Entomol.,* 18, 381–420, 1973.
4. Fraenkel, G., *Entomol. Exp. Appl.,* 12, 473–486, 1969.
5. House, H. L., Insect nutrition, in *Biology of Nutrition, International Encyclopaedia of Food and Nutrition,* Vol. 18, T-W-Fiennes, R. N., Ed., Pergamon Press, Oxford, Engl.,1972, 513–573.
6. House, H. L., Nutrition, in *The Physiology of Insecta,* Vol. 5, Rockstein, M., Ed., Academic Press, New York, 1974, 1–62.
7. Kasting, R. and McGinnis, A. J., *Ann. N. Y. Acad. Sci.,* 139, 98–110, 1966.
8. Schoonhoven, L. M., *Annu. Rev. Entomol.,* 13, 115–136, 1969.
9. Singh, P., Bibliography of artificial diets for insects and mites, *N. Z. Dep. Sci. Ind. Res. Bull.,* No. 209, p. 75, 1972.
10. Thorsteinson, A. J., *Annu. Rev. Entomol.,* 5, 193–298, 1960.
11. Wood, D. L., Silverstein, R. M., and Nakajima, M., Eds., *Control of Insect Behavior by Natural Products,* Academic Press, New York, 1970.

Table 1
AMINO ACID REQUIREMENTS

Most insects obtain the amino acids needed for tissue protein synthesis by digestion of protein in natural foods. Free amino acids constitute the major or only supply for homopterous plantsucking bugs which are specialized for feeding on phloem or xylem saps containing little or no protein; in accordance with such a natural diet, certain aphids lack digestive proteinases. Besides Homoptera, other insects such as termites, psocids, wood-infesting beetles and roaches, etc., thrive on foods apparently deficient in nitrogen, but in all such cases the insect is probably largely dependent for nitrogenous nutrition upon the synthetic abilities of microorganisms associated with the food or symbiotic within the insect.

Usually the first step in attempting to rear an insect on chemically defined food is to develop a semisynthetic diet with a refined protein as the nitrogen source. Casein is most often used because of its ready availability as fat- and vitamin-free preparations; it is generally satisfactory, though sometimes requiring augmentation with particular amino acids (glycine, glutamic acid, cystine) which it may provide inadequately. Egg albumen, lactalbumen, and soy protein are commonly used alternatives, and, for phytophagous insects in particular, are sometimes superior. Probably most proteins that are amenable to tryptic digestion and which comprise a complete array of the 20 or so nutritive amino acids would be satisfactory if free from toxic or feeding-inhibitant contaminants; however, if the purpose is to develop diets expressly for nutritional studies, as distinct from standardized rearing regimes, the usefulness of a protein depends on its freedom from vitamins and other covert micronutrients.

To transform a diet based on casein to one that is completely defined, the casein often is replaced by an amino acid mixture of a composition based on analysis of caseinamino acids, the rationale being that if amino acids digested from casein sufficed, so too should a mixture of free amino acids in the same relative proportions. This approach has not always succeeded, and with more analytical data available in recent years on the composition of proteins of both insects and their natural foods, one or another amino acid pattern is often taken as the model for concocting a dietary mixture.

Given a satisfactory amino-acid-based diet, the results of deleting individual amino acids are generally clear with respect to those that are essential. Confusion can arise with respect to nonessential amino acids; whether or not they influence growth may depend on the proportionality of other amino acids in the basal mix or on the amount of other energy-producing nutrients available. In practice, an understanding of insect amino acid requirements is most problematic for those species that normally harbor symbiotes or have a heavy microflora in the gut. If microorganisms, symbiotic or facultative, can readily be removed, then anomalies disappear, as several examples below show. However, it is generally difficult to obtain viable aposymbiotic insects for species that have evolved a close interdependence with intracellular symbiotes.

In the absence of a suitable synthetic diet, radiometric methods have been used extensively to indicate essential amino acids. Apart from general drawbacks to the method dealt with in the introduction, the following points apply with respect to amino acid requirements: Tryptophan is destroyed during hydrolysis of extracted protein, so information about it is usually lacking. Tyrosine can generally be synthesized from phenylalanine, which, being essential, acquires little or no isotope; tyrosine would therefore show no count unless radioactive phenylalanine specifically is used as a precursor. It needs to be emphasized that for insects with symbiotes, whatever interconversions are revealed by radiometric studies apply equally to symbiotes as to the insect, just as for chemical deletion studies.

Excluding insects known to harbor symbiotes, without exception all species studied using deletion techniques were found to require ten amino acids, namely arginine, histidine, isoleucine, leucine, lysine, methionine, phenylalanine, threonine, tryptophan, and valine. These ten are therefore tabulated as a group, and exceptions arising in the case of symbiote-containing insects, etc., are dealt with in footnotes. Of more interest are those amino acids generally considered inessential but which sometimes are required by particular species; these are individually tabulated, except that aspartic acid/asparagine, glutamic acid/glutamine, and cysteine/cystine are treated as single entities. For reasons discussed in the introduction, use of symbols denoting essentiality or otherwise is avoided for radioisotope studies, and instead, designations are made in terms of evidence for synthesis (i.e., appearance of radio label) from the isotopic precursor. Symbols used therefore have the following meanings:

Deletion method:
  R = required.
  r = slightly improves growth and therefore probably required.
  R̸ = not required.

Isotope method:
  $̸ = not synthesized to significant extent.
  s = low but appreciable synthesis.
  S = substantial synthesis.

Table 1 (continued)
AMINO ACID REQUIREMENTS

Diptera

	All ten essential amino acids	Generally nonessential								Others	Ref.
		Alanine	Aspartic acid	Cystine	Glutamic acid	Glycine	Proline	Serine	Tyrosine		
Aedes aegypti	R	R	R	r?[a]	R	r[a]	R	R	R[a]		22, 47, 76
A. aegypti — for egg production by adult♀	R	R	R	r	r?	R	R	R	R		15, 76
Agria housei (Pseudosareophaga affinis)	R	R[b]	R	R	R	r	R	R[b]	R[b]	c	34
Calliphora erythrocephala (Calliphora vicina)	R	R	R		R	R	R	R	R		72, 73
Cochliomyia hominivorax	R	R	R	R	R	r	R	R	R	c	21
Culex pipiens	R	R	R	R[d]	R	r	R	R	R		7
Drosophila melanogaster	R	R	R	R[d]	r[d]	r[d]	R	R	R	c,d	19, 20, 30, 31,
D. melanogaster — egg production	R	R	r	r	r	R	R	R	R		70
Hylemia antiqua	R	R	R	R	R	R	R	R	R	c	18
Hypoderma bovis — isotope method	S[e]	S	S	e	S	S?[e]	S	S?[e]	S?[e]		46
Musca vicina	R									f	5
Phormia regina	R[g]	R	r	R[g]	r	R	R?[g]	R	R		6, 32, 55, 65
P. regina — isotope method	S	S	S	S	S	S	S	S	S		43

a Studies disagree over cystine and glycine requirement. Lack of tyrosine affected pigmentation.[22]

b Slight growth retardation if absent.

c Hydroxyproline tested but not required.

d Cystine improved growth of some mutant strains. Either glycine or glutamic acid required. Citrulline but not ornithine partially spared arginine. D forms of several amino acids replaced L forms with variable efficiency but D-serine was toxic.

e Arginine had low labeling; cysteine was not retrieved; glycine, serine, and tyrosine had low labeling.

f Grows poorly with only ten essentials.

g Methionine said to be interchangeable with cystine,[6] but metabolic studies show no conversion of cystine to methionine.[27] Proline required in one study but not in another, perhaps a strain difference. D forms of several amino acids utilized with varying efficiency.

Table 1 (continued)
AMINO ACID REQUIREMENTS

	All ten essential amino acids	Generally nonessential									Ref.
		Alanine	Aspartic Acid	Cystine	Glutamic Acid	Glycine	Proline	Serine	Tyrosine	Others	
Hymenoptera											
Apis mellifera – isotope method, larvae	$	S	$	$	S	S	S	S	$		53
A. mellifera – adult worker growth	R	R	R	R	R	r?	r?h	r?h	R		4
Cephus cinctus – isotope method	$i	S	S	i	S	S	S	S	s		45
Exeristes roborator – isotope method	$	S	$j	j	S	S	S	S	$		79
Neodiprion pratti – isotope method	$j	S	$j	j	S	S	S	S	$		71
Coleoptera											
Anthonomus grandis	R	R	R	R	R k	R k	R	R	R	k	83
A. grandis – egg production	R	R	R	R	k	k	R	R	R		82
Ctenicera destructor – isotope method	$	S	S	l	$l	S	S	S	Sl		42
Lasioderma serricorne (symbiotes present)	R and R^m	R	r	r	r	R	R	r		m	58
Oryzaephilus surinamensis	R	R	R	r?	R	r?	R	R	r?		9, 10–12
Stegobium paniceum (with symbiotes)	R and R^n	R	R	R	R	r	R	R	R		59
S. paniceum (aposymbiotic)	R	r	R	R	R	R	R	R	R		59

h   Glycine, proline, and serine affected growth only if proportions of ten essential amino acids were suboptimal.

i   No isotope data for histidine; cystine not retrieved.

j   Low labeling for histidine. Very little aspartic acid. Cystine not retrieved.

k   Citrulline but not ornithine replaced arginine. With adequate glucose, there was virtually no reduction in larval growth when all inessentials were omitted. Individual deletion of nonessentials not tested for egg production, but optimal eggs with ten essentials plus glutamic acid and glycine showed other nonessentials not required.

l   Cystine not determined; glutamic acid had very low labeling; tyrosine not synthesized from glucose but readily synthesized from labeled phenylalanine.

m   Growth severely retarded only on omission of arginine or histidine; no reduction for phenylalanine, tryptophan, alanine, or tyrosine, and retardation various when all others, essential and inessential, were omitted individually. These irregular results are presumed due to symbiotes in gut caeca.

n   A clear requirement only for arginine in the normal insect with symbiotes.

Table 1 (continued)
AMINO ACID REQUIREMENTS

	All ten essential amino acids	Generally nonessential									Ref.
		Alanine	Aspartic Acid	Cystine	Glutamic Acid	Glycine	Proline	Serine	Tyrosine	Others	
Coleoptera (continued)											
Tenebrio molitor	R	r?	r?,o	r?	R	R	r?	R	R		13
Tribolium confusum	R	R	R?,o	R	R?,o	R	R	R	R		17, 51, 57, 78
Trogoderma granarium	R	R	R^p	R^p	R	R^p	R	R^p	R?,p	c	3
Lepidoptera											
Agrotis orthogonia – isotope method	s and $^q	S	S	s?	S	S	S	S	S^q		44
Argyrotaenia velutinana	R	R	R	S	R	R	R	R	R	r	63, 65—67, 69
A. velutinana – isotope method	s and $^r	S	S	S	S	S	$^r	S	$	s	68
Bombyx mori	R	R	R	R	r^s	R	R	R	R		1, 35, 36, 39—41
Chilo suppresalis (*Chilo simplex*)	R	R	R	R	R	R	R	R	R	c,t	37, 38

o   Grows with ten essentials only, but growth rate improved with all others included, or with just aspartic and glutamic acids.

p   Single omission of aspartic acid, cystine, glycine, or serine slightly retarded growth. Though tyrosine required for good growth, analysis of tissue amino acids showed it to be synthesized.

q   Some labeling in threonine; although cysteic acid was heavily labeled authors considered cysteine not labeled significantly and therefore essential; tyrosine synthesized from phenylalanine.

r   Reared through several consecutive generations with only ten essentials, but faster growth with others present. Slight labeling of methionine and threonine. Note disagreement between methods for proline. Cystine, homocysteine, cystathionine, and various methionine analogues can spare methionine, but inorganic sulfate cannot.[69,74,75] D forms of methionine, phenylalanine, and histidine replace L forms but with reduced growth.[63,65] Citrulline replaced arginine completely for three generations.[62]

s   Either aspartic or glutamic acid required, aspartic being best. Cystine and tyrosine partly spare methionine and phenylalanine, respectively. Citrulline but not ornithine partly spared arginine.[36] D forms of histidine and methionine substitute for L forms though growth thereby slowed.[41]

t   Chromatography of tissue extracts of larvae reared on diet lacking single nonrequired amino acids showed their presence in all cases except for hydroxyproline.

Table 1 (continued)
AMINO ACID REQUIREMENTS

	All ten essential amino acids	Generally nonessential								Ref.
		Alanine	Aspartic Acid	Cystine	Glutamic Acid	Glycine	Proline	Serine	Tyrosine	Others	
Lepidoptera (continued)											
Heliothis zea	R	R	R	R	R	R	R	R	R	u	64
H. zea – isotope method	s and $u	S	S	s u	S	S	s u	S	R		64
Pectinophora gossypiella	R	R	R	R	R	R	R	R	R		81
Neuroptera											
Chrysopa carnea	R	R	R	R	R	R	R	R	R		84
Hemiptera											
Acyrthosiphon pisum – with symbiotes	R	r?w	R	Rv	R	R	R	R	R	v	54, 61
Aphis fabae – with symbiotes	R and Rw	r	r	r	R	S	r?w	r?w	rw		48–50
Myzus persicae – with normal symbiotes	R and Rx	R	R	R	R	R	R	R	R		8
M. persicae – semiaposymbioticx	R	R	R	S	R	R	R	R	R		56
M. persicae – with normal symbiotes, isotope method	$	S	S	S	S	S	$	S	$		77
Oncopeltus fasciatus	R	R	R	R	R	R	R	R	R		2
Rhodnius prolixus – isotope method	$	S	S	s	S	$	S	S	$		60

u D forms of methionine, phenylalanine, and histidine utilized. Methionine and threonine appreciably labeled; cystine and proline had low labeling.

v Homoserine not required. Cysteine partially spared by homocysteine or dithiothreitol. D form of methionine can replace L form, but not so for cysteine. Inorganic sulfate partially spared both cysteine and methionine in several aphids; various degrees of sparing of sulfur amino acids by sulfate occur, the differences probably reflecting symbiote activities.[8,16,50,80]

w Only histidine and methionine clearly required. Omission of lysine, tryptophan, or valine had no effect. Of nonessentials whose omission slightly decreased growth, alanine, proline, and serine were considered phagostimulants. Cysteine is converted to methionine in absence of methionine, but methionine is not converted to cysteine when latter absent.[48]

x In first generation, only histidine, isoleucine, and methionine clearly required; possible need for lysine appeared in subsequent generation. Semiaposymbiotic larvae with few, deformed symbiotes were produced by antibiotic-fed mothers (completely aposymbiotic larvae could not survive); these larvae showed significant growth reductions on individual omission of essential amino acids but not for nonessentials.

Table 1 (continued)
AMINO ACID REQUIREMENTS

	All ten essential amino acids	Generally nonessential									Ref.
		Alanine	Aspartic Acid	Cystine	Glutamic Acid	Glycine	Proline	Serine	Tyrosine	Others	
Orthoptera											
Blattella germanica — with symbiotes[y]	R and R[y]	R?[y]	R	R?[y]	R	R	R?[y]	R?[y]	R	y	23, 28, 33

[y] All studies were confounded by either or both intracellular symbiotes and a heavy gut microflora; under these conditions isotopic tracer studies show that all amino acids become substantially labeled,[24-26] as also shown for *Periplaneta*.[52] Growth studies indicate requirements for all essential amino acids except methionine and phenylalanine. Inorganic sulfur can spare both methionine and cysteine, completely replacing them in the septic roach.[29] Indications that alanine, serine, and proline were required by aseptic roaches (but still with intracellular symbiotes) more likely reflect imbalance in a poor basal diet. Certain polypeptide fragments of some proteins improve the growth of aposymbiotic roaches.[4]

Table 1 (continued)

REFERENCES

1. Arai, N. and Ito, T., *Bull. Seric. Exp. Stn. Tokyo,* 21, 373–384, 1967.
2. Auclair, J. L., unpublished paper read at 1975 meeting of American Entomological Society.
3. Bhattacharya, A. K. and Pant, N. C., *J. Stored Prod. Res.,* 4, 249–257, 1968.
4. Brooks, M. A. and Kringen, W. B., in *Insect and Mite Nutrition,* Rodriquez, J. C., Ed., North Holland, Amsterdam, 1972, 353–364.
5. Chang, J. T. and Wang, M. Y., *Nature,* 181, 566, 1958.
6. Cheldelin, V. H. and Newburgh, R. W., *Ann. N.Y. Acad. Sci.,* 77, 373–383, 1959.
7. Dadd, R. H., unpublished.
8. Dadd, R. H. and Krieger, D. L., *J. Insect Physiol.,* 14, 741–764, 1968.
9. Davis, G. R. F., *Can. J. Zool.,* 34, 82–85, 1956.
10. Davis, G. R. F., *J. Nutr.,* 75, 275–278, 1961.
11. Davis, G. R. F., *J. Nutr.,* 91, 255–260, 1969.
12. Davis, G. R. F., *J. Insect Physiol.,* 18, 1287–1294, 1972.
13. Davis, G. R. F., *J. Nutr.,* 105, 1071–1075, 1975.
14. De Groot, A. P., *Physiol. Comp. Oecol.,* 3, 197–285, 1953.
15. Dimond, J. B., Lea, A. O., Hahnert, W. F., and De Long, D. M., *Can. Entomol.,* 88, 57–62, 1956.
16. Ehrhardt, P., *Biol. Zentralbl.,* 88, 335–348, 1969.
17. Fraenkel, G. and Printy, G. E., *Biol. Bull.,* 106, 149–157, 1954.
18. Friend, W. G., Backs, R. H., and Cass, L. M., *Can. J. Zool.,* 35, 535–543, 1957.
19. Geer, B. W., *Trans. Ill. State Acad. Sci.,* 59, 3–10, 1966.
20. Geer, B. W., *J. Nutr.,* 90, 31–39, 1966.
21. Gringrich, R. E., *Ann. Entomol. Soc. Am.,* 57, 351–360, 1964.
22. Golberg, L. and DeMeillon, B., *Biochem. J.,* 43, 379–387, 1948.
23. Gordon, H. T., *Ann. N.Y. Acad. Sci.,* 77, 290–351, 1959.
24. Henry, S. M., *Trans. N.Y. Acad. Sci.,* 24, 676–683, 1962.
25. Henry, S. M. and Cook, T. W., *Contr. Boyce Thompson Inst.,* 22, 507–508, 1964.
26. Henry, S. M. and Block, R. J., *Contr. Boyce Thompson Inst.,* 20, 317–329, 1960.
27. Henry, S. M. and Block, R. J., *Contr. Boyce Thompson Inst.,* 21, 447, 1962.
28. Hilchey, J. D., *Contr. Boyce Thompson Inst.,* 17, 203–219, 1953.
29. Hilchey, J. D., Block, R. J., Miller, L. P., and Weed, R. M., *Contr. Boyce Thompson Inst.,* 18, 109–123, 1955.
30. Hinton, T., *Ann. N.Y. Acad. Sci.,* 77, 366–372, 1959.
31. Hinton, T., Noyes, D. T., and Ellis, J., *Physiol. Zool.,* 24, 335–353, 1951.
32. Hodgson, E., Cheldelin, V. H., and Newburgh, R. W., *Can. J. Zool.,* 34, 527–532, 1956.
33. House, H. L., *Can. Entomol.,* 81, 133–139, 1949.
34. House, H. L., *Can. J. Zool.,* 32, 351–357, 1954.
35. Inokuchi, T., Horie, Y., and Ito, T., *Bull. Seric. Exp. Stn. Tokyo,* 22, 195–205, 1967.
36. Inokuchi, T., Horie, Y., and Ito, T., *Biochem. Biophys. Res. Commun.,* 35, 783–787, 1969.
37. Ishii, S. and Hirano, C., *Bull. Natl. Inst. Agric. Sci. Jpn. Ser. C,* No. 5, 35–48, 1955.
38. Ishii, S. and Hirano, C., in *Proc. 10th Int. Congr. Entomol. Montreal, 1956,* Vol. 2, Becker, E. C., Ed., 1958, 295–298.
39. Ito, T. and Arai, N., *J. Insect Physiol.,* 12, 861–869, 1966.
40. Ito, T. and Arai, N., *J. Insect Physiol.,* 13, 1813–1824, 1967.
41. Ito, T. and Inokuchi, T., in *Insect and Mite Nutrition,* Rodriguez, J. G., Ed., North Holland, Amsterdam, 1972, 517–529.
42. Kasting, R., Davis, G. R. F., and McGinnis, A. J., *J. Insect Physiol.,* 8, 589–596, 1962.
43. Kasting, R. and McGinnis, A. J., *Can. J. Biochem.,* 38, 1229–1234, 1960.
44. Kasting, R. and McGinnis, A. J., *J. Insect Physiol.,* 8, 97–103, 1962.
45. Kasting, R. and McGinnis, A. J., *Can. Entomol.,* 96, 1133–1137, 1964.
46. Kasting, R. and McGinnis, A. J., *Exp. Parasitol.,* 19, 249–253, 1966.
47. Lea, A. O. and DeLong, D. M., in *Proc. 10th Int. Congr. Entomol. Montreal, 1956,* Vol. 2, Becker, E. C., Ed., 1958, 299–302.
48. Leckstein, P. M., *Comp. Biochem. Physiol. B,* 49, 743–747, 1974.
49. Leckstein, P. M. and Llewellyn, M., *J. Insect Physiol.,* 19, 973–988, 1973.
50. Leckstein, P. M. and Llewellyn, M., *J. Insect Physiol.,* 20, 877–885, 1974.
51. Lemonde, A. and Bernard, R., *Can. J. Zool.,* 29, 80–83, 1951.
52. Lipke, H., Leto, S., and Graves, B., *J. Insect Physiol.,* 11, 1225–1232, 1965.
53. Lue, P. T. and Dixon, S. E., *Can. J. Zool.,* 45, 595–599, 1967.

Table 1 (continued)

54.   Markulla, M. and Laurema, S., *Ann. Agric. Fenn.*, 6, 77–80, 1967.
55.   McGinnis, A. J., Newburgh, R. W., and Cheldelin, V. H., *J. Nutr.*, 58, 309–324, 1956.
56.   Mittler, T. E., *J. Nutr.*, 101, 1023–1028, 1971.
57.   Naylor, A. F., *Can. J. Zool.*, 41, 1127–1132, 1963.
58.   Pant, N. C. and Gupta, P., *Indian J. Entomol.*, 23, 220–224, 1961.
59.   Pant, N. C., Gupta, P., and Nayar, J. K., *Experientia*, 15, 311, 1960.
60.   Pickett, C. and Friend, W. G., *J. Insect Physiol.*, 11, 1617–1673, 1965.
61.   Retnakaran, A. and Beck, S. D., *Comp. Biochem. Physiol.*, 24, 611–619, 1968.
62.   Rock, G. C., *J. Nutr.*, 98, 153–158, 1969.
63.   Rock, G. C., *J. Insect Physiol.*, 17, 2157–2168, 1971.
64.   Rock, G. C. and Hodgson, E., *J. Insect Physiol.*, 17, 1087–1097, 1971.
65.   Rock, G. C., Khan, A., and Hodgson, E., *J. Insect Physiol.*, 21, 693–703, 1975.
66.   Rock, G. C. and King, K. W., *J. Insect Physiol.*, 13, 59–68, 1967.
67.   Rock, G. C. and King, K. W., *J. Insect Physiol.*, 13, 175–186, 1967.
68.   Rock, G. C. and King, K. W., *J. Nutr.*, 95, 369–373, 1968.
69.   Rock, G. C., Ligon, B. G., and Hodgson, E., *Ann. Entomol. Soc. Am.*, 66, 177–179, 1973.
70.   Sang, J. H. and King, R. C., *J. Exp. Biol.*, 38, 793–809, 1961.
71.   Schaefer, C. H., *J. Insect Physiol.*, 10, 363–369, 1964.
72.   Sedee, D. J. W., *Acta Physiol. Pharmacol. Neerl.*, 3, 262–269, 1954.
73.   Sedee, D. J. W., *Entomol. Exp. Appl.*, 1, 38–40, 1958.
74.   Sharma, G. K., Hodgson, E., and Rock, G. C., *J. Insect Physiol.*, 18, 9–18, 1972.
75.   Sharma, G. K., Rock, G. C., and Hodgson, E., *J. Insect Physiol.*, 18, 1333–1341, 1972.
76.   Singh, K. R. P. and Brown, A. W. A., *J. Insect Physiol.*, 1, 199–220, 1957.
77.   Strong, F. E. and Sakamota, S. S., *J. Insect Physiol.*, 9, 875–879, 1963.
78.   Taylor, M. W. and Medici, J. C., *J. Nutr.*, 88, 176–180, 1966.
79.   Thompson, S. N., *J. Insect Physiol.*, 20, 1515–1528, 1974.
80.   Turner, R. B., *J. Insect Physiol.*, 17, 2451–2456, 1971.
81.   Vanderzant, E. S., *J. Econ. Entomol.*, 51, 309–311, 1958.
82.   Vanderzant, E. S., *J. Insect Physiol.*, 9, 683–691, 1963.
83.   Vanderzant, E. S., *J. Insect Physiol.*, 11, 659–670, 1965.
84.   Vanderzant, E. S., *J. Econ. Entomol.*, 66, 336–338, 1973.
85.   Vanderzant, E. S. and Chremos, J. H., *Ann. Entomol. Soc. Am.*, 64, 480–485, 1971.

Table 2

REQUIREMENTS FOR B VITAMINS AND OTHER
WATER-SOLUBLE GROWTH FACTORS

A glance at the table below suggests that insects generally require an exogenous supply of choline and all the B vitamins excepting B_{12}. Assuming this generalization to be correct, interest then centers on why particular species apparently are able to dispense with certain vitamins. Most cases of multiple vitamin dispensability are species known to harbor specific symbiotic microorganisms (e.g., aphids, cockroaches, and several beetles infesting stored products), and in several instances where symbiote defaunation was possible, the aposymbiotic insect was found to require an enlarged complement of vitamins. Insects that feed on vertebrate blood throughout all developmental stages also are thought to be in this category; the bloodsucking bug *Rhodnius prolixus,* which normally carried a symbiotic actinomycete, *Nocardia rhodnii,* cannot complete development if deprived of this organism unless the blood meal is augmented with several B vitamins.[70] When experiments conducted without asepsis indicate that B vitamins apparently are not needed by insects lacking specific symbiotes, there is often the possibility that facultative microorganisms in the gut or diet might be supplying the vitamins in question. Much work on stored-products insects and Orthoptera in which "dry" diets and nonsterile techniques were used was prone to this uncertainty, particularly with respect to folic acid and biotin, requirements for which are always especially minute. A requirement for these two vitamins is also particularly likely to be masked by their presence as contaminants of other dietary components such as the caseins or yeast fractions used in the meridic diets of much early work. Delayed manifestation of a deficiency can often result from substantial initial reserves in the egg or newly hatched larvae with which growth experiments are initiated, to be revealed only by growth failure late in development or in a second generation on deficient diet.

Of the growth factors listed, the status of cobalamine (vitamin B_{12}) is still equivocal; there is no clear demonstration of a need for it (nor is the possibility distinguished that an apparent requirement might be for cobalt rather than for the vitamin as such); on the other hand, if required, it would be needed in such low amounts as to be especially prone to the masking situations mentioned above. Carnitine appears to be essential only for beetles of the family Tenebrionidae; indications that a few species in other taxa may require it are of questionable biological significance. However, it is noteworthy that carnitine can spare choline for a number of species, and this would sometimes lead to the supposition that it was required if choline was suboptimally provided. Though carnitine and other choline-sparing compounds were sometimes considered fully adequate substitutes for choline, probably they would ultimately prove to be adequate only for certain functions of choline as a constituent of phospholids, as recently clarified in work with Diptera.[9,44-47] A requirement for nucleic acid or components thereof is virtually restricted to Diptera, in which all degrees of need, from nonrequirement to absolute essentiality, may be documented, in the case of *Drosophila melanogaster* as between various strains within the one species. Slight, positive effects of growth have been recorded for a few beetles, and RNA was considered essential to allow any growth at all of larvae of the moth, *Plodia interpunctella,*[84] but in the latter case the possibility was not excluded that effects might be due to impurities such as trace metals, etc.

Ascorbic acid (vitamin C) and inositol are essential for some insect species but not required by others. All species requiring ascorbic acid feed on plant tissues in nature, but not all plant feeders require it; evidently the evolution of a loss of ability to synthesize ascorbic acid has occurred many times and in various insect taxa as they become adapted to food having abundant supplies of the nutrient. In addition to the species tabulated below, the following Lepidoptera have been shown to require dietary vitamin C: *Alabama argillacea,*[115] *Carpocapsa pomonella,*[98] *Diatraea grandiosella,*[15] *Estigmene acrea,*[113] *Heliothis* spp.,[113,115] *Ostrinia (Pyrausta) nubilalis,*[16] *Spodoptera (Prodenia) littoralis,*[79] and *Trichoplusia ni.*[17]

Para-aminobenzoic acid was routinely tested as a possible vitamin in most studies before 1950, but since it has never been shown to be of value it is excluded from tabulation. Niacin is understood to include nicotinic acid or nicotinamide, and pyridoxine covers pyridoxamine and pyridoxal phosphate, since in all studies where these alternatives have been considered they proved completely interchangeable.

Symbols used in this table have the following meanings:

R = required.
r = development retarded, probably required.
Ɍ = apparently not required for one cycle of development.

Table 2 (continued)
REQUIREMENTS FOR B VITAMINS AND OTHER WATER-SOLUBLE GROWTH FACTORS

	Vitamin B complex								Lipogenic factors			Ascorbic acid	Nucleic acid	Others	Ref.
	Thiamine	Riboflavin	Niacin	Pyridoxin	Pantothenate	Folic acid	Biotin	B_{12}	Choline	Inositol	Carnitine				
Diptera															
Aedes aegypti	R	R	R	R	R	R	R	Ɍ?a	R	Ɍ	Ɍ^a	Ɍ	R^a	a	1, 52, 71, 72, 107, 109
Agria housei (*Pseudosarcophaga affinis*)	R	R	R	Ɍ?b	R	R?	R	Ɍ?b	R	Ɍ			R^b	b	4, 61—63
Calliphora erythrocephala (*Calliphora vicina*)	R	R	R	R	R	R	R	Ɍ	R	Ɍ			R		105
Cochliomyia hominivorax	R	R	R	Ɍ?	R	R	R	Ɍ	R	Ɍ	Ɍ		R^c		51
Culex pipiens	R	R	R	R	R	R	R	Ɍ	R	Ɍ	Ɍ	Ɍ	R^a		22
Drosophila melanogaster	R	R	R	R	R	R	R	Ɍ?d	R^d	Ɍ	Ɍ?d		R^d	d	13, 32, 42—50, 54, 55, 100—103
D. melanogaster — oviposition	R	R	R	R	R	R	R	Ɍ?d	R?d	Ɍ					104
Hylemya antiqua	R	R	R	R	R	R	Ɍ	Ɍ?e	R					e	39

a Early studies attributed slight growth improvements to vitamin B_{12}, carnitine, and glutathione, not substantiated by subsequent work RNA effective DNA not, RNA replaceable by mixture of adenylic, guanylic, cytidylic, uridylic, and thymidylic acids; for *Culex pipiens* adenylic, thymidylic and either cytidylic or uridylic acids replace RNA.[22]

b Subsequently shown that basal diets contained traces of pyridoxine which masked demonstration of a low requirement.[4] Pupation improved with B_{12}. Growth slow without nucleic acid though ultimately normal numbers of adults, both RNA and DNA utilized, and replaceable by nucleotides but not nucleosides or bases. Glutathione not required.

c No growth without RNA, replaceable by mixture of adenine, cytosine and guanine, uracil and uridylic acid inhibit growth.

d Reduced percentage of adults if B_{12} absent. Requirement for nucleic acid varies from not required to essential for different strains. RNA cannot be replaced by DNA but can be replaced by mixtures of nucleotides and bases, essentially those providing adenine and cytosine Carnitine β-methylcholine betaine and various other analogues spare choline variously, but without some choline male and female adults are infertile, choline may not be needed by adults from normally reared larvae, depending on reserves carried over pupation.

e Slightly reduced survival to pupae without B_{12}. Reduced mortality with thioctic (α-lipoic) acid. Slight positive effect of coenzyme A probably indicates inadequate pantothenate.

Table 2 (continued)
REQUIREMENTS FOR B VITAMINS AND OTHER WATER-SOLUBLE GROWTH FACTORS

	Vitamin B complex								Lipogenic factors			Ascorbic acid	Nucleic acid	Others	Ref.
	Thiamine	Riboflavin	Niacin	Pyridoxin	Pantothenate	Folic acid	Biotin	B$_{12}$	Choline	Inositol	Carnitine				
Diptera (continued)															
Musca domestica	R?[h]	r	R	R	R	R[f]	R	R	R[f]	R	R[f]		R?[g]	f	9–11, 64, 85, 94
Phormia regina	R	R	R	R	R	r[g]	r[g]	R	R[g]	R	R[g]				12, 14, 56–58, 81
Siphonaptera															
Xenopsylla cheopis	R?[h]	r	R	r	R	R?[h]	R?[h]		R?[h]						93
Hymenoptera															
Exeristes comstockii — egg production by adults	R	R	R	R	R	R	R	R	R	R			R		7, 8
Itoplectis conquisitor	R	R	R	R	R	R	R	R	R	R	R	R	R		117

f Folic acid and choline deficiency appear only late in development and not detected in one study. Development retarded without nucleic acid replaceable by adenylic and guanylic acids; adult flies require RNA for maximal fecundity, replaceable by adenylic and cytidylic acids. Choline is completely spared for larval growth by carnitine and acetyl-β-methylcholine and partially spared by methylaminoethanols and other analogues; however choline required for normal levels of acetylcholine in nervous issues.

g Folic acid and biotin probably required though effects of omission on one cycle of larval growth trivial. Choline spared by carnitine and dimethyaminoethanol but not by betaine. Nucleic acid not required but DNA and RNA have different complex effects on pupation and adult emergency.

h Dispensability of several vitamins not due to symbiotes but ascribed to relatively large reserves in an egg which is 1/10th the size of adult.

Table 2 (continued)
REQUIREMENTS FOR B VITAMINS AND OTHER WATER-SOLUBLE GROWTH FACTORS

	Vitamin B complex								Lipogenic factors		Carnitine	Ascorbic acid	Nucleic acid	Others	Ref.
	Thiamine	Riboflavin	Niacin	Pyridoxin	Pantothenate	Folic acid	Biotin	B₁₂	Choline	Inositol					
Coleoptera															
Anthonomus grandis	R	R	R	R	R	R	R?[i]		R[i]	R[i]		R[i]			110, 111, 113
Attagenus sp.	R	R	R	R	R	R	R		R	R					83
Carpophilus hemipterus	R	R	R	R	R	R?[k]	R?[k]		R	R					108
Dermestes maculatus (*Dermestes vulpinus*)	r	r	R	R	R	R	R		r	R					78
Lasioderma serricorne, symbiotic	R	R?[k]	r	R	r	r	R		r	R					6, 33, 68, 89, 90
Lasioderma serricorne, aposymbiotic	R	R	R	R	R	R	R		R	r?		R			
Oryzaephilus surinamensis, symbiotic (*Silvanus surinamensis*)					R		r		r	r?	r?[j]	R?	r?[j]	j	25—30, 33, 76
Palorus ratzeburgi	R	R	R	R	R	r	r		R	R?					18
Ptinus tectus	R	R	R	R	R	R	R		R	R?					33
Scolytus multistriatus	R	R	R	R	R	R	R		R	R					41
Sitophilus oryzae, symbiotic	R	R?[k]	R	r	R?[k]	R	R		R?[k]	R?					3
Stegobium paniceum, symbiotic (*Sitodrepa panicea*), aposymbiotic	R	r	R	R	R	r	R		r	r?					6, 33, 76, 89, 90
Tenebrio molitor	R	R	R	R	R	R	R		r?[l]	R?	R				35a, 36, 37, 73—75
Tribolium casteneum	R	R	R	R	R	R	R		R	R	R				2, 40, 80, 91
Tribolium confusum	R	R	R	R	R	R	R		r	r?	R				33, 35a, 38, 77
Trogoderma granarium	R	R	R	R[m]	R	R[m]	r[m]		R[m]	R					92, 106

i   Biotin known to be in casein of basal diet. Betaine, carnitine, and ethanolamine do not spare choline. Various bound forms of inositol utilized.

j   Carnitine slightly improved growth with certain basal diets. RNA slightly increased growth rate, replaceable by combinations of guanine and cytosine. Putrescine had tiny beneficial effect on growth. Factor G (a protein-associated growth factor for rats) slightly improved growth.

k   Believed that symbiotes *do not* provide these vitamins, but rather that they occurred as impurities in other dietary components. Suggested that failure to demonstrate choline requirement due to sparing by carnitine impurities.

l   Choline requirement may differ for different races.[75]

m   Effect of vitamin omission varied with type of casein in basal diet. Carnitine, butyrobetaine and methylaminoethanols partially spared choline.

Table 2 (continued)
REQUIREMENTS FOR B VITAMINS AND OTHER WATER-SOLUBLE GROWTH FACTORS

	Vitamin B complex								Lipogenic factors			Ascorbic acid	Nucleic acid	Others	Ref.
	Thiamine	Riboflavin	Niacin	Pyridoxin	Pantothenate	Folic acid	Biotin	B_{12}	Choline	Inositol	Carnitine				
Lepidoptera															
Agrotis orthogonia	R	R	R	R	R	R	R?	R	R	R		R?[n]			69
Anagasta kuehniella[o] (*Ephestia kuehniella*)	R	R	R	R	R	R	R		R	r?[o]		R?[o]			35, 35a
Argyrotaenia velutinana	R	R	R	R	R	R	R	r?	R	R		R			98, 99
Bombyx mori	R	R	R[p]	R	R	r	R		R[p]	R		R[p]			59, 60, 66, 67
Chilo simplex	R	R	R	R	R	R	R		R?[q]	R?[q]					65
Pectinophora gossypiella	R	R	R	R	R	R	R	R	R	R		R	R		88, 114, 115
Tineola bisselliella[r]	R	R	R	R	R	R	r		R	R			R		34

n   Ascorbic acid in basal diet but not tested specifically.

o   Other species of Lepidoptera related to *Anagasta* had similar vitamin requirements. Inositol requirement doubtful. Ascorbic acid improves growth probably by antioxidant protection of essential fatty acids.

p   Pyridine-3-aldehyde can substitute for niacin. Choline not spared by carnitine nor soybean phospholipids. Ascorbic acid also an important phagostimulant, can be replaced by arab-ascorbic acid, partly by dehydroascorbic acid, but not by gulonolactone or glucuronolactone.

q   Larvae on complete control diet only reached one fifth normal weight and none pupated, so cannot be concluded that choline and inositol not required.

r   Considerable slow growth to eventual pupation occurred in the absence of all vitamins except nicotinic acid, presumably due to traces of vitamins in casein, since symbiotes were excluded.

Table 2 (continued)
REQUIREMENTS FOR B VITAMINS AND OTHER WATER-SOLUBLE GROWTH FACTORS

	Vitamin B complex								Lipogenic factors						
	Thiamine	Riboflavin	Niacin	Pyridoxin	Pantothenate	Folic acid	Biotin	B_{12}	Choline	Inositol	Carnitine	Ascorbic acid	Nucleic acid	Others	Ref.
Hemiptera															
Myzus persicae, symbiotic[s]	R	R	R	R	R	R	R	R[s]	R	R	R[s]	R[s]	R[s]		21, 23, 24, 82
Neomyzus circumflexus, symbiotic[t]	R	R̸	R	r[t]	r[t]	R	r[t]		r[t]	r[t]		R			31
Orthoptera															
Acheta domestica	R	r	R	R	R	r	R		R	R		R̸[u]			97
Blatella germanica, symbiotic	R	r?[u]	R	r	R	r[u]	r[u]	r[u]	R[u]	R		R[u]			53, 86, 87, 116
Schistocerca gregaria[v]	R	R	R	R	R	R	R		R	R		R		v	5, 19, 20

s   In the absence of a vitamin poor growth becomes marked only after two consecutive generations of deficiency, presumably because of reserves carried over from plant-reared stock. Araboascorbic and dehydroascorbic acids can replace ascorbic acid but glucuronolactone and gulonolactone cannot. Apparent growth improvement with nucleic acids resulted from essential trace metal impurities. Vitamin B_{12} and carnitine can be assumed not required since 30 consecutive generations were reared in their absence.

t   Vitamin deficiencies became manifest only after two to four consecutive generations. Riboflavin not required for ten generations, and in its absence, pantothenate, biotin, and choline were not required. These anomalous relationships are probably the outcome of changes in activity of intracellular symbiotes.

u   Riboflavin shown to be synthesized by roach with normal microflora and symbiotes. Requirement for folic acid, biotin, and B_{12} apparent only as failure to produce eggs or death early in second generation. Choline spared by betaine, dimethyaminoethanol, and other analogues. Synthesis of ascorbic acid probably dependent on symbiotes as shown for some other roaches.[95,96]

v   *Locusta migratoria* has same vitamin requirements.

Table 2 (continued)

REFERENCES

1. Akov, S., *J. Insect Physiol.*, 8, 319–335, 1962.
2. Applebaum, S. W. and Lubin, Y., *Entomol. Exp. Appl.*, 10, 23–30, 1967.
3. Baker, J. E., *J. Insect Physiol.*, 21, 1337–1342, 1975.
4. Barlow, J. S., *Nature*, 196, 193–194, 1962.
5. Blackith, R. E. and Blackith, R. M., *Arch. Zool. Exp. Gen.*, 110, 303–340, 1969.
6. Blewett, M. and Fraenkel, G., *Proc. R. Soc. London Ser. B*, 132, 212–221, 1944.
7. Bracken, G. K., *Can. Entomol.*, 97, 1037–1041, 1965.
8. Bracken, G. K., *Can. Entomol.*, 98, 918–922, 1966.
9. Bridges, R. G., *J. Insect Physiol.*, 20, 2363–2374, 1974.
10. Bridges, R. G., Ricketts, J., and Cox, J. T., *J. Insect Physiol.*, 11, 225–236, 1965.
11. Brookes, V. J. and Fraenkel, G., *Physiol. Zool.*, 31, 208–223, 1958.
12. Brust, M. and Fraenkel, G., *Physiol. Zool.*, 28, 186–204, 1955.
13. Burnett, B. and Sang, J. H., *J. Insect Physiol.*, 9, 553–562, 1963.
14. Chelderlin, V. H. and Newburgh, R. W., *Ann. N.Y. Acad. Sci.*, 77, 373–383, 1959.
15. Chippendale, G. M., *J. Nutr.*, 105, 499–507, 1975.
16. Chippendale, G. M. and Beck, S. D., *Entomol. Exp. Appl.*, 7, 241–248, 1964.
17. Chippendale, G. M., Beck, S. D., and Strong, F. M., *J. Insect Physiol.*, 11, 211–223, 1965.
18. Cooper, M. I. and Fraenkel, G., *Physiol. Zool.*, 25, 20–28, 1952.
19. Dadd, R. H., *Proc. R. Soc. London Ser. B*, 153, 128–143, 1960.
20. Dadd, R. H., *J. Insect Physiol.*, 6, 1–12, 1961.
21. Dadd, R. H., *J. Insect Physiol.*, 13, 763–778, 1967.
22. Dadd, R. H., unpublished results.
23. Dadd, R. H. and Krieger, D. L., *J. Econ. Entomol.*, 60, 1512–1514, 1967.
24. Dadd, R. H., Krieger, D. L., and Mittler, T. E., *J. Insect Physiol.*, 13, 249–272, 1967.
25. Davis, G. R. F., *Arch. Int. Physiol. Biochim.*, 72, 70–75, 1964.
26. Davis, G. R. F., *Can. J. Zool.*, 44, 781–785, 1966.
27. Davis, G. R. F., *Can. Entomol.*, 98, 263–267, 1966.
28. Davis, G. R. F., *Comp. Biochem. Physiol.*, 19, 619–627, 1966.
29. Davis, G. R. F., *J. Insect Physiol.*, 13, 1737–1743, 1967.
30. Davis, G. R. F., *Comp. Biochem. Physiol. A*, 43, 927–933, 1972.
31. Ehrhardt, P., *Z. Vgl. Physiol.*, 60, 416–426, 1968.
32. Erk, F. C. and Sang, J. H., *J. Insect Physiol.*, 12, 43–51, 1966.
33. Fraenkel, G. and Blewett, M., *Biochem. J.*, 37, 686–692, 692–695, 1943.
34. Fraenkel, G. and Blewett, M., *J. Exp. Biol.*, 22, 156–161, 1946.
35. Fraenkel, G. and Blewett, M., *J. Exp. Biol.*, 22, 162–171, 1946.
35a. Fraenkel, G. and Blewett, M., *Biochem. J.*, 41, 469–475, 1947.
36. Fraenkel, G., Blewett, M., and Coles, M., *Physiol. Zool.*, 23, 92–108, 1950.
37. Fraenkel, G. and Chang, P.-I., *Physiol. Zool.*, 27, 40–56, 1954.
38. French, E. W. and Fraenkel, G., *Nature*, 173, 173, 1954.
39. Friend, W. G. and Patton, R. L., *Can. J. Zool.*, 34, 152–162, 1956.
40. Gabrani, K. and Pant, N. C., *Indian J. Entomol.*, 25, 266–268, 1963.
41. Galford, J. R., *J. Econ. Entomol.*, 65, 681–684, 1972.
42. Geer, B. W., *J. Exp. Zool.*, 154, 353–364, 1963.
43. Geer, B. W., *Genetics*, 49, 787–796, 1964.
44. Geer, B. W., *Biol. Bull.*, 133, 548–566, 1967.
45. Geer, B. W., *Arch. Int. Physiol. Biochim.*, 76, 797–805, 1968.
46. Geer, B. W. and Dolph, W. W., *J. Reprod. Fertil.*, 21, 9–15, 1970.
47. Geer, B. W., Dolph, W. W., McGuire, J. A., and Dates, R. J., *J. Exp. Zool.*, 176, 445–460, 1971.
48. Geer, B. W., Olander, R. M., and Sharp, P. L., *J. Insect Physiol.*, 16, 33–43, 1970.
49. Geer, B. W. and Vovis, G. F., *J. Exp. Zool.*, 158, 223–236, 1965.
50. Geer, B. W., Vovis, G. F., and Yund, M. A., *Physiol. Zool.*, 41, 280–292, 1968.
51. Gingrich, R. E., *Ann. Entomol. Soc. Am.*, 57, 351–360, 1964.
52. Golberg, L. and DeMeillon, B., *Biochem. J.*, 43, 372–379, 1948.
53. Gordon, H. T., *Ann. N.Y. Acad. Sci.*, 77, 290–351, 1959.
54. Hinton, T., *Physiol. Zool.*, 29, 20–26, 1956.
55. Hinton, T., Noyes, D. T., and Ellis, J., *Physiol. Zool.*, 24, 335–353, 1951.
56. Hodgson, E., Chelderlin, V. H., and Newburgh, R. W., *Can. J. Zool.*, 34, 527–532, 1956.
57. Hodgson, E. and Dauterman, W. C., *J. Insect Physiol.*, 10, 1005–1008, 1964.
58. Hodgson, E., Mehendale, H. M., Smith, E., and Khan, M. A. Q., *Comp. Biochem. Physiol.*, 29, 343–359, 1969.

Table 2 (continued)

59. Horie, Y. and Ito, T., *J. Insect Physiol.,* 11, 1585–1593, 1965.
60. Horie, Y., Watanabe, K., and Ito, T., *Bull. Seric. Exp. Stn. Tokyo,* 20, 393–409, 1966.
61. House, H. L., *Can. J. Zool.,* 32, 342–350, 1954.
62. House, H. L., *Can. J. Zool.,* 32, 358–365, 1954.
63. House, H. L., *Can. J. Zool.,* 42, 801–806, 1964.
64. House, H. L. and Barlow, J. S., *Ann. Entomol. Soc. Am.,* 51, 299–302, 1958.
65. Ishii, S. and Urushibara, H., *Bull. Nat. Inst. Agric. Sci. C* (Tokyo), 4, 109–133, 1954.
66. Ito, T., *Bull. Seric. Exp. Stn. Tokyo,* 17, 119–136, 1961.
67. Ito, T. and Arai, N., *Bull. Seric. Exp. Stn. Tokyo,* 20, 1–19, 1965.
68. Jurzitza, G., *Oecologia Berlin,* 3, 70–83, 1969.
69. Kasting, R. and McGinnis, A. J., *Can. J. Zool.,* 45, 787–796, 1967.
70. Lake, P. and Friend, W. G., *J. Insect Physiol.,* 14, 543–562, 1968.
71. Lea, A. O. and DeLong, D., in *Proc. 10th Int. Congr. Entomol. Montreal, 1956,* Vol. 2, Becker, E. C., Ed., 1958, 299–302.
72. Lea, A. O., Dimond, J. B., and DeLong, D., *J. Econ. Entomol.,* 49, 313–315, 1956.
73. Leclercq, J., *Biochim. Biophys. Acta,* 2, 329–332, 1948.
74. Leclercq, J., *Arch. Int. Physiol.,* 57, 67–70, 1959.
75. Leclercq, J., *Voeding,* 16, 785–790, 1955.
76. Lemonde, A. and Bernard, R., *Nat. Can.,* 80, 125–142, 1953.
77. Lemonde, A. and Barnard, R., *Rev. Can. Biol.,* 14, 8–13, 1955.
78. Levinson, Z. H., Barelkovsky, J., and Bar Ilan, A. R., *J. Stored Prod. Res.,* 3, 345–352, 1967.
79. Levinson, Z. H. and Navon, A., *J. Insect Physiol.,* 15, 591–595, 1969.
80. Magis, N., *Arch. Inst. Physiol.,* 62, 505–511, 1954.
81. McGinnis, A. J., Newburgh, R. W., and Cheldelin, V. H., *J. Nutr.,* 58, 309–324, 1956.
82. Mittler, T. E., Tsitsipis, J. A., and Kleinjan, J. E., *J. Insect Physiol.,* 16, 2315–2326, 1970.
83. Moore, W., *Ann. Entomol. Soc. Am.,* 39, 513–521, 1946.
84. Morere, J.-L., *Arch. Int. Physiol. Biochim.,* 82,625–630, 1974.
85. Morrison, P. E. and Davies, D. M., *Nature,* 201, 948–949, 1964.
86. Noland, J. L. and Baumann, C. A., *Proc. Soc. Exp. Biol. Med.,* 70, 198–201, 1949.
87. Noland, J. L., Lilly, J. H., and Baumann, C. A., *Ann. Entomol. Soc. Am.,* 42, 154–164, 1949.
88. Ouye, M. T. and Vanderzant, E. S., *J. Econ. Entomol.,* 57, 427–430, 1964.
89. Pant, N. C. and Fraenkel, G., *Science,* 112, 498–500, 1950.
90. Pant, N. C. and Fraenkel, G., *Biol. Bull.,* 107, 420–432, 1954.
91. Pant, N. C. and Gabrani, K., *Indian J. Entomol.,* 25, 110–115, 1963.
92. Pant, N. C. and Pant, J. C., *Indian J. Entomol.,* 22, 115–120, 1960.
93. Pausch, R. D. and Fraenkel, G., *Physiol. Zool.,* 39, 202–223, 1966.
94. Perry, A. S., Miller, G., and Buckner, A. J., *J. Insect Physiol.,* 11, 1277–1287, 1965.
95. Pierre, L. L., *Nature,* 193, 904–905, 1962.
96. Raychaudhuri, D. N. and Banerjee, M., *Sci. Cult.,* 34, 461–463, 1968.
97. Ritchot, C. and McFarlane, J. E., *Can. J. Zool.,* 39, 11–15, 1961.
98. Rock, G. C., *J. Econ. Entomol.,* 60, 1002–1005, 1967.
99. Rock, G. C., *Ann. Entomol. Soc. Am.,* 62, 611–613, 1969.
00. Royes, W. V. and Robertson, F. W., *J. Exp. Zool.,* 156, 105–135, 1964.
01. Sang, J. H., *J. Exp. Biol.,* 33, 45–72, 1956.
02. Sang, J. H., *Proc. R. Soc. Edinburgh Sect. B,* 66, 339–359, 1957.
03. Sang, J. H., *J. Nutr.,* 77, 355–368, 1962.
04. Sang, J. H. and King, R. C., *J. Exp. Biol.,* 38, 793–809, 1961.
05. Sedee, J. W., *Physiol. Zool.,* 31, 310–316, 1958.
06. Sehgal, S. S. and Agarwal, H. C., *J. Insect Physiol.,* 19, 419–425, 1973.
07. Singh, K. R. P. and Brown, A. W. A., *J. Insect Physiol.,* 1, 199–220, 1957.
08. Stride, G. O., *Trans. R. Entomol. Soc. London,* 104, 171–194, 1953.
09. Trager, W., *J. Biol. Chem.,* 176, 1211–1223, 1948.
10. Vanderzant, E. S., *J. Econ. Entomol.,* 52, 1018–1019, 1959.
11. Vanderzant, E. S., *J. Econ. Entomol.,* 56, 357–362, 1963.
12. Vanderzant, E. S., *J. Econ. Entomol.,* 68, 375–376, 1975.
13. Vanderzant, E. S., Pool, M. C., and Richardson, C. D., *J. Insect Physiol.,* 8, 287–297, 1962.
14. Vanderzant, E. S. and Reiser, R., *J. Econ. Entomol.,* 49, 454–458, 1956.
15. Vanderzant, E. S. and Richardson, C. D., *Science,* 140, 989–991, 1963.
16. Wollman, E., Giroud, A., and Ratsimananga, R., *C. R. Soc. Biol.,* 124, 434–435, 1937.
17. Yazgan, S., *J. Insect Physiol.,* 18, 2123–2141, 1972.

Table 3
LIPID GROWTH FACTORS

Dietary fats and oils (fatty acids and glycerol on digestion) undoubtedly are utilized for energy by many insects, perhaps as a major source in the normal food of some, but in no case have they been shown essential for this purpose. *Dermestes maculatus,* a beetle which prefers fatty animal foods, develops normally with carbohydrate and protein as the only energy source,[13,53] and larvae of the waxmoth, *Galleria mellonella,* utilize carbohydrate as readily as the beeswax (a mixture of long-chain fatty acids, alcohols, and esters) of old honeycomb, their natural food.[18,19] Although not required, fat or mixtures of fatty acids somewhat improve the growth rate of several species,[9,24-26,42,46,53,56,57] but it is generally unclear whether improvement stems from a phagostimulant or texturizing effect which optimizes the ingestion of food, or from the truly nutritional role of optimizing the proportions of tissue fatty acids which are otherwise synthesized by the insect. Apart from these widespread facultative uses of fat, small amounts of several lipid materials are essential to various insects as accessory growth factors. Besides sterol, dealt with in Table 4, these comprise certain polyunsaturated fatty acids (linoleic and linolenic acids), vitamin E (alpha-tocopherol), and vitamin A (or its provitamin, β-carotene). There is now no doubt that vitamin A is required by insects and other arthropods for normal vision,[14,32,36,37,73] and in its absence ommatidial malformations can be observed;[3,7,12] the evidence for a growth-promoting effect of vitamin A is more equivocal, and the effects observed could arguably result from behavioral changes consequent upon deficient vision.[17] Though a sterol is essential, the steroid D vitamins of vertebrate nutrition appear to have no significance for insects, nor have requirements for vitamins of the K group or the ubiquinones been detected. A phospholipid, lecithin, is sometimes listed as an essential dietary ingredient, but the presumption then is that it provides choline,[62a] or that it texturizes and renders diets more consumable.[34,68] Since fat is not required in the diets of insects, their substantial tissue fat evidently can be synthesized from nonlipid precursors. Metabolic studies (for reviews see References 27, 33) have established that insects can derive saturated and monoethenoid fatty acids from simple precursors such as glucose or acetate or from other fatty acids; in spite of an early study indicating that the mealworm could form linoleic acid from food presumed to lack it,[31] there is no convincing evidence that any insect, independently of symbiotes it may contain, can synthesize polyunsaturated fatty acids or form them by desaturation of other fatty acids that are synthesized or provided in food. This lack of synthetic ability accounts for the dietary essentiality of linoleic and/or linolenic acids for many insects, but poses an interesting question in the case of species apparently without the requirement (e.g., all Diptera studied); since these also lack synthetic ability, it would follow that they have no ultimate physiological need for polyunsaturated fatty acids. Broadly speaking, putative physiological functions of polyunsaturated fatty acids are of two kinds: As components of phospholipids they may be necessary to provide appropriate physical characteristics of cellular membranes, a function considered to be relatively unspecific and subserved adequately by various combinations of saturated and unsaturated fatty acids; or polyunsaturated fatty acid may be required specifically as the precursor of an essential metabolite, a role suggested by the fact that in Lepidoptera the most obvious deficiency symptom emerges as a characteristic dysfunction of adult emergence. In vertebrates, essential fatty acids are precursors of prostaglandins, now thought to be ubiquitous cellular regulators. However, prostaglandins derive mainly from arachidonic acid, which, if not in food, vertebrates can derive from dietary linoleic acid. By contrast, among insects, arachidonic acid has proved quite unable to ameliorate the adult molting failure of fatty acid deficiency (though occasionally improving growth slightly), and the weight of current evidence indicates that it is unimportant in insect nutrition. However, the following cautionary points should be noted. First, prostaglandin has recently been detected in insect tissues, being apparently involved in reproductive functions (work in progress reported by D. B. Destephano at 1975 meeting of the Entomological Society of America), and this raises the question of precursor material for its biosynthesis; second, since arachidonic acid is especially prone to oxidation, it is quite possible that toxic oxidation products rapidly developing in synthetic diets might obscure positive effects that could result from the presence of arachidonic acid as such. In the absence of lipid growth factors impaired development tends to occur only after considerable apparently normal growth, often first manifesting as a failure in some aspect of adult function, or in the F_1 generations. Since few studies are carried beyond one cycle of larval growth, it remains possible that requirements for lipid factors especially may be more widespread than is presently apparent. It was necessary to carry experiments beyond one larval cycle to reveal the linoleic acid requirement of *Blatella germanica*[38] and *Anthonomus grandis*[26] and the carotene requirement of *Locusta migratoria*.[17] With respect to lipid growth factors, it is therefore particularly important to recognize that a finding of no requirement based on a single larval cycle may be equivocal. With the foregoing caveats, where no entry appears in this listing for insects that are known to have developed well on defined diets, the presumption is that the unlisted factor was not required.

Symbols used in the table have the following meanings:

With respect to fatty acids:

    R = required (particularly for adult emergence in Lepidoptera).
    Ɍ = not required.
    r = growth retarded if lacking (adult emergence normal in Lepidoptera).
    $ = not synthesized.

With respect to vitamin E:
    R = required, generally for reproduction specifically.
    Ɍ = not required.

With respect to vitamin A, vision:

    R = reduced visual performance and/or reduced vitamin A in eye.

With respect to vitamin A, growth:

    R = required.
    r = growth retarded and/or pigmentation abnormal.

Table 3 (continued)
LIPID GROWTH FACTORS

	Polyunsaturated C18 fatty acids			Other fatty acids	Vitamin E	Vitamin A or β-carotene		Ref.
	Linoleic or linolenic	Linoleic	Linolenic			Growth	Vision	
Diptera								
Aedes aegypti	R				R[?a]	r[?a]	R	3, 35
Agria housei (Pseudosarcophaga affinis)	R			r[?b]	R	r		42—44
Cochliomyia hominivorax	R							34
Culex pipiens	R					r[?c]		20
Drosophila melanogaster	R, $					R	R	48, 62a, 73
Musca domestica	R, $							1, 4, 5, 14, 36, 37, 44
Musca vicina	R							51, 63
Phormia regina	R							6, 8
Hymenoptera								
Itoplectis conquisitor			R					72
Coleoptera								
Anthonomus grandis	R,[d] $				R[e]	R[?f]		26, 49, 71
Dermestes maculatus (vulpinus)	R?$[g]							13
Lasioderma serricorne	R							28
Oryzaephilus surinamensis (Sitodrepa surinamensis)	R				R			24

[a] Sometimes improved growth slightly, possibly because of contaminants or physical effects on media.[35]

[b] Growth rate improved by mixture of five fatty acids, of which oleic was the most important.[42]

[c] Growth rate and percent pupation increased by inclusion of β-carotene or various forms of vitamin A.[20]

[d] Deficiency in larval diet appeared mainly as reduced egg production by resulting adults.

[e] Not required for three consecutive generations.

[f] Eggs of female reared on carotene-deficient diet lacked usual yellowish color.

[g] Though omission produced no adverse effects during one cycle of larval growth, apparently no synthesis of polyunsaturated fatty acids occurs.

Table 3 (continued)
LIPID GROWTH FACTORS

	Polyunsaturated C18 fatty acids			Other fatty acids	Vitamin E	Vitamin A or β-carotene		Ref.
	Linoleic or linolenic	Linoleic	Linolenic			Growth	Vision	
Coleoptera (continued)								
Ptinus tectus	R							28
Stegobium paniceum (Sitodrepa panicea)	R							28
Tenebrio molitor	R?			h				25, 31, 50
Tribolium confusum	R							28
Lepidoptera								
Adoxophyes orana	R	R	R					67
Anagasta kuehniella (Ephestia kuehniella)	R			i	r^j			30, 31
Argyrotaenia velutinana	R, $							62
Bombyx mori	R, $							41, 46, 47, 64
Cadra cautella (Ephestia cautella)	R							30
Carpocapsa pomonella	R							61
Chilo suppressalis (Chilo simplex)	R							40
Diatraea grandiosella	R			l				11
Galleria mellonella	R				k			18, 19
Heliothis zea	$	R	R					2, 69
Hyalophora cecropia	$						R	7, 12, 65
Pectinophora gossypiella	R	R?m	R?m		R	R		70

h Though not required, and linoleic synthesis claimed,[31] fats or linoleic acid alone improved growth.[25]

i Arachidonic acid slightly improved growth but failed to allow adult emergence.

j Improved growth slightly, supposedly due to antioxidant protection of linoleic acid.

k Included in diet as protective antioxidant.

l Arachidonic acid gave no amelioration of pupal eclosion failure and retarded larval growth, the latter probably because of developing rancidity in diet.

m Suggested that linoleic and linolenic acids have separate functions, on basis of differences in growth/dosage curves.

Table 3 (continued)
LIPID GROWTH FACTORS

	Polyunsaturated C18 fatty acids			Other fatty acids	Vitamin E	Vitamin A or β-carotene		Ref.
	Linoleic or linolenic	Linoleic	Linolenic			Growth	Vision	
Lepidoptera (continued)								
Plodia interpunctella	R?n				R?n	R?n		30, 56—58
Sitotroga cerealella	R							9
Spodoptera exempta		R	R		R^o			23
Spodoptera littoralis (Prodenia litorina)		R	R					52
Tineola bisselliella	R							29
Trichoplusia ni	S	R	R					10, 39, 60
Orthoptera								
Acheta domesticus	R^p				R^p			54, 55
Blattella germanica	R^q							38
Locusta migratoria	R			r	R	r^r	R	15—17, 32
Melanoplus bivittatus	R				R	r^s		59
Schistocerca gregaria	R				R	r^t		15, 17
Hemiptera								
Myzus persicae^u	R				R	R		21, 22

n Early work showed fatty acids not required for normal adult emergence though they slightly increased growth rate, nor was tocopherol required for one generation,[30] recent studies over several consecutive generations indicate absolute requirements for a mixture of five fatty acids, tocopherol, and β-carotene, though the latter perhaps functioned as a critical phagostimulant rather than a required nutrient.[56-58]

o Deficiency during larval growth reduces egg viability over several generations, after which the surviving genetic strains continue with same low egg viability.

p Requirement for dietary oil or linoleic acid probably explicable in part by subsequent finding that vitamin E is essential for reproduction, in particular for sperm viability. Various fatty acids or methyl esters thereof are toxic or have indirect pheromone-like effects on growth.

q Deficiency during larval growth shows first as poor egg production or as weak, abnormally pale second generation larvae.

r Arachidonic acid slightly improved growth but could not avert abnormalities at the final molt.[16] Carotene was necessary for normal internal and external pigmentation, and deficiency imposed over two generations retarded growth in the second.[17] Vitamin A could not be detected in eyes of locusts reared on carotene-deficient diet, though present in eyes of locusts reared normally.[32]

s No effect on growth (one generation), but pigmentation abnormal.

t Pigmentation and growth affected as for Footnote r.

u Fifty generations reared in total absence of dietary lipid and similar results obtained with other aphids;[21,22] all aphids have intracellular symbiotes presumed to supply essential lipids.

Table 3 (continued)

REFERENCES

1. Barlow, J. S., *Can. J. Zool.,* 44, 775–779, 1966.
2. Barnett, J. W. and Berger, R. S., *Ann. Entomol. Soc. Am.,* 63, 917–924, 1970.
3. Brammer, J. D. and White, R. H., *Science,* 163, 821–823, 1969.
4. Bridges, R. G., *J. Insect Physiol.,* 17, 881–895, 1971.
5. Brookes, V. J. and Fraenkel, G., *Physiol. Zool.,* 31, 208–223, 1958.
6. Brust, M. and Fraenkel, G., *Physiol. Zool.,* 28, 186–204, 1955.
7. Carlson, S. D., Steeves, H. R., Vandeberg, J. S., and Robbins, W. E., *Science,* 158, 268–270, 1967.
8. Cheldelin, V. H. and Newburgh, R. W., *Ann. N.Y. Acad. Sci.,* 77, 373–383, 1959.
9. Chippendale, G. M., *J. Insect Physiol.,* 17, 2169–2177, 1971.
10. Chippendale, G. M., Beck, S. D., and Strong, F. M., *Nature,* 204, 710–711, 1964.
11. Chippendale, G. M. and Reddy, G. P. V., *J. Insect Physiol.,* 18, 305–316, 1972.
12. Clark, R. M., *J. Insect Physiol.,* 17, 1593–1598, 1971.
13. Cohen, E., *Entomol. Exp. Appl.,* 17, 433–438, 1974.
14. Cohen, C. F. and Barker, R. J., *J. Cell. Comp. Physiol.,* 62, 43–47, 1963.
15. Dadd, R. H., *J. Insect Physiol.,* 4, 319–347, 1960.
16. Dadd, R. H., *J. Insect Physiol.,* 6, 126–145, 1961.
17. Dadd, R. H., *Bull. Entomol. Res.,* 52, 63–81, 1961.
18. Dadd, R. H., *J. Insect Physiol.,* 10, 161–178, 1964.
19. Dadd, R. H., *J. Insect Physiol.,* 12, 1479–1492, 1966.
20. Dadd, R. H., unpublished observations.
21. Dadd, R. H. and Krieger, D. L., *J. Econ. Entomol.,* 60, 1512–1514, 1967.
22. Dadd, R. H. and Mittler, T. E., *Experientia,* 22, 832, 1966.
23. David, W. A. L. and Ellaby, S., *Entomol. Exp. Appl.,* 18, 269–280, 1975.
24. Davis, G. R. F., *Rev. Can. Biol.,* 26, 119–124, 1967.
25. Davis, G. R. F. and Sosulski, F. W., *Arch. Int. Physiol. Biochem.,* 81, 495–500, 1973.
26. Earle, N. W., Slatten, B., and Burks, M. L., *J. Insect Physiol.,* 13, 187–200, 1967.
27. Fast, P. G., *Prog. Chem. Fats Other Lipids,* 11, 181–242, 1970.
28. Fraenkel, G. and Blewett, M., *J. Exp. Biol.,* 20, 28–34, 1943.
29. Fraenkel, G. and Blewett, M., *J. Exp. Biol.,* 22, 156–161, 1946.
30. Fraenkel, G. and Blewett, M., *J. Exp. Biol.,* 22, 172–190, 1946.
31. Fraenkel, G. and Blewett, M., *Biochem. J.,* 41, 475–478, 1947.
32. Fisher, L. R. and Goldie, E. H., in *Progress in Photobiology, Proc. 3rd Int. Congr. Photobiol.,* Christensen, C. and Buchmann, B., Eds., Elsevier, Amsterdam, 1961, 153–154.
33. Gilbert, L. I., *Adv. Insect Physiol.,* 4, 69–211, 1967.
34. Gingrich, R. E., *Ann. Entomol. Soc. Am.,* 57, 351–360, 1964.
35. Golberg, L. and DeMeillon, B., *Biochem. J.,* 43, 372–379, 1948.
36. Goldsmith, T. H., Barker, R. J., and Cohen, C. F., *Science,* 146, 65–67, 1964.
37. Goldsmith, T. H. and Fernandez, H. R., in *The Functional Organization of the Compound Eye,* Bernhard, C. G., Ed., Pergamon, New York, 1966, 125–143.
38. Gordon, H. T., *Ann. N.Y. Acad. Sci.,* 77, 290–351, 1959.
39. Grau, P. A. and Terriere, L. C., *J. Insect Physiol.,* 17, 1637–1649, 1971.
40. Hirano, C., *Jpn. J. Appl. Entomol. Zool.,* 7, 59–62, 1963.
41. Horie, Y., Nakasone, S., and Ito, T., *J. Insect Physiol.,* 14, 971–981, 1968.
42. House, H. L. and Barlow, J. S., *J. Nutr.,* 72, 409–414, 1960.
43. House, H. L., *J. Insect Physiol.,* 11, 1039–1045, 1965.
44. House, H. L., *J. Insect Physiol.,* 12, 409–417, 1966.
45. Ishii, S. and Urushibara, H., *Bull. Nat. Inst. Agric. Sci. Tokyo C,* 4, 109–133, 1954.
46. Ito, T. and Nakasone, S., *Bull. Seric. Exp. Stn. Tokyo,* 20, 375–391, 1966.
47. Ito, T. and Nakasone, S., *Bull. Seric. Exp. Stn. Tokyo,* 23, 295–311, 1969.
48. Keith, A. D., *Comp. Biochem. Physiol.,* 21, 587–600, 1967.
49. Labremont, E. N., Stein, C. I., and Bennett, A. E., *Comp. Biochem. Physiol.,* 16, 289–302, 1965.
50. Leclercq, J. and DeBast, D., *Arch. Nutr. Aliment.,* 19, 19–25, 1965.
51. Levinson, Z. H. and Bergman, E. D., *Biochem. J.,* 65, 254–260, 1957.
52. Levinson, Z. H. and Navon, A., *J. Insect Physiol.,* 15, 591–595, 1969.
53. Levinson, Z. H., Barelkovsky, J., and Bar Ilan, A. R., *J. Stored Prod. Res.,* 3, 345–352, 1967.
54. McFarlane, J. E., Nielson, B., and Ghouri, A. S. K., *Can. J. Zool.,* 37, 913–916, 1959.
55. Meikle, J. E. S. and McFarlane, J. E., *Can. J. Zool.,* 43, 87–89, 1965.
56. Morere, J.-L., *Arch. Sci. Physiol.,* 24, 97–107, 1970.

Table 3 (continued)

57.   Morere, J.-L., *C.R. Acad. Sci.,* 272, 133–136, 1971.
58.   Morere, J.-L., *C.R. Acad. Sci.,* 272, 2229–2231, 1971.
59.   Nayar, J. K., *Can. J. Zool.,* 42, 11–22, 1964.
60.   Nelson, D. R. and Sukkestad, D. R., *J. Insect Physiol.,* 14, 293–300, 1968.
61.   Rock, E. C., *J. Econ. Entomol.,* 60, 1002–1005, 1967.
62.   Rock, E. C., Patton, R. L., and Glass, E. H., *J. Insect Physiol.,* 11, 91–101, 1965.
62a. Sang, J. H., *J. Exp. Biol.,* 33, 45–72, 1956.
63.   Silverman, P. H. and Levinson, Z. H., *Biochem. J.,* 58, 291–294, 1954.
64.   Sridhara, S. and Bhat, J. V., *Biochem. J.,* 94, 120–133, 1964.
65.   Stephen, W. F. and Gilbert, L. I., *J. Insect Physiol.,* 15, 1833–1854, 1969.
66.   Strong, F. E., *Science,* 140, 983–984, 1963.
67.   Tamaki, Y., *Jpn. J. Appl. Entomol. Zool.,* 5, 58–63, 1961.
68.   Thompson, S. N., *J. Insect Physiol.,* 20, 1515–1528, 1974.
69.   Vanderzant, E. S., *Ann. Entomol. Soc. Am.,* 61, 120–125, 1968.
70.   Vanderzant, E. S., Kerur, D., and Reiser, R., *J. Econ. Entomol.,* 49, 454–458, 1957.
71.   Vanderzant, E. S. and Richardson, C. D., *J. Insect Physiol.,* 10, 267–272, 1964.
72.   Yazgan, S., *J. Insect Physiol.,* 18, 2123–2141, 1972.
73.   Zimmerman, W. F. and Goldsmith, T. H., *Science,* 171, 1167–1169, 1971.

Table 4
STEROL UTILIZATION

Dietary sterol has been found essential for all insects studied nutritionally. The few apparent exceptions are among those species normally associated with symbiotic microorganisms, which probably synthesize sterol that can be used by the host insect.[13,21,43,45] With a few doubtful exceptions mentioned in footnotes, all steroids utilizable by insects are C27 — C29 3β-hydroxysterols. Utilizability is generally unimpaired by esterification with short-chain fatty acids such as acetate; esterification with long-chain fatty acids or benzoate may or not affect utilizability, but replacement of the 3β-hydroxyl group by Cl, Br, NH_2, oxymethyl, etc., prevents utilization. Sterol acetates are frequently used rather than plain sterols and are not distinguished in the table. Substances that have consistently failed to substitute for cholesterol and therefore are not mentioned in the table include mevalonate, farnesol, squalene, vertebrate steroid hormones and derivatives, bile acids, calciferols, and ketosterols. Cholesterol, the characteristic sterol of animal tissues, was until recently thought to satisfy completely the dietary sterol needs of all insects, but this view is no longer tenable since two species are now known for which dietary Δ_7 sterols are essential for continued development.[10,29] There is also evidence that some plant-feeding insects grow better with[31-33,39,48,59] or even require[9,19,47] plant sterols, notably β-sitosterol, rather than cholesterol; however, it is uncertain whether this is a strictly nutritional phenomenon since β-sitosterol is known also to be a phagostimulant in some of these cases. Since all insects require sterol, and most contain substantial tissue cholesterol even when their food contains phytosterols principally or exclusively, interest centers on the range of dietary sterols that can be utilized in particular cases, their structural affinities, and how they are variously converted to the tissue sterols of ultimate physiological importance. Several recent reviews[13,49,52,54,56] may be consulted for this burgeoning information on sterol conversions and metabolism. Here we note that determination of utilizability for insects is complicated by the fact that sterols are physiologically involved in at least two distinguishable functions: first, as relatively nonspecific structural components of membrane lipoproteins, a function that accounts for the bulk of sterol utilized; second, in a more sterol-specific "metabolic" function to which only a small fraction of the total sterol requirement is accountable. The "metabolic" function is thought to involve the need for sterol as precursor of the ecdysone moulting hormones, but there is evidence also for a sterol-specific membrane function.[13,35] With the few exceptions noted above, cholesterol subserves all these functions in most insects studied. In some cases other dietary sterols are equally well utilized for the nonspecific structural function but are nonutilizable for the specific metabolic and/or microstructural functions; they are then designated "sparing sterols" relative to the sterol that subserves all functions, and can be identified as such if their use alone gives poor growth but supports good growth when augmented with subliminal amounts of the wholly adequate sterol which, alone, would allow no growth. Since a well-utilized sparing sterol might support little growth in the total absence of the wholly adequate sterol which it can spare, experimentally it might appear nonutilized unless tested in a way specifically designed to reveal sparing activity, and this has generally not been done. It cannot therefore be unequivocally concluded that a sterol is not utilizable, at least as a sparing sterol, on the sole basis of negative results in tests using it singly. Conversely, it will also be evident that a sparing sterol may appear to be wholly adequate if it (or other components of a basal diet) carries, as an impurity, a small amount of the wholly adequate sterol which it spares. The problem of impure chemicals, common to all areas of nutrition involving micronutrients, is particularly acute with sterols; most commercial samples contain sterols other than those denominated (e.g., sitosterol commonly includes up to 7% campesterol), a difficulty particularly likely to have been intrusive in historically earlier work, and unsaturated sterols such as 7-dehydrocholesterol are notably unstable unless stored under special conditions. The uncertainty which consequently attaches to many attributions of utilizability or otherwise is well exemplified by the disagreements regarding cholestanol and 7-dehydrocholesterol documented by successive studies of *Dermestes vulpinus* (see Footnote j). Where different studies conflict, or doubt seems otherwise warranted, question marks are placed after symbols used in the table. Symbols are as follows:

U = well utilized for all sterol functions.

Ψ = apparently not utilized at all.

u = partially utilized, but no distinction as to whether in a sparing or general function, except as indicated in footnotes.

Table 4 (continued)
STEROL UTILIZATION

	Cholestanol	Cholesterol	7-Dehydro-cholesterol	Ergosterol	Zymosterol	Desmosterol	Campesterol	β-Sitosterol	Stigmasterol	Others	Ref.
Diptera											
Aedes aegypti	u	U	u	U	u?[a]			U	U	[a]	26
Cochliomyia hominivorax	Ψ	U	Ψ					Ψ			25
Culex pipiens	u[b]	U	u?[b]	Ψ		u	u	u	Ψ	[b]	18
Drosophila melanogaster		U	Ψ[c]	U		Ψ[c]		U	U	[c]	15,16
Drosophila pachea		Ψ	Ψ					Ψ	Ψ	[d]	29
Musca domestica	u[e]	U				u	u[e]	u[e]		[e]	20,34,41,55
Musca vicina	u	U	u	u				U?[f]	u	[f]	3,38
Phormia regina		U		u	Ψ			U			4
Sarcophaga bullata	u	U	U	Ψ				U	Ψ		27
Siphonaptera											
Xenopsylla cheopis		U		Ψ				U	u?[g]		46

[a] Zymosterol apparently utilized only when tested in certain basal diets. Of several other steroids tested, only coprosterol considered well utilized.

[b] With subliminal cholesterol, cholestanol acts as sparing sterol, as do β-sitosterol, campesterol, desmosterol, and lathosterol; with cholestanol as main sterol, 7-dehydrocholesterol allows good development (i.e., it appears to substitute for subliminal cholesterol), but this conclusion would be invalidated if cholesterol was a substantial impurity of the 7-dehydrocholesterol.

[c] Ostreasterol well utilized; 7-dehydrocholesterol, desmosterol, lathosterol, and 25-norcholesterol partially utilized by one mutant strain but not another. Several other steroids not utilized, and both utilized and nonutilized sterols had diverse effects on the expression of tumors in a tumor-prone mutant.[16]

[d] This monophagous species requires achottanol (probably Δ[7]-stigmastenol), the sterol of its host plant, senita cactus; complete development occurred also with Δ[7]-cholestenol (?lathosterol) and Δ[5,7]-cholestadienol (?7-dehydrocholesterol).

[e] Larval growth is normal with campesterol and β-sitosterol but resulting adults cannot produce viable eggs without cholesterol, which is preferentially absorbed if in the diet; the plant sterols are therefore sparing sterols, as is desmosterol, and, to a lesser extent, cholestanol.

[f] Many other steroids found nonutilizable; cholest-4-en-3β-ol was partially utilized; egg production not examined, so, by analogy with *Musca domestica,* β-sitosterol may be only a sparing sterol.

[g] Experimental data inconsistent for stigmasterol.

Table 4 (continued)
STEROL UTILIZATION

Coleoptera

	Cholestanol	Cholesterol	7-Dehydro-cholesterol	Ergosterol	Zymosterol	Desmosterol	Campesterol	β-Sitosterol	Stigmasterol	Others	Ref.
Anthonomus grandis		U						U	U		58
Callosobruchus chinensis	U	U						U	U	h	30
Dermestes maculatus (*Dermestes vulpinus*)	ψ?[j]	U	U?[j]	ψ	ψ	ψ?[j]		ψ?[j]	ψ	j	5,12–14,23, 37,55
Hylobius pales		u[k]	ψ	ψ				U	u	k	11,47
Lasioderma serricorne	u	U	U	U	u?[l]			U			22,40,44,45
Oryzaephilus surinamensis (*Silvanus surinamensis*)	u	U	u?	u	u?[l]			U			23,40
Ptinus tectus	u	U	U	U	ψ			u			23
Scolytus multistriatus	ψ	U		U				U	U		24
Sitophilus granarius (lacks symbiotes)	u	U	u	u				U			2
Sitophilus oryzae (with symbiotes)	U	U	U	U				U			2
Stegobium paniceum (*Sitodrepa panicea*)	u	U	U	U	ψ			U			22,40
Tenebrio molitor		U		U				U			36
Tribolium confusum	u	U	U	U	u?[l]	U	U	U	U		22,40,54,55
Trogoderma granarium	U	U	U	U				U	U	m	1,51
Xyleborus ferrugineus		u[n]	U	U						n	10,43

h Tetrahydrostigmasterol (stigmastanol) utilized; slight utilization claimed for epicholesterol doubtful.

j Early studies indicated cholesterol and 7-dehydrocholesterol to be fully utilized and all plant sterols nonutilized;[23,37] subsequently shown that cholestanol and other sterols had sparing action;[12] later claimed that very pure 7-dehydrocholesterol not utilized;[13,14] more recently shown that β-sitosterol, but not stigmasterol or ergosterol, partially utilizable,[5] and that 27-norcholesterol and 22-dehydrocholesterol well and partially utilized, respectively.[3] Most recently, desmosterol shown to be well utilized as sparing sterol.[55] One wonders whether these conflicting conclusions stem from a more widespread sparing ability than earlier recognized, perhaps dependent for expression on different levels of cholesterol reserves carried over via the egg to experimental larvae.

k A phagostimulant effect may be involved in the good utilization of β-sitosterol and the relatively poor utilization of cholesterol; lanosterol also not utilized.

l A small effect of zymosterol possibly due to 7-dehydrocholesterol impurities.

m No utilization of many other steroids tested.

n Ergosterol (in nature supplied by symbiotic fungus[43]) or 7-dehydrocholesterol required for pupation; cholesterol and lanosterol allow larval growth but not pupation and are thus sparing sterols.

Table 4 (continued)
STEROL UTILIZATION

	Cholestanol	Cholesterol	7-Dehydro-cholesterol	Ergosterol	Zymosterol	Desmosterol	Campesterol	β-Sitosterol	Stigmasterol	Others	Ref.
Lepidoptera											
Anagasta kuehniella (*Ephestia kuehniella*)	u?[o]	U	U	u	Ʋ			U			22
Argyrotaenia velutinana		U		u				U			50
Bombyx mori	u	U	?[p]	u			U	U[p]	U		31—33
Crambus trisectus		u[q]					U?[q]	U[q]			19
Diatraea grandiosella		U	u	Ʋ				U[r]	U		8,9
Ectomyelois ceratoniae		U						U			39
Manduca sexta (*Protoparce sexta*)		U				U	U	U	U	s	49,53
Pectinophora gossypiella		U	U					U	U		59
Sitotraga cerealella	U	U	U	U				U	U		7
Orthoptera											
Acheta domestica (*Gryllulus domesticus*)		U	U					U	U		6
Blattella germanica	U	U	Ʋ	Ʋ		U		U	U		28,42,55
Locusta migratoria	U	U	Ʋ	Ʋ				U	Ʋ		17
Schistocerca gregaria	U	U	Ʋ	Ʋ				U	Ʋ		17

[o] Slight effect may have been due to impurities.

[p] β-Sitosterol is also a feeding stimulant for silkworm; diet containing 7-dehydrocholesterol not eaten, so uncertain whether utilizable or not.

[q] β-Sitosterol (actually with 7% campesterol impurity) may be a required phagostimulant.

[r] β-Sitosterol better than cholesterol, possibly because of phagostimulation.

[s] Fucasterol, 24-methylenecholesterol, brassicasterol, and dihydrobrassicasterol also fully utilizable.

REFERENCES

1.   Agarwal, H. C., *J. Insect Physiol.*, 16, 2023—2026, 1970.
2.   Baker, J. E., *Ann. Entomol. Soc. Am.*, 67, 591—594, 1974.
3.   Bergmann, E. D. and Levinson, Z. H., *J. Insect Physiol.*, 12, 77—81, 1966.
4.   Brust, M. and Fraenkel, G., *Physiol. Zool.*, 28, 186—204, 1955.
5.   Budowski, P., Ishaaya, I., and Katz, M., *J. Nutr.*, 91, 201—207, 1967.
6.   Chauvin, R., *C.R. Acad. Sci.*, 229, 902, 1949.
7.   Chippendale, G. M., *J. Insect Physiol.*, 17, 2169—2177, 1971.
8.   Chippendale, G. M. and Reddy, G. P. V., *J. Insect Physiol.*, 18, 305—316, 1972.
9.   Chippendale, G. M. and Reddy, G. P. V., *Experientia*, 28, 485—486, 1972.
10. Chu, H.-M., Norris, D. M., and Kok, L. T., *J. Insect Physiol.*, 16, 1379—1387, 1970.
11. Clark, E. W., *J. Econ. Entomol.*, 66, 841—843, 1973.
12. Clark, A. J. and Bloch, K., *J. Biol. Chem.*, 234, 2583—2588, 1959.
13. Clayton, R. B., *J. Lipid Res.*, 5, 3—19, 1964.
14. Clayton, R. B. and Bloch, K., *J. Biol. Chem.*, 238, 586—591, 1963.
15. Cooke, J. and Sang, J. H., *J. Insect Physiol.*, 16, 801—812, 1970.
16. Cooke, J. and Sang, J. H., *Genet. Res.*, 20, 317—329, 1972.
17. Dadd, R. H., *J. Insect Physiol.*, 5, 161—168, 1960.
18. Dadd, R. H., unpublished data.
19. Dupnik, T. D. and Kamm, J. A., *J. Econ. Entomol.*, 63, 1578—1581, 1970.
20. Dutky, R. C., Robbins, W. E., Shortino, T. J., Kaplanis, J. N., and Vroman, J. N., *J. Insect Physiol.*, 13, 1501—1510, 1967.

Table 4 (continued)

21. Ehrhardt, P., *Experientia*, 24, 82–83, 1968.
22. Fraenkel, G. and Blewett, M., *Biochem. J.*, 37, 692–695, 1943.
23. Fraenkel, G., Reid, J. A., and Blewett, M., *Biochem. J.*, 35, 712–720, 1941.
24. Galford, J. R., *J. Econ. Entomol.*, 65, 681–684, 1972.
25. Gingrich, R. E., *Ann. Entomol. Soc. Am.*, 57, 351–360, 1964.
26. Golberg, L. and DeMeillon, B., *Biochem. J.*, 43, 372–379, 1948.
27. Goodfellow, R. D., Barnes F. J., and Graham, W. S., *J. Insect Physiol.*, 17, 1625–1635, 1971.
28. Gordon, H. T., *Ann. N.Y. Acad. Sci.*, 77, 290–351, 1959.
29. Heed, W. B. and Kircher, H. W., *Science*, 149, 758–761, 1965.
30. Ishii, S., *Botyu Kagaku*, 16, 83–90, 1951.
31. Ito, T., *Bull. Seric. Exp. Stn. Tokyo*, 17, 91–117, 1961.
32. Ito, T. and Horie, Y., *Annot. Zool. Jpn.*, 39, 1–6, 1966.
33. Ito, T., Kawashima, K., Nakahara, M., Nakanishi, K., and Terahara, A., *J. Insect Physiol.*, 10, 225–238, 1964.
34. Kaplanis, T. N., Robbins, W. E., Monroe, R. E., Shortino, T. J., and Thompson, M. J., *J. Insect Physiol.*, 11, 251–258, 1965.
35. Lasser, N. L. and Clayton, R. B., *J. Lipid Res.*, 7, 413–421, 1966.
36. Leclercq, J., *Biochim. Biophys. Acta*, 2, 614–617, 1948.
37. Levinson, Z. H., in *Verh. 11th Int. Kongr. Entomol. Wien 1960, Band 3, Symp. 3 and 4* (published separately), Pavan, M. and Eisner, T., Eds., Instituto di Entomologia Agraria dell', Universita di Pavia, Italy, 1960, 145–153.
38. Levinson, Z. H. and Bergmann, E. D., *Biochem. J.*, 65, 254–260, 1957.
39. Levinson, Z. H. and Gothilf, S., *Riv. Parassitol.*, 26, 19–26, 1965.
40. Magis, N., *Bull. Ann. Soc. Entomol. Belg.*, 90, 49–58, 1954.
41. Monroe, R. E., Kaplanis, J. N., and Robbins, W. E., *Ann. Entomol. Soc. Am.*, 54, 537–539, 1961.
42. Noland, J. L., *Arch. Biochem. Biophys.*, 48, 370–379, 1954.
43. Norris, D. M., Baker, J. M., and Chu, H. M., *Ann. Entomol. Soc. Am.*, 62, 413–414, 1969.
44. Pant, N. C. and Fraenkel, G., *Biol. Bull.*, 107, 420–432, 1954.
45. Pant, N. C. and Kapoor, S., *Indian J. Entomol.*, 25, 311–315, 1963.
46. Pausch, R. D. and Fraenkel, G., *Physiol. Zool.*, 39, 202–222, 1966.
47. Richmond, J. A. and Thomas, H. A., *Ann. Entomol. Soc. Am.*, 68, 329–332, 1975.
48. Riddisford, L. M., *Science*, 160, 1461–1462, 1968.
49. Robbins, W. E., Kaplanis, J. N., Svoboda, J. A., and Thompson, M. J., *Annu. Rev. Entomol.*, 16, 53–72, 1971.
50. Rock, G. C., *Ann. Entomol. Soc. Am.*, 62, 611–613, 1969.
51. Sehgal, S. S. and Agarwal, H. C., *J. Insect Physiol.*, 19, 419–425, 1973.
52. Svoboda, J. A., Kaplanis, J. N., Robbins, W. E., and Thompson, M. J., *Annu. Rev. Entomol.*, 20, 205–220, 1975.
53. Svoboda, J. A. and Robbins, W. E., *Experientia*, 24, 1131–1139, 1968.
54. Svoboda, J. A., Robbins, W. E., Cohen, C. F., and Shortino, T. J., in *Insect and Mite Nutrition*, Rodriguez, J. G., Ed., North Holland, Amsterdam, 1972, 505–516.
55. Svoboda, J. A., Thompson, M. J., and Robbins, W. E., *Life Sci.*, 6, 395–404, 1967.
56. Svoboda, J. A., Thompson, M. J., Robbins, W. E., and Elden, T. C., *Lipids*, 10, 524–527, 1975.
57. Thompson, M. J., Kaplanis, J. N., and Vroman, H. E., *Steroids*, 5, 551–553, 1965.
58. Vanderzant, E. S., *J. Econ. Entomol.*, 56, 357–362, 1963.
59. Vanderzant, E. S. and Reiser, R., *J. Econ. Entomol.*, 49, 454–458, 1956.

Table 5
SUGAR AND CARBOHYDRATE UTILIZATION

Natural foods of insects nearly always contain some form of carbohydrate, and therefore a purified carbohydrate or a sugar is generally included in synthetic diets for particular species at concentrations characteristic of their normal foods. Probably most insects can utilize the common sugars, and for many species it has been shown that omission of sugar or digestible polysaccharide from synthetic diets is to some degree deleterious. However, since fat or protein is known to be readily interchangeable with carbohydrate as an energy source for a number of species (see examples in Table 3), it should not be too readily assumed that carbohydrate is a required nutrient unless the possibility can be excluded that deleterious effects of omission might be obviated by adjustments to the protein or fatty content of the diet, a possibility not often checked. For example, glucose improves the growth of larvae of *Aedes aegypti* in defined media, but to an extent that depends on the level of amino acids provided,[10] perhaps accounting for divergent earlier findings regarding sugar in diets for this species.[1,19,33,34,47,49] Nevertheless, it may reasonably be generalized that the majority of insects require dietary carbohydrate, if only to optimize growth rates. A few species are definitely known not to require carbohydrate for larval growth;[6,8,15,16,18,46] however, several of these are Diptera, which require sugar to survive as adults, so in terms of continued generation they too require carbohydrate. It is also worth noting that little information is available on the nutritional requirements of insects that are predaceous both as larvae and adults and which therefore might be thought among the most likely to have evolved an independence from carbohydrate.

With regard to insects that do require carbohydrate, most interest concerns the utilizability or otherwise of particular polysaccharides, sugars, and related substances such as sugar alcohols and glycosides. Utilization studies fall into two categories, dealing respectively with (a) the ability of simple sugar solutions to support optimal longevity in those adult insects, mainly Diptera, for which carbohydrate (and water) are the essential and only maintenance nutrients, and (b) the ability of sugars and polysaccharides, as components of complete diets, to support normal growth of species having a marked larval requirement for dietary carbohydrate. In the table below, these categories are listed separately, but the following remarks apply to both.

A general assumption underlying carbohydrate utilization studies is that only monosaccharides and related simple molecules can readily be absorbed from the gut, and that, once absorbed, only certain ones can subsequently serve as substrates for the metabolic transformations whereby energy is produced. Hence, it follows that if in survival or growth studies oligo- and polysaccharides prove to be utilized, suitable digestive enzymes must be secreted into the gut to cleave these compounds into their constituent simple sugars. Utilization data for suitable arrays of mono- and polysaccharides thus afford a basis for inferring what digestive carbohydrases an insect possesses; conversely, knowing the utilizability of monosaccharides, a knowledge of the enzymes that are secreted can confirm utilization findings for polysaccharides. Such complementary studies are available for many insects, and on the whole are mutually supportive. When not in accord, this may stem from inadequate analysis of the carbohydrase complement, but more often discrepancies arise from uncertainties in the determination of carbohydrate utilizability by dietary methods.

The main cause of uncertainty arises from the fact that particular sugars are often among the most potent phagostimulants of natural insect foods, and on them then depends the induction and maintenance of feeding. In such cases, nonphagostimulant carbohydrates, though perhaps fully utilizable, might fail to induce feeding and hence be erroneously judged not utilizable by the criteria of survival or growth if the actual amounts of diet ingested were not checked and taken into account. If a carbohydrate supports moderate to good survival or growth it must be both ingested and utilized, and (excluding possible errors from impurities) its designation as having nutritive value is unequivocal. Also, if survival or growth is not supported by a carbohydrate in food that is shown to be well ingested, then it is unequivocally of no nutritive value (or even toxic, if survival is worse than in its absence). But if the carbohydrate so strongly inhibits feeding that even in combination with a powerful feeding stimulant no food is ingested, then its nutritive value is indeterminate. In a few of the studies tabulated below this latter situation has specifically been demonstrated and taken into account, but more often the various possibilities that could result in apparent nonutilization are not distinguished. Available information bearing on these distinctions is given in the footnotes, but in many cases the designation "not utilized" may equally mean "diet not eaten."

Data for sugars and other carbohydrates of central interest that were tested in most studies are tabulated, and the many additional materials used in particular cases are relegated to footnotes. Except where the contrary is indicated, the D form of sugars, etc., is to be assumed. Symbols used have the following meanings:

U = well utilized.

u = poorly utilized (this could include fully utilized materials which afforded only suboptimal feeding stimulation).

Ʋ = Apparently not utilized.

Table 5 (continued)
SUGAR AND CARBOHYDRATE UTILIZATION

Diptera
Adult Maintenance Studies

	Sugar alcohols		Pentoses		Hexoses					Disaccharides						Tri-saccharides		Poly-saccharides				
	Sorbitol	Mannitol	Xylose	Arabinose, L	Glucose	Fructose	Galactose	Mannose	Sorbose, L	Sucrose	Maltose	Trehalose	Cellobiose	Melibiose	Lactose	Melezitose	Raffinose	Starch	Glycogen	Inulin	Others	Ref.
Aedes aegypti[a]	U	U	Ψ	Ψ	U	U	u	Ψ	Ψ	U	U	u	Ψ	u	Ψ	U	u[a]	Ψ	Ψ	Ψ	a	17,44
Culiseta inornata[b]	u	U	u	Ψ	U	U	Ψ	u	Ψ	U	U		Ψ	u	Ψ	U	u	Ψ[b]	Ψ	Ψ	b	45
Calliphora erythrocephala[c]	U	U	u	Ψ	U	U	Ψ	U	Ψ	U	U	U	Ψ	U	Ψ[c]	U	U	Ψ[c]	Ψ	Ψ	c	13,22
Drosophila melanogaster[d]	u	u	u	Ψ	U	U	u	U	Ψ	U	U	U	Ψ	U	U	U	U	Ψ[d]	Ψ	Ψ	d	23
Erioischia brassicae[e]	U	Ψ?	Ψ?	Ψ	U	U	U	U	Ψ	U	U	U	U	u	U	U	U	Ψ[e]	Ψ	Ψ	e	12
Musca domestica[f]	U	Ψ	Ψ	Ψ	U	U	U	U	Ψ	U	U	U	Ψ?[g]	U	U	U	U	U[f]	U	Ψ	f	17
Phormia regina[g]	U	Ψ?[g]	Ψ?[g]	Ψ?[g]	U	U	U	U	Ψ?[g]	U	U	U	Ψ?[g]	u	Ψ?[g]	U	U	U?	Ψ	Ψ	g	24
Sarcophaga bullata[h]	U	Ψ	Ψ	Ψ	U	U	U	U	Ψ	U	U	U	Ψ	U	U	U	U	U?	Ψ	Ψ		17

a   Phagostimulatory effectiveness of all materials was determined[44] and digestive carbohydrases surveyed.[17] Others: dulcitol, inositol, glycerol, D-ribose, L-rhamnose, α-methyl-glucoside and β-methyl-mannoside not utilized.

b   Phagostimulatory effectiveness of materials tested. Others: dulcitol, inositol, glycerol, α-methyl-glucoside, and xylan not utilized.

c   Qualitative assessment of degree of feeding taken into account in determining utilizability, and digestive carbohydrases surveyed.[13] Others: dulcitol, erythritol, inositol, glycerol, α- and β-methyl-glucosides, -galactosides, and -mannosides, and fucose not utilized; slight utilization of lactose ascribed to bacterial breakdown.

d   Extent of feeding observed qualitatively. Others: inositol slightly and glycerol well utilized; several other sugar alcohols, fucose, xylan, and a large number of diverse metabolites not utilized. Suggested that cellobiose, sorbose, erythritol, and all pentoses except xylose are repellent.

e   Toxicity and/or repellency of nonutilized materials determined in combination with sucrose. Others: dulcitol, erythritol, inositol, glycerol, ribose, D-arabinose, and rhamnose not utilized; all these, and also sorbose, cellobiose, and dextrin, were toxic or repellent.

f   Others: dulcitol, inositol, glycerol, ribose, and rhamnose not utilized. When tested in combination with sucrose, all nonutilized materials with the exception of dulcitol proved toxic or repellent to some extent. A carbohydrase survey made.

g   Phagostimulant effect of all materials determined by tarsal/proboscis extension test. Authors regard substances marked Ψ? as very slightly utilized, as survival was marginally better than with water alone. Poor utilization for melibiose may result from poor phagostimulation. Others: ribose, rhamnose, lyxose, D-arabinose, fucose, glycerol, inositol, dulcitol, and erythritols not, or very poorly, utilized; α-methyl-glucoside utilized.

h   Others: dulcitol, inositol, glycerol, ribose, rhamnose, and α-methyl-mannoside not utilized. When tested in combination with sucrose all nonutilized materials were inhibitory. Carbohydrases of gut were surveyed.

Table 5 (continued)
SUGAR AND CARBOHYDRATE UTILIZATION

	Sugar alcohols		Pentoses		Hexoses					Disaccharides						Tri-saccharides		Poly-saccharides				
	Sorbitol	Mannitol	Xylose	Arabinose, L.	Glucose	Fructose	Galactose	Mannose	Sorbose, L.	Sucrose	Maltose	Trehalose	Cellobiose	Melibiose	Lactose	Melezitose	Raffinose	Starch	Glycogen	Inulin	Others	Ref.
Hymenoptera																						
Apis mellifera[i]	U	U	U?	U?	U	U	U?	U?	U?	U	U	U	U?	U?	U?	U	U?	u?			j	25,51,52
Larval Growth																						
Coleoptera																						
Anthonomus grandis	u				U	U	u	u?	U?j	U	U	U	U	u	U	U	u	U	U?j	U?j	j	50
Oryzaephilus surinamensis[k]	U	u?k	U?	U?	U	U	U?	u	u?	U	U	U	u	u	u	U	u	U	u	U	k	36
Stegobium paniceum[l]	u	u	u	u	U	U	u	U?	u?	U	U	U	U?	u	u	U	u	U	U	U	l	36
Tenebrio molitor	U	U	U?	U?m	U	U	U?	U?	U?	U	U	U	U?	U?	u?	U	U	U	U	U?	m	14,35
Tribolium castaneum	u	U	U?	U?	U	u?o	u	U	U?	u	U	U	U	u	u	U	U	U	u	U?	n	42
Tribolium confusum[o]	u	U	U?	U?	U		u	U		U	U	U	U	u	u	U	U	U		u?o	o	4
Trogoderma granarium	U?	U?	U?	U?	U	u	u	U	u	U	U	U	U	U	u	U	U	u			p	43

i   References cite contradictory studies. Xylose, arabinose, gelatose, mannose, melibiose and lactose are toxic; most of these, and raffinose, are phago-inhibitory also.

j   Others: ribose perhaps slightly utilized. Metabolic studies of conversion of ingested sugars to trehalose[40,41] indicate no utilization of sorbose and slight utilization of galactose, xylose, and sorbitol.

k   No tests or observations to indicate whether nonutilized substances might be feeding inhibitors or toxic. Experiments lacked a control diet without carbohydrate, making the designation "not utilized" uncertain. Others: dulcitol, ribose, rhamnose, and α-methyl-glucoside and -mannoside not utilized.

l   No control lacking carbohydrate and no tests for feeding inhibition. α-Methylmannoside and xylan utilized; dulcitol and rhamnose perhaps slightly utilized; ribose and α-methyl-glucoside not utilized.

m   Others: dulcitol, ribose, D-arabinose, rhamnose, α-methyl-glucoside and -mannose not utilized. The foregoing and also xylose, galactose, mannose, and sorbose were considered feeding inhibitors.

n   Dulcitol, rhamnose, and *soluble* starch not utilized.

o   Experiments lacked control diet without carbohydrate, making designations of no utilization uncertain. Others: rhamnose perhaps slightly utilized; fructose gave negative results because its hygroscopicity made diet wet.

p   Others: dulcitol and rhamnose not utilized.

Table 5 (continued)
SUGAR AND CARBOHYDRATE UTILIZATION

	Sugar alcohols		Pentoses		Hexoses					Disaccharides						Tri-saccharides		Poly-saccharides				
	Sorbitol	Mannitol	Xylose	Arabinose, L	Glucose	Fructose	Galactose	Mannose	Sorbose, L	Sucrose	Maltose	Trehalose	Cellobiose	Melibiose	Lactose	Melezitose	Raffinose	Starch	Glycogen	Inulin	Others	Ref.
Lepidoptera																						
Bombyx mori[r]	U	U	u?	Ʉ	U	U	u	U?		U	U	U	U	Ʉ	U	U	U	U?	U?	Ʉ	r	28,29,31,32
Chilo suppressalis[s]			Ʉ		U	U	Ʉ?	Ʉ?	Ʉ	u?	U	u	U	Ʉ	Ʉ	u	u	u	U	Ʉ		26
Orthoptera																						
Blattella germanica	U	u	Ʉ	u	U	U	U	U	Ʉ	U	U	U	U	U	U	U	U	U			t	20,20a
Locusta migratoria	u	u	Ʉ	Ʉ	U	U	Ʉ	U	Ʉu	U	U	U	U	U	U	U	U	U	U		u	9
Schistocerca gregaria	U	U	Ʉ?	Ʉ?	U	U	u	u		U	U	U	U	U	U	U	U	U	U		v	9
Hemiptera																						
Myzus persicae[w,x]	U		Ʉ?x	Ʉ?x	U	U	U	Ʉ?x	Ʉ?x	U	U	U	U	U	Ʉ	U	U				x	38
Oncopeltus fasciatus[w,y]	U	U	Ʉ?	Ʉ?	U	U	u	u	Ʉ	U	U	u	U	u	u?	U	U	?			y	3,21

r All sugars except sucrose, fructose, and raffinose are very poor phagostimulants,[30] so utilization determined in terms of (a) survival after force-feeding solution of carbohydrates[31,32] and (b) conversion of force-fed solution of sugars to glycogen and/or trehalose by starved larvae depleted of these metabolites.[28,-9] Digestive carbohydrates were surveyed.[27] Others: dulcitol, inositol, ribose, rhamnose, and α-methyl-glucoside and -mannoside not utilized.

s Substances were tested in synthetic diet on which growth was very poor even with the best sugar, fructose. Since there was no control diet lacking carbohydrate it is difficult to distinguish between nutritive nonutilization and feeding inhibition. Others: ribose and rhamnose not utilized.

t Others: α-methyl-glucoside slightly utilized; glycerol and turanose well utilized; dulcitol, inositol, ribose, D-arabinose, and rhamnose not utilized.

u Others: inositol slightly utilized; glycerol, ribose, rhamnose, and α-methyl-mannoside not utilized. At high concentration (26% of diet rather than 13%) xylose, rhamnose, and sorbose inhibit growth. Digestive carbohydrases surveyed in great detail.[39]

v Others: glycerol, ribose, and rhamnose not utilized; α-methyl-mannoside slightly utilized and inositol moderately well utilized. Xylose, rhamnose, and sorbose inhibit growth at high concentration. Digestive carbohydrases surveyed.[11]

w Because certain sugars are of prime importance in maintaining normal feeding in phytophagous Hemiptera, survival and growth data without ingestion studies are poor indicators of the nutritive value of carbohydrates, so some studies are omitted for this reason.[2,37] For the two species tabulated, correlated ingestion and growth data using mixed (stimulant and nonstimulant) sugars allow reasonable confidence in the designations listed.

x Others: ribose and rhamnose not utilized. Designations are uncertain for all pentoses, mannose, and sorbose because feeding inhibition could not be overcome by suboptimal levels of sucrose.

y Others: gentiobiose, turanose, and α-methyl-glycoside moderately well utilized; rhamnose, β-methyl-glucoside, and fucose not utilized. Digestive carbohydrases surveyed.[5,48]

Table 5 (continued)

REFERENCES

1. Akov, S., *J. Insect Physiol.,* 8, 319–335, 1962.
2. Auclair, J. L., *J. Insect Physiol.,* 13, 1247–1268, 1967.
3. Auclair, J. L., Srivastava, S. T., and Srivastava, P. N., *Entomol. Exp. Appl.,* 16, 525–540, 1973.
4. Bernard, R. and Lemonde, A., *Rev. Can. Biol.,* 8, 498–503, 1949.
5. Bongers, J., *Z. Vgl. Physiol.,* 70, 382–400, 1970.
6. Brookes, V. J. and Fraenkel, G., *Physiol. Zool.,* 31, 208–223, 1958.
7. Chippendale, G. M., *J. Nutr.,* 102, 187–194, 1972.
8. Cheldelin, V. H. and Newburgh, R. W., *Ann. N.Y. Acad. Sci.,* 77, 373–383, 1959.
9. Dadd, R. H., *J. Insect Physiol.,* 5, 301–316, 1960.
10. Dadd, R. H. and Sneller, V. P., unpublished observations.
11. Evans, W. A. L. and Payne, D. W., *J. Insect Physiol.,* 10, 657–674, 1964.
12. Finch, S. and Coaker, T. H., *Entomol. Exp. Appl.,* 12, 441–453, 1969.
13. Fraenkel, G., *J. Exp. Biol.,* 17, 18–29, 1940.
14. Fraenkel, G., *J. Cell. Comp. Physiol.,* 45, 393–408, 1955.
15. Fraenkel, G. and Blewett, M., *J. Exp. Biol.,* 20, 28–34, 1943.
16. Fraenkel, G. and Blewett, M., *J. Exp. Biol.,* 22, 156–161, 1946.
17. Galun, R. and Fraenkel, G., *J. Cell. Comp. Physiol.,* 50, 1–23, 1957.
18. Gingrich, R. E., *Ann. Entomol. Soc. Am.,* 57, 351–360, 1964.
19. Golberg, L. and DeMeillon, B., *Biochem. J.,* 43, 379–387, 1948.
20. Gordon, H. T., *Ann. N.Y. Acad. Sci.,* 77, 290–351, 1969.
20a. Gordon, H. T., *J. Insect Physiol.,* 14, 41–52, 1968.
21. Gordon, H. T., *Entomol. Exp. Appl.,* 17, 450–451, 1974.
22. Haslinger, F., *Z. Vgl. Physiol.,* 22, 614–640, 1936.
23. Hassett, C. C., *Biol. Bull.,* 95, 114–123, 1948.
24. Hassett, C. C., Dethier, V. G., and Gans, J., *Biol. Bull.,* 99, 446–453, 1950.
25. Haydak, M. H., *Annu. Rev. Entomol.,* 15, 143–156, 1970.
26. Hirano, C. and Ishii, S., *Bull. Nat. Inst. Agric. Sci. Jpn. C,* 7, 89–99, 1957.
27. Horie, Y., *Bull. Seric. Exp. Stn. Tokyo,* 15, 365–382, 1959.
28. Horie, Y., *Nature,* 188, 583–584, 1960.
29. Horie, Y., *Bull. Seric. Exp. Stn. Tokyo,* 16, 287–309, 1961.
30. Ito, T., *J. Insect Physiol.,* 5, 95–107, 1960.
31. Ito, T., *Nature,* 187, 527, 1960.
32. Ito, T. and Tanaka, M., *Bull. Seric. Exp. Stn. Tokyo,* 16, 267–285, 1961.
33. Lang, C. A., Basch, K. J., and Storey, R. S., *J. Nutr.,* 102, 1057–1066, 1972.
34. Lea, O. A. and DeLong, D., in *Proc. 10th Int. Congr. Entomol. Montreal, 1956,* Vol. 2, Becker, E. C., Ed., 1958, 299–302.
35. Leclercq, J., *Arch. Int. Physiol.,* 56, 130–133, 1948.
36. Lemonde, A. and Bernard, R., *Nat. Can.,* 80, 125–142, 1953.
37. Mitsuhashi, J. and Koyama, K., *Appl. Entomol. Zool.,* 4, 185–193, 1969.
38. Mittler, T. E., Dadd, R. H., and Daniels, S. C., *J. Insect Physiol.,* 16, 1873–1890, 1970.
39. Morgan, M. R. J., *J. Insect Physiol.,* 21, 1045–1053, 1975.
40. Nettles, W. C. and Burks, M. L., *J. Insect Physiol.,* 17, 1615–1623, 1971.
41. Nettles, W. C. and Burks, M. L., *J. Insect Physiol.,* 19, 1677–1687, 1973.
42. Pant, N. C. and Gabrani, K., *Indian J. Entomol.,* 25, 172–174, 1963.
43. Pant, N. C. and Oberoi, N. K., *Experientia,* 14, 71–72, 1958.
44. Salama, H. S., *J. Insect Physiol.,* 12, 1051–1060, 1966.
45. Salama, H. S., *Mosq. News,* 27, 32–35, 1967.
46. Sedee, P. D. J. W., *Entomol. Exp. Appl.,* 1, 38–40, 1958.
47. Singh, D. R. P. and Brown, A. W. A., *J. Insect Physiol.,* 1, 199–220, 1957.
48. Srivastava, S. T. and Auclair, J. L., *Entomol. Exp. Appl.,* 16, 301–304, 1973.
49. Trager, W., *J. Biol. Chem.,* 176, 1211–1223, 1948.
50. Vanderzant, E. S., *J. Insect Physiol.,* 11, 659–670, 1965.
51. Barker, R. J. and Lehner, Y., *J. Exp. Zool.,* 187, 277–286, 1974.
52. Barker, R. J. and Lehner, Y., *J. Exp. Zool.,* 188, 157–164, 1974.

Table 6
MINERAL REQUIREMENTS AND WATER

This is the least investigated area of insect nutrition. Because of difficulties in excluding particular inorganic ions from synthetic diets without grossly altering the balance of those remaining, few workers have tried to determine specific requirements by deletion experiments. Provision of an optimal level of a salt mixture is the most that is usually done. Except for potassium, magnesium, and phosphate, other inorganic nutrients probably are needed only in trace amounts by most insects, and it is therefore extremely difficult to avoid the masking of nominal deletions by contaminants in other dietary constituents, or by reserves carried over from the egg.

Water is always assumed to be required. In the plant or animal tissues which provide the natural food of most insects, water is usually the major component, and is therefore a major component in the formulation of most synthetic diets, or is supplied separately *ad libitum.* However, some insects which in nature feed on apparently dry foods or have been reared successfully on dry, powdery synthetic diets probably depend upon the hygroscopicity of their food to provide cryptic water,[6,32] or upon metabolic water resulting from the oxidation of fat and carbohydrate.[16,25]

On general principles it can be assumed that virtually all the inorganic entities for basic animal physiological functions would be found essential for insects (perhaps excepting iodine, since thyroid function is peculiar to the vertebrate line), so long as covert exogenous sources were excluded, and deficient dietary regimes imposed over an experimental span long enough to exhaust reserves passed on maternally.

Therefore, with mineral nutrients more than any others, experimental findings indicating nonrequirement are significant only in an operational sense. Because its theoretical connotations would be misleading, the designation "not required" is avoided in the table below. Symbols used have the following meanings:

R = required.
t = tested, but without effect.

Table 6 (continued)
MINERAL REQUIREMENTS AND WATER

	S[a]	Cl	P	K	Mg	Ca	Na	Fe	Zn	Mn	Cu	Others	Ref.
Diptera													
Agria housei (*Pseudosarcophaga affinis*)			R	R					t	t	t	t[b]	21,22
Culex pipiens			R	R	R	R[c]	t	R	R	R?	R?	t[d]	8
Drosophila melanogaster		R	R	R	R	t		R	t	t			38
Siphonaptera													
Xenopsylla cheopis	t	R	t	R	R	t		R	t	t	t	t[e]	34
Coleoptera													
Anthonomus grandis					R	t	t	t	t	t	t	t[f]	40
Tenebrio molitor			R						R				15,26
Tribolium confusum			R	R	R	R?[g]	t	R	R	R	t	t[g]	4,23,27, 29,30,33
Lepidoptera													
Argyrotaenia velutinana								R	R		R		37
Bombyx mori		R	R	R	R	R		R	R	R	R		20,24
Hemiptera													
Acyrthosiphon pisum[i]	R		R	R	R	R		R	R?	R?	R?	t&R?[h]	1,2,28 35,36
Myzus persicae[i]	R		R	R	R	R		R	R	R	R		7,9,11 31
Neomyzus circumflexus[i]	R				R			R	R	R	R		13,14
Orthoptera													
Blattella germanica				R	R	R?[j]	t		R?[j]	R?[j]	R		3,18,19

[a] As a component of the amino acids methionine and cystine, sulfur is essential to all insects; entries in this column indicate an additional requirement for sulfate.

[b] Iodine, cobalt.

[c] Several other mosquito larvae are very sensitive to calcium deficiency.[17,39]

[d] Cobalt.

[e] Aluminum, boron, cobalt, fluorine.

[f] Molybdenum, cobalt, iodine; phosphate and potassium not tested.

[g] Findings of different workers not in accord; cadmium toxic.

[h] Suggestion that molybdenum and boron might have had beneficial effect on reproduction, but no effect with cobalt.

[i] The sensitivity of these and other aphids[5,10,12] to trace metals may be related to requirements of their symbiotes for the metals.

[j] Requirements for calcium, zinc, and manganese involve a delicate balance which affects the viability of the intracellular symbiotes upon which survival also depends.[3]

Table 6 (continued)

REFERENCES

1. Akey, D. H. and Beck, S. D., *J. Insect Physiol.*, 18, 1901–1914, 1972.
2. Auclair, J. L. and Srivastava, P. N., *Can. Entomol.*, 104, 927–936, 1972.
3. Brooks, M. A., *Proc. Helminthol. Soc. Wash.*, 27, 212–220, 1960.
4. Chaudhary, K. D. and Lemonde, A., *Can. J. Zool.*, 40, 375–380, 1962.
5. Cress, D. C. and Chada, H. L., *Ann. Entomol. Soc. Am.*, 64, 1240–1244, 1971.
6. Dadd, R. H., *J. Insect Physiol.*, 10, 161–178, 1964.
7. Dadd, R. H., *J. Insect Physiol.*, 13, 763–778, 1967.
8. Dadd, R. H., unpublished.
9. Dadd, R. H., *Bull. Entomol. Soc. Am.*, 14, 22–26, 1968.
10. Dadd, R. H. and Krieger, D. L., *J. Econ. Entomol.*, 60, 1512–1514, 1967.
11. Dadd, R. H. and Mittler, T. E., *J. Insect Physiol.*, 11, 717–743, 1965.
12. Ehrhardt, P., *Z. Vgl. Physiol.*, 50, 293–312, 1965.
13. Ehrhardt, P., *Z. Vgl. Physiol.*, 58, 47–75, 1968.
14. Ehrhardt, P., *Biol. Zentralbl.*, 88, 335–348, 1969.
15. Fraenkel, G. S., *J. Nutr.*, 65, 361–396, 1958.
16. Fraenkel, G. and Blewett, M., *Bull. Entomol. Res.*, 35, 127–139, 1944.
17. Frost, F. M., Herms, W. B., and Hoskins, W. M., *J. Exp. Zool.*, 73, 461–479, 1936.
18. Gordon, H. T., *Ann. N.Y. Acad. Sci.*, 77, 290–351, 1959.
19. Henry, S. M. and Block, R. J., *Contrib. Boyce Thompson Inst.*, 21, 129–145, 1961.
20. Horie, Y., Watanabi, K., and Ito, T., *Bull. Seric. Exp. Stn. Tokyo*, 22, 181–193, 1967.
21. House, H. L. and Barlow, J. S., *Can. J. Zool.*, 34, 182–189, 1956.
22. House, H. L. and Barlow, J. S., *J. Insect Physiol.*, 11, 915–918, 1965.
23. Huot, L., Bernard, R., and Lemonde, A., *Can. J. Zool.*, 36, 7–13, 1958.
24. Ito, T. and Niimura, M., *Bull. Seric. Exp. Stn. Tokyo*, 20, 361–374, 1966.
25. Leclercq, J., *Arch. Int. Physiol.*, 55, 412–419, 1948.
26. Leclercq, J., *Arch. Int. Physiol. Biochim.*, 68, 500–503, 1960.
27. Lemonde, A. and Chaudhary, K. D., *Rev. Can. Biol.*, 25, 21–24, 1966.
28. Markkula, M. and Laurema, S., *Ann. Agric. Fenn.*, 6, 77–80, 1967.
29. Medici, J. C. and Taylor, M. W., *J. Nutr.*, 88, 181–186, 1966.
30. Medici, J. C. and Taylor, M. W., *J. Nutr.*, 93, 307–309, 1967.
31. Mittler, T. E., unpublished.
32. Murray, D. R. P., *Entomol. Exp. Appl.*, 11, 149–168, 1968.
33. Nelson, J. W. and Palmer, L. S., *J. Agric. Res.*, 50, 849–852, 1935.
34. Pausch, R. D. and Fraenkel, G., *Physiol. Zool.*, 39, 202–222, 1966.
35. Retnakaran, A. and Beck, S. D., *J. Nutr.*, 92, 43–52, 1967.
36. Retnakaran, A. and Beck, S. D., *Comp. Biochem. Physiol.*, 24, 611–619, 1968.
37. Rock, G. C., Glass, E. H., and Patton, R. L., *Ann. Entomol. Soc. Am.*, 57, 617–621, 1964.
38. Sang, J. H., *J. Exp. Biol.*, 33, 45–72, 1956.
39. Trager, W., *Biol. Bull.*, 17, 343–352, 1936.
40. Vanderzant, E. S., *J. Insect Physiol.*, 11, 659–670, 1965.

QUALITATIVE REQUIREMENTS AND UTILIZATION OF NUTRIENTS: ARTHROPODS OTHER THAN INSECTS

R. H. Dadd

INTRODUCTION

A recent review of arthropod nutrition[3] noted that compared with the information available for insects our knowledge of the qualitative nutrient requirements of other classes was meager and fragmentary. This is largely because of difficulties in developing the defined synthetic diets and aseptic rearing techniques necessary for deletion/growth studies. Because most arachnids are suctorial predators, and most crustaceans are aquatic detritus or filter feeders, formidable obstacles confront attempts to induce sustained feeding on stable artificial concoctions. This is still true, though some progress in the development of defined diets for several acarines and crustaceans promises to augment substantially the nutritional information for these classes. However, with few exceptions, these diets have yet to be used systematically for determining anything approaching the full range of qualitative specific requirements of any one species. Somewhat more information indicative of specific requirements is forthcoming from the application of radioisotope tracer methods, but nutritional conclusions based on such data are to be accepted cautiously in view of the ambiguities and uncertainties to which radiometric methods are prone as a means of determining nutritional need. The problems of nutritional interpretation of isotope data are discussed in the introduction to the tables in the section "Qualitative Requirements and Utilization of Nutrients: Insects" by R. H. Dadd elsewhere in this Handbook.

With regard to arthropod nutrition in general, one apparently characteristic feature is now reasonably well substantiated: A requirement for dietary sterol, demonstrated virtually without exception in numerous insect studies,* has been consistently found also in numerous other arthropods. Table 1 lists the species for which evidence of a dietary sterol requirement has been obtained. Several of the listed species readily convert phytosterols to cholesterol, which is the predominant tissue sterol in the Crustacea examined, regardless of food. Bioconversion of dietary phytosterols (ergosterol, stigmasterol, β-sitosterol, desmosterol, fucosterol, campesterol, and others) has been demonstrated variously in *Penaeus japonicus,*[21,39] *Palaemon serratus,*[34] and *Artemia salina.*[36-38] In contrast to these Crustacea, only a trace of cholesterol was detected in the phytophagous mite, *Tetranychus urticae,* the tissue sterols of which were almost entirely β-sitosterol, stigmasterol, and campesterol in proportions very similar to those of its host plant; if a mixture of β-sitosterol and stigmasterol was omitted from the synthetic diet, egg production was reduced, whereas the elimination of cholesterol slightly improved egg production.[6]

No other nutrient class has been studied over as wide a spectrum of the phylum as has sterol. The remaining rather fragmentary information is drawn entirely from work with Acarina and Crustacea, and will therefore be considered under these separate headings.

* See Table 4 in "Qualitative Requirements and Utilization of Nutrients: Insects."

Table 1

ARTHROPOD SPECIES FOR WHICH
EVIDENCE OF A DIETARY STEROL
REQUIREMENT HAS BEEN OBTAINED

	Ref.
Arachnida	
Avicularia avicularia[a]	47
Tetranychus urticae[c]	6
Diplopoda	
Graphidostreptus tumuliporus[a]	47
Crustacea	
Homarus vulgaris[a]	43
Homarus gammarus[a]	47
Panulirus japonica[a]	35
Astacus astacus[a]	45
Cancer pagurus[a]	40
Portunus trituberculatus[a]	35
Penaeus japonicus[a-c]	21, 22, 39
Palaemon serratus[a]	34
Palaemon aztecus	
Artemia salina[a-c]	25, 36, 37
Daphnia magna[a,c]	25
Moina macrocopa[c]	25

[a] Isotope method, usually with radio-labeled acetate.
[b] Comparative analyses of sterol content of food and tissues.
[c] Growth studies using artificial food with and without sterol.

ACARINA

Three species of mite have been reared with considerable success on aseptic, fully defined, synthetic diets: *Tetranychus urticae,* from protonymphs to egg-laying adults;[6,][30,32,42] and the cheese mites, *Tyrophagus putrescentiae*[31] and *Calophagus berlesei,*[29] from egg to adult stage. Data indicating the essentiality or otherwise of amino acids were determined radiometrically for the first two species, with confirmatory deletion/growth studies to clarify ambiguous results for *T. putrescentiae.* Amino acid analysis of *T. urticae* reared on a synthetic diet containing glucose-U-^{14}C revealed virtually no isotope in arginine, histidine, isoleucine, leucine, lysine, methionine, phenylalanine, tyrosine, and valine; except for tyrosine,* these can be considered essential. High specific activity in alanine, aspartic acid, cysteic acid/cystine, glutamic acid, glycine, and serine indicated that they were biosynthesized by the mite and would therefore probably not be required nutritionally. No data were obtained on tryptophan owing to its loss during tissue hydrolyzation. Threonine had a high count and was considered to be synthesized from glucose, an anomalous result since this amino acid is generally essential for animals. Proline had borderline isotopic activity, indicating it might be so slowly synthesized as to

* See the comment on the isotope method in the tables in "Qualitative Requirements and Utilization of Nutrients: Insects."

be a required nutrient; it is usually considered not essential for animals, but deletion studies with insects have revealed a number of exceptions that require it, sometimes in contradiction to the indications given by isotope studies.* Essentially similar results were obtained using the same radiometric methods with *T. putrescentiae;*[31] however, in this case, proline gained high radioactivity from the glucose and would unambiguously be indicated as nonessential. Threonine again had substantial activity, and was therefore examined in a deletion/growth experiment, together with phenylalanine, tyrosine, tryptophan, and valine; this showed that, like valine, threonine and tryptophan were nutritionally essential, whereas tyrosine was not required as long as phenylalanine was provided in the diet.[31] Allowing for the uncertainties of the radioisotope method, these data on amino acid requirements of mites show a basic similarity to data for the generality of insects.

Adult *T. urticae* that were fed a diet incorporating tritiated acetate were found to have substantial radioactivity in linoleic/linolenic acids (linolenic is the predominant fatty acid in this mite), though with only a quarter the specific activity for stearic/oleic acids (which combined, approach linolenic acid in amount). These data were taken to indicate that *T. urticae* can biosynthesize polyunsaturated fatty acids,[42] in contrast to another arachnid, the spider *Avicularia avicularia;* the myriapod, *Graphidostreptus tumuliporus;*[47] to insects generally; and, as will appear below, to several Crustacea. Fatty acid analysis of successive developmental stages of *T. urticae* indicated preferential accumulation of linolenic acid during feeding and growth, and rapid depletion of fatty acid reserves on starvation, particularly the more unsaturated they were. Though preferential accumulation might be considered as indicating a requirement, studies of egg production by mites maintained on a synthetic diet gave little indication that the absence of fatty acids was detrimental.[6] However, a doubling of the percentage viability of eggs resulted from the inclusion of vitamin E (α-tocopherol), and this can be interpreted in two ways: that vitamin E is required as such, or that unsaturated fatty acid is required but in the absence of an antioxidant such as tocopherol is lost from the diet.

Synthetic diets used in the foregoing mite studies contained comprehensive arrays of sugars, water and fat-soluble vitamins, nucleic acid, minerals, and ascorbic acid, but at this time no information on requirements specifically for these has been published.

CRUSTACEA

Because of the commercial value of large decapods as human food and the importance of small zooplanktonic forms in the food chains upon which fisheries ultimately depend, considerable work has focused on crustacean nutrition in the broadest sense. Mostly, this deals with natural dietaries, gross composition of crude foods, and their digestibility in relation to metabolism, etc., especially as influenced by seasonal and other environmental fluctuations and interactions, but offers little understanding of specific nutrient requirements.[1,8,17,23] Early attempts to rear crustacea artificially as a prelude to specific nutritional study used terrestrial isopods and small aquatic filter-feeders such as cladocerans and copepods. With the first type, problems associated with the provision of defined diets in water were avoided; with the second, where the problems of aquatic environment were grappled with, difficulties of using aseptic culture over long periods and on a large scale were reduced by working with small species. Nutrient needs suggested by such studies were reviewed recently,[3] but are of uncertain validity, either because the diets were too ill-defined, or, as in several copepod studies, because the diet comprised mono- or diaxenic microorganisms in a chemically defined medium, making it uncertain

---

* See Table 1 in "Qualitative Requirements and Utilization of Nutrients: Insects."

whether additions or deletions of nutrients from the basic medium acted directly on the crustacean or indirectly via an altered metabolism of the living food microorganisms.[25]

The first crustacean to be reared successfully in an aseptic, chemically defined dietary medium was the brine shrimp, *Artemia salina;*[28] it remains the species for which most information on qualitative requirements is available, the following points being established by deletion/growth experiments using many versions of the original defined medium.[26-28] The water-soluble B vitamins thiamine, riboflavin, niacin, pantothenate, pyridoxamine, folic acid, and biotin are essential; choline and inositol probably are required, since in their absence very few females matured to carry eggs. Vitamin B_{12}, carnitine, glutathione, and ascorbic acid were not required over one developmental cycle. An initially apparent need for glutathione was spared by cysteine, which is presumably inadequately supplied by the proteins of the diet, albumin and globulin. Of notable interest is an unusual requirement for putrescine, without which growth ceased in midlarval development. Like many dipteran insects, nucleic acid components (given as a mixture of thymidine plus four ribonucleotides) are required. In addition to sterol, other lipid growth factors (possibly vitamin E) are necessary for the production of fertile eggs.[25] A lipid fertility factor is also required by two other planktonic crustacea, *Daphnia magna* and *Moina macrocopa;* indeed, fertility in *Daphnia* was long ago shown to depend on the addition of vitamin E concentrate to its medium.[7] Using diets similar to those for *Artemia,* these two other species were shown to also require several B vitamins and nucleic acid components, though the specific nucleotides needed differ for the three species.[25] Further investigation of the nucleic acid requirement of *A. salina* indicates morphogenetic abnormalities when thymidine, pyrimidine bases, and purine nucleotides and other derivatives are absent from the diet;[13-15] a requirement for adenosine monophosphate varies with both the albumin concentration and salinity of the medium.[16]

A crucial feature of synthetic media for *Artemia, Daphnia,* and *Moina* was that both protein (albumin) and carbohydrate (starch) were needed as insoluble particulate suspensions, in optimal ratios that differed for the three species.[25] Nutritionally inert particulates could not be substituted for the starch, showing it was required nutritionally, as also indicated in recent work with *Artemia* using a simple microencapsulated diet of hemoglobin and starch.[18] It has not yet proved possible to replace these macromolecular nutrients with amino acid mixtures and sugars, probably because filter-feeding crustacea cannot ingest adequate amounts of simple nutrient solutes in liquid form without approaching concentrations exceeding some critically adverse osmotic level.[27] However, particulate albumin at grossly suboptimal concentrations could be satisfactorily fortified by dissolved amino acid mixtures, offering the possibility of determining amino acid requirements, but information on specific amino acids is not yet at hand.

Nonsterile, semisynthetic diets have recently come into use for nutritional studies with the prawn *Penaeus japonicus.*[4,19] With the aim of determining amino acid requirements, attempts were made to substitute amino acid mixtures for the major protein component, but without success.[5] Therefore, some understanding of crustacean amino acid nutrition currently depends upon isotopic tracer studies. A summary[17] of the earlier work with various large decapods (*Homarus* sp., *Astacus astacus* and *fluviatilis, Carcinus maenus, Micropodia rostraca*) and a cirripede, *Elminius modestus,* indicates that radioisotope from glucose or acetate was always found in alanine, aspartic acid, and glutamic acid, and generally also in glycine, serine, and proline, but not in histidine, isoleucine, leucine, lysine, phenylalanine, threonine, valine, or tyrosine, though tyrosine became labeled following injection of labeled phenylalanine in *Astacus.*[44] In the prawn, *Palaemon serratus,* injected with glucose-U-C^{14} and analyzed after 6 days of maintenance on an oligidic diet, all ten of the "rat essential" amino acids (and tyrosine) were unlabeled and therefore indicated as essential, whereas substantial labeling was found in alanine, aspartic acid, cysteic acid, glutamic acid, glycine, proline, and serine.[2] A similar study with *Astacus leptodactylus*[41] found label in alanine, aspartic acid, glutamic acid, glycine, proline, and serine, suggesting these were not essential; arginine, histidine, isoleucine, leucine, lysine, phenylalanine (and tyrosine), threonine, and valine were virtually unlabeled and thus probably essential. Methionine and cystine were labeled at a low but significant level, suggesting some biosynthesis; if methionine is accepted as essential on these data, as it usually is for insects and animals in general, then cystine should be likewise indicated as essential, as it is for a few insects. Of notable interest was the complete absence of label from asparagine, which this species could not apparently derive from the heavily labeled aspartic acid; this suggests that if asparagine as such were required as a protein constituent, it would probably be required exogenously as a nutrient.

Radioisotopic studies on fatty acid biosynthesis from acetate or palmitate by *Astacus astacus, Homarus gammarus,* and various planktonic copepods[24,33,46,47] show incorporation of isotope and biosynthesis of C 16-18 saturated and monounsaturated fatty acids and the long-chain C 20-22 polyunsaturated acids commonly found in crustacea, as also found in fatty alcohols and esters derived from them. By contrast, linoleic, when present, and especially linolenic acids were unlabeled, suggesting their derivation from food. Whether they are required nutrients can only be ascertained by growth/deletion studies, bearing in mind that many insects show no apparent requirement for growth even though unable to synthesize polyunsaturated fatty acids.*

The synthetic diets now developed for various crustaceans all include vitamin C (ascorbic acid) in their comprehensive vitamin mixtures, but whether it is actually required nutritionally has only just begun to be examined and clarified. Ascorbic acid content of the prawn *Palaemon serratus* and a crab, *Cancer pagurus,* undergoes major fluctuations in relation to developmental cycles, being low around molts, developing to high levels during the feeding and growth of intermoult periods, and accumulating preferentially in the eggs.[10,11] The eggs probably synthesize some ascorbic acid during early embryonic development, but, referring to studies in press at time of writing,[12] it was stated that the postembryonic prawn cannot synthesize vitamin C. It thus appears that vitamin C is likely to be an essential nutrient for some crustacea, as it is for a large number of phytophagous insects.

* See Table 3 in "Qualitative Requirements and Utilization of Nutrients: Insects."

REFERENCES

1. Conover, R. J., *Am. Zool.,* 8, 107–118, 1968.
2. Cowey, C. B. and Forster, J. R. M., *Mar. Biol.,* 10, 77–81, 1971.
3. Dadd, R. H., Arthropod nutrition, in *Chemical Zoology,* Vol. 5, Florkin, M. and Scheer, B. T., Eds., Academic Press, New York, 1970, 35–95.
4. Deshimaru, O. and Kuroki, K., *Nippon Suisan Gakkai,* 40, 413–419, 1974.
5. Deshimaru, O. and Kuroki, K., *Nippon Suisan Gakkai,* 40, 1127–1131, 1974.
6. Ekka, I., Rodriguez, J. G., and Davis, D. L., *J. Insect Physiol.,* 17, 1393–1399, 1971.
7. Fisher, L. R., Vitamins, in *The Physiology of Crustacea,* Vol. 1, Waterman, T. H., Ed., Academic Press, New York, 1960, 259–289.
8. Forster, J. R. M. and Gabbott, P. A., *J. Mar. Biol. Assoc. U.K.,* 51, 943–961, 1971.
9. Gosselin, L., *Arch. Int. Physiol. Biochim.,* 73, 543–544, 1965.
10. Guary, M. M., Ceccaldi, H. J., and Kanazawa, A., *Mar. Biol.,* 32, 349–355, 1975.
11. Guary, M. M. and Guary, J.-C., *Mar. Biol.,* 32, 357–363, 1975.
12. Guary, M., Kanazawa, A., Tanaka, N., and Ceccaldi, H. J., *Nippon Suisan Gakkai,* in press, 1976.
13. Hernandorena, A., *C. R. Acad. Sci. Ser. D,* 271, 1406–1410, 1970.
14. Hernandorena, A., *Arch. Zool. Exp. Gen.,* 113, 425–432, 1972.
15. Hernandorena, A., *Arch. Zool. Exp. Gen.,* 113, 489–498, 1972.
16. Hernandorena, A., *Biol. Bull.,* 146, 238–248, 1974.
17. Huggins, A. K. and Munday, K. A., *Adv. Comp. Physiol. Biochem.,* 3, 271–378, 1968.
18. Jones, D. A., Munford, J. G., and Gabbott, P. A., *Nature,* 247, 233–235, 1974.
19. Kanazawa, A., Shimaya, M., Kawasaki, K., and Kashiwada, K., *Nippon Suisan Gakkai,* 36, 949–954, 1970.
20. Kanazawa, A., Tanaka, N., and Kashiwada, K., *Nippon Suisan Gakkai,* 38, 1067–1071, 1972.
21. Kanazawa, A., Tanaka, N., Teshima, S., and Kashiwada, K., *Nippon Suisan Gakkai,* 37, 211–215, 1971.
22. Kanazawa, A., Tanaka, N., Teshima, S., and Kashiwada, K., *Nippon Suisan Gakkai,* 37, 1015–1019, 1971.
23. Marshall, S. M. and Orr, A. P., Feeding and nutrition, in *The Physiology of Crustacea,* Vol. 1, 227–258, Waterman, T. H., Ed., Academic Press, New York, 1960, 227–258.
24. Morris, R. J. and Sargent, J. R., *Mar. Biol.,* 22, 77–83, 1973.
25. Provasoli, L., Conklin, D. E., and D'Agostino, A. S., *Helgol. Wiss. Meersunters.,* 20, 443–454, 1970.
26. Provasoli, L. and D'Agostino, A., *Am. Zool.,* 2, 439, 1962.
27. Provasoli, L. and D'Agostino, A., *Biol. Bull.,* 136, 434–453, 1969.
28. Provasoli, L. and Shiraishi, K., *Biol. Bull.,* 117, 347–355, 1959.
29. Rodriguez, J. G., Inhibition of acarid mite development by fatty acids, in *Insect and Mite Nutrition,* Rodriguez, J. G., Ed., North Holland, Amsterdam, 1972, 637–650.
30. Rodriguez, J. G. and Hampton, R. E., *J. Insect Physiol.,* 12, 1209–1216, 1966.
31. Rodriguez, J. G. and Lasheen, A. M., *J. Insect Physiol.,* 17, 979–985, 1971.
32. Rodriguez, J. G., Singh, P., Seay, T. N., and Walling, M. V., *J. Insect Physiol.,* 13, 925–932, 1967.
33. Sargent, J. R. and Lee, R. F., *Mar. Biol.,* 31, 15–23, 1975.
34. Teshima, S. I., Ceccaldi, H. J., Patrois, J., and Kanazawa, A., *Comp. Biochem. Physiol. B,* 50, 485–490, 1975.
35. Teshima, S. and Kanazawa, A., *Comp. Biochem. Physiol.,* 13, 461–467, 1964.
36. Teshima, S. and Kanazawa, A., *Nippon Suisan Gakkai,* 37, 720–723, 1971.
37. Teshima, S. and Kanazawa, A., *Comp. Biochem. Physiol. B,* 38, 603–607, 1971.
38. Teshima, S. and Kanazawa, A., *Nippon Suisan Gakkai,* 38, 1305–1310, 1972.
39. Teshima, S., Kanazawa, A., and Okamoto, H., *Nippon Suisan Gakkai,* 40, 1015–1019, 1971.
40. Van den Oord, A., *Comp. Biochem. Physiol.,* 13, 461–467, 1964.
41. Van Marrewijk, W. J. A. and Zandee, D. I., *Comp. Biochem. Physiol. B,* 50, 449–455, 1975.
42. Walling, M. V., White, D. C., and Rodriguez, J. G., *J. Insect Physiol.,* 14, 1445–1458, 1968.
43. Zandee, D. I., *Nature,* 202, 1335–1336, 1964.
44. Zandee, D. I., *Arch. Int. Physiol. Biochim.,* 74, 35–44, 1966.
45. Zandee, D. I., *Arch. Int. Physiol. Biochim.,* 74, 435–441, 1966.
46. Zandee, D. I., *Arch. Int. Physiol. Biochim.,* 74, 614–626, 1966.
47. Zandee, D. I., *Comp. Biochem. Physiol.,* 20, 811–822, 1967.

NUTRIENT REQUIREMENTS AND UTILIZATION: MOLLUSCS (GASTROPODA AND BIVALVIA)

A. A. Wright

I. MINERALS

Aquatic molluscs, especially marine bivalves, are notorious for their ability to concentrate heavy metals present in water.[5,24] For those metals with a physiological function, this may be a device that ensures the availability of adequate amounts when the element itself is in relatively low concentration in the medium.[15] It appears to lead to excessive quantities with many of the metals, however, and also takes place with those of no known physiological function.[18] This has important implications for animals in polluted waters, especially where contamination from radionuclides exists, but it will not be further considered here.

Figures have been given for the concentration of many trace elements in the tissues of molluscs: *Cepaea vindobonensis,*[23] *Chlamys opercularis,*[6] *Crassostrea virginica,*[19] *Pecten maximus,*[6] and clams and mussels;[3] the concentrations of some common and trace elements have also been determined in eight species of marine molluscs[29] and in *Lymnaea stagnalis.*[31] Studies on the ionic composition of tissues and body fluids, especially in relation to salt and water balance, have given figures for the concentrations of some common elements for *Littorina littorea,*[27] *Arion ater,*[26] and a wide range of terrestrial molluscs.[7] A simple salt mixture was used in a diet supporting growth in *Arion ater,*[40] but a more complex mixture was successful in carrying *Biomphalaria glabrata* through its whole life cycle.[37] A complex mixture was also used to rear *Helix pomatia.*[16]

Table 1 lists elements of nutritional importance; R (= required) indicates that the element has been shown to have beneficial effects for the animal concerned, and U (= utilized) indicates that the element has been shown to be taken up actively, but in the case of trace elements the data may have been obtained from experiments using high concentrations that are of toxic rather than of nutritional importance.

Table 1
MINERALS

Species	Ca	Cl	Cu	Fe	Mg	Mn	P	K	Na	Zn	Other	Reference
Gastropoda												
Agriolimax reticulatus	U											39
Biomphalaria glabrata	R								R			10, 34
Busycon canaliculatum			U									4
Goniobasis clavaeformis							U					9
Helix aspersa	R											8, 38
Lymnaea auricularia	U										U[a]	35
Lymnaea peregra	U											42
Lymnaea stagnalis	U		U								U[a]	33, 35
Planorbarius corneus	U										U[a]	35, 42
Strophocheilus oblongus musculus	U		U		U		U	U	U		U[b]	17
Taphius glabratus			U							U	U[c]	41
Bivalvia												
Argopecten irradians							U					30
Crassostrea virginica	U	R	U			R?	U			U		11, 21, 22, 25, 32
Elliptio complanata			U									15
Mya arenaria							U				U[d]	2, 13
Mytilus edulis			U			U	U			U	U[d]	1, 14, 20, 36
Tellina tenuis			U									28
Venus striatula							U					1, 2

[a] Sr, and perhaps Ba.
[b] Sulfur utilized from cabbage or lettuce.
[c] Radioactive elements, also Cd, taken up.
[d] Accumulation of Co.

REFERENCES

1. **Allen, J. A.**, *Comp. Biochem. Physiol.,* 36, 131–141, 1970.
2. **Allen, J. A.**, *J. Mar. Biol. Assoc. U.K.,* 42, 609–623, 1962.
3. **Bertine, K. K. and Goldberg, E. D.**, *Limnol. Oceanogr.,* 17, 877–884, 1972.
4. **Betzer, S. B. and Pilson, M. E. Q.**, *Biol. Bull.,* 148, 1–15, 1975.
5. **Brooks, R. R. and Rumsby, M. G.**, *Limnol. Oceanogr.,* 10, 521–527, 1965; see also **Vinogradov, A. P.**, *Sears Foundation for Marine Research Memoirs,* Vol. 2, Yale University, New Haven, 1953.
6. **Bryan, S. W.**, *J. Mar. Biol. Assoc. U.K.,* 53, 145–166, 1973.
7. **Burton, R. F.**, *Comp. Biochem. Physiol. A.,* 39, 267–275, 875–878, 1971; 43, 655–664, 1972.
8. **Crowell, H. H.**, *Proc. Malacol. Soc., London,* 40, 491–503, 1973.
9. **Elwood, J. W. and Goldstein, R. A.**, *Freshwater Biol.,* 5, 397–406, 1975.
10. **Frank, G. H.**, *Bull. W.H.O.,* 29, 531–537, 1963.
11. **Galtsoff, P. S.**, *Fishery Bulletin of the Fish and Wildlife Service,* Vol. 64, U.S. Department of the Interior, Washington, D.C., 1964.
12. **Greenaway, P.**, *J. Exp. Biol.,* 54, 609–620, 1971.
13. **Harrison, F. L.**, in *Radioactive Contamination of the Marine Environment* (Proceedings of a Symposium), Krippner, M., Ed., International Atomic Energy Agency, Vienna, 1972.
14. **Hobden, D. J.**, *J. Mar. Biol. Assoc. U.K.,* 49, 661–668, 1969.
15. **Hobden, D. J.**, *Can. J. Zool.,* 48, 83–86, 1970.
16. **Howes, N. H. and Whellock, R. B.**, *Biochem. J.,* 31, 1489–1498, 1937.
17. **de Jorge, F. B., Peterson, J. A., and Ditadi, A. S. F.**, *Comp. Biochem. Physiol.,* 33, 837–843, 1970.
18. **Keckes, S., Pucar, Z., and Marazovic, Lj.**, in *Radioecological Concentration Processes,* Aberg, B. and Hungate, F. P., Eds., Pergamon Press, Oxford, 1966, pp. 993–994.
19. **Kopfler, F. C. and Mayer, J.**, *Proc. Natl. Shellfish. Assoc.,* 63, 27–34, 1972.
20. **Pentreath, R. J.**, *J. Mar. Biol. Assoc. U.K.,* 53, 127–144, 1973.
21. **Pomeroy, L.**, *Proc. Natl. Shellfish. Assoc.,* 43, 167–170, 1952.
22. **Pomeroy, L. R. and Haskin, H. A.**, *Biol. Bull.,* 107, 123–129, 1954.
23. **Ponomarenko, A. B., Trufanov, G. B. and Golubev, C. H.**, *Ekologiya,* 5, 94–96, 1974.
24. **Pringle, B. H., Hissong, D. E., Kantz, E. L., and Mulawka, S. T.**, *J. Sanit. Eng. Div. Am. Soc. Civ. Eng.,* 94, 455–475, 1968.
25. **Prytherch, H. F.**, *Science,* 73, 429–431, 1931; *Ecol. Monogr.,* 4, 47–107, 1934.
26. **Roach, D. K.**, *J. Exp. Biol.,* 40, 613–623, 1963.
27. **Rumsey, T. J.**, *Comp. Biochem. Physiol. A,* 45, 327–344, 1973.
28. **Saward, D., Stirling, A., and Topping, G.**, *Mar. Biol.,* 29, 351–361, 1975.
29. **Segar, D. A., Collins, J. D., and Riley, J. P.**, *J. Mar. Biol. Assoc. U.K.,* 51, 131–136, 1971.
30. **Schelske, C. L.**, in *Radioactive Contamination of the Marine Environment* (Proceedings of a Symposium), Krippner, M., Ed., International Atomic Energy Agency, Vienna, 1972.
31. **Spronk, N., Brinkman, F. G., Van Hoek, R. J., and Knook, D. L.**, *Comp. Biochem. Physiol. A,* 38, 387–405, 1971.
32. **Swift, M. L., Conger, K., Exler, J., and Lakshmanan, S.**, *Life Sci.,* 17, 1679–1684, 1975.
33. **Spronk, N., Tilders, F., and Van Hoek, R. J.**, *Comp. Biochem. Physiol. A,* 45, 257–272, 1973.
34. **Thomas, J. D., Benjamin, M., Lough, A., and Aram, R. H.**, *J. Anim. Ecol.,* 43, 839–860, 1974.
35. **Van der Borght, O. and Van Puymbroeck, S.**, *Nature London,* 204, 533–534, 1964; 210, 791–793, 1966.
36. **Van Weers, A. W.**, in *Radioactive Contamination of the Marine Environment* (Proceedings of a Symposium), Krippner, M., Ed., International Atomic Energy Agency, Vienna, 1972.
37. **Vieira, E. C.**, *Am. J. Trop. Med. Hyg.,* 16, 792–796, 1967.
38. **Wagge, L. E.**, *J. Exp. Zool.,* 120, 311–342, 1952.
39. **Walker, G.**, *Proc. Malacol. Soc. London,* 40, 33–44, 1972.
40. **Wright, A. A.**, *Comp. Biochem. Physiol. A,* 46, 593–603, 1973.
41. **Yager, C. M. and Harry, H. W.**, *Malacologia,* 1, 339–353, 1964.
42. **Young, J. O.**, *Proc. Malacol. Soc. London,* 41, 439–445, 1975.

II. VITAMINS

Little information is available. Vitamin K has not been tested, and the results of earlier work need confirmation. The term "required" is used where beneficial effects have been noted, though animals can be reared in the apparent absence of the vitamin. An entry followed by a question mark indicates that the vitamin has been used in a diet or is taken up by the organism, but no nutritional significance has been established.

Table 2

VITAMINS

Species	Stage[a]	Thiamine	Riboflavine	Nicotinic acid	B$_6$	Biotin	Pantothenic acid	Folic acid	p-Aminobenzoic acid	Inositol	B$_{12}$	Choline	Other	Reference
Gastropoda														
Arion ater		R	R	R	R	R?	E	R			R?	R?	R?[b]	3
Biomphalaria glabrata		E	R̸	E	E	R̸	E	E		R̸	E	R̸	R̸,R[c]	5, 6
Helix aspersa													R[d]	7
Helix pomatia													R[e]	2
Bivalvia														
Crassostrea virginica	L	R[f]	R[f]		R[f]		R[f]							1
Ostrea lurida	L	R[g]	R[g]		R[g]	R[g]	R[g]							1
Tapes japonica	A			R̸[g]							R?			4
Venus mercenaria	L	R̸[g]	R̸[g]	R̸[g]	R̸[g]	R̸[g]	R̸[g]				R̸[g]		R̸[h]	1

Abbreviations: E = essential; R = required; R̸ = not required.

a L = larvae; A = adult.
b Ascorbic acid.
c Ascorbic acid and vitamin A are nonessential; vitamin E is required for reproduction.
d Vitamin D supports uptake and transport of calcium and increases length of life.
e Vitamin A and vitamins of the B group (yeast used) are required; vitamin D may be required.
f Tested alone and in combination.
g Tested only in combination, not separately.
h Vitamin A.

REFERENCES

1. Davis, H. C. and Chanley, P. E., *Proc. Natl. Shellfish. Assoc.,* 45, 59–74, 1956.
2. Howes, N. H. and Whellock, R. B., *Biochem. J.,* 31, 1489–1498, 1937.
3. Ridgway, J. W. and Wright, A. A., *Comp. Biochem. Physiol. A,* 51, 727–732, 1975.
4. Tozawa, H. and Sagara, J., *Bull. Jpn. Soc. Sci. Fish.,* 27, 785–788, 1961.
5. Vieira, E. C., *Am. J. Trop. Med. Hyg.,* 16, 792–796, 1967.
6. Vieira, E. C., Senna, I. A., Rogana, S. M. G., and Tupynamba, M. L. V. C., in *Germfree Research,* Academic Press, New York, 1973, pp. 657–660.
7. Wagge, L. E., *J. Exp. Zool.,* 120, 311–342, 1952.

III. SUGARS

Entries in parentheses indicate that digestive enzymes capable of hydrolyzing these sugars are present or that in vitro utilization has been observed. Some entries refer only to observed uptake of the sugars. Information relates to adult or metamorphosed form for species with larval stages.

Table 3
SUGARS

Species	Cellobiose	Fructose	Galactose	Glucose	Lactose	Maltose	Mannose	Ribose	Sucrose	Trehalose	Xylose	Reference
Gastropoda												
Agriolimax reticulatus			U	U								29
Ariolimax columbianus			U	U		U						18
Arion ater	(U)			(U)	(U)	(U)			U	(U)		6, 26, 31
Biomphalaria glabrata				U					U			23
Elysia viridis			U[a]	U[a]			U[a]					27
Haliotis rufescens	(U)			(U)	(U)							1, 5
Helix pomatia			(U)	U	(U)	(U)			(∅)			2, 11, 13, 22
Patella sp.				U								4
Tridachia crispata			U[a]	U[a]							(U)	27
Bivalvia												
Crassostrea gigas			U	U								3
Crassostrea virginica				U								8, 9, 12, 19, 25
Mytilus edulis				U								15, 16, 21, 33
Ostrea edulis		(U)		U	(U)	(U)			(U)			7, 17, 32
Placopecten magellanicus					(U)	(U)				(U)		30
Rangia cuneata				U								14, 24

Abbreviations: U = utilized; ∅ = not utilized.

[a] Products formed by fixation of ^{14}C-bicarbonate photosynthetically in symbiotic chloroplasts obtained from dietary algae and released to the host animal. A similar process is seen in *Placobranchus ianthobapsus*[28] and in species of *Tridacna*, where the symbionts are zooxanthellae.[10,20]

REFERENCES

1.   Allen, W. V. and Kilgore, J., *Comp. Biochem. Physiol. A,* 50, 771–775, 1975.
2.   Baldwin, E., *Biochem. J.,* 32, 1225–1237, 1938.
3.   Bamford, D. R. and Gingles, R., *Comp. Biochem. Physiol. B,* 49, 637–646, 1974.
4.   Barry, R. J. C. and Munday, K. A., *J. Mar. Biol. Assoc. U.K.,* 38, 81–95, 1959.
5.   Bennet, R., Jr., Thamassi, N., and Nakada, H. I., *Comp. Biochem. Physiol. B,* 40, 807–811, 1971.
6.   Evans, W. A. L. and Jones, E. G., *Comp. Biochem. Physiol.,* 5, 149–160, 1962.
7.   L-Fando, J. J., Garcia-Fernandez, M. C., and R–Caudela, J. L., *Comp. Biochem. Physiol. B,* 43, 807–814, 1972.
8.   Galtsoff, P. S. and Whipple, D. V., *Bull. U.S. Bur. Fish.,* 46, 489–508, 1930.
9.   Gillespie, L., Ingle, R. M., and Havens, W. K., *Q. J. Fla. Acad. Sci.,* 27, 279–288, 1964.
10. Goreau, T. F., Goreau, N. I., and Yonge, C. M., *J. Zool.,* 169, 417–454, 1973.
11. Got, R., Marnay, A., and Jarrige, P., *Experientia,* 21, 653–654, 1965.
12. Haven, D. S., *Chesapeake Sci.,* 6, 43–51, 1965.
13. Holtz, F. and von Brand, T., *Biol. Bull.,* 79, 423–431, 1940.
14. Hopkins, S. H., Anderson, J. W., and Horvath, K., *Proc. Natl. Shellfish. Assoc.,* 64, 4, 1973.
15. Loxton, J. and Chaplin, A. E., *Biochem. Soc. Trans.,* 2, 41–43, 1973.
16. Malanga, C. J. and Crouthamel, W. G., *Am. Zool.,* 12, xxxix, 1972.
17. Mathers, N. F., *Proc. Malacol. Soc. London,* 40, 359–367, 1973.
18. Meenakshi, V. R. and Scheer, B. T., *Comp. Biochem. Physiol.,* 26, 1091–1097, 1968.
19. Mitchell, P. H., *Bull. U.S. Bur. Fish.,* 35, 153–161, 1915.
20. Muscatine, L., *Science,* 156, 516–519, 1967.
21. Péquignat, E., *Mar. Biol.,* 19, 227–234, 1973.
22. Schwartz (1934), see Goddard, C. K. and Martin, A. W., in *Physiology of Mollusca,* Vol. 2, Wilbur, K. M. and Yonge, C. M., Eds., Academic Press, New York, 1966, pp. 275–308.
23. Senna, I. A. and Vieira, E. C., *Am. J. Trop. Med. Hyg.,* 19, 568–570, 1970.
24. Stokes, T. and Awapara, J., *Comp. Biochem. Physiol.,* 25, 883–892, 1968.
25. Swift, M. L., Conger, K., Exler, J., and Lakshmanan, S., *Life Sci.,* 17, 1679–1684, 1975.
26. Thompson, G. A., Jr., *J. Biol. Chem.,* 240, 1912–1918, 1965; *Biochemistry,* 5, 1290–1296, 1966.
27. Trench, R. K., Boyle, J. E., and Smith, D. C., *Proc. R. Soc. London Ser. B,* 185, 453–464, 1974.
28. Trench, M. E., Trench, R. K., and Muscatine, L., *Comp. Biochem. Physiol.,* 37, 113–117, 1970.
29. Walker, G., *Proc. Malacol. Soc. London,* 40, 33–44, 1972.
30. Wojtowicz, M. B., *Comp. Biochem. Physiol. A,* 43, 131–141, 1972.
31. Wright, A. A., *Comp. Biochem. Physiol. A,* 46, 593–603, 1973.
32. Yonge, C. M., *J. Mar. Biol. Assoc. U.K.,* 15, 643–653, 1928.
33. de Zwaan, A. and van Marrewijk, W. J. A., *Comp. Biochem. Physiol. B,* 44, 429–439, 1973.

IV. CARBOHYDRATES OTHER THAN SUGARS

Most entries have been obtained from the interpretation of studies on digestive enzymes, and these have been put in parentheses, together with a few based on in vitro studies of utilization. Information on the activities of enzymes attacking certain polysaccharides has been given for a wider range of species than shown in Table 4 (see References 6 and 19).

Table 4
CARBOHYDRATES OTHER THAN SUGARS

Species	Cellulose	Dextrin	Glycerol	Glycogen	α-Methyl glucoside	Sorbitol	Starch	Other	Reference
Gastropoda									
Aplysia vaccaria	(U)							(U)[a]	14 (see also 23)
Arion ater	U						U	U[b]	7, 25
Biomphalaria glabrata	U			(U)	(U)		U		20
Eulota mackii								(U)[c]	22
Haliotis rufescens	(U)			(U)	(U)			(U)[d]	2
Helix pomatia	(U)						U	(U)[c] (U)[e]	1, 10, 13, 15, 22, 23
Lymnaea stagnalis	(U)						U	(U)[f]	9, 23
Succinea putris							U	(U)[c]	22
Bivalvia									
Crassostrea virginica	U?	Ψ					U	U[g]	4, 5, 6, 12, 17, 19
Macoma baltica								U[h]	3
Ostrea edulis		(U)	(U)		(U)		(U)	U[i]	8, 16
Portlandia arctica								U[h]	3
Placopecten magellanicus					(U)		(U)	(U)[i]	24
Scrobicularia plana	(U)						(U)	(U)[j]	18, 23
Spisula solidissima								(U)[k]	21
Teredo norvegica	(U)								11

Abbreviations: U = utilized; Ψ = not utilized.

a Digestive enzymes attack 1:4-β-polyglucosides, α- and β-glucosides, α-galactosides, including raffinose and melibiose, and β-galactosides.

b Pectin. No significant difference for growth between diets containing starch alone, cellulose alone, starch + cellulose, or starch + cellulose + pectin. (Vieira, personal communication; see Senna, I. A., M.Sc. thesis, Belo Horizonte).

c Zosterinase or polygalacturonase activity.

d Enzymes attack alginic acid, fucoidan, α-D(+)-melibiose, α-L-fucosides, α- and β-D-glucosides, α- and β-D-galactosides, α-D-mannosides, β-D-glucuronides, N-acetyl-β-D-glucosaminosides, and β-D-xylulosides, but not β-L-fucosides or pectic acid.

e Galactogen stimulates oxygen uptake by hepatopancreas in vitro; also see Reference 24.

f Enzymes attack laminarin.

g Corn starch and wheat flour,[14] and a "carbohydrate of biological origin."[16]

h Carbohydrates of marine sediments.

i Enzymes attack α- and β-glucosides, β-galactosides, laminarin, and chitobiose, but not chitin.

j Enzymes attack poly-β-glucosides and β-glucosides.

k Laminarin.

l Enzymes attacking 1:4-β-polysaccharides, 1:6-glucosides, α- and β-glucosides, β-galactosides and β-fructosides.

REFERENCES

1.   Baldwin, E., *Biochem. J.,* 32, 1225–1237, 1938.
2.   Bennett, R., Jr., Thamassi, N., and Nakada, H. I., *Comp. Biochem. Physiol. B,* 40, 807–811, 1971.
3.   Bubnova, N. P., *Okeanologiya,* 14, 743–747, 1974.
4.   Castell, J. D. and Trider, D. J., *J. Fish. Res. Board Can.,* 31, 95–99, 1974.
5.   Collier, A., Ray, S. M., Magnitzy, A. W., and Bell, J. O., *U.S. Fish. Wildl. Serv. Fish. Bull.,* 54, 167–185, 1953.
6.   Dunathan, J. P., Ingle, R. M., and Havens, W. K., Jr., *Fla. Dep. Nat. Resourc. Mar. Res. Lab. Tech. Ser.,* 58, 1–39, 1969.
7.   Evans, W. A. L. and Jones, E. G., *Comp. Biochem. Physiol.,* 5, 149–160, 1962.
8.   L-Fando, J. J., Garcia-Fernandez, M. C., and R-Caudela, J. L., *Comp. Biochem. Physiol. B,* 43, 807–814, 1972.
9.   Friedl, F. E., *Comp. Biochem. Physiol. A,* 39, 605–610, 1971; *Comp. Biochem. Physiol.,* 52, 377–380, 1975.
10. Goudsmit, E. M., *Malacol. Rev.,* 6, 58–59, 1973.
11. Greenfield, L. J. and Lane, C. E., *J. Biol. Chem.,* 204, 668–672, 1953.
12. Haven, D. S., *Chesapeake Sci.,* 6, 43–51, 1965.
13. Howes, N. H. and Whellock, R. B., *Biochem. J.,* 31, 1489–1498, 1937.
14. Koningsor, R. L., Jr., McLean, N., and Hunsaker, D., *Comp. Biochem. Physiol. B,* 43, 237–240, 1972.
15. Marshall J. J., *Comp. Biochem. Physiol. B,* 44, 981–988, 1973.
16. Mathers, N. F., *Proc. Malacol. Soc. London,* 40, 359–367, 1973.
17. Mitchell, P. H., *Bull. U.S. Bur. Fish.,* 35, 153–161, 1915.
18. Payne, D. W., Thorpe, N. A., and Donaldson, E. M., *Proc. Malacol. Soc. London,* 40, 147–160, 1972.
19. Sayce, C. S. and Tufts, D. F., *Proc. Natl. Shellfish. Assoc.,* 58, 14, 1967.
20. Senna, I. A. and Vieira, E. C., *Am. J. Trop. Med. Hyg.,* 19, 568–570, 1970.
21. Shallenberger, R. S., Searles, C., and Lenis, B. A., *Experientia,* 30, 597–598, 1974.
22. Shibaeva, V. I., Ovodova, R. G., and Ovodov, Y. S., *Comp. Biochem. Physiol. B,* 46, 561–565, 1973.
23. Stone, B. A. and Morton, J. E., *Proc. Malacol. Soc. London,* 33, 127–141, 1958.
24. Wojtowicz, M. B., *Comp. Biochem. Physiol. A,* 43, 131–141, 1972.
25. Wright, A. A., *Comp. Biochem. Physiol. A,* 46, 593–603, 1973.

V. PROTEINS

Casein has been used in artificial diets for the rearing of *Arion ater,*[11] *Biomphalaria glabrata,*[9] and *Helix pomatia*[6] at approximate levels of 16.8%, 17.8%, and 25.8% of the dry matter respectively. The growth of *Arion ater* is proportional to the protein content of the diet over a wide range, but a level of 33.6% had toxic effects after a few weeks.[8] For *Biomphalaria,* ovalbumin and lactalbumin have been substituted for casein; human globin A, trypsin, chymotrypsinogen, and lysosyme, each fortified with amino acids in an attempt to mimic the composition of casein, have also been used, but none were able to replace casein with the same efficiency.[10] Neither *Biomphalaria*[10] nor *Arion ater*[12] tolerated the replacement of casein with an amino acid mixture simulating its composition. The uptake of an algal hydrolysate and of leucine, both labeled with tritium, by *Biomphalaria*[6] may, therefore, have little nutritional significance. Eating their natural foods, *Archidoris pseudoargus* absorbs 93% of nitrogen ingested, with a high absorption of all amino acids; *Dendronotus frondosus* also absorbs 93% of the nitrogen, but with more variation in the absorption of individual amino acids; *Aplysia punctata* absorbs 74% of the nitrogen, with still wider variation in the absorption of individual amino acids, from 65% for lysine to 82% for methionine.[1] No clear relation was found between the composition of the food and growth rate in *Aplysia punctata.*[2]

No studies have been carried out on the protein nutrition of bivalves, although phytoplankton and suspensoids have been shown to have a high essential amino acid index for *Meretrix meretrix lusoria.*[7] Nitrogen fixation has been reported to occur in young adults of *Teredora malleolus*[3] to an extent that may be nutritionally significant. When artificial foods were used to fatten *Crassostrea virginica,* cereal meals were superior to the corresponding pure starch, but it is uncertain whether protein or some other constituent of the meals led to this effect.[4]

REFERENCES

1. Carefoot, T. H., *Comp. Biochem. Physiol.,* 21, 627–652, 1967.
2. Carefoot, T. H., *J. Mar. Biol. Assoc. U.K.,* 47, 565–589, 1967.
3. Carpenter, E. J. and Culliney, J. L., *Science,* 187, 551–552, 1975.
4. Dunathan, J. P., Ingle, R. M., and Havens, W. K., Jr., *Fla. Dep. Nat. Resourc. Mar. Res. Lab. Tech. Ser.,* 58, 1–39, 1969.
5. Gilbertson, D. E. and Jones, K. G., *Comp. Biochem. Physiol. A,* 42, 621–626, 1972.
6. Howes, N. H. and Whellock, R. B., *Biochem. J.,* 31, 1489–1498, 1937.
7. Okaichi, T., *Bull. Jpn. Soc. Sci. Fish.,* 40, 471–478, 1974.
8. Ridgway, J. W., Ph.D. thesis, University of Bradford, 1971.
9. Senna, I. A. and Vieira, E. C., *Am. J. Trop. Med. Hyg.,* 19, 568–570, 1970.
10. Vieira, E. C., Senna, I. A., Rogana, S. M. G., and Tupynamba, M. L. V. C., in *Germfree Research,* Academic Press, New York, pp. 657–660.
11. Wright, A. A., *Comp. Biochem. Physiol. A,* 46, 593–603, 1973.
12. Wright, A. A., unpublished results.

VI. AMINO ACIDS

Information on the requirement for any individual amino acid depends, almost entirely, on whether it is formed or not formed from ^{14}C-labeled precursors. These have usually been bicarbonate or glucose, but acetate and succinate have also been used. Work has been carried out on whole animals and on tissue preparations. In some cases, uptake from food or water has been recorded. None of these experiments give any indication whether a particular amino acid is taken up, or formed, at a sufficient rate or in sufficient quantities to satisfy the nutritional requirement of the animal.

Table 5
AMINO ACIDS

Species	Alanine	Arginine	Aspartic acid	Cysteine	Cystine	Glutamic acid	Glycine	Histidine	Hydroxyproline	Isoleucine	Leucine	Lysine	Methionine	Phenylalanine	Proline	Serine	Threonine	Tryptophan	Tyrosine	Valine	Others	Reference
Gastropoda																						
Agriolimax reticulatus	U	U	U			U	U															25
Aplysia punctata	U	U	U	U		U	U	U		U	U	U	U	R̸[a]	U	U	R̸[a]	U	U	R̸[a]	U[b]	10, 11
Archidoris pseudoargus	R̸	U	R̸			R̸	U	U				U	U	U	U	R̸	U	U	U	U	U[b]	11
Austrocochlea obtusa	R̸		R̸			R̸					U										R̸[c]	7
Biomphalaria glabrata											U											13
Bulimulus alternatus	R̸	R̸	R̸			R̸	R̸														R̸[c,d]	9
Cepaea hortensis	R̸		R̸			R̸				U[e]												8
Cepaea nemoralis	R̸		R̸			R̸														U[e]		8, 24
Crepidula fornicata	U	U	U	U		U	U	U		U	U	U	U	U	U	U	U	U	U	U	U[b]	20
Dendronotus frondosus	R̸	R	R̸	R̸		R̸	R̸	R		R	R	R	R	R	R̸	R̸	R	R	U	U		11
Elysia viridis	R̸	R̸	R̸			R̸	R̸			R	R	R	R	R	R̸	R̸	R	R	R[f?] R	R	R̸[c]	23
Haliotis rufescens	R̸					R̸	R̸															2
Helix aspersa	R̸						U															9
Helix pomatia																						8
Littorina littorea																				U		20
Melanerita melanotragus	R̸	R̸	R̸			R̸	R̸								R̸	R̸					R̸[c]	7
Otala lactaea	R̸	R̸	R̸			R̸	R̸									R̸					R̸[c,g]	9
Pomatias elegans	R̸	R̸	R̸			R̸																8
Strophocheilus oblongus	R̸	R̸	R̸			R̸	R̸								R̸	R̸					R̸[c]	22
Tridacna maxima											U											14

Abbreviations: R = required; U = utilized, i.e., absorbed from food or water.

[a] Isoleucine, phenylalanine, threonine, and valine were utilized by this species, but in some experiments the method used indicated that no uptake of these amino acids had occurred. This may be a deficiency in the method rather than a true lack of requirement.

[b] Taurine.

[c] Citrulline.

[d] Glutamine.

[e] Injected valine and, to lesser extent, isoleucine are utilized in the biosynthesis of hydrocarbons.

[f] Possible synthesis from phenylalanine has not been investigated.

[g] Ornithine.

Table 5 (continued)
AMINO ACIDS

Species	Alanine	Arginine	Aspartic acid	Cysteine	Cystine	Glutamic acid	Glycine	Histidine	Hydroxyproline	Isoleucine	Leucine	Lysine	Methionine	Phenylalanine	Proline	Serine	Threonine	Tryptophan	Tyrosine	Valine	Others	Reference
Bivalvia																						
Aequipecten irradians	U						U															20
Cerastoderma edule	R																					6
Crassostrea virginica	R		R			R						U										4, 15
Mercenaria mercenaria	R		R				U															20
Modiolus demissus	R		R																			18
Modiolus modiolus	U																					18
Mya arenaria	R	R	R	R		R	R	R		R			U	R	R	R	R	R	R[f?]		R[h]	1, 21
Mytilus californianus	R	R	R	R[i]		R	U						R	R		R	R	R	R			16
Mytilus edulis	R		R			R	U							U							U[j]	5, 9, 17, 18, 19, 20
Mytilus edulis											U[k]								U			26, 27
Ostrea edulis	R		R	R		R	U															12
Rangia cuneata	R	U				R	U						U	U							R[b]	3, 4, 9
Spisula solidissima						U	U						U	U					U			20
Yoldia limatula						U	U															20

h Methionine is rapidly turned over and gives rise to taurine and cysteic acid.

i ^{35}S methionine forms labeled cysteine.

j A mixture of labeled amino acids equivalent to a protein hydrolysate of *Chorella* is taken up.

k Glucose is formed at a low rate from labeled leucine.

REFERENCES

1. Allen, J. A. and Garrett, M. R., *Comp. Biochem. Physiol. A,* 41, 307–317, 1972.
2. Allen, W. V. and Kilgore, J., *Comp. Biochem. Physiol. A,* 50, 771–775, 1975.
3. Anderson, J. W. and Bedford, W. B., *Biol. Bull.,* 144, 229–247, 1973.
4. Awapara, J. and Campbell, J. W., *Comp. Biochem. Physiol.,* 11, 231–235, 1964.
5. Bamford, D. R. and Campbell, E., *Comp. Biochem. Physiol. A,* 53, 295–299, 1976.
6. Bamford, D. R. and McCrea, R., *Comp. Biochem. Physiol. A,* 50, 811–817, 1974.
7. Bryant, C., *Comp. Biochem. Physiol.,* 14, 223–230, 1965.
8. Bryant, C., Hines, W. J. W., and Smith, M. J. H., *Comp. Biochem. Physiol.,* 11, 147–153, 1964.
9. Campbell, J. W. and Bishop, S. H., Nitrogen metabolism in molluscs, in *Comparative Biochemistry of Nitrogen Metabolism,* Vol. 1, Campbell, J. W., Ed., Academic Press, New York, 1970, pp. 103–206; see also Campbell, J. W. and Speeg, K. V., *Comp. Biochem. Physiol.,* 25, 3–32, 1969.
10. Carefoot, T. H., *J. Mar. Biol. Assoc. U.K.,* 47, 565–589, 1967.
11. Carefoot, T. H., *Comp. Biochem. Physiol.,* 21, 627–652, 1967.
12. L-Fando, J. J., Garcia-Fernandez, M. C., and R-Caudela, J. L., *Comp. Biochem. Physiol. B,* 43, 807–814, 1972.
13. Gilbertson, D. E. and Jones, K. G., *Comp. Biochem. Physiol. A,* 42, 621–626, 1972.
14. Goreau, T. F., Goreau, N. I., and Yonge, C. M., *J. Zool.,* 169, 417–454, 1973.
15. Hammen, C. S. and Wilbur, K. M., *J. Biol. Chem.,* 234, 1268–1271, 1959.
16. Harrison, C., *Veliger,* 18, 189–193, 1975.
17. Loxton, J. and Chaplin, A. E., *Trans. Biochem. Soc.,* 2, 41–43, 1972.
18. Malanga, C. J. and Crouthamel, W. G., *Am. Zool.,* 12, xxxix, 1972.
19. Péquignat, E., *Mar. Biol.,* 19, 227–244, 1973.
20. Stephens, G. C. and Schinske, R. A., *Limnol. Oceanogr.,* 6, 175–181, 1961.
21. Stewart, M. G. and Bamford, D. R., *Comp. Biochem. Physiol. A,* 52, 67–74, 1975.
22. Tramell, P. R. and Campbell, J. W., *Comp. Biochem. Physiol. B,* 42, 439–450, 1972.
23. Trench, R. K., Boyle, J. E., and Smith, D. C., *Proc. R. Soc. London Ser. B,* 185, 453–464, 1974.
24. van der Horst, D. J. and Oudejans, R. C. H. M., *Comp. Biochem. Physiol. B,* 41, 823–829, 1972.
25. Walker, G., *Proc. Malacol. Soc. London,* 40, 33–44, 1972.
26. de Zwaan, A. and van Marrewijk, W. J. A., *Comp. Biochem. Physiol. B,* 44, 429–439, 1973.
27. de Zwaan, A. and Zandee, D. I., *Comp. Biochem. Physiol. B,* 43, 47–54, 1972.

VII. FATTY ACIDS AND STEROLS

Few dietary experiments have been carried out, and most information has been derived from the investigation of biochemical competence in these animals. No fatty acid requirements have been reported, even potential ones, but some molluscs appear not to synthesize sterols, or to do so at a low rate. This represents a potential requirement for sterols in these species. The information available is summarized below.

Gastropoda

Olive oil is taken up by *Helix aspersa*[9] and by *Arion ater;*[25] oleic acid is also taken up by *Helix pomatia,*[8] but in the two latter species there is evidence that body lipids are increased and that the neutral lipids incorporate the administered material itself.[8,16] *Cepaea nemoralis* utilizes injected palmitic acid for the synthesis of hydrocarbons,[13] whereas *Arion ater* synthesizes lipids from this acid when fed with it;[11] it is also absorbed by *Agriolimax reticulatus.*[23] A purified diet for *Biomphalaria glabrata*[5] contains no lipid, and the addition of triolein and linoleic acid to it had no effect on growth.

The lack of requirement for fatty acids is supported by evidence that molluscs readily synthesize these from acetate. Synthesis has been reported in *Aplysia depilans,*[6,7,21] *Archidoris tuberculata,*[17,18] *Arianta arbustorum,*[17] *Arion* sp.,[17] *Buccinum undatum,*[17] *Crepidula fornicata,*[17] *Cepaea nemoralis,*[12,14] *Dendronotus frondosus,*[17,18] *Helix pomatia,*[17] *Limnaea peregra,*[17] *Limnaea stagnalis,*[17] *Littorina littorea,*[17] *Monodonta turbinata,*[17] *Murex brandaris,*[17] *Nassa* sp.,[22] *Natica cataena,*[17] *Neptunea antiqua,*[17] *Patella coerulea,*[17] *Planorbarius corneus,*[17] *Purpura lapillus,*[17] *Succinea putris,*[17] and *Viviparus fasciatus,*[17] and may be presumed to occur in *Lamellidoris bilamellata* and *Peltodoris atromaculata.*[18]

Sterol synthesis from acetate was also reported in these species, except for *Buccinum undatum, Murex brandaris, Natica cataena, Neptunea antiqua* and *Purpura lapillus,*[17] where synthesis from mevalonate does occur. *Haliotis gurneri* also synthesizes sterols from mevalonate.[10] Sterol synthesis may be a general property of Gastropoda,[18] and hence such compounds would not be essential dietary constituents, although earlier work had suggested that cholesterol, or a contaminant of it, might be essential[5] or at least dietarily valuable[23] to *Helix.* Synthesis appears very slow in *Buccinum,*[17] *Murex,*[17] *Nassa,*[22] *Natica,*[19] *Neptunea,*[17] and *Purpura,*[17] and dietary supplementation may then be necessary.

Bivalvia

Crassostrea virginica has been shown to utilize labeled palmitic acid from a solution in which its concentration was similar to that in the sea (2×10^{-7} M).[1] When adsorbed onto celite, the acid was still used. Its uptake from solution has also been recorded for *Ostrea gryphea.*[3] Cod-liver oil is of higher nutritional value than corn oil for *Crassostrea virginica,*[23] which also digests and absorbs olive oil and peanut oil.[4] Other species, especially *Modiolus demissus,* were used in the latter experiments.[4] The lipids of *Chlorella* are utilized by the scallop, *Chlamys hericia.*[15]

No generalizations regarding sterol synthesis in this group can be made.[17] Synthesis from acetate-2-^{14}C has been reported for *Cardium edule,*[17] *Mytilus californianus,*[17] and *Saxidomus giganteus;*[17] *Crassostrea virginica*[17] appears to synthesize cholesterol, but a C_{29} sterol, tentatively identified as 20-isofucosterol, may be an essential sterol.[17] *Ostrea edulis*[17,19] apparently synthesized sterols from acetate-1-^{14}C in one experiment, but not in another, whereas the sterols of *Anodonta cygnea*[17,20] were not, or hardly, labeled when this material was administered. No synthesis from labeled acetate was detected in *Atrina fragilis,*[19] *Mya arenaria,*[17,20] or *Mytilus edulis;*[17,19] *Cyprina islandica*[17,20] did not use labeled mevalonate for this purpose either, but *Mytilus edulis* is reported to do so.[10] Neither labeled mevalonate nor methionine-CH$_3$-^{14}C gave rise to labeled sterols in *Ostrea gryphea.*[17] Alkylation of existing sterols may account for the incorporation of isotope from acetate-2-[14]C.[20]

REFERENCES

1.   Bunde, T. A. and Fried, M., *Fed. Proc.,* 33, 1337, 1974.
2.   Castell, J. D. and Trider, D. J., *J. Fish. Res. Board Can.,* 31, 95–99, 1974
3.   Fevrier, A., Barbier, M., and Saliot, A., *C. R. Acad. Sci. Ser. D,* 281, 239–241, 1975.
4.   George, W. C., *Biol. Bull.* 102, 118–127, 1952.
5.   Howes, N. H. and Whellock, R. B., *Biochem. J.,* 31, 1489–1498, 1937.
6.   Lupo di Prisco, C., Dessi, F., Tomasucci, M. and Basile, C., *Gen. Comp. Endocrinol.,* 18, 605, 1972.
7.   Lupo di Prisco, C., Dessi Fulgheri, F., and Tomasucci, M., *Comp. Biochem. Physiol. B,* 45, 303–310, 1973.
8.   Oudejans, R. C. H. M. and van der Horst, D. J., *Lipids,* 9, 798–803, 1974.
9.   Sumner, A. T., *J. R. Microsc. Soc.,* 84, 415–421, 1965.
10. Teshima, S. I. and Kanazawa, A., *Comp. Biochem. Physiol. B,* 47, 555–561, 1974.
11. Thompson, G. A., Jr. *J. Biol. Chem.,* 240, 1912–1918, 1965.
12. van der Horst, D. J., *Comp. Biochem. Physiol. B,* 46, 551–560, 1973.
13. van der Horst, D. J. and Oudejans, R. C. H. M., *Comp. Biochem. Physiol. B,* 41, 823–839, 1972.
14. van der Horst, D. J. and Voogt, P. A., *Comp. Biochem. Physiol. B,* 42, 1–6, 1972.
15. Vassallo, M. T., *Comp. Biochem. Physiol. A,* 44, 1169–1175, 1973.

16.   **Vieira, E. C., Senna, I. A., Rogana, S. M. G., and Tupynamba, M. L. V. C.,** in *Germfree Research,* Academic Press, New York, 1973, pp. 657–660.
17.   **Voogt, P. A.,** in *Chemical Zoology,* Vol. 7, Mollusca, Florkin, M. and Scheer, B. T., Eds., Academic Press, New York, 1972.
18.   **Voogt, P. A.,** *Int. J. Biochem.,* 4, 479–488, 1973.
19.   **Voogt, P. A.,** *Comp. Biochem. Physiol. B,* 50, 499–504, 1975.
20.   **Voogt, P. A.,** *Comp. Biochem. Physiol. B,* 50, 505–510, 1975.
21.   **Voogt, P. A. and van Rheenen, J. W. A.,** *Experientia,* 29, 1070–1071, 1973.
22.   **Voogt, P. A. and van Rheenen, J. W. A.,** *Neth. J. Sea Res.,* 6, 409–416, 1973.
23.   **Wagge, L. E.,** *J. Exp. Zool.,* 120, 311–342, 1952.
24.   **Walker, G.,** *Proc. Malacol. Soc. London,* 40, 33–44, 1972.
25.   **Wright, A. A.,** *Comp. Biochem. Physiol. A,* 46, 593–603, 1973, and unpublished observations.

VIII. MISCELLANEOUS ORGANIC COMPOUNDS

No compounds are known to be essential, or even notably beneficial, but a number of different ones have been shown to be available to molluscs or used by them. Dissolved organic matter, in the form of a hydrolysate of *Platimonas viridis* cells, is used by the gastropods *Acmaea testudinalis, Cadlina laevis,* and *Dendronotus frondosus,* and by the bivalve *Mytilus edulis.*[2] For the latter species, the relationship has been further analyzed.[3] In the relation between the gastropod *Tridacna crocea* and its symbiotic zooxanthellae, glycerol appears to be made available by the alga, as well as some glycollic acid,[4] which is also found in *Elysia viridis* as a result of the activity of its symbiotic chloroplasts.[8]

In the gastropod *Arion ater,* chimyl alcohol, long-chain alcohols, and glyceryl ethers are used for the synthesis of various components of the lipid material in the body. Dietary glyceryl ethers suppress the synthesis of this type of compound in the body, as do unsaturated glyceryl ethers, which are incorporated into body lipids, although they are not normal components.[5-7] This effect was suggested in the preceding Part VII, where use of vegetable and animal oils is recorded. The bivalve *Ostrea gryphea* shows uptake of cetylic alcohol and of dotriacontane from solution in sea water.[1]

REFERENCES

1.   **Fevrier, A., Barbier, M., and Saliot, A.,** *C. R. Acad. Sci. Ser. D,* 281, 239–241, 1975.
2.   **Khailov, K. M.,** *Dokl. Biol. Sci.,* 198, 341–344, 1971.
3.   **Khailov, K. M., Finenko, G. A., Burlakova, Z. P., and Smirnov, V. A.,** *Dokl. Akad. Nauk SSSR Ser. Biol.,* 209, 1210–1212, 1973.
4.   **Muscatine, L.,** *Science,* 156, 516–519, 1967.
5.   **Thompson, G. A., Jr.,** *J. Biol. Chem.,* 240, 1912–1918, 1965.
6.   **Thompson, G. A., Jr.,** *Biochemistry,* 5, 1290–1296, 1966.
7.   **Thompson, G. A., Jr.,** *Biochem. Biophys. Acta,* 152, 409–412, 1968.
8.   **Trench, R. K., Boyle, J. E., and Smith, D. C.,** *Proc. R. Soc. London Ser. B,* 185, 453–464, 1974.

IX. NUTRITIONAL VALUE OF ALGAE FOR BIVALVES

The rearing of commercially valuable bivalve shellfish has led to the extensive investigation of the nutritional value of many species of alga so that intensive hatchery techniques could be developed. A complete review of this field is not attempted, but tables indicating the relative values of different algae for this purpose have been given by Loosanoff and Davis,[1] summarizing principally American work, and by Walne,[2] summarizing British work in this field.

The species of alga listed in Table 6 are arranged in approximately descending order of value as foods for the larvae of the clam *Mercenaria mercenaria* and the oyster *Crassostrea virginica.* Several other species of alga were tested, but are not listed, since they were either poor foods or were toxic.

The growth of *Ostrea edulis* fed on various foods was compared to the growth of controls fed on *Isochrysis*. The index (see Table 7) for each experimental series was calculated by dividing the largest mean size in the series at the measurement made on the day nearest to 21 days from the commencement of the test by the mean size of the control on that day. Mean values are given where more than one experiment was carried out.

The growth of *Mercenaria mercenaria* fed on various foods was compared to the growth of controls fed on *Isochrysis* (some fed on *Tetraselmis*). The index (see Table 8) for each experimental series was calculated by dividing the largest mean size in the series at the measurement made on the day nearest to 21 days from the commencement of the test by the mean size of the control on that day. Mean values are given where more than one experiment was carried out.

Table 6
NUTRITIONAL VALUE OF ALGAE FOR *MERCENARIA MERCENARIA* AND *CRASSOSTREA VIRGINICA*

M. mercenaria	*C. virginica*

A. Good Foods

Monochrysis lutheri	*Monochrysis lutheri*
Isochrysis galbana	*Isochrysis galbana*
Dicrateria sp. (B II)	*Chromulina pleiades*[a]
Chlorococcum sp.	*Dicrateria inornata*[a]
Platymonas sp. (1)	*Pyramimonas grossi*[a]
	Hemiselmis refescens[a]

B. Medium Foods

Carteria sp.	
Chlamydomonas sp.	
Cyclotella sp. (0-3A)	*Dunaliella euchlora*
Chlorella sp. (580)	*Platymonas* sp. (1)
Stichococcus sp. (0–18)	*Cyclotella* sp. (0-3A)
Phaeodactylum tricornutum	*Dunaliella* sp.
Skeletonema costatum	*Chlorococcum* sp.
Chlamydomonas sp. (D)	*Chlorella* sp. (UHMC)
Rhodomonas sp.	*Phaeodactylum tricornutum*
Dunaliella sp.	*Cryptomonas* sp.
Olisthodiscus sp.	
Dunaliella euchlora	

[a] Not tested on clam larvae.

From Loosanoff, V. L. and Davis, H. C., *Adv. Mar. Biol.*, 1, 1–136, 1963. With permission.

Table 7
INDEX OF THE NUTRITIONAL VALUE OF ALGAE FOR *OSTREA EDULIS*

Species	Index of food value (No. of Experiments)
Monochrysis lutherii	1.36 (2)
Chaetoceros calcitrans	1.28 (1)
Tetraselmis suecica	1.20 (3)
Skeletonema costatum	1.01 (2)
Isochrysis galbana	1.00
Dicrateria inornata	0.94 (2)
Cryptomonas sp.	0.64 (2)
Cricosphaera carterae	0.62 (3)
Chlorella stigmatophora	0.60 (2)
Phaeodactylum tricornutum	0.59 (3)
Olisthodiscus sp.	0.56 (2)
Nannochloris atomus	0.54 (3)
Chlorella autotrophica	0.52 (2)
Pavlova gyrans	0.50 (2)
Micromonas pusilla	0.44 (3)
Dunaliella euchlora	0.40 (1)
Dunaliella tertiolecta	0.39 (2)
Chlamydomonas coccoides	0.30 (2)

From Walne, P. R., *Fish. Invest. Minist. Agric. Fish. Food G.B. Ser. II,* 26(5), 38, 1970. With permission of the Controller of Her Britannic Majesty's Stationery Office.

Table 8
INDEX OF THE NUTRITIONAL VALUE OF ALGAE FOR *MERCENARIA MERCENARIA*

Species	Index of food value (No. of Experiments)
Skeletonema costatum	3.30 (1)
Pyramimonas grossii	1.19 (1)
Tetraselmis suecica	1.11 (3)
Isochrysis galbana	1.00
Nannochloris atomus	0.92 (3)
Olisthodiscus sp.	0.75 (3)
Micromonas pusilla	0.74 (2)
Cricosphaera carterae	0.70 (3)
Dicrateria inornata	0.67 (3)
Monochrysis lutherii	0.59 (3)
Phaeodactylum tricornutum	0.44 (1)
Chlorella stigmatophora	0.31 (2)
Chlamydomonas coccoides	0.19 (1)
Dunaliella tertiolecta	0.14 (1)

From Walne, P. R., *Fish. Invest. Minist. Agric. Fish. Food G.B. Ser. II,* 26(5), 39, 1970. With permission of the Controller of Her Britannic Majesty's Stationery Office.

REFERENCES

1. **Loosanoff, V. L. and Davis, H. C.,** *Adv. Mar. Biol.,* 1, 1–136, 1963.
2. **Walne, P. R.,** *Fish. Invest. Minist. Agric. Fish. Food G.B. Ser. II,* 26(5), 1970.

QUALITATIVE REQUIREMENTS AND UTILIZATION OF NUTRIENTS: ECHINODERMATA

J. M. Lawrence and M. Jangoux

Part I
Food Used to Maintain Post-metamorphic Echinoderms in the Laboratory

General statements on the subject may be found in References 30, 35, 37, 46, 68, and 72.

Species	Food	Comments	Reference
Crinoidea			
Antedonidae			
Antedon pesatus	Plankton	Pebbles and shells are provided for attachment of cirri.	24
Echinoidea			
Cidaridae			
Eucidaris tribuloides	*Cliona* sp.	Used for absorption and assimilation studies; see Part II, Tables 1 and 2.	53
Diadematidae			
Diadema antillarum	Kenl-Ration dog food		44
Echinothrix calamaris	*Laurencia*	Not natural food.	10
Arbaciidae			
Arbacia punctulata	*Ulva* sp.		14
	Frozen shrimp		35
	Algae, lettuce, spinach		68
	Fucus, Laminaria, Cliona celata		30
Toxopneustidae			
Lytechinus pictus	Sessile diatoms, *Nitzschia* sp.	The algae are grown in plastic dishes, kept in continuous light; the urchins are transferred when the algae have been consumed or have overgrown the dish.	5
	Nitzschia sp.		36

QUALITATIVE REQUIREMENTS AND UTILIZATION OF NUTRIENTS: ECHINODERMATA

Part I (continued)

Species	Food	Comments	Reference
Toxopneustidae (continued)			
Lytechinus variegatus	*Thalassia testudinum*	Used for assimilation studies; see Part II, Table 2.	60
Tripneustes ventricosus	*Thalassia testudinum*	Used for absorption and assimilation studies; see Part II, Tables 1 and 2.	61
Echinidae			
Psammechinus miliaris	*Cardium edule*, *Polydora*-infested oyster shells, *Elminius*		27
	Juveniles are fed *Dunaliella* and *Nitzschia*; fed lettuce 55 days after metamorphosis		51
	Codium dicotomun, *Delesseria*, *Fucus*		23
	Fine suspensions of crab liver, fine algal growth	Juvenile urchins.	9
	Laminaria saccharina, with and without attached bryozoans		40
Paracentrotus lividus	*Laminaria*		18
Echinometridae			
Evechinus chloroticus	Macroscopic algae	Juvenile urchins.	17
Strongylocentrotidae			
Strongylocentrotus purpuratus	*Hedophyllum sessile*		1
	Macrocystis pyrifera		47
	Brown algal species		33
	Nitzschia sp.	Juvenile urchins. The diatoms are grown on plastic dishes; the urchins are transferred to fresh dishes every 5 days or when the algae have been consumed.	37
	Macrocystis, Egregia, Laminaria, Ulva	Adult urchins.	37
	Frozen shrimp		35
	Variety of algal species	Used for absorption and assimilation studies; see Part II; Tables 1 and 2.	49
	Macrocystis sp.		38
			63

QUALITATIVE REQUIREMENTS AND UTILIZATION OF NUTRIENTS: ECHINODERMATA

Part I (continued)

Species	Food	Comments	Reference
Strongylocentrotidae (continued)			
Strongylocentrotus franciscanus	*Macrocystis pyrifera*		48
Strongylocentrotus droebachiensis	*Macrocystis* sp.		50
	Ascophyllum nodosum		64
	Laminaria longicuris	400 fc, seasonal light–dark regime. Used for absorption and assimilation studies; see Part II, Tables 1 and 2.	59
	Variety of algal species	Used for absorption and assimilation studies; see Part II, Tables 1 and 2.	74
Strongylocentrotus intermedius	Variety of algal species	Used for absorption and assimilation studies; see Part II, Tables 1 and 2.	22
	Laminaria ochotensis		43
Strongylocentrotus pulcherrimus	A green alga and a brown alga, various food products	Used for assimilation studies; see Part II, Table 2.	62
Apatopygidae			
Apatopygus recens	Organic matter in bottom sediment		31
Asteroidea			
Luidiidae			
Luidia ciliaris	*Ophiothrix fragilis, Ophiura albida, Ophiocomina nigra Ophiura albida*		6
Luidia sarsi			20
Luidia clathrata	*Donax* sp.; Ralston-Purina dog or cat food		15

QUALITATIVE REQUIREMENTS AND UTILIZATION OF NUTRIENTS: ECHINODERMATA

Part I (continued)

Species	Food	Comments	Reference
Astropectinidae			
Astropecten irregularis	*Spisula subtruncata, Nucula nitida Mya arenaria*		12
Astropecten iranciacus	Mactridae (*Mactra corallina, Spisula subtruncata*), Cardiidae (*Cardium tuberculatum, C. papillosum, Laevicardium* sp.), Donacidae (*Donax venestus, D. semistriatus*).		54
Oreasteridae			
Protoreaster nodosus	Fresh or frozen fish		34
Protoreaster lincki	Dry mussels		39
Culcita schmideliana	Dry mussels		39
Ophidiasteridae			
Gomophia egyptica	Solitary ascidians (e.g., *Ascidia*); sponge (*Tethya*)	Compound ascidians, other sponges, actinians, and corals are not eaten.	80
Pterasteridae			
Pteraster tesselatus	Sponges (*Ectyodoryx parasitica, Mycale adhaerens, Choanites latus*)		66
Asterinidae			
Asterina gibbosa	From metamorphosis to 8 months: egg powder; older than 8 months: *Mytilus* or fish meat		16
	Meat of bivalves; bryozoans; sponges; ascidians (Didemnidae, Botryllidae) *Botryllus* sp.; *Mytilus* meat		75
			39
Patiriella regularis	Crabs or fish.	Used for assimilation studies; see Table II.	13
Echinasteridae			
Echinaster sepositus	Sponges (*Grantia compressa, Adocia simulans, Desmacidon fruticosum, Axinella* sp.)		75

QUALITATIVE REQUIREMENTS AND UTILIZATION OF NUTRIENTS: ECHINODERMATA

Part I (continued)

Species	Food	Comments	Reference
Echinasteridae (continued)			
Henricia sanguinolenta	See *Echinaster sepositus*		75
	Isochrysis galbana, Monochrysis lutheri, Nitzschia closterium, Phaeodactylum tricornutum, Cyclotella nana	Does not eat sponges, but rather the same food as sponges.	65
Acanthasteridae			
Acanthaster planci	Encrusting algae and sedentary microfauna; corals	Early juveniles are herbivorous; late juveniles are carnivorous.	78
	Encrusting algae, *Acropora, Pocillopora, Pavona*	Early juveniles are herbivorous; late juveniles are carnivorous. Used for assimilation studies; see Part II, Table 2.	79
Asteriidae			
Asterias rubens	*Mytilus edulis, Tellina crassa*		76
	Mytilus edulis; juveniles are fed barnacles (*Elminius modestus, Balanus* sp.)		28
	Ciona intestinalis	Used for assimilation studies; see Part II, Table 2.	26
	Mytilus edulis, Balanus balanoides		39
	Mussels		8
Asterias vulgaris	*Mytilus*		42
	Oysters and mussels		70
Asterias forbesi	Barnacles, clams; juveniles eat *Mulinia lateralis*	See Part II, Table 2.	57
	Venus mercenaria		3
	Cracked snails and mussels		4
	Oysters	Used for assimilation studies; see Part II, Table 2.	52
Asterias amurensis	*Venerupsis japonica*		32
	Scapharca broughtoni and *Tapes japonica*		41

QUALITATIVE REQUIREMENTS AND UTILIZATION OF NUTRIENTS: ECHINODERMATA

Part I (continued)

Species	Food	Comments	Reference
Asteriidae (continued)			
Crossaster papposus	*Asterias rubens*		28
Leptasterias pusilla	Cracked snails and mussels		4
Leptasterias hexactis	Barnacles, mussels, chitons, limpets, gastropods		11
Marthasterias glacialis	*Mytilus, Buccinum, Echinocardium*		39
Pisaster ochraceus	*Mytilus californianus*		19
	Mussels		45
	Mytilus californianus		2
Pycnopodia helianthoides	Juveniles are fed on material associated with bottom debris		25
Ophiuroidea			
Ophiolepidinae			
Ophiolepis elegans	Juveniles are fed *Dunaliella* and *Tetraselmis*, and probably also protozoans and bacteria in the culture		72
Ophiocomidae			
Ophiocomina nigra	Chopped mussel and herring, or *Phaeodactylum* sp. and harpacticoid species		21
Ophiactidae			
Ophiophous aculeata	*Sceletonema costatum*		67
Amphiuridae			
Amphiura filiformis	*Dunaliella marina*	Current speed: 5 – 10 cm/sec.	69
	Phaeodactylum sp.		7
Axiognathus squamata	Unicellular algae	Current speed: 2 – 3 cm/sec.	56
Ophiothricidae			
Ophiothrix fragilis	*Nitzschia* sp.		77
	Sceletonema costatum		67

QUALITATIVE REQUIREMENTS AND UTILIZATION OF NUTRIENTS: ECHINODERMATA

Part I (continued)

Holothuroidea

Species	Food	Comments	Reference
Synaptidae			
Leptosynapta crassipatina	Organic material in mud and detritus		71
Rhabdomolgus ruber	Suspended detritus		58
Holothuriidae			
Holothuria forskali	Organic matter in bottom sand; encrusting algae		55

QUALITATIVE REQUIREMENTS AND UTILIZATION OF NUTRIENTS: ECHINODERMATA

Part I (continued)

REFERENCES

1. Allen, W. V., *Comp. Biochem. Physiol. A,* 47, 1297–1311, 1974.
2. Allen, W. V. and Giese, A. C., *Comp. Biochem. Physiol.,* 17, 23–38, 1966.
3. Anderson, J. M., *Biol. Bull.,* 107, 157–173, 1954.
4. Anderson, J. M., *Biol. Bull.,* 128, 1–23, 1965.
5. Brandriff, B., Hinegardner, R. T., and Steinhardt, R., *J. Exp. Zool.,* 192, 13–24, 1975.
6. Brun, E., *J. Mar. Biol. Assoc. U.K.,* 52, 225–236, 1972.
7. Buchanan, J. B., *J. Mar. Biol. Assoc. U.K.,* 44, 565–576, 1964.
8. Bull, H. D., *Rep. Dove Mar. Lab. Ser. 3,* 2, 60–65, 1934.
9. Bull, H. O., *Rep. Dove Mar. Lab. Ser. 3,* 6, 39–42, 1938.
10. Castro, P., in *Aspects of the Biology of Symbiosis,* Chang, T. C., Ed., University Park Press, Baltimore, 1971, 229–247.
11. Chia, F.-S., *Acta Zool.,* 49, 321–364, 1968.
12. Christinsen, A. M., *Ophelia,* 8, 1–134, 1970.
13. Crump, R. G., *J. Exp. Mar. Biol. Ecol.,* 7, 137–162, 1971.
14. Davies, T. T., Crenshaw, M. A., and Heatfield, B. M., *J. Paleontol.,* 46, 874–883, 1972.
15. Dehn, P. F. and Lawrence, J. M., *Fla. Sci.,* 38(Suppl. 1), 6, 1975.
16. Delavault, R., *Bull. Lab. Marit. Dinard,* 47, 3–7, 1962.
17. Dix, T. G., *N.Z. J. Mar. Freshwater Res.,* 6, 48–68, 1972.
18. Fechter, H., *Mar. Biol.,* 19, 285–289, 1973.
19. Feder, H., *Ophelia,* 8, 161–185, 1970.
20. Fenchel, T., *Ophelia,* 2, 223–236, 1965.
21. Fontaine, A. R., *J. Mar. Biol. Assoc. U.K.,* 45, 373–385, 1965.
22. Fuji, A., *Mem. Fac. Fish. Hokkaido Univ.,* 15, 83–160, 1967.
23. Gezelius, G., *Zool. Bijdr.* (Uppsala), 35, 329–337, 1962.
24. Gislen, T., *Zool. Bijdr.* (Uppsala), 9, 1–316, 1924.
25. Greer, D. L., *Pac. Sci.,* 16, 280–285, 1962.
26. Gulliksen, B. and Skjaeveland, S. H., *Sarsia,* 52, 15–20, 1973.
27. Hancock, D. A., *J. Mar. Biol. Assoc. U.K.,* 34, 255–262, 1955.
28. Hancock, D. A., *J. Mar. Biol. Assoc. U.K.,* 37, 565–589, 1958.
29. Hancock, D. A., *Ophelia,* 13, 1–30, 1974.
30. Harvey, E. B., *The American Arbacia and Other Sea Urchins,* Princeton University Press, Princeton, N.J., 1956.
31. Higgins, R. C., *J. Zool.* (London), 173, 505–516, 1974.
32. Hatanaka, M. and Kosaka, M., *Tohoku J. Agric. Res.,* 9, 159–178, 1958.
33. Heatfield, B. M., *Biol. Bull.,* 139, 151–163, 1970.
34. Hildeman, W. H. and Dix, T. G., *Transplantation,* 15, 624–633, 1972.
35. Hinegardner, R. T., in *Methods in Developmental Biology,* Wilt, F. H. and Wessels, N. K., Eds., T. Y. Crowell Co., New York, 1967, 139–155.
36. Hinegardner, R. T., *Biol. Bull.,* 137, 465–475, 1969.
37. Hinegardner, R. T., *Am. Zool.,* 15, 679–689, 1975.
38. Holland, N. D. and Nimitz, Sr. M., *Biol. Bull.,* 137, 280–293, 1964.
39. Jangoux, M., unpublished observations, 1976.
40. Jensen, M., *Ophelia,* 7, 65–78, 1969.
41. Kim, Y. S., *Bull. Fac. Fish. Hokkaido Univ.,* 19, 244–249, 1969.
42. King, H. D., *Wilhelm Roux Arch. Entwicklungsmech. Org.,* 17, 351–363, 1898.
43. Kobayashi, S. and Taki, J., *Calcif. Tissue Res.,* 4, 210–223, 1969.
44. Kristensen, I., *Caribb. J. Sci.,* 4, 441–443, 1964.
45. Landenberger, D. E., *Anim. Behav.,* 14, 414–418, 1966.
46. Lawrence, J. M., *Oceanogr. Mar. Biol. Ann. Rev.,* 13, 213–286, 1975.
47. Lawrence, J. M., Lawrence, A. L., and Giese, A. C., *Physiol. Zool.,* 39, 281–290, 1966.
48. Lawrence, J. M., Lawrence, A. L., and Giese, A. C., unpublished observations, 1963.
49. Leighton, D. L., Doctoral dissertation, University of California, San Diego, 1968.
50. Leighton, D. L., *Nova Hedwigia,* 32, 421–453, 1971.
51. Lundin, L. G., *Hereditas,* 75, 151–153, 1973.
52. MacKenzie, C. L., *Fish. Bull. Fish Wildl. Serv. U.S.,* 68, 67–72, 1969.
53. McPherson, B. F., Doctoral dissertation, University of Miami, Fla., 1968.

QUALITATIVE REQUIREMENTS AND UTILIZATION OF NUTRIENTS: ECHINODERMATA

Part I (continued)

54. Massé, H., *Cah. Biol. Mar.,* 16, 495–510, 1975.
55. Massin, C., unpublished observations, 1976.
56. Martin, R. B., *Tane* (New Zealand), 14, 65, 1968.
57. Mead, A. D., *Bull. U.S. Fish Comm.,* 19, 203–224, 1899.
58. Menker, D., *Mar. Biol.,* 6, 167–186, 1970.
59. Miller, R. J. and Mann, K. H., *Mar. Biol.,* 18, 99–114, 1973.
60. Moore, H. B., Jutare, T., Bauer, J. C., and Jones, J. A., *Bull. Mar. Sci. Gulf Carribb.,* 13, 23–53, 1963.
61. Moore, H. B., Jutare, T., Jones, J. A., McPherson, B. F., and Roper, C. F. E., *Bull. Mar. Sci. Gulf Carribb.,* 13, 267–281, 1963.
62. Nagai, Y. and Kaneko, K., *Mar. Biol.,* 29, 105–108, 1975.
63. Pearse, J. S. and Pearse, V. B., *Am. Zool.,* 15, 731–753, 1975.
64. Percy, J. A., Doctoral dissertation, Memorial University of Newfoundland, St. Johns, 1971.
65. Rasmussen, B. N., *Medd. Dan. Fisk. Havunders.,* 4, 157–213, 1965.
66. Rodenhouse, I. Z. and Guberlet, J. E., *Univ. Wash. Publ. Biol.,* 12, 23–45, 1946.
67. Roushdy, H. M. and Hansen, V. K., *Nature,* 188, 517–518, 1960.
68. Ruggieri, G. D., in *Culture of Marine Invertebrate Animals,* Clark, W. C. and Chanley, M. H., Eds., Plenum Press, New York, 1975, p. 229.
69. Salzwedel, H., *Veroeff. Inst. Meeresforsch. Bremerhaven,* 14, 161–167, 1974.
70. Smith, G. F. M., *J. Fish. Res. Bd. Can.,* 5, 84–103, 1940.
71. Smith, G. N., *J. Exp. Zool.,* 177, 319–330, 1971.
72. Stancyk, S., *Mar. Biol.,* 21, 7–12, 1973.
73. Tyler, A. and Tyler, B. S., in *Physiology of Echinodermata,* Boolootian, R. A., Ed., John Wiley & Sons, New York, 1966, pp. 639–682.
74. Vadas, R. L., Doctoral dissertation, University of Washington, Seattle, 1968.
75. Vasserot, J., *Bull. Soc. Zool. Fr.,* 86, 796–809, 1961.
76. Vevers, H. G., *J. Mar. Biol. Assoc. U.K.,* 28, 165–187, 1949.
77. Warner, G. F. and Woodley, J. D., *J. Mar. Biol. Assoc. U.K.,* 55, 199–210, 1975.
78. Yamaguchi, M., in *Biology and Geology of Coral Reefs: Geology,* Vol. 1, Jones, O. A. and Endean, R., Eds., Academic Press, New York, 1973.
79. Yamaguchi, M., *Pac. Sci.,* 18, 123–138, 1974.
80. Yamaguchi, M., *Micronesica,* 10, 57–64, 1974.

QUALITATIVE REQUIREMENTS AND UTILIZATION OF NUTRIENTS: ECHINODERMATA

J. M. Lawrence and M. Jangoux

Part II

Absorption and Assimilation Efficiencies of Post-metamorphic Echinoderms

Table 1
ABSORPTION EFFICIENCIES

Echinoidea

Food	Absorption efficiency of food[g]*
Cliona sp.	15—50

Cidaridae
Eucidaris tribuloides. Absorption efficiency of food.[9]

Diadematidae
Echinothrix calamaris. Absorption efficiency of food.[2]

Plant	Absorption efficiency of food[a]	Comments
Laurencia sp.	$\bar{x}$ = 83.8 (77.0 – 89.0)	Alga probably never ingested in nature

Toxopneustidae
Lytechinus variegatus. Absorption efficiency of food.[11]

Plant	Absorption efficiency of food[a]	Comments
Thalassia testudinum	54	Winter animals
	57	Summer animals

* Footnotes appear at the end of the table.

QUALITATIVE REQUIREMENTS AND UTILIZATION OF NUTRIENTS: ECHINODERMATA

Part II (continued)

Table 1 (continued)
ABSORPTION EFFICIENCIES

Toxopneustidae (continued)

Lytechinus variegatus. Comparison of absorption efficiency of various marine plants.[8]

Plant	Absorption efficiency of food[c]	Absorption efficiency of lipid[h]	Absorption efficiency of protein[h]	Absorption efficiency of carbohydrate[h]	Comments
Halimeda incrassata	$\bar{x} = 43 \pm 10$	$\bar{x} = 71 \pm 7$	$\bar{x} = 65 \pm 12$	$\bar{x} = 35 \pm 11$	Summer animals
Thalassia testudinum	$\bar{x} = 19 \pm 7$	$\bar{x} = 67 \pm 6$	$\bar{x} = 47 \pm 2$	$\bar{x} = 13 \pm 1$	Summer animals, no epiphytes
Ulva lactuca	$\bar{x} = 13 \pm 2$	$\bar{x} = -53 \pm 28$	$\bar{x} = 9 \pm 14$	$\bar{x} = 15 \pm 3$	Winter animals
Sargassum sp.	$\bar{x} = 6 \pm 7$	$\bar{x} = 43 \pm 6$	$\bar{x} = -35 \pm 2$	$\bar{x} = 7 \pm 8$	Summer animals
Eucheuma isiforme	$\bar{x} = -35 \pm 16$	$\bar{x} = -56 \pm 42$	$\bar{x} = -42 \pm 36$	$\bar{x} = -34 \pm 17$	Summer animals
Syringodium filiforme	$\bar{x} = -3 \pm 11$	$\bar{x} = 45 \pm 10$	$\bar{x} = -1 \pm 10$	$\bar{x} = -5 \pm 12$	Summer animals

Tripneustes ventricosus. Absorption efficiency of food.[11]

Plant	Absorption efficiency of food[a]	Comments
Thalassia testudinum	52	Winter animals
	56	Summer animals

Echinidae

Psammechinus miliaris. Effect of starvation on absorption efficiency.[3]

Plant	Absorption efficiency of food[a]	Absorption efficiency of food[f]	Comments
Fucus vesiculosus	$\bar{x} = 81.1$	$\bar{x} = 85.1$	Fed animals
	$\bar{x} = 63.0$	$\bar{x} = 61.2$	Animals starved before the experiment

QUALITATIVE REQUIREMENTS AND UTILIZATION OF NUTRIENTS: ECHINODERMATA

Part II (continued)

Table 1 (continued)
ABSORPTION EFFICIENCIES

Strongylocentrotidae

Strongylocentrotus intermedius. Comparison of absorption efficiencies for various marine plants.[5]

Plant	Absorption efficiency of food[a]	Absorption efficiency of nitrogen[b]	Comments (month in which efficiencies were measured)
Laminaria japonica	56.7	64.6	January
	66.0	72.1	June
Alaria crassifolia	69.4	70.2	June
Agarum cribrosum	72.6	73.5	June
Sargassum tortile	62.8	63.0	January
Sargassum thunbergii	58.7	69.7	January
Scytosiphon lomentaria	83.4	87.5	January
Ulva pertusa	74.4	86.8	January
	81.9	88.2	June
Chondrus ocellatus	56.8	69.9	January
	61.5	75.9	June
Pachymeniopsis yendoi	68.3	83.5	June
Rhodymenia palmata	74.4	78.9	June
Rhodoglossum pulchrum	76.3	95.1	January
Phyllospadix iwatensis	32.4	67.2	June

Strongylocentrotus intermedius. Comparison of absorption efficiencies for different sizes of echinoids (size group indicates age, with I indicating first year, etc.) and H.D (mm).[5]

Plant	Absorption efficiency of food[a]	Absorption efficiency of nitrogen[b]	Size group	H.D. (food)	H.D. (nitrogen)
Laminaria japonica	60.1	68.1	I	24.6	18.61
	60.8	68.7	II	34.9	28.7
	64.1	68.2	III	41.7	38.7
	58.7	68.6	IV	49.1	46.8
	57.6	71.7	V	60.2	54.8
	57.7	—	VI	69.8	—

QUALITATIVE REQUIREMENTS AND UTILIZATION OF NUTRIENTS: ECHINODERMATA

Part II (continued)

Table 1 (continued)
ABSORPTION EFFICIENCIES

Strongylocentrotidae (continued)

Strongylocentrotus intermedius. Comparison of absorption efficiencies for echinoids fed different amounts of food.[5]

Plant	Absorption efficiency of food[a]	Amount of food ingested (mg/day)
Ulva pertusa	79.8	83.0
	78.2	83.6
	79.8	37.1
	82.6	27.7
	83.2	20.1
Alaria crassifolia	72.4	228.6
	77.1	131.1
	82.1	50.6
	83.2	29.2
	85.1	15.9
Laminaria japonica	65.7	204.9
	68.1	157.3
	72.0	136.8
	75.9	116.2
	80.0	50.2
	81.8	31.8

Strongylocentrotus intermedius. Comparison of absorption efficiencies at different seasons.[5]

Plant	Absorption efficiency of food[a]	Months
Laminaria japonica	55—60	December to June
	73	July and August
	63	September

QUALITATIVE REQUIREMENTS AND UTILIZATION OF NUTRIENTS: ECHINODERMATA

Part II (continued)

Table 1 (continued)
ABSORPTION EFFICIENCIES

Strongylocentrotidae (continued)

Strongylocentrotus purpuratus. Comparison of absorption efficiencies for various marine plants, several at different seasons.[7]

Plant	Absorption efficiency of food[c]	Absorption efficiency of lipid[d]	Absorption efficiency of protein[d]	Absorption efficiency of carbohydrate[d]	Comments (months in which efficiencies were measured)
Macrocystis pyrifera	$\bar{x}$ = 70 ± 7.2 / 62.2	1.47	47.6	66.1	Between September and March / April
Egregia laevigata	$\bar{x}$ = 56.0 ± 13.2 / 34.4	19.0	35.4	35.4	Between September and March / April
Laminaria farlowii	$\bar{x}$ = 74.7 ± 5.5				Between September and March
Eisenia arborea	$\bar{x}$ = 57.1 ± 8.3				Between September and March
Pterygophora californica	$\bar{x}$ = 49.8 ± 6.0 / 44.2	− 22.6	4.8	53.6	Between September and March / April
Cystoseira osmundacea	$\bar{x}$ = 46.2 ± 14.0 / 13.1	−101.4	32.2	−33.4	Between September and March / April
Halidrys dioica	$\bar{x}$ = 53.3 ± 5.5				Between September and March
Pelagophycus porra	$\bar{x}$ = 77.8 ± 4.7				Between September and March
Agarum fimbriatum	$\bar{x}$ = 52.7 ± 11.6				Between September and March
Ulva sp.	$\bar{x}$ = 64.7 ± 4.7				Between September and March
Codium fragile	$\bar{x}$ = 68.0 ± 13.5				Between September and March
Rhodymenia sp.	$\bar{x}$ = 33.5 ± 17.5				Between September and March
Gelidium cartilagneum	$\bar{x}$ = 58.1 ± 33.1				Between September and March
Gigartina arinata	$\bar{x}$ = 51.2 ± 12.9				Between September and March
Gigartina canaliculata	$\bar{x}$ = 30.3 ± 24.1				Between September and March
Phyllospadix scouleri	$\bar{x}$ = 51.5 ± 17.0				Between September and March
Bossliella sp.	$\bar{x}$ = 29.2				Between September and March
Corallina officinalis	$\bar{x}$ = 16.5				Between September and March

Strongylocentrotus purpuratus. Absorption efficiency of food.[4]

Plant	Absorption efficiency of food[e]	Comments
Iridaea flaccidum	89—93	Values obtained 1 to 9 days after feeding

QUALITATIVE REQUIREMENTS AND UTILIZATION OF NUTRIENTS: ECHINODERMATA

Part II (continued)

Table 1 (continued)
ABSORPTION EFFICIENCIES

Strongylocentrotidae (continued)

Strongylocentrotus purpuratus. Comparison of absorption efficiencies for various marine plants.[1]

Plant	Absorption efficiency of food[a]
Macrocystis pyrifera	$\bar{x} = 80 \pm 7$
Egregia laevigata	$\bar{x} = 62 \pm 11$
Petalonaia fascia	$\bar{x} = 51$
Halidrys dioica	$\bar{x} = 45$

Strongylocentrotus purpuratus. Comparison of absorption efficiencies for various marine plants.[1,4]

Plant	Absorption efficiency of food[a]	Comments (months in which efficiencies were measured)
Nereocystis luetkeana	$\bar{x} = 85.2 \pm 5.3$	December, February
Costaria costata	$\bar{x} = 77.4 \pm 9.9$	November
Callophyllis flabellulata	$\bar{x} = 67.9 \pm 16.6$	October
Laminaria saccharina	$\bar{x} = 64.4 \pm 9.9$	December, February
Agarum cribrosum	$\bar{x} = 44.3 \pm 20.7$	July, January
Agarum fimbriatum	$\bar{x} = 36.4 \pm 13.8$	October, January
Monostroma fuscum	$\bar{x} = 28.0 \pm 16.0$	February

Strongylocentrotus droebachiensis. Comparison of absorption efficiency at different seasons.[1,3]

Plant	Absorption efficiency of food[a]	Month
Laminaria digitata	71.2	February
	58.9	July
	42.1	August
	65.1	November

QUALITATIVE REQUIREMENTS AND UTILIZATION OF NUTRIENTS: ECHINODERMATA

Part II (continued)

Table 1 (continued)
ABSORPTION EFFICIENCIES

Strongylocentrotidae (continued)

Strongylocentrotus droebachiensis. Comparison of absorption efficiencies for various marine plants by different size groups of echinoids.[6]

Plant	H.D. (mm)	Absorption efficiency of food[a]	Comments (month in which efficiencies were measured)
Laminaria spp.	9.0—10.9	41.6	February
	19.0—20.9	47.1	
	29.0—30.9	51.9	
	39.0—40.9	42.3	
	49.0—50.9	32.5	
	58.3—60.9	20.8	
	66.9—72.7	9.0	
Laminaria spp.	8.0—11.9	—	September
	28.0—31.9	35.0	
	48.0—51.9	25.1	
Alaria esculenta	8.0—11.9	5.5	September
	28.0—31.9	52.4	
	48.0—51.9	49.6	
Ulva sp.	8.0—11.9	56.4	September
	28.0—31.9	70.5	
	48.0—51.9	66.9	
Fucus vesiculosus	8.0—11.9	—	September
	28.0—31.9	69.0	
	48.0—51.9	76.7	
Ascophyllum nodosum	8.0—11.9	—	September
	28.0—31.9	75.8	
	48.0—51.9	67.4	
Desmarestia viridis	8.0—11.9	23.8	September
	28.0—31.9	69.4	
	48.0—51.9	51.1	
Agarum cribrosum	8.0—11.9	—	September
	28.0—31.9	22.0	
	48.0—51.9	63.4	
Ptilota serrata	8.0—11.9	—	September
	28.0—31.9	10.6	

QUALITATIVE REQUIREMENTS AND UTILIZATION OF NUTRIENTS: ECHINODERMATA

Part II (continued)

Table 1 (continued)
ABSORPTION EFFICIENCIES

Strongylocentrotidae (continued)

Strongylocentrotus droebachiensis. Comparison of absorption efficiencies for various marine plants.[14]

Plant	Absorption efficiency of food[a]	Comments (months in which efficiencies were measured)
Nereocystis luetkeana	$\bar{x} = 83.6 \pm 5.6$	December, February
Costaria costata	$\bar{x} = 77.7 \pm 3.3$	November
Laminaria saccharina	$\bar{x} = 76.8 \pm 11.9$	December, February
Callophyllis flabellulata	$\bar{x} = 62.3 \pm 6.4$	October
Monostroma fuscum	$\bar{x} = 56.3 \pm 8.6$	February
Agarum fimbriatum	$\bar{x} = 45.5 \pm 16.3$	October, January
Agarum cribrosum	$\bar{x} = 40.0 \pm 16.2$	July, January

Strongylocentrotus droebachiensis. Comparison of absorption efficiencies at different seasons for echinoids fed *Laminaria longicruris*.[10]

Time interval	Temperature (°C)	Absorption efficiency[f]
February–March	2	58 ± 9
April–May	4	59 ± 4
June–July	11	61 ± 5
August–September	17	71 ± 2
October–November	8.5	65 ± 3
December–January	7.5	49 ± 6

Strongylocentrotus franciscanus. Comparison of absorption efficiencies for various marine plants.[14]

Plant	Absorption efficiency of food[a]	Comments (months in which efficiencies were measured)
Nereocystis luetkeana	$\bar{x} = 91.2 \pm 1.8$	December, February
Costaria costata	$\bar{x} = 82.6 \pm 1.1$	November
Laminaria saccharina	$\bar{x} = 77.8 \pm 14.8$	December, February
Callophyllis flabellulata	$\bar{x} = 70.8 \pm 19.1$	October
Agarum cribrosum	$\bar{x} = 55.8 \pm 19.9$	July, January
Agarum fimbriatum	$\bar{x} = 51.5 \pm 11.9$	October, January
Monostroma fuscum	$\bar{x} = 47.6 \pm 10.6$	February

QUALITATIVE REQUIREMENTS AND UTILIZATION OF NUTRIENTS: ECHINODERMATA

Part II (continued)

Table 1 (continued)
ABSORPTION EFFICIENCIES

Strongylocentrotidae (continued)

Strongylocentrotus pulcherrimus. Digestibility[b] of artificial diet.[1][2] (See Table 2 for composition of diet.)

Dry matter	54.7%
Nitrogen	64.7%
Lipid	25%

[a] Absorption efficiency of food = $\dfrac{\text{dry weight of food eaten} - \text{dry weight of feces}}{\text{dry weight of food eaten}} \times 100$.

[b] Absorption efficiency of component = $\dfrac{\text{component in food eaten} - \text{component in feces}}{\text{component in food eaten}} \times 100$.

[c] Absorption efficiency of food = $\dfrac{\text{fraction of organic matter in food} - \text{fraction of organic matter in feces}}{(1 - \text{fraction of organic matter in feces}) (\text{fraction of organic matter in food})} \times 100$.

[d] Calculated by Lowe and Lawrence.[8] See footnote h.

[e] Algae grown in ^{14}C-labeled bicarbonate. Absorption efficiency of food = $\dfrac{\text{activity of gut contents}}{\text{activity of alga}} \times 100$.

[f] Absorption efficiency of food = $\dfrac{\text{calories of food eaten} - \text{calories of feces}}{\text{calories of food eaten}} \times 100$.

[g] Absorption efficiency of food = $\dfrac{\text{ash-free dry weight absorbed}}{\text{ash-free dry weight ingested}} \times 100$.

[h] Absorption efficiency of organic component of food = $\dfrac{(\text{organic component}) (\text{fraction of ash in food}) - (\text{organic component}) (\text{fraction of ash in feces})}{(\text{organic component}) (\text{fraction of ash in food})} \times 100$.

QUALITATIVE REQUIREMENTS AND UTILIZATION OF NUTRIENTS: ECHINODERMATA

Part II (continued)

REFERENCES

1. Boolootian, R. A. and Lasker, R., *Comp. Biochem. Physiol.*, 11, 273–289, 1964.
2. Castro, P., in *Aspects of the Biology of Symbiosis,* Chang, T. C., Ed., University Park Press, Baltimore, 1971, 229–247.
3. Faller-Fritsch, R. and Emson, R., unpublished manuscript, 1972.
4. Farmanfarmaian, A. and Phillips, J. H., *Biol. Bull. Mar. Biol. Lab. Woods Hole, Mass.,* 123, 105–120, 1962.
5. Fuji, A., *Mem. Fac. Fish. Hokkaido Univ.,* 15, 83–160, 1967.
6. Himmelman, J. H., Master's thesis, Memorial University of Newfoundland, St. Johns, 1969.
7. Leighton, D. L., Doctoral dissertation, University of California, San Diego, 1968.
8. Lowe, E. and Lawrence, J. M., *J. Exp. Mar. Biol. Ecol.,* 21, 223–234, 1975.
9. MacPherson, B. F., Doctoral dissertation, University of Miami, Florida, 1968.
10. Miller, R. J. and Mann, K. H., *Mar. Biol.,* 18, 99–114, 1973.
11. Moore, H. B. and MacPherson, B. F., *Bull. Mar. Sci.,* 15, 855–871, 1965.
12. Nagai, Y. and Kaneko, K., *Mar. Biol.,* 29, 105–108, 1975.
13. Percy, J. A., Doctoral dissertation, Memorial University of Newfoundland, St. Johns, 1971.
14. Vadas, R. L., Doctoral dissertation, University of Washington, Seattle, 1968.

QUALITATIVE REQUIREMENTS AND UTILIZATION OF NUTRIENTS: ECHINODERMATA

Part II (continued)

Table 2
ASSIMILATION EFFICIENCIES AND UTILIZATION OF NUTRIENTS

Echinoidea

Cidaridae
Eucidaris tribuloides.[13]

Food	Net assimilation efficiency of organic matter[k]*	Gross assimilation efficiency of organic matter[l]
Cliona sp.	17–44%	5–7%

Toxopneustidae
Tripneustes ventricosus.[16]

Plant fed	Gross assimilation efficiency of organic matter[d]
Thalassia testudinum	3.8%

Lytechinus variegatus. Mean population assimilation efficiency, feeding on *Thalassia.*[15]

H.D. (mm)	Gross assimilation efficiency[j]
23	16%
65	5%

Lytechinus variegatus.[16]

Plant fed	Gross assimilation efficiency of organic matter[d]
Thalassia testudinum	3.0%

* Footnotes appear at the end of the table.

QUALITATIVE REQUIREMENTS AND UTILIZATION OF NUTRIENTS: ECHINODERMATA

Part II (continued)

Table 2 (continued)
ASSIMILATION EFFICIENCIES AND UTILIZATION OF NUTRIENTS

Echinidae

Psammechinus miliaris. Comparison of the coefficient of growth anabolism, E^m, of young sea urchins.[9]

Food	E
Laminaria saccharina without epizoic bryozoans	3.4
Laminaria saccharina with epizoic bryozoans	7.6

Strongylocentrotidae

Strongylocentrotus intermedius. Comparison of somatic and gonadal growth (mg dry weight/day) and of net assimilation efficiency (%) of echinoids fed various marine plants. H.D. = 46.83–51.96 mm.[5]

Plant fed	Somatic growth	Gonadal growth	Assimilation efficiency[a]
Alaria crassifolia	23.8	15.9	28.7
Scytosiphon lomentaria	15.5	23.1	18.9
Laminaria japonica	9.5	10.4	14.8
Rhodymenia palmata	12.0	6.7	25.9
Ulva pertusa	12.2	5.9	27.6
Sargassum tortile	6.6	6.5	16.9
Pachymeniopsis yendoi	0.4	6.5	9.9
Agarum cribrosum	2.9	0.2	4.5

Strongylocentrotus intermedius. Comparison of somatic and gonadal growth (mg dry weight/day and mg protein/day), net assimilation efficiency (% dry weight), and net assimilation efficiency of protein (% protein) of different sizes of echinoids fed *Laminaria japonica.*[5]

Size group	H.D. (mm)	Somatic growth		Gonadal growth		Assimilation efficiency	
		Dry weight	Protein	Dry weight	Protein	Dry weight[a]	Protein[b]
I	18.6	18.4	1.1	6.1	1.6	51.2	61.2
II	28.7	21.5	1.2	11.8	3.2	39.4	57.0
III	38.7	9.8	0.4	14.8	4.0	21.8	42.9
IV	46.8	6.1	0.12	17.5	4.8	17.6	39.4
V	54.8	2.4	-0.21	21.1	5.6	14.5	37.3

QUALITATIVE REQUIREMENTS AND UTILIZATION OF NUTRIENTS: ECHINODERMATA

Part II (continued)

Table 2 (continued)

ASSIMILATION EFFICIENCIES AND UTILIZATION OF NUTRIENTS

Strongylocentrotidae (continued)

Strongylocentrotus intermedius. Comparison of somatic and gonadal growth (mg dry weight/day), and of net assimilation efficiency (%) of echinoids fed different amounts of algae (mg dry weight/day).[5]

Plant fed	Amount of food	Somatic growth	Gonadal growth	Assimilation efficiency[a]
Ulva pertusa	83.0	11.7	9.4	32.0
	83.6	12.2	5.9	27.6
	37.1	3.7	2.6	21.4
	27.7	3.2	1.2	19.5
	20.1	0.3	1.0	7.5
Alaria crassifolia	228.6	22.6	28.1	30.6
	131.1	16.4	10.5	26.4
	50.6	7.6	2.0	23.1
	29.2	0.7	0.3	3.8
	15.9	0.1	-0.5	-2.7
Laminaria japonica	204.7	7.1	16.7	17.6
	157.3	5.9	13.2	17.8
	136.8	9.9	6.8	16.9
	116.2	9.9	4.9	16.8
	50.7	3.5	2.0	13.7
	31.8	0.2	1.5	6.5

Strongylocentrotus purpuratus. Comparison of gross assimilation efficiencies (%) of total matter consumed and of organic matter consumed of echinoids fed different algae in January.[10]

Plant	Gross assimilation efficiency of total food[c]	Gross assimilation efficiency of organic matter[d]	Organic matter in plant (% dry weight)
Macrocystis sp.	30.2	8.8	83.3
Egregia sp.	33.6	12.0	67.8
Pterygophora sp.	17.1	5.3	77.9
Cystoseira sp.	29.0	9.6	73.2

QUALITATIVE REQUIREMENTS AND UTILIZATION OF NUTRIENTS: ECHINODERMATA

Part II (continued)

Table 2 (continued)

ASSIMILATION EFFICIENCIES AND UTILIZATION OF NUTRIENTS

Strongylocentrotidae (continued)

Strongylocentrotus purpuratus. Comparison of gross assimilation efficiencies (%), gonadal development, and increase in H.D. (mm) of echinoids fed different algae in autumn (period of gonadal development).[10]

Plant	Gross assimilation efficiency of total food[c]	Gonad index[f]	Increase in H.D.
Macrocystis sp.	12.7	18.9	1.1
Egregia sp.	13.3	10.3	1.2
Pterygophora sp.	6.3	12.8	0.5
Bossiella sp.	0.4	7.3	0.4

Strongylocentrotus purpuratus. Comparison of gross assimilation efficiencies (%), gonadal development, and increase in H.D. (mm) of echinoids fed different algae in spring.[10]

Plant	Gross assimilation efficiency[c]	Gonad index[f]	Increase in H.D.
Macrocystis sp.	24.0	4.1	1.2
Pterygophora sp.	2.42	3.0	2.5
Ulva sp.	24.3	1.4	1.9
Codium sp.	1.4	1.8	0.2
Rhodymenia sp.	15.9	1.9	1.1
Gigartina sp.	18.8	2.0	1.1

Strongylocentrotus purpuratus. Comparison of gross assimilation efficiencies (%) of echinoids fed different components of *Macrocystis*.[11]

Plant component	Gross assimilation efficiency[e]
Blades	16.2
Haptera	13.9

Strongylocentrotus franciscanus. Comparison of gross assimilation efficiencies (%) of echinoids fed different components of *Macrocystis*.[11]

Plant component	Gross assimilation efficiency[e]
Blades	10.8
Haptera	4.5

QUALITATIVE REQUIREMENTS AND UTILIZATION OF NUTRIENTS: ECHINODERMATA

Part II (continued)

Table 2 (continued)

ASSIMILATION EFFICIENCIES AND UTILIZATION OF NUTRIENTS

Strongylocentrotidae (continued)

Strongylocentrotus droebachiensis. Comparison of the growth of echinoids fed different algae from January 1965 to March 1966.[19]

Plant fed	Mean initial wet weight (g)	Mean initial H.D. (mm)	Mean net gain in wet weight/month	Mean net gain in H.D./month
Nereocystis luetkeana	2.6	18	11.9	2.3
	1.1	12	14.0	2.4
Laminaria saccharina	3.6	22	11.0	1.9
	1.1	13	10.5	1.8
Agarum fibriatum	3.2	20	4.5	1.2
	1.0	12	4.9	1.4
Agarum cribrosum	2.8	19	4.0	1.2
	1.1	12	4.9	1.4

Strongylocentrotus droebachiensis. Comparison of continued growth and gonadal development of echinoids fed for an additional 5 months either the original food or another food.[19]

Plant fed	Original plant fed for 15 months	Gonad index[g] at end of experiment	Mean net gain in wet weight/month during the 5 months
Laminaria saccharina	*L. saccharina*	20.6	15.9
Laminaria saccharina	*A. fimbriatum*	16.8	12.0
Nereocystis luetkeana	*N. luetkeana*	27.1	10.1
Nereocystis luetkeana	*A. cribrosum*	22.4	14.9
Agarum cribrosum	*A. cribrosum*	7.2	8.3
Agarum cribrosum	*N. luetkeana*	13.7	1.8
Agarum fimbriatum	*A. fimbriatum*	8.7	6.1
Agarum fimbriatum	*L. saccharina*	11.6	6.2

Strongylocentrotus droebachiensis. Mean population assimilation efficiency (%).[14]

	Not including estimated loss by dissolved organic material	Including estimated loss by dissolved organic material
Gross efficiency[h]	13	4.3
	22	7.0

QUALITATIVE REQUIREMENTS AND UTILIZATION OF NUTRIENTS: ECHINODERMATA

Part II (continued)

Table 2 (continued)

ASSIMILATION EFFICIENCIES AND UTILIZATION OF NUTRIENTS

Strongylocentrotidae (continued)

Strongylocentrotus pulcherrimus. Seasonal growth of sea urchins reared on a diet of 18% (of dry weight) white-fish meal, 18% soybean meal, 18% yellow corn, 18% yeast, 5% soy bean oil, 22.9% agar-agar, and 0.1% vitamin mixture. One gram of the vitamin mixture contained the following: A, 2,500 IU; D_2, 200 IU; B_1 (nitrate), 1 mg; B_2, 15 mg; B_6 (nitrate), 1 mg; B_{12}, 1 μg; C, 37.5 mg; E, 1 mg; nicotinic acid, 10 mg; folic acid, 0.5 mg; pantothenic acid (Ca salt), 5 mg.[17]

Month	Average individual wet weight (g)
August	9.2
September	10.3
October	10.8
November	11.3
December	11.4
January	11.4

Asteroidea

Luidiidae

Luidia sarsi.[4] Growth efficiency[e] of animals fed *Ophiura albida*: 33%

Astropectinidae

Astropecten irregularis. Net growth efficiency[k] of animals fed different diets.[1]

Diet	Growth efficiency
Spisula subtruncata	38%
Nucula nitida	48%

QUALITATIVE REQUIREMENTS AND UTILIZATION OF NUTRIENTS: ECHINODERMATA

Part II (continued)

Table 2 (continued)
ASSIMILATION EFFICIENCIES AND UTILIZATION OF NUTRIENTS

Asterinidae

Patiriella regularis. Comparison of wet weights of the entire animal and of indices of body components[n] of starved and of fed animals. Means ±1 S.D. are given.[2]

	Starved	Fed crabs (*Hemiplax hirtipes*)	Fed fish (red mullet or cod)
Wet body weight	8.9	19.7	17.3
Pyloric caeca index	4.79 ± 0.74	25.23 ± 4.35	21.47 ± 4.61
Gonad index	1.36 ± 1.45	26.13 ± 10.69	17.14 ± 6.68

Acanthasteridae

Acanthaster planci. Comparison of the time (days) necessary for different sizes of juveniles to double their total diameter. The animals were fed encrusting algae during the herbivorous stage, and *Acropora*, *Pocillopora*, and *Pavona* during the carnivorous stage.[20]

Stage	Doubling time
Small, herbivorous juveniles	38 days
Large, carnivorous juveniles	61—63 days

Acanthaster planci. Comparison of the time (days) necessary for carnivorous juveniles to double their weight[o] on two different coral diets.[20]

Diet	Doubling time
Acropora nasuta	24.24
Pocillopora damicornis	22.92

QUALITATIVE REQUIREMENTS AND UTILIZATION OF NUTRIENTS: ECHINODERMATA

Part II (continued)

Table 2 (continued)

ASSIMILATION EFFICIENCIES AND UTILIZATION OF NUTRIENTS

Asteriidae

Asterias amurensis. Comparison of the "efficiency of conversion"[e] of adult animals fed *Veneruptis japonicus* at different seasons.[8]

Date	Average efficiency of conversion
April 9 to May 8	55.5
May 9 to June 7	37.8
June 8 to July 7	11.9
July 23 to August 21	−119.0
August 22 to September 20	−81.6
September 21 to October 19	33.1
October 20 to November 11	47.9
November 12 to November 26	7.1
November 27 to December 11	16.0
December 2 to December 18	−209.4

Asterias vulgaris. Comparison of disc diameter growth in animals of different sizes, fed oysters and mussels.[18]

Diet	Original mean size	Mean % increase in 4 months
Oysters	5.4	35
	7.7	13
	10.5	4
Mussels	5.5	73
	7.2	52

Asterias rubens. Gross assimilation efficiency[c] of animals fed *Ciona intestinalis.*[6]

$\bar{x}$ = 4.3%; range: 3.5−5.2%

QUALITATIVE REQUIREMENTS AND UTILIZATION OF NUTRIENTS: ECHINODERMATA

Part II (continued)

Table 2 (continued)

ASSIMILATION EFFICIENCIES AND UTILIZATION OF NUTRIENTS

Asteriidae (continued)

Asterias rubens. Increase in radius of animals fed mussels from September 1 to December 7, and barnacles from December 7 to December 30.[7]

Date	Radius (mm)
September 17	12—13
September 30	10—17
October 31	15—24
December 31	16—29

Asterias forbesi. Effect of temperature on feeding rate and weight gain.[12]

Temperature (°C)	Number of oysters consumed in 28 days	Gain in wet weight (g) in 28 days
5	2.3	+ 4.49
10	3.0	+ 7.19
15	4.1	+13.79
22.8	2.8	+ 5.89

Pisaster ochraceus.[3]   Growth efficiency[e] of animals fed *Mytilus californianus*: 43%.

QUALITATIVE REQUIREMENTS AND UTILIZATION OF NUTRIENTS: ECHINODERMATA

Part II (continued)

[a] Net assimilation efficiency = growth (dry weight/day) × 100/absorbed food (dry weight/day).

[b] Net assimilation efficiency of protein = growth (protein weight/day) × 100/absorbed protein (protein weight/day).

[c] Gross assimilation efficiency = growth (dry weight) × 100/ingested food (dry weight).

[d] Gross assimilation efficiency = growth (organic matter) × 100/ingested food (dry weight).

[e] Gross assimilation efficiency = growth (wet weight) × 100/ingested food (wet weight).

[f] Gonad index = ml of gonad × 100/wet weight of animal.

[g] Gonad index = wet weight of gonad × 100/wet weight of drained testis.

[h] Gross assimilation efficiency = growth (kcal/m²/year) × 100/ingested food (kcal/m²/year).

[i] Net assimilation efficiency = growth (kcal/m²/year) × 100/absorbed food (kcal/m²/year).

[j] Gross assimilation efficiency = growth (dry weight/m²/30 days) × 100/ingested food (dry weight/m²/30 days).

[k] Net assimilation efficiency = growth (ash-free weight) × 100/absorbed nutrients (ash-free weight).

[l] Gross assimilation efficiency = growth (ash-free weight) × 100/ingested food (ash-free weight).

[m] E = dl/dt, horizontal test diameter measured.

[n] Body component index = g wet weight of component × 100/g wet total body weight.

[o] Doubling time = 1/growth coefficient.

REFERENCES

1. Christinsen, A. M., *Ophelia,* 8, 1–134, 1970.
2. Crump, R. G., *J. Exp. Mar. Biol. Ecol.,* 7, 137–162, 1971.
3. Feder, H. M., *Ophelia,* 8, 161–185, 1970.
4. Fenchel, T., *Ophelia,* 2, 223–236, 1965.
5. Fuji, A., *Mem. Fac. Fish. Hokkaido Univ.,* 15, 83–160, 1967.
6. Gulliksen, B. and Skjaeveland, S. H., *Sarsia,* 52, 15–20, 1973.
7. Hancock, D. A., *J. Mar. Biol. Assoc. U.K.,* 37, 565–589, 1958.
8. Hatanaka, M. and Kosaka, M., *Tohoku J. Agric. Res.,* 9, 159–178, 1958.
9. Jensen, M., *Ophelia,* 7, 65–78, 1969.
10. Leighton, D. L., Doctoral dissertation, University of California, San Diego, 1968.
11. Leighton, D. L., *Nova Hedwigia,* 32, 421–453, 1971.
12. MacKenzie, C. L., *U.S. Fish Wildl. Serv. Fish. Bull.,* 68, 67–72, 1969.
13. MacPherson, B. F., Doctoral dissertation, University of Miami, Florida, 1968.
14. Miller, R. J. and Mann, K. H., *Mar. Biol.,* 18, 99–114, 1973.
15. Moore, H. B., Jutare, T., Bauer, J. C., and Jones, J. A., *Bull. Mar. Sci. Gulf Caribb.,* 13, 23–53, 1963.
16. Moore, H. B., Jutare, T., Jones, J. A., MacPherson, B. F., and Roper, C. F. E., *Bull. Mar. Sci. Gulf Caribb.,* 13, 267–281, 1963.
17. Nagai, Y. and Kaneko, K., *Mar. Biol.,* 29, 105–108, 1975.
18. Smith, G. F. M., *J. Fish. Res. Bd. Can.,* 5, 84–103, 1940.
19. Vadas, R. L., Doctoral dissertation, University of Washington, Seattle, 1968.
20. Yamaguchi, M., *Pac. Sci.,* 28, 123–138, 1974.

QUALITATIVE NUTRITIONAL REQUIREMENTS OF ECHINODERM LARVAE

J. M. Lawrence, L. Fenaux, and M. Jangoux

General statements on the subject may be found in References 4, 13, 15, 21—23, 26, 33—35, and 41—43. Echinoderm larvae with indirect development require feeding when the larval stage is reached; references concerned with the food fed the larval stages are listed here. Echinoderm larvae with direct development do not require feeding and will not be considered.

Species	Food	Comments	Ref.
Holothuroidea			
Stichopodidae			
Stichopus japonicus	*Monas* sp.		24, 25
Parastichopus californicus	*Phaeodactylus tricornutum* Bohlin; *Dunaliella tertiolecta* Butcher; *Isochrysis galbana* Parke.	Through metamorphosis.	42
Echinoidea			
Phymosomatidae			
Glyptocidaris crenularis	*Peridinium* sp.		12
Diadematidae			
Diadema setosum	Diatoms	To pluteus.	38
Diadema antillarum	*Nitzschia closterium*	First larval stage.	33
Arbaciidae			
Arbacia punctulata	*Nitzschia closterium*; *Lichmophora* sp.	Through metamorphosis; 12 plutei/15 ml.	19
	N. closterium and diatoms.	Through metamorphosis.	16
	Dunaliella tertiolecta (Best). *Rhodomonas* sp. and *Pyranimonas* sp. also used.	Through metamorphosis. Fed once a day (3000 algal cells/ml for young plutei); one individual/ml medium for mature larvae. Unsatisfactory algae: *Amphidinium operculayum, Coccolithus huxleyi, Cryptomonas* sp., *Cyclotella nana, Cylindrotheca closterium, Eutreptiella* sp., *Isochrysis galbana, Melosira nummuloides, Monochrysis lutheri, Nitzschia brevirostris, Phaeodactylum tricornutum.*	22

QUALITATIVE NUTRITIONAL REQUIREMENTS OF ECHINODERM LARVAE (continued)

Species	Food	Comments	Ref.
Arbaciidae (continued)			
Arbacia lixula	*D. tertiolecta*	Through metamorphosis. See comments for *Lytechinus pictus*.	3
	Planktonic coccus	Suggests nutritional role for bacteria in pluteus stage.	8
	Platymonas	Through metamorphosis.	40
Temnopleuridae			
Temnopleurus toreumaticus	Diatoms	To final pluteus stage.	38
Mespilia globulus	Diatoms	Through metamorphosis.	38
Toxopneustidae			
Toxopneustes pileolus	Diatoms	Through metamorphosis.	38
Tripneustes gratilla	Diatoms	To pluteus.	38
Lytechinus variegatus	*Nitzschia closterium*	To echinopluteus.	10
	Fresh-water green algal species grown in sea water.	Through metamorphosis; 100 embryos/125 ml.	31
	Rhodomonas sp.	Through metamorphosis. See *Arbacia punctulata*.	22
Lytechinus pictus	*Rhodomonas* sp.; *Pyranimonas* sp.	Through metamorphosis. See *Arbacia punctulata*.	22
	Dunaliella tertiolecta (Best)	Through metamorphosis; 50 units of penicillin/ml.	1
	Rhodomonas sp.	Through metamorphosis; 400 plutei/6 l. Fed three times per week.	3
Pseudocentrotus depressus	*Skeletonema costatum, Chaetoceros simplex, Chlamydomonas* sp. *C. simplex, Chlamydomonas* sp.	Little difference in growth on different algae. Growth greatest at algal concentrations of 10^4 cells/ml or 10^5 cells/ml.	24
			45
Sphaerechinus granularis	Planktonic coccus.	Suggests nutritional role of bacteria in pluteus stage.	8
Echinidae			
Echinus esculentus	Unicellular green algae.	150 larvae/10 gal in later stages.	30
Paracentrotus lividus	Planktonic coccus.	Suggests nutritional role of bacteria in pluteus stage.	8

QUALITATIVE NUTRITIONAL REQUIREMENTS OF ECHINODERM LARVAE (continued)

Species	Food	Comments	Ref.
Echinidae (continued)			
Parechinus angulosus	*Dunaliella primolecta* (Plymouth Coll. No. 81), *Monochrysis lutheri* (Plymouth Coll. No. 75), *Phaeodactylum tricornutum* (Plymouth Coll. No. 100), unidentified flagellate species.	Through metamorphosis; 20—50 larvae/ml.	5
Psammechinus miliaris	*Dunaliella* sp.; dry fish food mixture of Tetra min (Tetra Werke product, W. Germany) and Hykro Shrimpmeal (Hykro, Denmark).	Through metamorphosis.	29
Echinometridae			
Echinometra mathaei	Diatoms	Through metamorphosis.	38
	Rhodomonas sp.	Through metamorphosis.	22
Echinostrephus molaris	Diatoms	To pluteus.	38
Evechinus chloroticus	"Growth" on container walls; *Phaeodactylus tricornutum*.	Through metamorphosis.	9
Strongylocentrotidae			
Strongylocentrotus purpuratus	*Rhodomonas* sp.	Through metamorphosis.	22
	Phaeodactylum tricornutum Bohlin; *Dunaliella tertiolecta* Butcher; *Isochrysis galbana* Parke.	Through metamorphosis.	42
Strongylocentrotus pallidus	*P. tricornutum* Bohlin; *D. tertiolecta* Butcher; *I. galbana* Parke.	Through metamorphosis.	42
Strongylocentrotus franciscanus	*P. tricornutum* Bohlin; *D. tertiolecta* Butcher; *I. galbana* Parke.	Through metamorphosis.	42
Strongylocentrotus droebachiensis	*P. tricornutum* Bohlin; *D. tertiolecta* Butcher; *I. galbana* Parke.	Through metamorphosis.	42
Strongylocentrotus pulcherrimus	Diatoms	To pluteus.	38
Allocentrotus fragilis	*P. tricornutum* Bohlin; *D. tertiolecta* Butcher; *I. galbana* Parke.	Through metamorphosis.	42

QUALITATIVE NUTRITIONAL REQUIREMENTS OF ECHINODERM LARVAE (continued)

Species	Food	Comments	Ref.
Clypeasteridae *Clypeaster humilis*	*Nitzschia closterium*; *Chlorella sparchi*; *Chlamydomonas* sp.; *Dunaliella* sp.	Through metamorphosis. Growth on algal cultures less reliable than replacing medium with fresh sea water each day.	34
Dendrasteridae *Dendraster excentricus*	*Phaeodactylum tricornutum* Bohlin; *Dunaliella tertiolecta* Butcher; *Isochrysis galbana* Parke.	Through metamorphosis.	42
Mellitidae *Mellita quinquiesperforata*	*Dunaliella* sp.	Through metamorphosis.	2
Echinolampadidae *Echinolampas (Palaeolampas) crassa*	*Phaeodactylum tricornutum* (Plymouth Coll. No. 100);*Monochrysis lutheri* (Plymouth Coll. No. 75); unidentified flagellates and diatoms.	Through metamorphosis; 20—50 larvae/mł.	6
Schizasteridae *Brisaster latifrons*	*Phaeodactylum tricornutum* Bohlin; *Dunaliella tertiolecta* Butcher; *Isochrysis galbana* Parke.	Through metamorphosis.	42
Loveniidae *Echinocardium cordatum*	*Dunaliella marina*	Through metamorphosis.	11
Spatangidae *Spatangus purpureus*	*Dunaliella marina*	Through metamorphosis.	11

QUALITATIVE NUTRITIONAL REQUIREMENTS OF ECHINODERM LARVAE (continued)

Species	Food	Comments	Ref.
Ophiuroidea			
Ophiactadae			
Ophiactis balli	Diatoms	To 26-day-old ophiopluteus.	32
Ophiopholis aculeata	Phytoplankton from sea water; cultures from sea urchin stomachs; *Dunaliella* sp.	Through metamorphosis. *Chlamydomonas* sp. not digested.	37
	Phaeodactylum tricornutum Bohlin; *Dunaliella tertiolecta* Butcher; *Isochrysis galbana* Parke.	Through metamorphosis.	42
Ophiothricidae			
Ophiothrix quinquemaculata	*Phaeodactylum tricornulatum* West; *Platymonas* sp. Bohlin.	Through metamorphosis; 12 or more larvae/150 ml.	18
Asteroidea			
Luidiidae			
Luidia foliolata	*Phaeodactylum tricornutum* Bohlin; *Dunaliella tertiolecta* Butcher; *Isochrysis galbana* Parke.	Through metamorphosis.	42
Asterinidae			
Asterina glacialis	*Nitzschia* sp.	To 4-week-old bipinnaria. *Nitzschia* apparently not digested.	32
Patiria miniata	Diatoms	Slow regression. No growth.	36
	Phaeodactylum tricornutum Bohlin; *Dunaliella tertiolecta* Butcher; *Isochrysis galbana* Parke.	To brachiolarian stage.	42
Acanthasteridae			
Acanthaster planci	*Isochrysis galbana; Gymnodinium* sp.; *Amphidinium* sp.; *Cyclotella nana; Dunaliella primolecta.*		20
	Amphidinium; D. primolecta or *Monochrysis lutheri*	*Amphidinium* necessary for development; used with either of the other two species.	28
	Nitzschia closteridium; Chlamydomonas sp.	Through metamorphosis.	46
	D. primolecta	Through metamorphosis.	39

QUALITATIVE NUTRITIONAL REQUIREMENTS OF ECHINODERM LARVAE (continued)

Species	Food	Comments	Ref.
Asteriidae			
Asterias rubens	*Nitzschia*	Better with bacterial and flagellate infection.	14
Asterias forbesi	*Phaeodactylum tricornutum*		44
	Mixed cultures of *Chlorella, Nitzschia, Dunaliella,* and phytomonads; *Nitzschia* plus bacteria and flagellates.		27
Marthasterias glacialis	Filtered suspension of yolk or *Chlorella.*	*Chlorella* better as food.	7
Pisaster ochraceus	*Phaeodactylum tricornutum* Bohlin; *Dunaliella tertiolecta* Butcher; *Isochrysis galbana* Parke.	Through metamorphosis.	42
Pycnopodia helianthiodes	*Nitzschia* and *Dunaliella.* Marine protozoans and diatoms added to diet of older larvae.		17
	P. tricornutum Bohlin; *D. tertiolecta* Butcher; *I. galbana* Parke.	Through metamorphosis.	42

QUALITATIVE NUTRITIONAL REQUIREMENTS OF ECHINODERM LARVAE (continued)

REFERENCES

1. Brandiff, B., Hinegardner, R. T., and Steinhardt, R., *J. Exp. Zool.,* 192, 13–24, 1975.
2. Caldwell, J., *Bull. Assoc. Southeast. Biol.,* 20, 43, 1973.
3. Cameron, R. A. and Hinegardner, R. T., *Biol. Bull.,* 146, 335–342, 1974.
4. Costello, D. P., Davidson, M. E., Eggers, A., Fox, M. H., and Henley, C., *Methods of Obtaining and Handling Marine Eggs and Embryos,* Marine Biological Laboratory, Woods Hole, Massachusetts, 1957.
5. Cram, D. L., *Trans. R. Soc. S. Afr.,* 39, 321–337, 1971.
6. Cram, D. L., *Trans. R. Soc. S. Afr.,* 39, 339–352, 1971.
7. Delage, Y., *Arch. Zool. Exp. Gen.,* 2, 27–46, 1904.
8. Deveze, L., *Recl. Trav. Stn. Mar. Endoume Fac. Sci. Marseilles,* No. 8, 55–59, 1953.
9. Dix, T. G., *N.Z. J. Mar. Freshwater Res.,* 3, 13–16, 1969.
10. Eastwood, J. L. J., Doctoral dissertation, Lehigh University, Bethlehem, Pa., 1972.
11. Fenaux, L., *Bull. Mus. Natl. Hist. Nat. Zool.,* 3(31), Zool, 25, 297–304, 1972.
12. Fukushi, T., *Bull. Mar. Biol. Stn. Asamushi Tohoku Univ.,* 10, 57–63, 1970.
13. Needham, J. G., Ed., *Culture Methods for Invertebrate Animals,* Dover, New York, 1937.
14. Gemmil, J. F., *Phil. Trans. R. Soc. London Ser. B,* 205, 213–234, 1914.
15. Giudice, G., *Developmental Biology of the Sea Urchin Embryo,* Academic Press, New York, 1973.
16. Gordon, I., *Phil. Trans. R. Soc. London Ser. B,* 217, 289–334, 1929.
17. Greer, D. L., *Pac. Sci.,* 16, 280–285, 1962.
18. Guille, A., *Vie et Milieu,* 15, 243–308, 1964.
19. Harvey, E. B., *The American Arbacia and Other Sea Urchins,* Princeton University Press, Princeton, N.J., 1956.
20. Henderson, J. A. and Lucas, J. S., *Nature,* 232, 255–257, 1971.
21. Hinegardner, R. T., in *Methods in Developmental Biology,* Wilt, F. H. and Wessels, N. K., Eds., Crowell, New York, 1967, 139–155.
22. Hinegardner, R. T., *Biol. Bull.,* 137, 465–475, 1969.
23. Hinegardner, R. T., *Am. Zool.,* 15, 679–689, 1975.
24. Hirano, R. and Oshima, Y., *Bull. Jpn. Soc. Sci. Fish.,* 29, 282–297, 1963.
25. Imai, T., Inabra, D., Sato, R., and Hatanaka, M., *Bull. Inst. Agric. Res. Tohoku Univ.,* 2, 269–276, 1950.
26. Just, E. E., *Basic Methods for Experiments on Eggs of Marine Animals,* P. Blatiston's Sons and Co., Philadelphia, 1939.
27. Larson, E. J., in *Culture Methods for Invertebrate Animals,* Needham, J. G., Ed., Dover, New York, 1937, 552.
28. Lucas, J. E., *Micronesica, J. Coll. Guam,* 9, 197–203, 1973.
29. Lundin, L. G., *Hereditas,* 75, 151–153, 1973.
30. MacBride, E. W., *Phil. Trans. R. Soc. London Ser. B,* 195, 285–327, 1903.
31. Mazur, J. E. and Miller, J. W., *Ohio J. Sci.,* 71, 30–36, 1971.
32. Mortensen, Th., *J. Mar. Biol. Assoc. U. K.,* 10, 1–18, 1913.
33. Mortensen, Th., *Studies of the Development and Larval Forms of Echinoderms,* Gad, G. E. C., Ed., Copenhagen, 1921.
34. Mortensen, Th., *K. Dan. Vidensk. Selsk. Skr. Naturvidensk. Math. Afd.,* 9, Raekke 7(1), 1–65, 1937.
35. Mortensen, Th., *K. Dan. Vidensk. Selsk. Skr. Naturvidensk. Math. Afd.,* 9, Raekke 7(3), 1–59, 1938.
36. Newman, H. H., *Biol. Bull.,* 40, 118–125, 1921.
37. Olsen, H., *Bergens Mus. Årbok Naturvitensk. Rekke,* No. 6, 1–107, 1942.
38. Onoda, K., *Jpn. J. Zool.,* 6, 637–654, 1936.
39. Ormond, R. F. and Campbell, A. C., *Proc. 2nd Int. Coral Reef Symp.,* Great Barrier Reef Committee, Brisbane, Australia, 1974, 595–619.
40. Pressoir, L., *Bull. Inst. Oceanogr.,* 1142, 1–22, 1959.
41. Ruggieri, G. D., in *Culture of Marine Invertebrate Animals,* Clark, W. C., and Chanley, M. H., Plenum, New York, 1975, 229.
42. Strathman, R. R., *J. Exp. Mar. Biol. Ecol.,* 6, 109–160, 1971.
43. Tyler, A. and Tyler, B. S., in *Physiology of Echinodermata,* Boolootian, R., Ed., Interscience, New York, 1966, 639–682.
44. Vanden Bossche, J. P., unpublished.
45. Yamabe, A., *Aquaculture,* 10, 213–220, 1962.
46. Yamaguchi, M., in *Biology and Geology of Coral Reefs,* Jones, O. A. and Endean, R., Eds., Academic Press, New York, 1973, 369.

Vertebrates

QUALITATIVE REQUIREMENTS AND UTILIZATION OF NUTRIENTS: FISHES

H. George Ketola

Table 1
AMINO ACIDS

An amino acid is listed as required if the rate of growth is increased by dietary supplementation with the crystalline form of the nutrient. Reports have not been found to demonstrate the requirement of fish for cystine (or cysteine) and tyrosine in the presences of their potential precursors methionine and phenylalanine. *Abbreviations*: R = required, Ɍ = not required.

Species	Alanine	Arginine	Aspartic acid	Cyst(e)ine	Glutamic acid	Glycine	Histidine	Isoleucine	Leucine	Lysine
Salmo gairdneri	Ɍ	R	Ɍ	Ɍ	Ɍ	Ɍ	R	R	R	R
Oncorhynchus tshawytscha	Ɍ	R	Ɍ	Ɍ	Ɍ	Ɍ	R	R	R	R
Oncorhynchus kisutch	–	R	–	–	–	–	R	–	–	–
Oncorhynchus nerka	Ɍ	R	Ɍ	Ɍ	Ɍ	–	R	R	R	R
Ictalurus punctatus	Ɍ	R	Ɍ	Ɍ	Ɍ	Ɍ	R	R	R	R
Cyprinus carpio	Ɍ	R	Ɍ	Ɍ	Ɍ	Ɍ	R	R	R	R
Anguilla japonica	Ɍ	R	Ɍ	Ɍ	Ɍ	Ɍ	R	R	R	R
Solea solea	Ɍ	R	Ɍ	Ɍ[a]	Ɍ	Ɍ	R	R	R	R
Pleuronectes platessa	Ɍ	R	Ɍ	Ɍ[a]	Ɍ	Ɍ	R	R	R	R

Species (continued)	Methionine	Phenyl-alanine	Proline	Serine	Threonine	Tryptophan	Tyrosine	Valine	Ref.
Salmo gairdneri	R	R	–	–	R	R	Ɍ	R	1, 2
Oncorhynchus tshawytscha	R	R	Ɍ	R	R	R	Ɍ	R	3–6
Oncorhynchus kisutch	–	–	–	–	–	R	–	–	7, 8
Oncorhynchus nerka	R	R	Ɍ	–	R	R	Ɍ	R	9
Ictalurus punctatus	R	R	Ɍ	–	R	R	Ɍ	R	10
Cyprinus carpio	R	R	Ɍ	Ɍ	R	R	Ɍ	R	11
Anguilla japonica	R	R	Ɍ[b]	Ɍ	R	R	Ɍ	R	12
Solea solea	R	R	Ɍ[b]	Ɍ	R	–	Ɍ[c]	R	13
Pleuronectes platessa	R	R	Ɍ[b]	Ɍ	R	–	Ɍ[c]	R	13

[a] Data indicated biosynthesis of cysteine, presumably from methionine.
[b] Not conclusive.
[c] Data were interpreted to indicate that tyrosine was not essential in presence of sufficient dietary phenylalanine.

REFERENCES

1.   Kloppel, T. M. and Post, G., *J. Nutr.,* 105, 861–866, 1975.
2.   Shanks, W. E., Gahimer, G. D., and Halver, J. E., *Prog. Fish Cult.,* 24, 68–73, 1962.
3.   Chance, R. E., Mertz, E. T., and Halver, J. E., *J. Nutr.,* 83, 177–185, 1964.
4.   Halver, J. E., DeLong, D. C., and Mertz, E. T., *Fed. Proc. Fed. Am. Soc. Exp. Biol.,* 18, 527, 1959.
5.   DeLong, D. C., Halver, J. E., and Mertz, E. T., *J. Nutr.,* 76, 174–178, 1962.
6.   Halver, J. E., DeLong, D. C., and Mertz, E. T., *J. Nutr.,* 63, 95–105, 1957.
7.   Halver, J. E., *Fed. Proc. Fed. Am. Soc. Exp. Biol.,* 24, 169, 1965.
8.   Klein, R. G. and Halver, J. E., *J. Nutr.,* 100, 1105–1110, 1970.
9.   Halver, J. E. and Shanks, W. E., *J. Nutr.,* 72, 340–346, 1960.
10. DuPree, H. K. and Halver, J. E., *Trans. Am. Fish. Soc.,* 99, 90–92, 1970.
11. Nose, T., Arai, S., Lee, D. L., and Hashimoto, Y., *Nihon Suisan Gakkai-Shi,* 40, 903–908, 1974.
12. Nose, T., *Tansuiku Suisan Kenkyusho Kenkyu Hokoku,* 19, 31–36, 1969.
13. Cowey, C. B., Adron, J., and Blair, A., *J. Mar. Biol. Assoc. U.K.,* 50, 87–95, 1970.

Table 2
WATER-SOLUBLE VITAMINS

Abbreviations: R = required, R̶ = not required.

Species (synonym)	Thiamine	Riboflavin	Pantothenic acid	Niacin	Vitamin B₆	Biotin	Choline	Folacin	Vitamin B₁₂	myo-Inositol	L-Ascorbic acid	p-Amino benzoic acid	Ref.
Salmo gairdneri (*S. irideus*)	R	R	R	—	R	R	R	R	—	R[a]	R	R	1–6
Salmo trutta	—	R	—	R	R	R	R	R	R	R	—	—	7, 8
Salmo salar	—	R	R	—	R	—	—	—	—	R̶	—	—	9
Salvelinus fontinalis	—	R	R	R	R	R	R	R	—	R	R	R[a,b]	10–16
Salvelinus namaycush	R	R	R	R	R	R	R[c]	R	—	R	—	—	17–20
Oncorhynchus tshawytscha	R	—	R	R	R	R	R	R	—	R	—	—	21
Oncorhynchus kisutch	R	R	R	—	R	R	R	R	—	R	R	—	5, 22, 23
Ictalurus punctatus	R	R	R	R	R	—	R	R	R	R̶	R[d]	R̶	24–28
Cyprinus carpio	—	R	R	R	R	—	R	—	—[e]	R	—	—	29–34
Carassius auratus	R	—	—	—	—	R	—	—	—	—	—	—	35
Chrysophrys major	R	R	R	R	R	—	—	—	R	R	R	R̶	36, 37
Anguilla japonica	R	R	R	R	R	R	R	R	R	R	R	R̶	38
Scophthalmus maximus	R	—	—	—	—	—	—	—	—	—	—	—	39

[a] References 2, 3, 4, and 6 cite conflicting data.

[b] A requirement was demonstrated in the presence of dietary folacin.

[c] N-Methylaminoethanol and N,N-diemthylaminoethanol were utilized as effective substitutes for choline, whereas aminoethanol and betaine were not.[19]

[d] Also, a requirement for ascorbic acid by *Ictalurus frucatus* is indicated.[28]

[e] Reference 34 indicates a nutritional requirement for vitamin B₁₂ by *Catla catla, Labeo rohita,* and *Cirrhina mrigala.*

REFERENCES

1. Phillips, A. M., Jr., Podoliak, H. A., Brockway, D. R., and Vaughn, R. R., *Fish. Res. Bull. N.Y. Conserv. Dep.,* 21, 86–87, 1957.
2. Wolf, L. E., *Prog. Fish Cult.,* 13, 21–24, 1951.
3. McLaren, B. A., Keller, E., O'Donnell, D. J., and Elvehjem, C. A., *Arch. Biochem. Biophys.,* 15, 169–178, 1947.
4. Kitamura, S., Ohara, S., Suwa, T., and Nakagawa, K., *Nihon Suisan Gakkai-Shi,* 31, 818–826, 1965.
5. Halver, J. E., Ashley, L. M., and Smith, R. R., *Trans. Am. Fish. Soc.,* 98, 762–771, 1969.
6. Kitamura, S., Suwa, T., Ohara, S., and Nakagawa, K., *Nihon Suisan Gakkai-Shi,* 33, 1120–1125, 1967.
7. Phillips, A. M., Jr., Lovelace, F. E., Podoliak, H. A., Brockway, D. R., and Balzer, G. C., Jr., *Fish. Res. Bull. N.Y. Conserv. Dep.,* 18, 46–47, 1954.
8. Phillips, A. M., Jr., Lovelace, F. E., Podoliak, H. A., Brockway, D. R., and Balzer, G. C., Jr., *Fish. Res. Bull. N.Y. Conserv. Dep.,* 19, 51–53, 1955.
9. Phillips, A. M., Jr., Podoliak, H. A., Dumas, R. E., and Thoesen, R. W., *Fish. Res. Bull. N.Y. Conserv. Dep.,* 22, 79–82, 1958.
10. Phillips, A. M., Jr., Lovelace, F. E., Brockway, D. R., and Balzer, G. C., Jr., *Fish. Res. Bull. N.Y. Conserv. Dep.,* 16, 9–11, 1952.
11. Poston, H. A. and DiLorenzo, R. N., *Proc. Soc. Exp. Biol. Med.,* 144, 110–112, 1973.
12. Poston, H. A. and McCartney, T. H., *J. Nutr.,* 104, 315–322, 1974.
13. Phillips, A. M., Jr., *Prog. Fish Cult.,* 25, 132–134, 1963.
14. Poston, H. A., *Fish. Res. Bull. N.Y. Conserv. Dep.,* 30, 46–51, 1967.
15. Phillips, A. M., Jr., Podoliak, H. A., Poston, H. A., Livingston, D. L., Booke, H. E., and Pyle, E. A., *Fish. Res. Bull. N.Y. Conserv. Dep.,* 27, 66–70, 1964.
16. Phillips, A. M., Jr., Lovelace, F. E., Brockway, D. R., and Balzer, G. C., Jr., *Fish. Res. Bull. N.Y. Conserv. Dep.,* 17, 5–7, 1953.
17. Phillips, A. M., Jr., Podoliak, H. A., Brockway, D. R., and Balzer, G. C., Jr., *Fish. Res. Bull. N.Y. Conserv. Dep.,* 20, 16–19, 1957.
18. Poston, H. A., *Fed. Proc. Fed. Am. Soc. Exp. Biol.,* 34, 884, 1975.
19. Ketola, H. G., *J. Anim. Sci.,* 43, 474–477, 1976.
20. Phillips, A. M., Jr., Podoliak, H. A., Dumas, R. E., and Thoesen, R. W., *Fish. Res. Bull. N.Y. Conserv. Dep.,* 22, 82–84, 1958.
21. Halver, J. E., *J. Nutr.,* 62, 225–243, 1957.
22. Coates, J. A. and Halver, J. E., *U.S. Fish Wildl. Serv. Spec. Sci. Rep. Fish.,* 281, 1–9, 1958.
23. Smith, C. E. and Halver, J. E., *J. Fish. Res. Board Can.,* 26, 111–114, 1969.
24. DuPree, H. K., *U.S. Dep. Inter. Bur. Sp. Fish. Wildl. Tech. Pap.,* 7, 3–12, 1966.
25. Murai, T. and Andrews, J. W., *Trans. Am. Fish. Soc.,* 104, 313–316, 1975.
26. Lovell, R. T., *J. Nutr.,* 103, 134–138, 1973.
27. Andrews, J. W. and Murai, T., *J. Nutr.,* 105, 557–561, 1975.
28. Wilson, R. P., *Comp. Biochem. Physiol. B,* 46, 635–638, 1973.
29. Aoe, H. and Masuda, I., *Nihon Suisan Gakkai-Shi,* 33, 674–685, 1967.
30. Ogino, C., *Nihon Suisan Gakkai-Shi,* 33, 351–354, 1967.
31. Aoe, H., Masuda, I., Saito, T., and Komo, A., *Nihon Suisan Gakkai-Shi,* 33, 355–360, 1967.
32. Ogino, C., *Nihon Suisan Gakkai-Shi,* 31, 546–551, 1965.
33. Ogino, C., Uki, N., Watanabe, T., Iida, Z., and Ando, K., *Nihon Suisan Gakkai-Shi,* 36, 1140–1146, 1970.
34. Das, B. C., *Can. J. Biochem. Physiol.,* 38, 453–458, 1960.
35. Tomiyama, T. and Ohba, N., *Nihon Suisan Gakkai-Shi,* 33, 448–452, 1967.
36. Yone, Y. and Fujii, M., *Kyushu Daigaku Nogakubu Fuzoku Suisan Jikkensho Hokoku,* 2, 25–32, 1974.
37. Shitanda, K., Furuichi, M., and Yone, Y., *Kyushu Daigaku Nogakubu Fuzoku Suisan Jikkensho Hokoku,* 1, 29–36, 1971.
38. Arai, S., Nose, T., and Hashimoto, Y., *Tansuiku Suisan Kenkyusho Kenkyu Hokoku,* 42, 69–79, 1972.
39. Cowey, C. B., Adron, J. W., Knox, D., and Ball, G. T., *Br. J. Nutr.,* 34, 383–390, 1975.

Table 3
FAT-SOLUBLE VITAMINS

Abbreviations: R = required, U = utilized, Ψ = not utilized.

Species	Vitamin A	Vitamin D	Vitamin E[a]	Vitamin K	β-Carotene	Ref.
Salmo gairdneri	R[b]	R[c]	–	R[d]	–	1–3
Salmo trutta	–	–	R	–	–	4
Salmo salar	–	–	R	–	–	5
Salvelinus fontinalis	–	–	–	R[e]	Ψ[f]	6, 7
Salvelinus namaycush	–	–·	–	R[g]	–	8
Oncorhynchus tshawytscha	–	–	R	–	–	9
Perca fluviatilus	–	–	–	–	U[h]	10
Gadus callarias	–	–	–	–	U[h]	11
Ictalurus punctatus	–	–	R	–	–	12
Cyprinus carpio	–	–	R	–	–	13
Lebistes reticulatus	–	–	–	–	U[h,i]	14
Xiphophorus variatus	–	–	–	–	U[h,i]	14

[a] α-Tocopherol.

[b] Forms of vitamin A studied include retinyl acetate and retinyl palmitate.

[c] Reference 3 showed that cholecalciferol was required and indicated that calciferol was utilized to partially satisfy the vitamin D requirement. Utilization of calciferol was not confirmed in a later study.[1]

[d] Form of vitamin K studied was menaquinone.

[e] Form of vitamin K studied was phylloquinone.

[f] Dietary β-carotene failed to increase tissue storage of vitamin A.

[g] Forms of vitamin K studied were menaquinone sodium bisulfite and menaquinone dimethylpyrimidinol bisulfite (trade name Hetrazeen®).

[h] Dietary β-carotene increased tissue storage of vitamin A. Degree of utilization was not determined.

[i] Utilization of other pro-vitamin A substances was examined.[1,4]

REFERENCES

1. Kitamura, S., Suwa, T., Ohara, S., and Nakagawa, K., *Nihon Suisan Gakkai-Shi,* 33, 1120–1125, 1967.
2. Kitamura, S., Suwa, T., Ohara, S., and Nakagawa, K., *Nihon Suisan Gakkai-Shi,* 33, 1126–1131, 1967.
3. McLaren, B. A., Keller, E., O'Donnell, D. J., and Elvehjem, C. A., *Arch. Biochem. Biophys.,* 15, 169–178, 1947.
4. Poston, H. A., *Fish. Res. Bull. N.Y. Conserv. Dep.,* 28, 6–9, 1964.
5. Poston, H. A., Combs, G. F., Jr., and Leibovitz, L., *J. Nutr.,* 106, 892–904, 1976.
6. Poston, H. A., *Prog. Fish. Cult.,* 26, 59–64, 1964.
7. Poston, H. A., *Fish. Res. Bull. N.Y. Conserv. Dep.,* 32, 41–43, 1969.
8. Poston, H. A., *J. Fish. Res. Board Can.,* 33, 1791–1793, 1976.
9. Woodall, A. N., Ashley, L. M., Halver, J. E., Olcott, H. S., and VanDerVeen, J., *J. Nutr.,* 84, 125–135, 1964.
10. Morton, R. A. and Creed, R. H., *Biochem. J.,* 33, 318–324, 1939.
11. Neilands, J. B., *Arch. Biochem. Biophys.,* 13, 415–419, 1947.
12. Murai, T. and Andrews, J. W., *J. Nutr.,* 104, 1416–1431, 1974.
13. Aoe, H., Abe, I., Saito, T., Fukawa, H., and Koyama, H., *Nihon Suisan Gakkai-Shi,* 38, 845–851, 1972.
14. Gross, J. and Budowski, P., *Biochem. J.,* 101, 747–754, 1966.

Table 4
MINERALS

Abbreviations: R = required, U = Utilized.

Species	Ca	P	Fe	I	Se	Mg	Ref.
Salmo gairdneri	–	–	–	–	–	R	1
Salmo salar	–	R[a]	–	–	R[b]	–	2, 3
Salmo trutta	U	–	–	–	–	–	4
Salvelinus fontinalis	U	–	R	R	–	–	5–7
Oncorhynchus tshawytscha	–	–	–	U	–	–	8
Ictalurus punctatus	U	R[a,c]	–	–	–	–	9
Chrysophrys major	R[d]	R[e]	–	–	–	–	10
Esox sp.	–	–	–	R	–	–	11
Xiphophorus helleri	–	–	R	–	–	–	12
Xiphophorus maculatus	–	–	R	–	–	–	12

[a] As phosphate.
[b] As selenite.
[c] Represents responses to dietary supplements of $NaH_2PO_4 \cdot H_2O$.
[d] Indicated by data.
[e] Represents a growth response to dietary supplements of $NaH_2PO_4 \cdot 2H_2O$.

REFERENCES

1. Cowey, C. B., *J. Fish. Res. Board Can.,* 33, 1040–1045, 1976.
2. Ketola, H. G., *Trans. Am. Fish. Soc.,* 104, 548–551, 1975.
3. Poston, H. A., Combs, G. F., Jr., Leibovitz, L., *J. Nutr.,* 106, 892–904, 1976.
4. Podoliak, H. A. and Holden, H. K., Jr., *Fish. Res. Bull. N.Y. Conserv. Dep.,* 28, 64–70, 1964.
5. McCay, C. M., Tunison, A. V., Crowell, M., and Paul, H., *J. Biol. Chem.,* 114, 259–263, 1936.
6. Kawatsu, H., *Tansuiku Suisan Kenkyusho Kenkyu Hokoku,* 22, 59–67, 1972.
7. Marine, D. and Lenhart, C. H., *J. Exp. Med.,* 10, 311–337, 1910.
8. Woodall, A. N. and LaRoche, G., *J. Nutr.,* 82, 475–482, 1964.
9. Andrews, J. W., Murai, T., and Campbell, C., *J. Nutr.,* 103, 766–771, 1973.
10. Sakamoto, S. and Yone, Y., *Nihon Suisan Gakkai-Shi,* 39, 343–348, 1973.
11. Marine, D. and Lenhart, C. H., *Bull. Johns Hopkins Hosp.,* 21, 95–98, 1910.
12. Roeder, M. and Roeder, R. H., *J. Nutr.,* 90, 86–90, 1966.

Table 5
FATTY ACIDS

Abbreviations: R = required, U = utilized.

Species	18:2ω6[a]	18:3ω3[b]	Ref.
Salmo gairdneri	R[c]	R[c]	1—6
Oncorhynchus tshawytscha	R[d]	R[d]	7
Ictalurus punctatus	U	U	8
Cyprinus carpio	R[e]	R[e]	9, 10

[a] 9,12-Octadecadienoic acid (linoleic acid).

[b] 9,12,15-Octadecatrienoic acid (linolenic acid).

[c] References cite conflicting data.

[d] Experimental purified diets contained β-carotene as the source of supplemental vitamin A.

[e] Tentative; description of experimental diets used includes no record of supplementation with fat-soluble vitamins A and D.

REFERENCES

1. Higashi, H., Kaneko, T., Ishii, S., Ushiyama, M., and Sugihashi, T., *J. Nutr. Sci. Vitaminol.,* 12, 74—79, 1966.
2. Watanabe, T., Ogino, C., Koshiishi, Y., and Matsunaga, T., *Nihon Suisan Gakkai-Shi,* 40, 493—499, 1974.
3. Yu, T. C. and Sinnhuber, R. O., *Lipids,* 10, 63—66, 1975.
4. Castell, J. D., Sinnhuber, R. O., Wales, J. H., and Lee, D. J., *J. Nutr.,* 102, 77—86, 1972.
5. Watanabe, T., Takashima, F., and Ogino, C., *Nihon Suisan Gakkai-Shi,* 40, 181—188, 1974.
6. Yu, T. C. and Sinnhuber, R. O., *Lipids,* 7, 450—454, 1972.
7. Nicolaides, N. and Woodall, A. N., *J. Nutr.,* 78, 431—437, 1962.
8. Stickney, R. R. and Andrews, J. W., *J. Nutr.,* 102, 249—258, 1972.
9. Watanabe, T., Utsue, O., Kobayashi, I., and Ogino, C., *Nihon Suisan Gakkai-Shi,* 41, 257—262, 1975.
10. Watanabe, T., Takeuchi, T., and Ogino, C., *Nihon Suisan Gakkai-Shi,* 41, 263—269, 1975.

CARBOHYDRATES

Although fish need glucose physiologically, a nutritional requirement for carbohydrate per se has not been clearly demonstrated. A nutritional response of fish to supplements of carbohydrates depends upon the species and age of fish, water temperature, and type of carbohydrate, and may be confounded or obscured by the level of supplementation, experimental choice of the dietary ingredient(s) and amount(s) removed during supplementation, and degree of balance maintained between energy and critical nutrients in the diet.

A major physiological function of dietary carbohydrate is to serve as a source of energy, and the requirements of fish for energy are poorly defined. Furthermore, only limited information is available on quantitative efficiencies of digestion and utilization of energy from various feed ingredients and nutrient classes as they vary with the level of supplementation in the diet.

In general, various studies demonstrate poor utilization of dietary carbohydrates such as dextrin, fructose, galactose, lactose, glucosamine, glucose, maltose, starch, or sucrose by *Oncorhynchus tshawytscha*,[1] *Salmo gairdneri*,[2,3] and *Salvelinus fontinalis*.[4,5]

REFERENCES

1. Buhler, D. R. and Halver, J. E., *J. Nutr.*, 74, 307–318, 1961.
2. McLaren, B. A., Herman, E. F., and Elvehjem, C. A., *Arch. Biochem. Biophys.*, 10, 433–441, 1946.
3. Luquet, P., Léger, C., and Bergot, F., *Ann. Hydrobiol.*, 6, 61–70, 1975.
4. McCartney, T. H., *Fish. Res. Bull. N.Y. Conserv. Dep.*, 34, 43–52, 1971.
5. Phillips, A. M., Jr., Tunison, A. V., and Brockway, D. R., *Fish. Res. Bull. N.Y. Conserv. Dep.*, 11, 1–44, 1948.

NUTRITION OF AMPHIBIA

G. W. Brown, Jr.

I. HABITS AND ADAPTATIONS

M. Gross

Amphibia are remarkable in their adaptation to a variety of environmental conditions. They occupy aquatic, terrestrial, and arboreal habitats all over the world, except in extremely arid or cold regions. Amphibia are grouped in three orders:

Order Salientia (Anura): frogs and toads
Order Caudata (Urodela): newts and salamanders
Order Caecilia (Apoda): limbless amphibia (caecilians)

Since the nature of the diet of amphibia depends upon their habitats, Table 1 provides a summary of these habitats. For literature on habits, adaptations, and water relationships, consult References 1 to 15.

REFERENCES

1. **Barbour, T.,** *Reptiles and Amphibians: Their Habits and Adaptations,* Houghton Mifflin Co., Boston, New York, 1926.
2. **Bentley, P. J.,** *Science,* 152, 619–623, 1966.
3. **Boernke, W. E.,** *Comp. Biochem. Physiol. B,* 44, 1035–1042, 1973.
4. **Cochran, D. M.,** *Living Amphibians of the World,* Doubleday and Co., Garden City, N.Y., 1961.
5. **Cunningham, J. D.,** *Herpetologica,* 19, 56–61, 1963.
6. **Greenwald, L.,** *Physiol. Zool.,* 45, 229–237, 1972.
7. **Heatwole, H. and Lim, K.,** *Ecology,* 42, 814–819, 1961.
8. **Inger, R. F. and Greenberg, B.,** *Ecology,* 47, 746–759, 1966.
9. **Noble, G. K.,** *Biology of the Amphibia,* Dover Publications, New York, 1954.
10. **Norris, K. S. and Lowe, C. H.,** *Ecology,* 45, 565–580, 1964.
11. **Organ, J. A.,** *Ecol. Monogr.,* 31, 189–220, 1961.
12. **Tevis, L., Jr.,** *Ecology,* 47, 766–775, 1966.
13. **Thorson, T. B. and Svihla, A.,** *Ecology,* 24, 374–381, 1943.
14. **Thorson, T. B.,** *Ecology,* 36, 100–116, 1955.
15. **Brown, G. W., Jr.,** in *Physiology of the Amphibia,* Moore, J. A., Ed., Academic Press, New York, London, 1964, pp. 1–98.

Table 1
HABITATS OF AMPHIBIAN SPECIES[1-4]

(Nonbreeding Season)

Species	Habitat	Species	Habitat
Terrestrial Species		**Terrestrial Species (continued)**	
Anurans		**Aquatic species**	
Rana sylvatica	Terrestrial		
Rana temporaria	Terrestrial	**Anurans**	
Rana pipiens	Terrestrial-semiaquatic	*Rana aurora*	Semiaquatic
Hyla regilla	Terrestrial	*Rana esculenta*	Semiaquatic
Hyla arborea	Terrestrial-arboreal	*Rana grylio*	Aquatic
Hyla moorei	Terrestrial-arboreal	*Rana clamitans*	Aquatic
Hyla rubella	Terrestrial-arboreal	*Rana catesbeiana*	Aquatic
Hyla cinerea	Terrestrial-arboreal	*Rana boylii*	Aquatic
Eleutherodactylus gollmeri	Terrestrial	*Hyla arenicolor*	Aquatic
Eleutherodactylus maussi	Terrestrial	*Xenopus laevis*	Aquatic
Bufo boreas	Terrestrial	*Xenopus tropicalis*	Aquatic
Bufo terrestris	Terrestrial	*Pipa pipa*	Aquatic
Bufo canorus	Terrestrial	*Telmatobius microphthalmus*	Aquatic
Bufo americanus	Terrestrial	*Telmatobius culeus*	Aquatic
Bufo bufo	Dry-terrestrial		
Bufo calamita	Dry-terrestrial	**Urodeles**	
Bufo punctatus	Dry-terrestrial	*Desmognathus quadramaculatus*	Aquatic
Scaphiopus couchii	Terrestrial-fossorial	*Plethodon dunni*	Terrestrial-aquatic
Scaphiopus holbrookii	Terrestrial-fossorial	*Ambystoma mexicanum*	Aquatic
Scaphiopus hammondii	Terrestrial-fossorial	*Triturus granulosus*	Aquatic
Cyclorana platycephalus	Terrestrial		
Urodeles			
Desmognathus wrighti	Terrestrial-arboreal		
Desmognathus o. carolinensis	Terrestrial-arboreal		
Desmognathus fuscus	Terrestrial-aquatic		
Desmognathus monticola	Terrestrial-aquatic		
Aneides ferreus	Terrestrial-arboreal		
Aneides lugubris	Terrestrial-arboreal		
Dicamptodon ensatus	Terrestrial		
Plethodon cinereus	Terrestrial		
Triturus cristatus	Terrestrial		
Salamandra salamandra	Terrestrial		

REFERENCES

1. **Noble, G. K.,** *Biology of the Amphibia,* Dover Publications, New York, 1954.
2. **Thorson, T. B. and Svihla, A.,** *Ecology,* 24, 374–381, 1943.
3. **Moore, J. A., Ed.,** *Physiology of the Amphibia,* Vol. 1, Academic Press, New York, London, 1964.
4. Miscellaneous sources.

II. WATER ASPECTS

T. E. Bucsko

Introduction

Amphibians do not normally drink water, and thus regulation of moisture content of the body, even in times of stress, depends upon (1) dermal absorption, (2) evaporative loss through the skin and if present, the lungs, (3) loss through the kidneys, and (4) production of metabolic water.[1] Amphibians have no extensively developed integumentary modifications that tend to conserve body water. Even in the most terrestrial of species the integument itself offers little resistance to water loss, but on the contrary loses water freely by evaporation.[2] Because of this adaptation, Amphibia are restricted in both their possible habitats and in their behavioral patterns, including feeding.

Poikilothermic Nature

Amphibia are poikilothermic. Hence, the temperature of the environment has a direct effect upon their metabolism and behavior. Water is such a good conductor that immersed Amphibia follow closely the temperature of their aquatic environment. Water is also a stable medium, without sudden fluctuations; thus, immersed Amphibia will respond seasonally to thermal changes. Each species, however, has its own temperature level at which it shows optimal physiological response.

Both types of Amphibia — semiterrestrial and terrestrial — are behaviorally influenced by humidity.[2] Since the skin is moist, it depresses the body temperature below that of the environment. The loss of heat on land due to the cooling by evaporation offers protection against overheating, but the skin itself is subject to rapid desiccation. Few Amphibia other than the rough-skinned toads and salamanders will remain long in a dry atmosphere.[3] These latter species depend largely on a pulmonary evaporation mechanism to regulate body temperature.

Desiccation and Urea Retention

The ornithine–urea cycle enzymes, specifically arginase, were studied[4] to see if adaptive responses to environmental shifts in water availability, such as drought, are exhibited in certain Amphibia. Arginase has been implicated in the active transport of urea in the kidney tubules of several species of Amphibia.[5,6] Dehydration results in significant increases in arginase activity.[4] The adaptive value of arginase activity is reflected in three ways: (1) increased production of urea, which may act as a buffer against osmotic stress;[4,7] (2) large amounts of water are needed to excrete ammonia, which is highly toxic, but under conditions of dehydration, ammonia can be converted to urea and thus be excreted with a less abundant water supply; (3) arginase may be involved in active transport of urea, compensating for a lower glomerular filtration rate, which decreases urea excretion.[4]

Desiccation and Skin Permeability

Estivating *Xenopus laevis* was found by Balinsky[8] to have increased levels of ornithine–urea cycle enzymes. In response to osmotic stress of a dehydrating character, an increase in blood level for urea and a decrease in urea excretion in anuran Amphibia were found.[9] Amphibia commonly have mucous glands, which secrete a transparent substance that serves as a lubricant in the water and keeps the skin moist on land.[3] Anuran species occupying habitats of different aridity often show variations in their cutaneous permeability, due to different osmotic problems. Bentley and Main[10] suggest that there are regional differences in permeability of the integument. *Hyla moorei*, an arboreal frog, has skin in its pelvic region that is much more permeable than the skin in the pectoral or dorsal areas. This pelvic specialization is not found in *Neobatrachus pelobatiodes,* a fossorial type.

Evidence was presented[11] suggesting that the skin water balance response in two species of toads is confined to the pelvic patch. Changes in this pelvic patch permeability alone are able to account for about half the dermal uptake rate of dehydrated animals.

Amphibia in deserts are particularly liable to die as a result of desiccation, because they cannot prevent loss of moisture via the skin. Among the morphological adaptations many of these Amphibia have in common are a hydration patch of very thin pink skin and a bladder that stores water.[12] Anurans (frogs and toads) are seen to have the widest distribution, thus possessing most of the small modifications in the physiological pattern necessary to survive in a hot, arid region. Bentley[13] has summarized some of the major adaptations as (1) localized skin permeability to collect water from damp surfaces and assist rapid rehydration, (2) extra water storage and resorption capabilities in a large urinary bladder, (3) controlled water transfer across the skin and bladder by hormones, (4) ability of amphibian tissues to continue functioning in the presence of high osmotic concentrations of solutes, and (5) adaptations of behavior to find microenvironments where conditions are favorable for survival and reproduction.

Salts and gases are controlled osmotically in the amphibian's integument. Thus, studies have been made to see if Amphibia make physiological adjustments to live in an environment that may cause sodium loss.

In addition to normal sodium losses in urine, Amphibia living in fresh water have a diffusive loss across their body surfaces, which would not be experienced by Amphibia on land or in dilute sea water. Greenwald[14] performed studies on terrestrial, semiterrestrial, freshwater, and marine Amphibia. Freshwater Amphibia placed in fresh water showed lower sodium uptake and loss rates than those from other environments. All of the Amphibia studied showed compensatory responses to sodium loss rate or increasing uptake rate.

Studies have been made with some anurans from a saline habitat and with others not adapted to saline conditions to determine salt tolerances and osmoregulatory responses.[15] Concentrations of salts in the lymph, serum, and urine were studied at various external concentrations. Tolerances to salinity in the six species tested were found to be no higher than those known previously for other species. The adaptive significance of hypertonic urine in a terrestrial anuran inhabiting saline areas is discussed by Ruibal.[15] Urine serves as a water reserve, and bladder water is utilized during dehydration. The unique ability of *Pleurodema nebulosa* to produce hypotonic urine when immersed in a medium of relatively high salinity allows it to live in saline habitats that are occupied by no other anuran,[15] except possibly *Rana cancrivora.*[16]

An optimal rate of water loss in Amphibia exists in the air.[17] At slower rates, tolerance becomes low, possibly due to increased metabolic expenses elsewhere. Temperature extremes are a major factor in terrestrial tolerance to dehydration. At low temperatures, slower evaporation occurs, with the consequent prolongation of the exposure to unfavorable physiological conditions decreasing the tolerance in the Amphibia.[2] Frogs die after a loss of only 15% of their body weight if evaporation is rapid, but can tolerate nearly twice as great a loss if the rate of evaporation is slower. In 1857, Kunde found that some toads even withstand a loss of 50% of their body weight when evaporation is slow.[3]

Rehydration

Amphibia commonly take up water through the skin, but secrete a corresponding amount of urine. Aquatic frogs, when returned to water after dehydration, will absorb water through the skin by osmosis until the original weight is restored — no urine is secreted until osmosis is complete.[18] The rate of dehydration and rehydration are temperature-dependent, occurring four to five times faster at 30°C than at 0°C. A frog weighing 60 g will lose 25% of its weight in 8 to 11 hr at room temperature. This produces an increase of nearly twice the original osmotic pressure in the blood. The rate of absorption of water will then become proportional to the difference in osmotic pressure between the blood and its surrounding fluid.[18]

The rate of rehydration is greater in species from arid regions than in those from wet ones. This is due to the localization of permeable integumental regions on such amphibians' bodies, a necessary adaptation for their habitat.[10] Species such as *Neobatrachus pelobatiodes* and other fossorial Amphibia utilize cutaneous permeability to rapidly regain water lost and to actively accumulate sodium by the addition of the neurohypophysial hormone vasotocin.

Sodium Balance and Salinity

Upon exposure to slightly hyperosmotic saline solutions, *Xenopus laevis,* normally a freshwater species, becomes desiccated. Both the rate of urea production and the level of carbamoylphosphate synthetase activity increase.[19] Krogh[18] noted that anurans make up water loss in hypertonic solution by a shift in the osmotic gradient, accomplishing this by drinking water and by passing quantities of concentrated urine. This same drinking behavior is also seen in *Bufo bufo.*[20] Some species of anurans decrease the permeability of the skin when faced with increased salinity.[21]

The main elements contributing to survival in hyperosmotic media are elevated levels of sodium and chloride ions, reduced urine output, and reduced water uptake, with the urine always remaining hypoosmotic to the blood.[20] The completely aquatic toad *Xenopus laevis* can survive for extended periods in a hyperosmotic solution, where it accumulates urea converted from ammonia.[22] Most specimens of *Xenopus laevis* can survive for two weeks in a 33% hyperosmotic solution, but all die within 33 hr in a 40% solution.[20]

The survival capacity of *Xenopus laevis* in a hyperosmotic medium is greater in winter; this reflects a difference in body water content, which is greater in winter than in summer.[23] The biological significance of this active uptake of water and salts is seen in frogs, some of which spend 7 months of the year under water without food.[18]

Effect of Substrate

The exchange of water between soil and Amphibia is found both in wet and in dry communities, and species can be characterized by their water requirements as wholly aquatic, semiterrestrial, or wholly terrestrial.[24] Different effects of substrate have been identified and labeled:[1] (1) the *absorption threshold,* which is the level of substrate moisture above which there is a net gain in body water in dehydrated species, and below which there is a net loss; (2) the *critical level,* the substrate moisture level below which amphibian water loss increases markedly; and (3) the *limiting range,* the substrate moisture levels between the absorption threshold and the critical level.

Above the absorption threshold, rates of moisture uptake are correlated with soil moisture if comparisons are between samples with sufficiently differing moisture tensions. The soil moisture tension scale is defined as a scale of availability of soil water to Amphibia.[1] As the length of time a salamander spends in soil below the absorption threshold increases, the rate of water loss to the soil drops, because the water is absorbed in the soil immediately adjacent to the amphibian's skin, and a melting front is formed. Here the humidity increases, and the dehydration rate decreases.[24]

The critical level for *Plethodon cinereus* coincides with the point at which there is a sharp increase in soil moisture tension and a decrease in relative humidity of the soil interface.[1] The rate of water exchange then is a function of soil moisture tension. This can be correlated with the body weight and dehydration deficit of the Amphibian.[24]

REFERENCES

1.   Heatwole, H. and Lim, K., *Ecology,* 42, 814–819, 1961.
2.   Thorson, T. B., *Ecology,* 36, 100–116, 1955.
3.   Noble, G. K., *Biology of the Amphibia,* Dover Publications, New York, 1954.
4.   Boernke, W. E., *Comp. Biochem. Physiol. B,* 44, 647–655, 1973.
5.   Forster, R. P., *Am. J. Physiol.,* 179, 372–377, 1954.
6.   Schmidt-Nielsen, B. and Schrauger, C. R., *Am. J. Physiol.,* 205, 483–488, 1963.
7.   Brown, G. W., Jr. and Brown, S. G., *Science,* 155, 570–573, 1967.
8.   Balinsky, J. B., Choritz, E. L., Coe, C. G. L., and Van der Schans, G. S., *Comp. Biochem. Physiol.,* 22, 59–68, 1967.
9.   Scheer, B. T. and Markel, R. P., *Comp. Biochem. Physiol.,* 7, 289–297, 1962.
10.   Bentley, P. J. and Main, A. R., *Am. J. Physiol.,* 223, 361–363, 1972.
11.   Baldwin, R. A., *Comp. Biochem. Physiol. A,* 47, 1285–1295, 1974.
12.   Tevis, L., Jr., *Ecology,* 47, 766–775, 1966.
13.   Bentley, P. J., *Science,* 152, 619–623, 1966.
14.   Greenwald, L., *Physiol. Zool.,* 45, 229–237, 1972.
15.   Ruibal, R., *Physiol. Zool.,* 35, 133–147, 1962.
16.   Gordon, M. S., Schmidt-Nielsen, K., and Kelley, H. M., *J. Exp. Biol.,* 38, 659–678, 1961.
17.   Thorson, T. B., *Copeia,* 1956, 230–237, 1956.
18.   Krogh, A., *Osmotic Regulation in Aquatic Animals,* Cambridge University Press, Cambridge, 1939.
19.   McBean, R. L. and Goldstein, L., *Science,* 157, 931–932, 1967.
20.   Ireland, M. P., *Comp. Biochem. Physiol. A,* 46, 469–476, 1973.
21.   Dicker, S. E. and Elliot, A. B., *J. Physiol. London,* 207, 119–132, 1970.
22.   Balinsky, J. B., Cragg, M. M., and Baldwin, E., *Comp. Biochem. Physiol.,* 3, 236–244, 1961.
23.   Deyrup, I. J., in *Physiology of the Amphibia,* Moore, J. A., Ed., Academic Press, New York, 1964, pp. 251–328.
24.   Spight, T. M., *Biol. Bull.,* 132, 126–132, 1967.

III. FEEDING BEHAVIOR

D. K. Baker

Most naturalists will agree that frogs and toads will eat just about any small animal that they can get into their mouths.[1-3] Besides this requirement, there are few, if any, limiting factors in the diets of these amphibians. Even the senses of taste and smell can be of little importance in prey selection, as in the case of *Rana pipiens.*[3]

Since frogs and toads are such fortuitous feeders, any trends or patterns found in food selection are subject to the unique conditions under which the investigator is working. Any data on food habits would be subject to many factors, the most important being the time of year, locality, and methods of collection. For example, if the investigator collects green frogs on the border of a pond during July, a large portion of the food may consist of transforming smaller frogs.[1] Sampling during another season or at another location might indicate a completely different diet for the same species of frog.

The frequency of certain organisms in the diet of Amphibia can also depend on several types of environmental factors. For example, the leopard frog, *Rana pipiens,* varies its insect diet seasonally, depending on the availability of certain species in any given month.[4] This availability is related to the life cycles of the insects. Dineen[5] showed that larval tiger salamanders exhibited dual feeding habits; they ate plankton when light levels were high, and large insects when there was less available light. In the case of the cave-dwelling salamander, *Gyrinophilus palleuccus,* the availability of troglobitic isopods depends largely upon the amount of rainfall received in the area.[6] The isopods live in the organic debris that is washed into the caves by the intermittent surface streams that accompany extended periods of rain. These examples serve to show that a broad spectrum of sampling situations must be examined before any generalizations are made about the diets of Amphibia.

Salientia exhibit a preference for predatory species.[2] The theoretical food chain is as follows: the invertebrate phytoplankton are ingested by the saprophagous fauna, which in turn are eaten by the invertebrate predators, who then fall prey to the frogs and toads. This preference for predators was attributed to their greater mobility and greater daytime activity. Since a great many of the Salientia are nocturnal, the fact that the prey is active during the day may be only part of the reason they are being taken as a food item. In fact, there may be a great many invertebrate predators that are also nocturnal. Thus, movement would seem to be the factor that determines whether or not an animal will be attacked.

Frogs seem to be especially sensitive to movement. Dickerson[7] states, "Frogs will blindly attack any moving object without hesitation, and can be easily captured using a fish hook baited with a colored bit of cloth." Frogs strike at inanimate objects that have been set in motion. Hamilton[1] observed a *Rana clamitans* that was sitting on a rock in the middle of a small stream and stuffing itself with elm seeds drifting by in the riffles. Despite this sensitivity to movement, frogs tend to be slightly far-sighted in comparison to toads, and they will sometimes miss a potential meal that passes directly in front of them.[7]

Toads do not attack their prey as blindly as do frogs; they are more conservative in their actions. A toad's first reaction to a moving object is to orient itself toward it and fix its eyes upon it.[7] The toad will not strike at the first movement, as a frog would, but will wait for a second movement and then pounce upon its prey. Bragg[8] observed a toad that was ingesting mulberries as they fell from a nearby bush and rolled along the ground. Toads will even distinguish the forward-moving end of its prey, so that it can be grasped head first, which makes swallowing it much easier. Despite such advantages of discriminatory feeding, there are also some disadvantages. Maturing frogs display a gradual preference for larger and larger prey, and a medium-sized individual may have trouble deciding whether or not to attack a medium-to-large-sized invertebrate. Sometimes the toad may consider the prey too large to be ingested when, in fact, it could be easily swallowed.[8] Toads tend to avoid vertebrates, possibly due to an inability to digest bones.[7] Although a frog will greedily eat smaller toads and frogs, a toad will not eat small toads. Toads somehow distinguish the movements of a young toad from those of an invertebrate.

Frogs are omnivorous, and no small animal is safe from a large specimen (especially *Rana catesbeiana*) unless it is too large to be ingested.[9] A bullfrog (*Rana catesbeiana*) raised during one summer ate 427 g of various foods, which included 56 amphibians and 4 birds.[9]

Barrington[10] made several observations on the digestion time of frogs, specifically *Rana catesbeiana*. He concluded that the digestive action of this frog is quite slow compared to that of mammals. For example, one specimen required 24 hr to digest a small worm, and another frog digested a smaller frog in 48 to 68 hr.[10] Digestion time is dependent on the quantity of food present in the stomach and may extend to 3 days or longer (Frost, cited in Reference 10). Digestion may not be complete in a full stomach, and incompletely digested insects may be discharged in the feces. Parts of a beetle have been recovered from a frog as long as 17 days after feeding.[10]

The digestive tract of tadpoles is quite different from that of the adult toad or frog. In his investigations of *Rana temporaria* tadpoles, Savage[11] found that the gastrointestinal tract is never empty, except at metamorphosis. He also found the variety of food eaten to be so great that he thought it unlikely that tadpoles are confined to any particular species of plant or animal. Tadpoles eat almost continually, and will eat debris if there is nothing else to consume; they consume at random organisms contained in bottom ooze.[12]

The digestion time of tadpoles is in direct contrast to that of the adult frog. The rate of digestion is extremely short — an average of 6.5 hr for *Rana temporaria,* and 4.5 hr for *Bufo bufo.*[11] This would not allow enough time for the breakdown of the cellulose walls of plants to occur by enzymatic action due to either bacteria or the tadpole itself. In general, all classes of satisfactory food have one thing in common — thick cell walls are absent.[11] Higher plants are thought to be poor food sources, due to their thick cell walls,[11] but they may be utilized by tadpoles after bacterial breakdown has occurred.[12] Algae have thin cell walls, protozoans have no cell wall, and Entomastraca appear to swell osmotically and burst in the alimentary canal; these characteristics make all of these organisms good food sources.

Due to the very short digestion time observed in tadpoles, Savage[11] felt that the cell walls of algae must be ruptured by mechanical means. He proposed that the tadpoles use their teeth to rupture cell walls, and he attempted to support this theory by examining photographs of tadpoles scraping algae off of the glass sides of an aquarium. He found that the algae are scraped off an object by means of the back and labial teeth, and that the contents of the cells are made available after the labial teeth pierce the cell walls. Sand might serve as a grinding medium in addition to these teeth.[12]

After the cells are loose or ruptured, two ways are used by tadpoles to extract the cells and their contents from the water that passes through the pharynx. Small particles are trapped in the mucous cords that form in an eddy behind the dorsal velum, and are then conveyed by ciliary action into the esophagus. Larger particles are thrown into the esophageal funnel by the centrifugal action induced by water passing in a circular path at that point.[11] The cell constituents are then degraded further through enzymatic action, and the nutrients are absorbed as they pass along the intestine.

Several investigators have experimented with special diets in order to further clarify certain aspects of tadpole digestion and metabolism. Tadpoles fed on a meat-only diet (liver) developed shorter intestines than either wild tadpoles or those fed a plants-only diet (lettuce).[13] Lettuce, because of its cellulose content, probably produces mechanical irritation, which presumably results in a longer intestine.[13] In larvae fed the lettuce diet, the intestines generally contained a greater bulk of food than did those of tadpoles on the liver diet, which may also have an effect on intestinal length. To further support the theory of mechanical stimulation as an agent responsible for longer intestines, Janes[13] cited the fact that control animals, which were fed a complete diet, had an intermediate-length gut during early development; later on, when they ingested small amounts of sand with their food, their intestinal length was greater than that of animals raised on lettuce. A less digestible food tends to accumulate in the intestine and thus causes greater mechanical stimulation, which in turn leads to lengthening of the intestine.

Burke[14] fed tadpoles with bacteria washed from agar slants. He observed the larvae actively feeding on the bacteria as the food settled in the water. By using control groups and various dilutions of bacteria, he found that they serve as a satisfactory food supply for the tadpoles. His conclusions indicated that water bacteria contain all the food factors necessary for the metamorphosis of *Rana pretiosa*. Later investigators have disagreed with this conclusion; they felt that other factors, found in the mud bottoms of ponds, stimulate development.[12]

Tadpoles have been known to resort to cannibalism when alternate food sources are depleted. Bleakney[15] found a small pond containing hundreds of tadpoles (*Rana sylvatica*) – all with chewed tails. The pond was grown over with conifers, so that no light penetrated to the pond; therefore, the aquatic plants could not grow fast enough to meet the nutritional needs of the tadpole population. Tadpoles will attack tadpoles of another species instead of their own siblings.[16] Tadpoles of *Hyla septintrionalis* attacked and consumed tadpoles of *Bufo americanus* when ten of each species were placed together in 200 ml of distilled water;[16] even though all the tadpoles were of equal size, only the hylids were left after 24 hr. The hylids attacked the bufonids with fervor; frequently, several would attack a single bufonid and rend it apart. If hylids were placed by themselves in an aquarium without food, actively feeding hylids would eventually resort to cannibalism. From hatching to metamorphosis, tadpoles seem to be nonpalatable to their siblings (in an actively feeding population). However, once injured or debilitated through starvation, tadpoles are attacked by their cohorts. An injured tadpole may become the object of a violent feeding aggregation, in which the noninjured individuals remain immune to attack.[16]

Salamanders, as a group, are omnivorous. However, the diets of separate species may be very restricted. Many salamanders prefer some foods to others and often will ignore potential food items, even when those items may be in greater supply than the preferred ones. *Rhyacotriton olympicus* is such a selective feeder; 95% of its diet consists of arthropods, even though freshwater snails and earthworms abound on the stream banks and under rocks, where the salamander forages.[17] Dineen[5] found that larval *Ambystoma tigrinum* ate a few molluscs and tadpoles, but they did not eat the terrestrial insects that were floating on the surface of the pond in great numbers. Adult salamanders of the species *Triturus viridescens viridescens* inhabiting the high-altitude lakes of North Carolina eat freshwater clams exclusively, though other invertebrates are undoubtedly present in these lakes.[18] These examples indicate that salamanders are not fortuitous feeders, even though the combined diets of all species of salamanders include many of the same foods that frogs and toads eat.

REFERENCES

1. Hamilton, W. J., Jr., *Copeia,* 1948, 203—207, 1948.
2. Zimka, J., *Ekol. Pol. Ser.,* 14, 589—605, 1966.
3. Drake, C. J., *Ohio Nat.,* 14, 257—269, 1914.
4. Linzey, D. W., *Herpetologica,* 23, 11—17, 1967.
5. Dineen, C. F., *Proc. Indiana Acad. Sci.,* 65, 231—233, 1955.
6. Brandon, R. A., *Herpetologica,* 23, 52—53, 1967.
7. Dickerson, M. C., *The Frog Book,* Doubleday, Page Co., New York, 1906.
8. Bragg, A. N., *Herpetologica,* 13, 189—191, 1957.
9. Noble, G. K., *The Biology of Amphibia,* Dover Publications, New York, 1954.
10. Barrington, E. J. W., *Biol. Rev.,* 17, 1—27, 1942.
11. Savage, R. M., *Proc. Zool. Soc. London,* 122, 467—514, 1952.
12. Nathan, J. M. and James, V. G., *Copeia,* 1972, 669—679, 1972.
13. Janes, R. G., *Copeia,* 1939, 134—140, 1939.
14. Burke, V., *Science,* 78, 194—195, 1933.
15. Bleakney, S., *Herpetologica,* 14, 34, 1958.
16. Spielman, A. and Sullivan, J. J., *Am. J. Trop. Med. Hyg.,* 23, 704—709, 1974.
17. Bury, R. B. and Martin, M., *Copeia,* 1967, 487, 1967.
18. Behre, E. H., *Copeia,* 1953, 60, 1953.

IV. FOOD FOR CAPTIVE AMPHIBIA

W. A. Cooke

Amphibian larvae are provided with a large yolk sac, which is functional even after mouth parts are well developed and useable. These larvae can then exist without food for a short while, even though their digestive tract is complete and can accept food. Once the yolk sac is absorbed, the animal requires food, and if kept in captivity, it must be provided with an ample supply. Researchers have provided a great variety of foods for their captive amphibians (see Table 2).

REFERENCES

1.   Rugh, R., *Experimental Embryology. A Manual of Techniques and Procedures,* rev. ed., Burgess Publishing Co., Minneapolis, 1948.
2.   Moore, J. A., *Genetics,* 27, 408–416, 1942.
3.   Briggs, R. W., *Science,* 93, 256–257, 1941.
4.   Volpe, E. P., *J. Hered.,* 51, 151–155, 1960.
5.   Etkin, W. and Huth, T., *J. Exp. Zool.,* 82, 463–495, 1939.
6.   Borland, J., *J. Exp. Zool.,* 94, 115–143, 1943.
7.   Nace, G. W., *Bioscience,* 18, 767–775, 1968.
8.   Ting, H., *Science,* 112, 539–540, 1950.
9.   Adolph, E. F., *Biol. Bull.,* 61, 376–386, 1931.
10. Richard, 1958; Rose, 1960; West, 1960; cited in Merrell, D. J., *Turtox News,* 41, 263–265, 1963.
11. Cairns, A. M., Bock, J. W., and Bock, F. G., *Nature London,* 213, 191–193, 1967.
12. Hamburger, V., *A Manual of Experimental Embryology,* rev. ed., University of Chicago Press, Chicago, 1960.
13. Merrell, D. J., *Turtox News,* 41, 263–265, 1963.

Table 2A
FOOD FOR LARVAE

Food	Preparation	Reference
Anura		
Lettuce or spinach	Boiled	1–5
Cabbage	Boiled	6
Romaine or escarole	Boiled	7
Sphagnum moss	Dried	8
Algae	Intact	1, 9
Oatmeal	Dry	5
Pablum		3
Milk	Dehydrated	3
Purina Rabbit Chow		10
Purina Guinea Pig Chow		11
Corn meal		10
Beef liver	Cooked, dehydrated, and powdered	1, 3, 6
Liverwurst		1
Egg yolk	Dried and powdered	1, 8
Bacto-Beef Extract	Mixed with whole-wheat flour, dried, and pulverized	1
Protozoan	Intact	1
Shrimp	Dried and powdered	5, 6
Liver	Raw and Minced	7
Urodela		
Protozoan	Intact	1
Daphnia	Intact	1, 12
Red worm (*Tubifex*)	Intact	1, 12
White worm (*Enchytrea*)	Intact	1
Mealflies	Intact	1, 12
Wax moth (*Galleria*)	Intact	1
Drosophila (vestigial mutant)	Intact	1
Plant lice	Intact	1
Small ants	Intact	1
Culex pipiens (wriggler)	Intact	7
Brine shrimp	Intact	7
Ostracods	Intact	12
Amphibian larvae	Intact	1
Earthworm	Cut-up	12
Beef liver	Thin strips	12

Table 2B
FOOD FOR THE POST-METAMORPHIC STAGES

Food	Preparation	Reference
Anura		
Earthworms	Cut-up	1
Blow flies	Intact	1
Mealworms	Intact	1, 7, 11
Ants	Intact	1
Spiders	Intact	1
Roaches	Intact	1
Caterpillars	Intact	1
Grasshoppers	Intact	1
Culex pipiens	Intact	1, 7
Mealfly (*Sarcophaga bullata*)	Intact	7
Green bottlefly (*Phoenicia sericata*)	Intact	7
Field cricket (*Acheta domestica*)	Intact	7, 11
Mammalian liver or muscle	Brewer's yeast and cod-liver oil	1
Liver	Raw strips	1
Crayfish	Intact	1
Minnows	Intact	1
Small frogs	Intact	1
Urodela		
White worms	Intact	1
Earthworms	Intact	1
Beef or calf's liver	Thin strips	1

V. CALORIC REQUIREMENTS

T. Hansen

Table 3
THE METABOLIC RATES OF SOME AMPHIBIANS

Species[a]	Oxygen consumption[b] (μl·hr^{-1}·g^{-1})	Temperature (°C)	Weight (g)	Caloric equivalent[c] (cal·hr^{-1}·g^{-1})	Ref.[f]
Acris crepitans (F)	30 260	5 25	N.D.[d] N.D.	0.141 1.22	1[e] 1[e]
Ambystoma jeffersonianum (S)	131.7	N.D.	N.D.	0.619	2
Ambystoma maculatum (S)	25 130	5 25	N.D. N.D.	0.117 0.611	3[e] 3[e]
Ambystoma microstomum (S)	333.5	N.D.	N.D.	1.57	2
Ambystoma punctatum (S)	163.3	N.D.	N.D.	0.768	2
Ambystoma trigrinum (S)	111.2	N.D.	N.D.	0.522	2
Amphiuma tridactylum (S)	3.55	15	611	0.0167	4
Aneides lugubris (S)	387.0	20	N.D.	1.82	5
Batrachoseps attenuatus (S)	75.7	20	0.931	0.356	5
Bufo bufo (T)	112.2	17	N.D.	0.528	6
Bufo marinus (T)	48.0	22	N.D.	0.226	7
Desmognathus fuscus (S)	39.0	16.2—16.8	2.50	0.1833	8
Desmognathus ochrophaeus (S)	47.7 13.94 34.08	16.4—16.8 5 15	1.33 1.17 1.24	0.25 0.07 0.16	8 9 9

[a] (F) = frog; (S) = salamander; (T) = toad.
[b] Grams body weight.
[c] If the author did not calculate the caloric equivalent of the O_2 consumption, the value 4.7 kcal/l O_2 was used.
[d] Not determined.
[e] Not precise, as they were read from a graph.
[f] See Reference 25.

Table 3 (continued)
THE METABOLIC RATES OF SOME AMPHIBIANS

Species[a]	Oxygen consumption[b] (μl·hr^{-1}·g^{-1})	Temperature (°C)	Weight (g)	Caloric equivalent[c] (cal·hr^{-1}·g^{-1})	Ref.[f]
Desmognathus quadramaculatus (S)	24	5	N.D.	0.113	3[e]
	78	25	N.D.	0.367	3[g]
Eurycea bislineata (S)	55.4	15.8—16.7	1.13	0.26	8
Eurycea nana (S)	40	15	0.216—0.431	0.188	10
	55	20	0.280—0.592	0.259	10
	81	25	0.216—0.431	0.381	10
Eurycea neotones (S)	36	15	0.756—0.849	0.169	10
	51	20	0.280—0.592	0.240	10
	58	25	0.756—0.849	0.273	10
	79	30	0.756—0.879	0.371	10
Eurycea pterophila (S)	66	15	0.280—0.592	0.310	10
	85	20	0.280—0.592	0.40	10
	65	25	0.280—0.592	0.31	10
	62	30	0.280—0.592	0.291	10
Gyrinophilus porphyriticus (S)	17.6	15.8—16.7	12.1	0.083	8
Hyla cinerea (F)	102	25	4.5	0.479	11
Hyla versicolor (F)	102	25	4.5	0.479	11
Necturus maculosus (S)	27.5	22	N.D.	0.129	7
Plethodon cinereus (S)	41.4	15.5—16.0	0.98	0.195	8
Pseudacris nigrita (F)	162	25	1.0	0.761	11
Rana catesbeiana (F)	43	25	350	0.20	11
Rana clamitans (F)	93	25	35	0.44	11

g Taken at maximum activity.

Table 3 (continued)
THE METABOLIC RATES OF SOME AMPHIBIANS

Species[a]	Oxygen consumption[b] ($\mu l \cdot hr^{-1} \cdot g^{-1}$)	Temperature (°C)	Weight (g)	Caloric equivalent[c] ($cal \cdot hr^{-1} \cdot g^{-1}$)	Ref.[f]
Rana	95.8	17	N.D.	0.46	6
esculenta (F)	97	25	N.D.	0.456	12[h]
	82.5	10—20	N.D.	0.387	13
	212.0	19	N.D.	0.996	13
	79.0	N.D.	N.D.	0.371	14
	120.0	20—25	N.D.	0.564	12
	136.0	N.D.	N.D.	0.639	14
	50.5	17.7	N.D.	0.237	15
	116.0	20	N.D.	0.545	16
	152	20—21	N.D.	0.715	12
	74	22—27	N.D.	0.348	17
	16.3	10	N.D.	0.077	17
	44.2	15—19	N.D.	0.208	18
	73.4	15—19	N.D.	0.345	18
	87.0	25	N.D.	0.409	12
	54.3	20	N.D.	0.256	19
Rana	194.0	20	N.D.	0.912	12
fusca (F)	107	20	N.D.	0.503	12
	77.8	26.3	N.D.	0.366	20
Rana	93	25	35	0.44	11
pipiens (F)	91	17	N.D.	0.43	6
	127.6	25	N.D.	0.60	21[g]
	70.8	22	N.D.	0.333	7
	44.0	22	N.D.	0.207	15
	166.0	25—28	N.D.	0.78	22
Rana sylvatica (F)	108	25	6.0	0.508	11
Rana	78.5	17	N.D.	0.37	6
temporaria (F)	68.9	24.7	N.D.	0.324	23
	68.9	24.7	N.D.	0.989	23
	127.0	24.7	N.D.	0.059	23
	200.0	19.3	N.D.	0.94	23
	173.0	24.7	N.D.	0.81	23
	69.9	19—25	N.D.	0.328	24
	65.6	22.5	N.D.	0.308	19
Rana	5.70	5	N.D.	0.027	17
viridis (F)	8.40	5	N.D.	0.039	17
Siren lacertina (S)	52.9	22	N.D.	0.249	7
Taricha	40	5	N.D.	0.188	3[e]
granulosa (S)	132	25	N.D.	0.620	3[e]
Xenopus	63.1	17	N.D.	0.297	6
laevis (T)	45.9	22	N.D.	0.216	7
Xenopus mulleri (T)	104.6	17	N.D.	0.492	6

[h] Acclimated at 5°C.

Table 3 (continued)
THE METABOLIC RATES OF SOME AMPHIBIANS

REFERENCES

1. Dunlap, D. G., *Comp. Biochem. Physiol. A,* 38, 1–16, 1971.
2. Helff, O. M., *J. Exp. Zool.,* 49, 353–361, 1923.
3. Whitford, W. G. and Hutchison, V. H., *Physiol. Zool.,* 40, 127–133, 1967.
4. Toews, D. P., *Can. J. Zool.,* 51, 664–666, 1973.
5. Cook, S. F., *Univ. Calif. Berkeley Publ. Zool.,* 53, 367–376, 1949.
6. Jones, D. R., *Comp. Biochem. Physiol.,* 20, 691–707, 1967.
7. Bentley, P. J. and Shield, J. W., *Comp. Biochem. Physiol. A,* 46, 29–38, 1973.
8. Evans, G., *Ecology,* 20, 74–95, 1939.
9. Fitzpatrick, L. C., *Ecol. Monogr.,* 43, 43–58, 1973.
10. Norris, W. E., Grandy, P. A., and Davis, W. K., *Biol. Bull.,* 125, 523–533, 1963.
11. Davison, J., *Biol. Bull.,* 109, 407–419, 1955.
12. Krogh, A., *Skand. Arch. Physiol.,* 15, 328–419, 1903.
13. Bohr, C., *Skand. Arch. Physiol.,* 10, 74–91, 1900.
14. Galperin, L., Okun, M., Simonson, E., and Sirkinia, G., *Arbeitsphysiologie,* 8, 407–423, 1935.
15. Oertmann, E., *Pfluegers Arch. Gesamte Physiol.,* 15, 381–398, 1877.
16. Schultz, H., *Pfluegers Arch. Gesamte Physiol.,* 14, 78–91, 1877.
17. Dontcheff, L. and Kayser, C., *C. R. Acad. Sci.,* 201, 474–476, 1935.
18. Precht, H., *Biol. Zentralbl.,* 70, 71–85, 1951.
19. Vernon, H. M., *J. Physiol.,* 21, 443–496, 1897.
20. Joel, A., *Hoppe Seyler's Z. Physiol. Chem.,* 107, 231–263, 1919.
21. Turney, L. D. and Hutchison, V. H., *Comp. Biochem. Physiol. A,* 49, 583–601, 1974.
22. Long, W. D. and Johnson, R. E., *Am. J. Physiol.,* 171, 744–745, 1952.
23. Dolk, H. E. and Postma, N., *Z. Vergl. Physiol.,* 5, 417–444, 1927.
24. Kayser, C., *Les Echanges Respiratoires des Hibernants,* Lons-LeSaunier, Imprimerie M. Declume, Paris, 1940, pp. 141–203.
25. Fromm, P. O. and Johnson, R. E., *J. Cell. Comp. Physiol.,* 45, 343–359, 1935.

Table 4
METABOLIC RATES OF SOME AMPHIBIA AT SPECIFIC TIMES OF YEAR

Species[a]	Oxygen consumption[b] (μl·hr^{-1}·g^{-1})	Temperature (°C)	Time of year	Caloric equivalent[c] (cal·hr^{-1}·g^{-1})	Ref.
Eurycea	38.10	10	October–November	0.179	1
bislineata (S)	40.20	10	February–March	0.189	1
	44.75	10	May–June	0.210	1
	11.74	1	October–November	0.055	1
	13.68	1	February–March	0.064	1
	16.86	1	February–March	0.079	1
	15.34	1	May–June	0.072	1
Plethodon	30.32	10	October–November	0.143	1
cinereus (S)	33.46	10	February–March	0.157	1
	37.43	10	May–June	0.176	1
	18.58	1	October–November	0.087	1
	15.44	1	February–March	0.073	1
	15.71	1	February–March	0.074	1
	14.25	1	May–June	0.067	1
Rana	97.0	25	June	0.456	2
esculenta (F)	120.0	20–25	April	0.564	2
	87.0	25	December	0.409	2
	74.0	22–27	June	0.348	3
Rana	194.0	20	April	0.912	2
fusca (F)	182.0	20	May	0.855	2
	107.0	20	October	0.503	2
Rana	39.6	22–28	March–May	0.832[d]	4
pipiens (F)	33.8	22–28	June–August	0.777[d]	4
	21.8	22–28	September–November	0.673[d]	4
	23.9	22–28	December–February	0.712[d]	4
	44.0	22	March	0.207	5
	166.0	25–28	Summer	0.78	6
Rana	5.7	5	October	0.027	3
viridis (F)	8.41	5	December	0.039	3

[a] (F) = frog; (S) = salamander.
[b] Grams body weight.
[c] If the author did not calculate the caloric equivalent for the O_2 consumption, the value 4.7 kcal/l was used.
[d] Fromm's data.

REFERENCES

1. Vernberg, F. J., *Physiol. Zool.*, 25, 243–249, 1952.
2. Krogh, A., *Skand. Arch. Physiol.*, 15, 328–419, 1904.
3. Dontcheff, L. and Kayser, C., *C. R. Soc. Biol.*, 121, 1453–1455, 1935.
4. Fromm, P. O., *Physiol. Zool.*, 29, 234–240, 1956.
5. Oertmann, E., *Pfluegers Arch. Gesamte Physiol.*, 15, 381–398, 1877.
6. Long, W. D. and Johnson, R. E., *Am. J. Physiol.*, 171, 744–745, 1952.

VI. MINERAL REQUIREMENTS

B. J. Coffey and G. W. Brown, Jr.

Most of the information on the mineral requirements of Amphibia is inferential in nature and is based on the qualitative requirements as known in other vertebrates. Table 5 indicates mineral elements in addition to carbon, hydrogen, oxygen, and nitrogen that are considered to be essential for Amphibia. The following brief list of references provides some information on the role of mineral elements in Amphibia.

Batra, S., The effects of zinc and lanthanum on calcium uptake by mitochondria and fragmented sarcoplasmic reticulum of frog skeletal muscle, *J. Cell. Physiol.*, 82, 245–256, 1973.

Beaugé, L. A., Medici, A., and Sjodin, R. A., The influence of external caesium ions on potassium efflux in frog skeletal muscle, *J. Physiol.*, 228, 1–11, 1973.

Beaugé, L. A. and Ortiz, O., Further evidences for a potassium-like action of lithium ions on sodium efflux in frog skeletal muscle, *J. Physiol.*, 226, 675–697, 1972.

Bentley, P. J. and Main, A. R., Zonal differences in permeability of the skin of some anuran Amphibia, *Am. J. Physiol.*, 223, 361–363, 1972.

Binet, L. and Magrou, J., Sulfur and growth, *C. R. Acad. Sci.*, 193, 115–117, 1931.

Borchard, V. and Schneider, K. V., Intoxication, detoxication and copper storage of central nervous tissue at different external Cu(II)-concentrations, *Arch. Toxicol.*, 33, 17–30, 1974.

Chiarandini, D. J. and Stefani, E., Effects of manganese on the electrical and mechanical properties of frog skeletal muscle fibres, *J. Physiol.*, 232, 129–147, 1973.

Crim, J. W., Studies on the possible regulation of plasma sodium by prolactin in Amphibia, *Comp. Biochem. Physiol. A*, 43, 349–357, 1972.

Cofré, G. and Crabbé, J., Active sodium transport by the colon of *Bufo marinus:* stimulation by aldosterone and antidiuretic hormone, *J. Physiol.*, 188, 177–190, 1967.

Das, P. K., Sinha, P. S., Srivastava, R. K., and Sanyal, A. K., Studies on ciliary movement. II. Effects of certain physical and chemical factors on ciliary movement in frog's oesophagus, *Arch. Int. Pharmacodyn.*, 153, 367–378, 1965.

Dietz, T. H., Kirschner, L. B., and Porter, D., The roles of sodium transport and anion permeability in generating transepithelial potential differences in larval salamanders, *J. Exp. Biol.*, 46, 85–96, 1967.

Guerrero, S. and Riker, W. K., Effects of some divalent cations on sympathetic ganglion function, *J. Pharmacol. Exp. Ther.*, 186, 152–159, 1973.

Krylov, O. A., The role of haloids (bromine and iodine) in metamorphosis of Amphibia, *Byull. Eksp. Biol. Med.*, 49, 85–89, 1960.

Kuusisto, A. N. and Telkka, A., The effect of sodium fluoride on the metamorphosis of tadpoles, *Acta Odontol. Scand.*, 19, 121–127, 1961.

Pasanen, S. and Koskela, P., Seasonal changes in calcium, magnesium, copper and zinc content in the liver of the common frog, *Rana temporaria* L., *Comp. Biochem. Physiol. A*, 48, 27–36, 1974.

Schoffeniels, E. and Tercafs, R. R., Potential difference and net flux of water in the isolated amphibian skin, *Biochem. Pharmacol.*, 11, 769–778, 1962.

Schwartz, M., Kashiwa, H. K., Jacobson, A., and Rehm, W. S., Concentration and localization of calcium in frog gastric mucosa, *Am. J. Physiol.*, 212, 241–246, 1967.

Simkiss, K., *Calcium in Reproductive Physiology: A Comparative Study of Vertebrates*, Reinhold Publishing Corp., New York, and Chapman and Hall, London, 1967.

Simkiss, K., Calcium and carbonate metabolism in the frog (*Rana temporaria*) during respiratory acidosis, *Am. J. Physiol.*, 214, 627–534, 1968.

Underwood, E. J., *Trace Elements in Human and Animal Nutrition*, 3rd ed., Academic Press, New York, 1971.

Walton, K. G., DeLorenzo, R. J., Curran, P. F., and Greengard, P., Regulation of protein phosphorylation and sodium transport in toad bladder, *J. Gen. Physiol.*, 65, 153–177, 1975.

Weakly, J. N., The action of cobalt ions on neuromuscular transmission in the frog, *J. Physiol. London*, 234, 597–612, 1973.

Woerdeman, M. W., The influence of potassium upon the isolated heart of frog larva, *Arch. Neerl. Physiol.*, 9, 153–158, 1924.

Table 5
MINERAL IONS AND ELEMENTS ESSENTIAL TO AMPHIBIA[a]

(Other than C, H, O, and N)

Element or ion	Role or function
Ca^{2+}	Bone formation, blood coagulation, capillary integrity, enzyme activation, permeability, nervous system.
Cl^-	Chief anion of extracellular fluid, esophageal or gastric HCl, acid–base balance, blood transport of CO_2 ("chloride shift").
Co (micronutrient)	Stimulation of hematopoiesis, vitamin B_{12}, enzyme activation.
Cu (micronutrient)	Erythropoiesis, Cu-flavin enzymes (polyphenol oxidase, etc.).
F^- (micronutrient)	Bone.
Fe (micronutrient)	Part of heme group in hemoglobin, catalase, cytochromes, Fe-flavins
I_2 or I^- (micronutrient)	Part of thyroxine, metamorphosis, metabolic regulation through thyroxine.
K^+	Chief cation of intracellular fluid, nervous system, enzyme activation.
Mg^{2+}	Neuromuscular system, bone formation, activation of enzymes (especially involving phosphate transfer).
Mn^{2+} (micronutrient)	Metal ion enzyme activation (e.g., arginase).
Mo (micronutrient)	Mo-flavin enzymes (aldehyde oxidase, xanthine oxidase).
Na^+	Chief cation of extracellular fluid, nerve conduction, enzyme activation.
P (HPO_4^{2-}/$H_2PO_4^-$)	Bone formation, organic phosphates, tissue buffer.
S (organic)	Amino acids, proteins, coenzymes, redox reactions.
Zn^{2+} (micronutrient)	Some enzymes (carbonic anhydrase, glutamic dehydrogenase).
Se (micronutrient)	Requirement not demonstrated.
Cr (micronutrient)	Requirement not demonstrated.

[a] Adapted from Brown, G. W., Jr., in *Physiology of the Amphibia,* Moore, J. A., Ed., Academic Press, New York, 1964, p. 80.

VII. VITAMINS

J. C. Hoeman

In very few cases have vitamin requirements been established for Amphibia. Most of the literature on vitamins in Amphibia is concerned with the distribution and function of vitamins rather than with the role of vitamins in growth and nutrition. Vitamins, in general, are precursors of coenzymes and prosthetic groups for enzymes. In the absence of information to the contrary, the role of vitamins in amphibian nutrition can be expected to be similar to that of vitamins in any vertebrate. Table 6 gives some information on the utilization of vitamins by Amphibia.

REFERENCES

1.   Love, R. M., Collins, F. D., and Morton, R. A., *Biochem. J.,* 51, 37–38, 1952.
2.   Allende, I. L. C., Caligaris, L. S., and Orias, O., *Rev. Soc. Argent. Biol.,* 24, 301–306, 1948.
3.   Günder, I., *Z. Naturforsch. Teil B,* 10, 173–177, 1955.
4.   Koller, M., *Boll. Soc. Ital. Biol. Sper.,* 15, 718–719, 1941.
5.   Yagi, K., *J. Jpn. Biochem. Soc.,* 23, 72–74, 1951.
6.   Catolla-Cavalcanti, A., *Acta Vitaminol.,* 5, 162–164, 1951.
7.   Lindahl, P. E., and Lennerstrand, A., *Ark. Kemi Mineral. Geol.,* 15B, 1–6, 1942.
8.   Nakamura, S., *Trans. Jpn. Pathol. Soc.,* 22, 97–101, 1932.
9.   Tomita, M., Hamada, K., Yoshiko, M., and Sasaki, M., *J. Biochem. Tokyo,* 39, 299–303, 1952.
10. DeBastiani, G. and Zatti, P., *Boll. Soc. Ital. Biol. Sper.,* 25, 1323–1327, 1949.
11. Wald, G., *Harvey Lect.,* 41, 117–160, 1946.
12. Buck, M. G. and Zadunaisky, J., *Biochim. Biophys. Acta,* 389, 251–260, 1975.
13. Collins, F. D., Love, R. M., and Morton, R. A., *Biochem. J.,* 53, 632–636, 1953.
14. Radu, V. G. and Oniceanu, H., *Ann. Sci. Univ. Jassy,* 26, 815–822, 1940.
15. Hara, T., *Seiri Seitai,* 3, 102–110, 1949.
16. Yagi, K., *J. Jpn. Biochem. Soc.,* 22, 143–147, 1950.
17. Hubbard, R. and Colman, A. D., *Science,* 130, 977–978, 1959.
18. Morton, R. A. and Rosen, D. G., *Biochem. J.,* 45, 612–627, 1949.
19. Nollet, H. and Raffy, A., *C. R. Acad. Sci.,* 210, 269–270, 1940.
20. Collins, F. D., Love, R. M., and Morton, R. A., *Biochem. J.,* 53, 629–632, 1953.
21. Wald, G., *Biol. Symp.,* 7, 43–71, 1942.
22. Kasyako, K. S., *Fiziol. Zhur.,* 35, 124–127, 1949.
23. Bruce, H. M. and Parkes, A. J., *J. Endocrinol.,* 7, 64–81, 1950.

Table 6
VITAMIN REQUIREMENTS AND UTILIZATION IN AMPHIBIA

Species	Stage	Vitamins												Reference
		A	A_1	A_2	B_c	A_p	B_1	B_2	B_3	B_5	C	D	D_p	
Ambystoma tigrinum	A	U	–	–	–	∅	–	–	–	–	–	–	–	1
Bufo arenarum	A	–	–	–	–	–	–	–	–	–	U	–	–	2
Bufo bufo	A	–	–	–	–	–	–	U	–	–	–	–	–	3
Bufo vulgaris	A	–	–	–	–	–	U	U	U	U	U	–	–	4,5
	L	–	–	–	R	–	–	–	–	–	–	–	–	6,7,8
Hyla arborea	A	–	–	–	–	–	U	–	–	–	–	–	–	4
Megalobatrachus japonicus	A	–	–	–	–	–	U	U	–	–	–	–	–	9
Rana agilis	A	–	–	–	–	–	U	–	–	–	–	–	–	10
Rana catesbeiana	A	–	–	U	–	–	–	–	–	–	U	–	–	11,12
	L	–	U	–	–	–	–	–	–	–	–	–	–	11
Rana esculenta	A	–	–	–	–	–	–	–	–	–	U	–	–	13,14
	L	–	U	–	–	–	–	–	–	–	–	–	–	13
Rana mugiens	A	–	–	–	–	–	–	U	–	–	–	–	–	5
Rana nigromaculata	A	–	–	–	–	–	U	U	–	–	–	–	–	15,16
Rana pipiens	A	U	–	–	–	–	–	–	U	–	–	–	–	17,7
Rana temporaria	A	U	–	–	–	U	–	–	–	–	–	–	U	13,18
	L	–	U	–	–	–	–	–	–	–	–	–	–	13
Rana viridis	A	–	–	–	–	–	–	U	–	–	–	–	–	19
Triton cristatus	A	–	–	–	–	–	U	–	–	–	–	–	–	4
Triturus carniflex	A	–	U	U	–	–	–	–	–	–	–	–	–	20
Triturus cristatus	A	–	U	U	–	–	–	–	–	–	–	–	–	20
Triturus cristatus	A	–	U	U	–	–	–	–	–	–	–	–	–	13
Triturus viridescens	A	–	–	U	–	–	–	–	–	–	–	–	–	21
Triturus vulgarus	A	–	–	–	–	–	–	–	–	–	U	–	–	22
Xenopus laevis	A	–	–	–	–	–	–	–	–	–	–	R	–	23

Abbreviations. Vitamins: B_c = vitamin B complex; A_p = β-carotene, vitamin A precursor; D_p = vitamin D precursors, 7-dehydrocholesterol and ergosterol. Stage: A = adult; L = larva. Table: R = required; U = utilized; ∅ = not utilized.

VIII. PROTEINS AND AMINO ACIDS

G. W. Brown, Jr. and G. Dalton

Most adult Amphibia are carnivorous; hence, they have a relatively rich protein diet. The essential amino acid requirements have not been established for Amphibia. On the basis of requirements for other vertebrates and for fish (for salmon, see Reference 1), the following amino acids are probably essential: L-leucine, L-isoleucine, L-lysine, L-methionine, L-phenylalanine, L-threonine, L-tryptophan, and L-valine. Larval Amphibia presumably require L-arginine in addition. The ornithine—urea cycle becomes induced during metamorphosis of the frog *Rana catesbeiana*,[2] which provides a route for the synthesis of arginine in the adult. Neotenic forms, such as the mudpuppy, *Necturus*, may also require L-arginine, because the level of ornithine—urea cycle enzymes in adult *Necturus* is not very active.[3]

Utilization of the protein of reserve materials in embryonic and larval development of the salamander *Ambystoma mexicanum* is shown in Figure 1.

Larval Amphibia excrete their waste nitrogen as ammonia. Upon induction of the ornithine—urea cycle at metamorphosis (see Figure 2), this ammonia is largely converted to urea for excretion. The partition of waste nitrogen between ammonia and urea is dependent upon habitat (see Table 7).

Table 8 provides information on the occurrence and levels of free amino acids in tissues of Amphibia. Hydrolysis of dietary protein to amino acids is effected by gastric pepsin, by pancreatic trypsin, and presumably by peptidases of the small intestine.[4]

REFERENCES

1. Halver, J. E. and Shanks, W. E., *J. Nutr.*, 72, 340–346, 1960.
2. Brown, G. W., Jr. and Cohen, P. P., in *Symposium on the Chemical Basis of Development*, McElroy, W. D. and Glass, B., Eds., Johns Hopkins Press, Baltimore, Maryland, 1958, pp. 495–513.
3. Brown, G. W., Jr. and Cohen, P. P., *Biochem. J.*, 75, 82–91, 1960.
4. Reeder, W. G., in *Physiology of the Amphibia*, Vol. 1, Moore, J. A., Ed., Academic Press, New York and London, 1964, pp. 99–149.

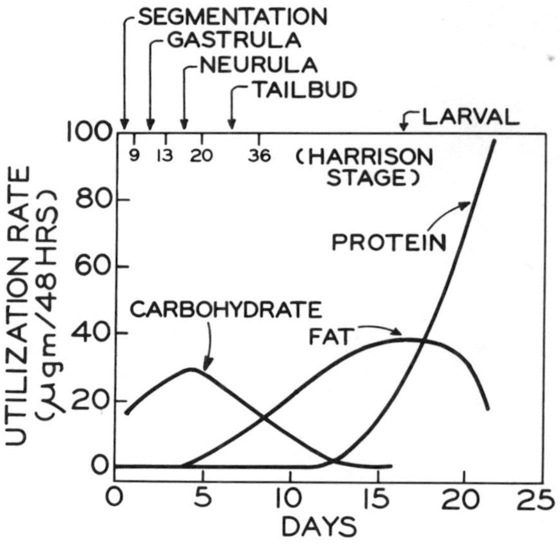

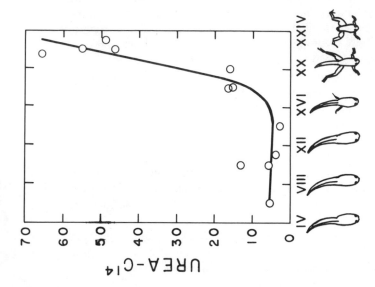

FIGURE 2. Incorporation of bicarbonate-^{14}C into urea-^{14}C by tadpole liver slices. Ordinate: thousands of counts/min incorporated into urea. Abscissa: larval stage number after Taylor and Kollros. Tissue (0.1 g) from *Rana catesbeiana* tadpoles incubated in buffer medium at pH 7.5, containing 6.8×10^5 counts/min of bicarbonate-^{14}C. Incubation for 2.0 hr at 37°C, with oxygen in the gas phase. (From Brown, G. W., Jr., *Biochim. Biophys. Acta*, 60, 185–186, 1962. With permission.)

FIGURE 1. Rate of utilization of reserve materials in the embryonic and larval development of *Ambystoma mexicanum* (salamander). (From Løvtrup, S., *C. R. Trav. Lab. Carlsberg Ser. Chim.*, 28, 426–443, 1953. With permission).

Table 7[1,2]

PARTITION OF EXCRETORY NITROGEN BY AMPHIBIA

Species	Habitat (non-breeding season)	Microatoms N/g body weight/24 hr as		Ratio Urea-N Ammonia-N
		Ammonia	Urea	
Urodela				
Salamandra salamandra	Terrestrial	0.24	4.5	19
Triturus cristatus	Terrestrial	0.20	4.8	24
Ambystoma mexicanum (neo-tenic larvae)	Aquatic	5.3	3.5	0.66
Anura				
Pipidae				
Xenopus laevis	Aquatic	12	7.7	0.64
Xenopus tropicalis	Aquatic	9.2	5.7	0.62
Hymenochirus sp. (*boettgeri?*)	Aquatic	6.9	2.0	0.29
Pipa pipa	Aquatic	7.2	0.54	0.75
Ranidae				
Rana esculenta	Semi-aquatic	0.35	3.2	9.1
Rana temporaria	Terrestrial	0.63	6.6	11
Hylidae				
Hyla arborea	Terrestro-arboreal	0.50	11	22
Bufonidae				
Bufo bufo	Dry terrestrial	0.44	7.6	17
Bufo calamita	Dry terrestrial	0.41	6.1	15

From Moore, J. A., *Physiology of the Amphibia,* Vol. 1, Academic Press, New York, 1964, p. 61. With permission.

REFERENCES

1. **Cragg, M. M., Balinsky, J. B., and Baldwin, E.,** *Comp. Biochem. Physiol.,* 3, 227–235, 1961.
2. **Brown, G. W., Jr.,** in *Physiology of the Amphibia,* Vol. 1, Moore, J. A., Ed., Academic Press, New York and London, 1964, p. 61.

Table 8
FREE AMINO ACID CONTENT OF AMPHIBIAN TISSUES

(Values in μmol/kg wet weight or liter of plasma)

Amino acid	Rana catesbeiana				Rana pipiens			Xenopus laevis Plasma[a]		
	Adult liver	Tadpole liver	Muscle	Kidney	Brain	Liver	Muscle	Fed previous day	Starved 28 days	Estivating
Asp	340	460	206	243	857	850	345	65 (1.0)	4 (0.9)	128 (2.8)
Thr	190	160	31	389	—	252	308	670 (10.7)	40 (9.2)	242 (5.5)
Ser	380	474	68	823	462	231	505	597 (9.5)	34 (7.8)	383 (8.7)
Gln	—	—	—	—	1,960	112	1,248	465 (7.4)	47 (10.8)	233 (5.3)
Glu	724	1,022	185	1,540	5,360	2,580	644	170 (2.7)	9 (2.1)	196 (4.4)
Pro	169	249	70	401	—	0	0	270 (4.3)	0 (0)	117 (2.6)
Cit	—	—	—	—	—	0	0	49 (0.8)	2 (0.5)	17 (0.4)
Gly	487	409	273	1,220	826	229	378	366 (5.9)	13 (3.0)	392 (8.9)
Ala	590	781	483	1,298	197	1,674	1,605	1,141 (18.2)	121 (27.8)	612 (13.9)
α-ABA	—	—	—	—	—	0	0	26 (0.4)	0.3 (0.1)	0 (0)
Val	267	112	93	348	88	719	101	538 (8.6)	29 (6.6)	233 (5.3)
Cys	330	70	8	432	—	632	—	98 (1.6)	11 (2.5)	18 (0.4)
Met	91	33	32	—	71	26	38	151 (2.4)	0.4 (0.1)	64 (1.5)
Ile	170	79	86	230	38	70	51	216 (3.5)	15 (3.4)	178 (4.0)
Leu	333	181	123	550	91	122	97	469 (7.5)	25 (5.7)	326 (7.4)
Tyr	163	57	58	249	80	153	176	101 (1.6)	16 (3.7)	106 (2.4)
Phe	163	51	63	278	—	59	40	139 (2.2)	17 (3.9)	123 (2.8)
Orn	—	—	—	—	33	185	49	171 (2.7)	12 (2.7)	211 (4.8)
Lys	394	374	367	886	140	274	305	288 (4.6)	16 (3.7)	440 (10.0)
His	140	72	2,560	826	165	134	804	130 (2.1)	19 (4.4)	116 (6.3)
Try	—	—	—	—	—	—	—	50 (0.8)	—	—
Arg	120	21	173	224	220	27	74	137 (2.2)	11 (2.5)	29 (1.6)
Taurine	—	—	—	—	2,720	—	—	—	—	—
γ-ABA	—	—	—	—	2,500	—	—	—	—	—
Ammonia	—	—	—	—	—	1,890	800	408	59	408
Urea	—	3,020	2,620	5,180	—	760	970	1,630	—	29,500

[a] Values in parentheses: percentage of total amino acids.

From Balinsky, J. B., in Comparative Biochemistry of Nitrogen Metabolism, Vol. 2, Campbell, J. W., Ed., Academic Press, London and New York, 1970. p. 521. With permission.

IX. CARBOHYDRATES

G. D. Cortner

Although many studies have been concerned with carbohydrates in relation to the physiology of amphibians and for medical experimentation, little emphasis has been placed by research workers on carbohydrates in relation to amphibian nutrition. The information is sparse and scattered, particularly concerning adult Amphibia. Figure 1 of Section VIII illustrates the utilization of reserve carbohydrate in embryonic and larval development of *Ambystoma mexicanum.*

Adult Amphibia ordinarily do not encounter high levels of carbohydrate in their natural diet, and probably are not well equipped to handle high percentages of it in their diet. Larval Amphibia are largely herbivorous, encountering higher levels of carbohydrate in the diet than do the adults. In larval Amphibia, the gut is long and coiled, reflecting their herbivorous diet. After metamorphosis, the gut is proportionally shorter in relation to body length. Cellulose is probably hydrolyzed in tadpoles by intestinal flora of the small intestine, although this concept has been challenged.[1]

Amylase of pancreatic juice functions to hydrolyze glycogen and starch in the intestine.[1] Pancreatic juice also contains diastase and glucosidase.[1] The following sugars are transported by "active transport" across the intestinal wall of *Rana pipiens:*[2,3] D-arabinose, D-fructose, D-galactose, D-glucose, D-mannose, D-xylose, and 3-methyl-D-glucose. Some blood sugar values are given in Table 9.

REFERENCES

1. **Reeder, W. G.,** in *Physiology of the Amphibia,* Vol. 1, Moore, J. A., Ed., Academic Press, New York and London, 1964, pp. 99–149.
2. **Lawrence, A. L.,** *Comp. Biochem. Physiol.,* 9, 69–73, 1962.
3. **Csaky, T. Z. and Ho, P. M.,** *Life Sci.,* 5, 1025–1030, 1966.

Table 9
NORMAL BLOOD SUGAR VALUES FOR SOME AMPHIBIA[a]

(Values in mg/100 ml)

Species	Blood sugar
Bombina variegata	41 (28–75)
Xenopus laevis[b]	25–30
Bufo arenarum	20–80
Bufo d'Orbigny	62
Bufo marinus	25
Bufo viridis	36
Ceratophrys ornata	52
Leptodactylus occelata	129?
Rana ridibunda	30
Rana esculenta	29–100
Rana temporaria	20–50 (as high as 150)
Rana catesbeiana[c]	13
Rana pipiens	29–74
Rana vignomaculata	14–48
Rana sp.	10–65

[a] Adapted from Motelica, I. and Vladescu, C., *Rev. Roum. Biol. Ser. Zool.,* 10, 447–450, 1965.
[b] If depancreatized, *X. laevis* develops diabetes (230 mg/100 ml for blood sugar). See Slome, D., *J. Exp. Biol.,* 13, 1–6, 1936.
[c] *R. catesbeiana* appear to be refractory to injections of alloxan. See Wright, P. A., *Endocrinology,* 64, 551–558, 1959.

X. LIPIDS

B. A. Manz

Very little is known about lipid requirements of Amphibia. The nature of any essential fatty acids required for growth and development has not been established, although linolenic and linoleic acids are probably as essential as they are for other vertebrates. Little is known about the intermediary metabolism of lipids in Amphibia, but the basic cellular metabolic processes of Amphibia are believed to be similar to those of other vertebrates.[1] Lipase is secreted by the pancreas as in other vertebrates.[2]

Utilization of reserve lipid material during embryonic and larval development of the salamander *Ambystoma mexicanum* is shown in Figure 1 of Section VIII. Table 10 gives the percentages of several fatty acids of the depot fat of three frogs.

REFERENCES

1. **Brown, G. W., Jr.,** in *Physiology of the Amphibia,* Vol. 1, Moore, J. A., Ed., Academic Press, New York and London, 1964, pp. 1—98.
2. **Reeder, W. G.,** in *Physiology of the Amphibia,* Vol. 1, Moore, J. A., Ed., Academic Press, New York and London, 1964, pp 99—149.

Table 10

FATTY ACID COMPOSITION[a] OF AMPHIBIAN DEPOT FAT

Species	Lauric (C_{12})	Myristic (C_{14})	Palmitic (C_{16})	Stearic (C_{18})	Arachidonic (C_{20})	Unsaturated C_{12}	C_{14}	C_{16}	C_{18}	C_{20}	Reference
Rana tigrina	—	2.7	25.0	9.0	—	—	1.5	19.4	38.6	3.8	1
Rana temporaria	—	4.0	11.0	3.0	—	—	—	15.0	52.0	15.0	2
Bufo arenarum	0.5	3.4	18.2	3.8	0.5	0.1	1.1	13.1	57.9	1.4	2

[a] Values given are percentages.

REFERENCES

1. Pathak, S. P. and Agarwal, B. C. L., Indian J. Exp. Biol., 4, 248–249, 1966.
2. Hilditch, T. P. and Williams, P. N., The Chemical Constitution of Natural Fats, 4th ed., John Wiley & Sons, New York, 1964.

XI. DISEASES[1]

K. D. Edwards

Amphibia are subject to numerous diseases. Some diseases can be diagnosed readily by external signs. Sometimes the only sign of disease may be a refusal to eat or a condition of lethargy. Amphibia are generally hardy animals, but they may be handicapped through errors in development or by injuries from accidents or from fighting. Many Amphibia spend much of their lives either in or very near the water, and thus are exposed to numerous parasites that live a part of their life cycles in water. Such parasites cause the more common disease problems of Amphibia. Investigation of diseases in Amphibia is made difficult by several factors: the animals are highly adapted to their environment, and few display best the signs of health or disease when kept in captivity; also, many species are nocturnal and basically secretive, and when ill, they hide in the most available and most inaccessible place, often dying there unnoticed. Problems in raising and maintaining Amphibia in colonies for research purposes are discussed by Nace.[2] Because of susceptibility to diseases such as "red-leg" under captive conditions, commercial frog farming is a difficult undertaking.

Some common diseases of Amphibia are tabulated in Table 11. Susceptibility and resistance to disease are influenced by the nutritional status of an organism, hence the inclusion of this list of diseases of Amphibia in this volume.

REFERENCES

1. **Reichenbach-Klinke, H. H. and Elkan, E.,** *The Principal Diseases of Lower Vertebrates,* Academic Press, London and New York, 1965.
2. **Nace, G. W.,** *Bioscience,* 18, 767–775, 1968.

Table 11
COMMON DISEASES OF AMPHIBIA[a]

Region	Disease	Signs and symptoms	Cause	Organism(s)	Reference
Skin	"Red-leg"	Haemorrhages, inflammation, pink to red discoloration of skin	*Aeromonas hydrophila*	Anurans, Urodeles	1
	"Molchpest"	Secreting ulcers, gray pustules, skin shedding, edema, apathy	Unknown	Urodeles, especially newts	1
		Haemorrhages inflammation	Infection, injury, a vitaminosis	Anurans, Urodeles	1
	Avitaminosis	Skin shedding, delayed metamorphosis	Feeding faults, disturbance in carbohydrate metabolism	*Bufo vulgaris*	2
	Albinism	Pale skin	Failure of the pituitary gland, heredity, feeding faults	Urodeles, especially axolotls	1
	Fungus infection	Gray patches, disturbed skin shedding	Fungi	Urodeles, Anurans	1
	Tuberculosis	Skin inflamed and ulcerated	*Mycobacterium*	Anurans, *Bufo bufo*	3
	Epithelioma	Swelling	Herpesvirus, cancer	*Rana pipiens*	4
Digestive organs	Bacterial infection	Peritonitis, intestinal inflammation	*Pseudomonas*	Anurans, Urodeles	1
	Tuberculosis	Intestinal inflammation and ulceration	*Mycobacterium*	Anurans, Urodeles	1
	Protozoan infection	Intestinal inflammation	Protozoa, flagellates	Anurans, Urodeles	1

[a] Expanded from original table by Reichenbach-Klinke and Elkan.[1]

Table 11 (continued)
COMMON DISEASES OF AMPHIBIA[a]

Region	Disease	Signs and symptoms	Cause	Organism(s)	Reference
Digestive organs (continued)	Parasitism	Intestinal inflammation and obstruction	*Acanthocephala*, helminths	Anurans, Urodeles	1
	Liver	Liver abscesses, congestion, and degeneration	Tuberculosis, avitaminosis	Anurans, Urodeles	1
	Hydrops	Excessive accumulation of fluid in lymph sacs	Destruction of lymph, heart failure?	*Xenopus laevis*	1
Lungs		Breathing difficulties	Tuberculosis, carcinoma, nematodes, trematodes	Anurans, Urodeles	1
Urogenital organs	Kidney tumors	Swelling	Carcinoma	*Rana pipiens*	5
	Kidney enlargement	Swelling	Obstruction of uriniferous tubules, tuberculosis, nematodes	Anurans, Urodeles	1
	Kidney stones		Diet of spinach	*Rana pipiens*	6
Nervous system	Paralysis	Loss of bodily control	Parasitic infection	Anurans, Urodeles	1
			Tumors	*Rana catesbeiana*	7
Nervous receptors	Cercarial infestation	Swellings around lateral line, loss of equilibrium, refusal to feed, paleness	Absorption of an endotoxin produced by the cercariae	*Xenopus laevis*	8
Skeletal system	Rickets and osteoporosis	Skeletal deformities	Shortage of vitamin D in the diet	*Xenopus laevis*	9
Endocrine glands		Delayed metamorphosis	Congenital faults	Anurans, Urodeles	1
		Sex reversal	Hormonal imbalance	Anurans	1
Thyroid	Goiter	Hyperactivity, thyroid malfunction, swelling	Cabbage feeding and MeCN	*Rana pipiens*	10

Table 11 (continued)
COMMON DISEASES OF AMPHIBIA[a]

REFERENCES

1.   **Reichenbach-Klinke, H. H. and Elkan, E.,** *The Principal Diseases of Lower Vertebrates,* Academic Press, London and New York, 1965.
2.   **Nakamura, S.,** *Trans. Jpn. Pathol. Soc.,* 22, 97–101, 1932.
3.   **Elkan, E.,** *Proc. Zool. Soc. London,* 134, 275–296, 1960.
4.   **Lucke, B.,** *Ann. N.Y. Acad. Sci.,* 54, 1093–1109, 1952.
5.   **Lucke, B.,** *J. Exp. Med.,* 68, 457–467, 1934.
6.   **Briggs, R. W.,** *Science,* 93, 256–257, 1941.
7.   **Schlumberger, H. G. and Lucke, B.,** *Cancer Res.,* 8, 657–707, 1948.
8.   **Elkan, E. and Murray, R. W.,** *Proc. Zool. Soc. London,* 122, 121–126, 1952.
9.   **Bruce, H. M. and Parkes, A. S.,** *J. Endocrinol.,* 7, 64–81, 1950.
10. **Borland, J. R.,** *J. Exp. Zool.,* 94, 115–143, 1943.

QUALITATIVE REQUIREMENTS AND UTILIZATION OF NUTRIENTS: REPTILES

R. A. Coulson

QUALITATIVE REQUIREMENTS

Knowledge of the food requirements of reptiles is almost nonexistent, and most of the available information was gleaned from opinions of naturalists and zookeepers. If a reptile had been in captivity for more than 1 year, and if it had at least maintained its body weight, it may be assumed that the diet was reasonably satisfactory. The more rigorous requirements for nutritional adequacy in mammalian studies have not been applied, and, considering the nature of some of the experimental limitations, it may prove almost impossible to do some types of studies.

In any discussion of reptilian nutrition and physiology, metabolic rate is a prime consideration. Body size varies from less than a gram for an immature skink to 700,000 g for a grown male crocodilian, and the difference in their metabolic rates on a gram basis is of the order of 100 to 1. In effect, then, a deficiency would be apparent in the tiny reptile with its high metabolic rate in a fraction of the time required to detect it in a grown crocodile, Galapagos turtle, or large reticulated python. A very large python can eat enough at one time to suffice for almost a year, while the smallest lizards show signs of inanition when deprived of food for only 1 or 2 days. Imposed upon the inherent differences in metabolic rates is another important factor, temperature. Within limits, an increase in temperature is accompanied by an increase in metabolic rate, but the exact relationship in those reptiles studied is complex. Figure 1 shows the effect of a gradual increase in body temperature on a small alligator (*Alligator mississippiensis*). Between 5 and 15°C the difference was slight, while between 25 and 30°C the oxygen consumption increased greatly. If one performed a series of feeding experiments at 20°C and another series at 30°C, the animals in the second series with about four times the metabolic rate of those in the first would develop deficiencies in one fourth the time. The problem is even more complicated because at 20°C many reptiles will not feed in nature, and force-feeding often harms and occasionally kills the animal.

Although the oxygen requirements of most common reptiles are unknown, it is possible to construct a crude curve of the relationship between body weight and metabolic rate using information gathered from the literature. If one plots data from turtles, lizards, snakes, and crocodilians indiscriminately, a curious fact emerges: At 30°C a given weight of reptile has about the same metabolic rate, whether the reptile is a lizard, turtle, snake or crocodile. As might be expected, the very large reptiles have lower metabolic rates than the small ones. Perhaps the most remarkable finding is that in those animals with weights below 100 g, there is a great increase in metabolic rate for each small decrease in body weight (Table 1). If this relationship holds for the newborn skink or for other tiny lizards, it would appear that their metabolic rates at 30°C would be about ten times as high as an adult man at 37°C. The food requirements would then be at least 10 times those of man on an equivalent weight basis, and more probably 20 times as great, since a large share of the food would be utilized for growth.

Carbohydrate Requirements

It appears that absorbed carbohydrates are utilized by reptiles at a rate in keeping with their oxygen consumption. Injected monosaccharides are catabolized readily by the alligator, with fructose disappearing the fastest (Table 2). The inability to catabolize sucrose might have been expected, as sucrose is not present in the extracellular or

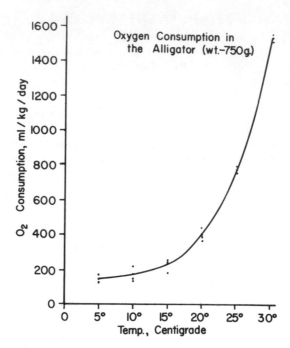

FIGURE 1.   Effect of gradual increase in body temperature on *Alligator mississippiensis.*

intracellular fluids in any animal so far as is known. It was surprising to find that alligators, caimans, and at least one turtle (*Pseudemys scripta elegans*) were unable to hydrolyse lactose, maltose, or starch when they were force-fed. Analysis of the gut washings showed the compounds to be present in the large intestine 3 days after feeding, and by the 4th day the feces contained them in quantity. Evidently, crocodilians do not derive energy from the starch-containing weeds etc. which they occasionally ingest, largely by accident. Failure of *P. scripta* to hydrolyze di- and polysaccharides was unexpected in view of the fact that they appear to be omnivorous, at least as adults.

All immature reptiles are carnivorous, existing on insects, crustaceans, and various whole vertebrates.[1] In the smallest reptiles, small amounts of carbohydrate may be derived from the extracellular glucose of the eaten animal, but most of the glycogen present in the tissues would have disappeared before digestion was complete. In the larger reptiles with their slower rate of digestion, no trace of carbohydrate would survive fermentation or glycolysis to lactic acid. There is little question that there is no carbohydrate requirement.

Fatty Acid Requirements

The quantity of fat eaten by most reptiles is considerable. The fact that the feces are largely fat free is an indication that fats are digested well and that the products are absorbed with little difficulty. There does not seem to be any information on the requirements of individual fatty acids but it is unlikely that deficiencies occur in nature. Excessively fatty diets result in steatitis in crocodilians, a disorder characterized by fatty livers and diffuse yellow lipid deposits; if severe, the disease is fatal.[2,3]

Amino Acid Requirements

Once again, it would not be reasonable to expect young carnivorous animals to develop amino acid deficiencies. The proteins available to them in nature are animal in origin and contain the full complement of amino acids. If one were to attempt to actually determine the amino acid requirements of a large reptile by feeding experiments in the same way done in the rat, the time and effort involved would be very great, but it is possible to determine which amino acids are essential in an indirect way. Turtles (*Pseudemys scripta elegans*), chameleons (*Anolis carolinensis*), alligators (*Alligator mississippiensis*), and caimans (*Caiman crocodilus crocodilus*) were all fed or injected with a single amino acid at a time, and each of the four species received 20 or more different (single) amino acids over a prolonged period. After the administration of each amino acid, the plasma and homogenates of various tissues and organs were analyzed for their amino acid contents and the results compared with control values from animals that had not received the amino acid. If amino acid A was injected and if there were significant rises in amino acids B and C, it was assumed that amino acid A was converted in whole or in part into the other two. Where the carbon of A was believed to be converted, ^{14}C isotopes of A were administered to the animals to verify the reaction. Some amino acids were subject to interconversions and others were not. Those amino acids that were not synthesized from any others apparently represent the essential amino acids. The results from all four species were so similar that they may be summarized together; Table 3 divides the amino acids into three groups: those never synthesized, those synthesized slowly or only in traces, and those synthesized in large yield. The first group would appear to be truly essential, the second group semiessential, and the third group nonessential.

It is interesting to note that those amino acids that appeared to be essential based on their apparent lack of synthesis were the same ones known to be essential in mammals, with the partial exception of histidine which seems not to be required by an adult mammal. Information on essential amino acids in snakes appears to be lacking.

In addition to the above, ornithine and α-aminobutyrate were synthesized by all, but citrulline was apparently not synthesized by the crocodilians or chameleon.

The presence of a balanced protein in the diet of a reptile is not necessarily an indication that the animal will thrive on it. Several vegetable proteins were force-fed to turtles (*P. scripta elegans*), alligators, and caimans (*C. crocodilus crocodilus*). The proteins from corn (zein), wheat (gluten), soy beans, rice, and peas (edestin) were not digested and appeared in the feces a few days later. These animals appear to need animal protein.

Too much protein in the diet during periods of slow growth produces gout of considerable severity. In crocodilia, gram quantities of crystalline uric acid are deposited in the joints and particularly on the ventral surface of the sternum. In time, the legs become paralyzed and the kidneys are blocked with solid urates. In the early stages, withholding food reverses the condition. The condition is caused by urate production exceeding urate excretion.[4] Similar disorders have been reported in other reptiles.[3]

Vitamin Requirements

As was the case for other types of nutritional studies, the difficulty of determining a vitamin requirement in a reptile cannot be overemphasized. Proper experiments would require the force-feeding of a purified experimental diet for months or even for years in the case of the largest reptiles. It would be reasonable to assume that they require vitamins in quantities commensurate with their metabolic rates. The small amount of vitamin C obtained from the animal tissues eaten by carnivorous reptiles would suggest that this vitamin is not required, but proof is lacking.

A definite requirement for vitamin D exists, and it is equally certain that sunlight (ultraviolet) can provide it. Conditions resembling rickets are common in reptiles kept indoors as pets or under zoo conditions. In crocodilia, avitaminosis D is manifest by a shortening of the head in growing animals and by malocclusion of the jaws and teeth and occasionally by distortion of the leg bones or even the spine. The syndrome may be complicated by a concomitant lack of calcium and phosphorus in the diet.

Table 4 lists the avitaminoses reported in various reptiles. The evidence for the development of a specific deficiency is not as strong as it is for mammals. In general, if a vitamin alone or in combination with others appeared to be effective in curing a disorder, it was assumed that the disorder occurred through lack of a vitamin. Unfortunately, the possibility that vitamins have a pharmacological effect when given in quantity cannot be ruled out.

It should be stated that the use of vitamin supplements in zoos is a common practice, and many believe the results to be beneficial. Unfortunately, scientific proof of the need for the supplements has not been demonstrated in animals receiving diets similar to those available to wild specimens.

Mineral Requirements

Reptiles vary in their need for calcium and phosphate, with those equipped with bony shells, as in the turtles, or with massive skull or skeletal bones, as in the crocodiles, requiring large amounts. Even the smallest lizards require appreciable amounts, and the usual pet feedings of meal worms (*Tenebrio molotor*) result in poor growth unless calcium phosphate supplements are provided. In the absence of more definite information, the requirements of the other minerals may be assumed to resemble those of the mammals. Obviously, iron is required for hemoglobin, zinc for carbonic anhydrase, iodine for thyroglobulins, etc. These would be well supplied in the food of carnivorous species.

UTILIZATION OF NUTRIENTS

In the absence of evidence to the contrary, it would be reasonable to assume that absorbed products of digestion are incorporated in macromolecules or degraded to CO_2 and the various nitrogenous excretion products by reactions familiar to better-known species. Saccharides are converted to glycogen, fats and proteins are synthesized from the usual precursors, etc. It would be unreasonable to expect that all species followed exactly the same schemes in intermediary metabolism, and, indeed, minor differences have been recorded. A detailed study of amino acid interconversions was conducted, with the results summarized in Table 5.

Table 1
OXYGEN CONSUMPTION AND BODY WEIGHT: REPTILES, TEMPERATURE 30°C[a]

Reptile	Body weight, (g)[b]	O$_2$ consumption (liters/kg/day)[c]	Ref.
Anolis carolinensis	7	5.3	1
Eumeces obsoletus	10	4.0	2
Uta stansburiana	10	14.9	3
Sceloporus occidentalis	15	13.2	3
Crotophytus collaris	18	4.8	4
Dipsosaurus dorsalis	40	2.3	5
Drymarchon corais c.	50	1.1	6
Caiman crocodilus c.	116	2.1	7
Alligator mississippiensis	50	2.4	7
A. mississippiensis	90	2.2	7
A. mississippiensis	100	2.2	7
Caiman crocodilus c.	160	1.9	7
Amphibolus barbatus	300	3.8	8
Crocodilus niloticus	700	1.6	9
Alligator mississippiensis	750	1.5	7
Pseudemys scripta e.	1,000	1.3	10
Iguana tuberculata	1,000	3.0	11
Alligator mississippiensis	3,000	0.8	7
Crotalus atrox	3,500	0.9	12
Python sebae	70,000	0.6	13
Constrictor constrictor	70,000	0.5	13

[a] This figure is not a constant since it is used to indicate the ambient temperature in some cases and the body temperature in others.

[b] These are estimates of the approximate body weight of representative specimens. The actual body weight of the animals tested was not always reported.

[c] The validity of some of these values is questionable. They were obtained on animals of various body weights and at different seasons of the year. While attempts were made to study the "resting" animals, it is difficult to control movement in a confined reptile, and therefore the amount of O$_2$ consumed by various specimens within a species was variable.

REFERENCES

1. Dessauer, H. C., personal communications, 1970.
2. Dawson, W. R., *Physiol. Zool.,* 33, 87–103, 1960.
3. Dawson, W. R. and Bartholomew, G., *Physiol. Zool.,* 29, 40–51, 1956.
4. Dawson, W. R. and Templeton, J. R., *Physiol. Zool.,* 36, 219–236, 1963.
5. Moberly, W. R., *Physiol. Zool.,* 36, 152–160, 1963.
6. Thill, H., *Z. Wiss. Zool.,* 150, 51, 1937.
7. Hernandez, T. and Coulson, R. A., *Proc. Soc. Exp. Biol. Med.,* 79, 145–149, 1952.
8. Bartholomew, G. A. and Tucker, V. A., *Physiol. Zool.,* 36, 199–218, 1963.
9. Coulson, R. A. and Hernandez, T., *Biochemistry of the Alligator, A Study of Metabolism in Slow Motion,* Louisiana State University Press, Baton Rouge, 1964, 10.
10. Coulson, R. A. and Hernandez, T., in *Chemical Zoology IX,* Florkin, M. and Scheer, B. T., Eds., Academic Press, New York, 1974, 217–246.
11. Vernberg, F. J., *Biol. Bull.,* 117, 163–184, 1959.
12. Sumner, F. B. and Lanham, U. N., *Biol. Bull.,* 82, 313–327, 1942.
13. Benedict, F. G., *Carnegie Inst. Washington Publ.,* p. 425, 1932.

Table 2
THE RATE OF DISAPPEARANCE OF SACCHARIDES (1 g/kg) INJECTED i.p. INTO 1-kg ALLIGATORS AT 28°C

Saccharide	Rate (%/hr)	Probable route of disposal
D-Glucosamine	0.7	Renal excretion
6-*O*-methylglucose	0.7	Renal excretion
Sucrose	0.7	Renal excretion
L-Sorbose	1.5	Excretion + catabolism
D-Galactosamine	1.7	Excretion + catabolism
D-Mannoheptulose	2.3	Excretion + catabolism
3-*O*-methylglucose	3.0	Catabolism
L-Rhamnose	4.5	Catabolism
D-Xylose	4.3	Catabolism
D-Galactose	5.0	Catabolism
6-Acetylglucose	5.0	Catabolism
D-Glucose	5.1	Catabolism
L-Xylose	6.9	Catabolism
D-Talose	7.1	Catabolism
D-Arabinose	7.9	Catabolism
D-Mannose	8.0	Catabolism
D-Ribose	8.1	Catabolism
D-Fructose	12.0	Catabolism

REFERENCE

Coulson, R. A. and Hernandez, T., *Biochemistry of the Alligator, A Study of Metabolism in Slow Motion,* Louisiana State University Press, Baton Rouge, 1964, 34.

Table 3
ESSENTIAL AND NONESSENTIAL AMINO ACIDS IN *ALLIGATOR MISSISSIPPIENSIS, CAIMAN CROCODILUS CROCODILUS, PSEUDEMYS SCRIPTA ELEGANS,* AND *ANOLIS CAROLINENSIS*

Essential	Semiessential	Nonessential
Thr	Pro	Hyp
Val	Cy_2S_2	Asp
Met	Arg	Ser
Ile		Asn
Leu		Gln
Phe		Glu
Lys		Gly
His		Ala
Trp		Tyr

REFERENCE

Coulson, R. A. and Hernandez, T., in *Chemical Zoology IX,* Florkin, M. and Scheer, B. T., Eds., Academic Press, 1974, 217.

Table 4
REPORTED VITAMIN REQUIREMENTS

Vitamin deficiency	Reptile	Nature of disorder	Ref.	Comments
A	Turtles	As in man	1—4	When fed lettuce in captivity
C	Snakes Lizards	Epidermal hemorrhages	2—5	In captivity
D	All reptiles	Osteoporosis, rickets	6	In captivity without ultraviolet light Often complicated by Ca and P deficits
E	Crocodilians	Steatitis	3, 4, 6, 7	Perhaps serves as an antioxidant for fatty acids
K	Crocodilians	Gingival bleeding	6	In captivity

REFERENCES

1. Doyle, R. E. and Moreland, A. F., *Lab. Anim. Dig.,* 4, 3, 1968.
2. Reichenbach-Klinke, H. and Elkan, E., *The Principal Diseases of Lower Vertebrates,* Academic Press, New York, 1965.
3. Wallach, J. D., in *Current Veterinary Therapy IV,* Kirk, R. W., Ed., W. B. Saunders, Philadelphia, 1971.
4. Wallach, J. D., *J. Am. Vet. Med. Assoc.,* 155, 1017, 1969.
5. Nelson, L., *J. Zoo Anim. Med.,* 2, 19, 1971.
6. Frye, F. L., *Husbandry, Medicine and Surgery in Captive Reptiles,* V. M. Publishing, Bonner Springs, Kans., 1973.
7. Wallach, J. D. and Hoessle, C., *J. Am. Vet. Med. Assoc.,* 153, 845, 1968.

Table 5
AMINO ACID INTERCONVERSIONS

Amino acid given	Amino acid derivatives detected[b]	Apparent yield[a]			
		Caiman[c][1]	Alligator[d][1]	Turtle[e][1]	Anolis[f][1,2]
Gly	Ser(C,N)	+	+	+	+++
	Ala(C,N)	±	0	0	0
	Gln(N)	+	0	0	+
Ala	Glu	+	+	++	+++
	Gly(C,N)	+	0	+	+
	Asp(C,N)	±	+	+	++
	Gln(C,N)	+	+	+	+
	Ser(C,N)	0	0	0	+++

[a] ±: probable trace; +: significant, ++: considerable yield; +++: almost complete conversion.

[b] (C): carbon incorporation; (C,N): carbon and nitrogen incorporation; (N): nitrogen incorporation; (U): unknown.

[c] *Caiman crocodilus crocodilus.*

[d] *Alligator mississippiensis.*

[e] *Pseudemys scripta elegans.*

[f] *Anolis carolinensis.*

[g] α-Aminobutyrate.

[h] Citrulline.

Table 5 (continued)
AMINO ACID INTERCONVERSIONS

Amino acid given	Amino acid derivatives detected[b]	Apparent yield[a]			
		Caiman[c1]	Alligator[d1]	Turtle[e1]	Anolis[f1,2]
Ser	Gly(C,N)	++	+	±	+
	Ala(C,N)	0	±	0	++
	Gln(U)	±	±	0	
Thr	Glu(C,N)	0	0	0	+
	Ser(C,N)	0	0	0	+
	Gly(C,N)	+++	+++	++	+
	Ala(C,N)	0	0	0	+
	Gln(C,N)	+	0	±	±
Asp	Glu(N)	++	++	++	++
	Ala(N)	++	+	±	+
	Gln(C,N)	+	±	0	
	Gly(U)	±	0	0	
Asn	Asp(C,N)	++	+	+	++
	Glu(N)	+	+	+	+
	Ala(C,N)	+	±	±	+
	Gln(N)	+	+	+	±
	Gly(U)	±	0	0	±
Glu	Asp(C,N)	+	+	++	+
	Gln(C,N)	+	±	+	+
	Ala(C,N)	++	+	+	++
	Gly(C,N)	+	0	0	+
Gln	Asn(U)	+	±	±	±
	Gly(U)	±	0	0	+
	Ala(C,N)	+	+	0	±
	Glu(C,N)	++	+	++	+
	Asp(C,N)	+	+	+	±
Val	Gly(U)	±	0	±	0
	Ala(U)	±	0	0	0
	Glu(U)	±	0	0	0
	Gln(U)	+	0	0	0
Leu	Gly(U)	+	0	+	0
	Ala(U)	±	0	0	0
	Glu(U)	±	0	0	0
	Gln(U)	+	0	0	±
lle	Ala(U)	±	0	0	0
	Glu(U)	±	0	±	0
	Gln(U)	±	0	0	0
	Asp(U)	±	0	0	0
	Ile(*allo*)(U)	±	0	±	0

Table 5 (continued)
AMINO ACID INTERCONVERSIONS

Amino acid given	Amino acid derivatives detected[b]	Apparent yield[a]			
		Caiman[c1]	Alligator[d1]	Turtle[e1]	Anolis[f1,2]
Orn	Arg(C,N)	+	+	±	0
	Gly(N)	++	0	0	0
	Ala(U)	+	0	0	0
	Glu(C,N)	+++	++	+++	+
	Gln(C,N)	+	±	0	++
Arg	Orn(C,N)	++	++	+++	++
	Ala(U)	±	0	0	0
	Glu(C,N)	±	0	++	±
	Gln(U)	0	0	0	+
Cit	Orn(U)	+	+	+	0
	Arg(U)	+	+	+	0
	Gly(U)	0	0	±	+
	Glu(U)	0	0	0	+
	Gln(U)	±	0	0	+
Met	Cys-Cys(C,N)	±	+	±	+
	αAB[g](C,N)	0	++	±	±
	Gly(U)	±	0	0	0
	Glu(U)	0	0	±	0
Lys	Gly(U)	+	0	±	+
	Ala(U)	0	0	±	0
	Glu(C,N)	+	±	+++	+
	Gln(C,N)	+	0	+	++
His	FlGlu(C,N)	++	++	++	+
	"Cit"[h](C,N)	0	±		±
	Asn(U)	±	0	±	+
	Gly(U)	±	0	±	+
	Ala(U)	0	0	+	0
	Glu(N)	+	+	0	+
	Gln(N)	+++	++	+	+++
	1-CH$_3$ His(U)				
	3-CH$_3$-His(U)				
Phe	Tyr(C,N)	±	+	+	+
	Gly(U)	0	0	0	+
	Glu(U)	0	0	±	0
	Gln(U)	±	0	0	±

Table 5 (continued)
AMINO ACID INTERCONVERSIONS

Amino acid given	Amino acid derivatives detected[b]	Apparent yield[a]			
		Caiman[c1]	Alligator[d1]	Turtle[e1]	Anolis[f1,2]
Trp	5-OH Trp (C,N)	±	±	0	±
Pro	Gly(C,N)	+	±	+	±
	Glu(C,N)	+++	++	+++	+
	Gln(C,N)	++	+	+	±
	Asp(U)	0	0	±	0
Hyp	Gly(C,N)	±	±	0	+

REFERENCES

1. **Coulson, R. A. and Hernandez, T.,** in *Comparative Biochemistry of Nitrogen Metabolism,* Campbell, J. W., Ed., Academic Press, New York, 1970, 639–710.
2. **Herbert, J. D. and Coulson, R. A.,** *Comp. Biochem. Physiol.,* 42B, 463–473, 1972.

REFERENCES

1. **Wallach, J. D.,** *J. Am. Vet. Med. Assoc.,* 159, 1632–1643, 1971.
2. **Wallach, J. D. and Hoessle, C.,** *J. Am. Vet. Med. Assoc.,* 153, 845–847, 1968.
3. **Frye, F. L.,** *Husbandry, Medicine and Surgery in Captive Reptiles,* V. M. Publishing, Bonner Springs, Kans., 1973.
4. **Coulson, T. D., Coulson, R. A., and Hernandez, T.,** *Zoologica,* 58, 47–52, 1973.

QUALITATIVE NUTRIENT REQUIREMENTS OF BIRDS

P. Vohra and L. C. Norris

All life in our ecosystem depends upon the flow of energy through it. Sun is the powerhouse of this energy. Autotrophic organisms meet their requirements for high energy in the form of adenosine triphosphate (ATP) by synthesizing their nutrients from low energy inorganic raw materials through the process of photosynthesis. The solar energy is captured by the photosynthetic autotrophic organisms on land and in water which synthesize carbohydrates, fats, proteins, and other essential organic nutrients from carbon dioxide, water, ammonia, and some sulfur compounds. Chlorophyll-containing plants and plankton are examples of autotrophic organisms that provide food for land and ocean inhabitants.

As a contrast, the heterotrophic organisms, which comprise all the animals, including humans, are incapable of meeting their energy requirements by synthesizing the necessary nutrients from low energy inorganic materials, and depend on other autotrophic organisms to do this work for them. They extract the energy for their life processes from prefabricated energy-rich nutrients stored by plants.

In contrast to ruminants, birds share with humans the attribute of a single stomach and belong to the group of monogastric animals. The qualitative requirements of birds for nutrients are far more critical than those of ruminants because of their monogastric nature.

There is very little experimental evidence on the qualitative requirements of birds other than domestic poultry. It is very time consuming and expensive to develop such information, and unless, the birds are of great economic importance, scientists lack financial means to carry on such studies. This difficulty is further compounded by the fact that the class *Aves* consists of almost 27 orders of living birds, 170 families, and 28,500 species and subspecies, and their body weights can vary from a few grams (hummingbird) to about 145 kg (ostrich).[1,2] Again, the birds may be capable of flying long distances, soaring over short distances, or incapable of doing either.

Birds vary greatly in their development at hatching time. Nidicolous birds have to look after their altricial nestlings, while the precocial young of nidifugous birds are ready to fend for themselves as soon as they hatch. The parents have to collect food for the altricial birds and feed them, sometimes in a predigested regurgitated form. Precocial chicks grow at a slower rate than altricial chicks, and the relative rates of development of various organs are not the same in any of the species.

Birds also differ in their food habits. They can be highly restrictive (stenophagic) of their food or eat a variety (euryphagic). They may be herbivorous, carnivorous, or omnivorous. Even the herbivorous birds may feed their baby chicks insects and other invertebrates during the early period of rapid growth.

The food habits of some birds are given in Table 1. In many instances, birds have suitable anatomical adaptations, especially in the beaks, feet, and digestive tracts, to cope with their feeding habits. The information on these topics has been reviewed by Storer,[3] Ziswiler and Farner,[4] and Hill.[5,6]

The main differences in the digestive apparatus of the birds are modifications in esophagus to form a crop, division of stomach into muscular and glandular parts, and in the posterior portion of the tract to form blind tubular sacs called ceca. The role of avian ceca has been reviewed by McNab.[7] The ceca have the highest microbial as well as protozoan concentration. Volatile fatty acids such as acetic, propionic, and butyric acid (but mostly acetic acid) are produced in ceca as well as in the crop of some birds. The ceca appear to decompose cellulose in birds such as wild grouse and ptarmigans.[8]

Table 1
FOOD HABITS OF SOME BIRDS

Granivorous:	Tinamidae (tinamous); Anatidae (swans, geese, fresh-water ducks); Tetraonidae (grouse, ptarmigans); Phasianidae (quail, partridges, pheasants, jungle fowl, peafowl, guineafowl); Gruidae (cranes); Rallidae (rails, coots); Laridae (gulls, terns); Columbidae (pigeons, doves); Alaudidae (larks); Corvidae (crows, magpies, jays); Paridae (titmice); Sittidae (nuthatches); Certhiidae (creepers); Icteridae (blackbirds); Fringillidae (grosbeaks, finches, buntings, sparrows)
Frugivorous:	Tetraonidae; Gruidae; Picidae (woodpeckers), Paridae; Mimidae (thrashers); Turdidae (thrushes); Bombycillidae (waxwings); Vireonidae (vireos)
Insectivorous:	Gaviidae (loons); Falconidae (falcons); Tetraonidae; Phasianidae; Gruidae; Laridae; Cuculidae (cuckoos); Caprimulgidae (goatsuckers); Apodidae (swifts); Trochilidae (hummingbirds); Tryannidae (flycatchers); Alaudidae; Hirundinidae (swallows); Paridae; Sittidae; Certhiidae; Troglodytidae (wrens); Sylviidae (warblers); Bombycillidae; Vireonidae; Parulidae (wood warblers); Icteridae; Fringillidae
Piscivorous:	Gaviidae; Podicipedidae (grebes); Pelecanidae (pelicans); Phalacrocoracidae (cormorants); Ardeidae (herons, egrets, bittern); Accipitridae (kites, hawks, eagles, vultures); Pandionidae (osprey); Laridae; Alcedinidae (kingfishers)
Amphibians and aquatic inverte-brates, crustaceans, molluscs:	Gaviidae; Podicipedidae; Phalacrocoracidae; Ardeidae; Laridae; Alcedinidae
Predators (birds, animals, and other vertebrates):	Cathartidae (vultures); Accipitridae; Falconidae; Rallidae; Charadriidae (plovers); Scolopacidae (sandpipers); Cuculidae; Strigidae (owls); Corvidae
Cerophagous:	(wax eating): Indicatoridae (honeyguides)
Nectar:	Trochilidae; Coerebidae (honeycreepers); Drepanididae (honeycreepers); Miliphagidae (honeyeaters); Nectariniidae (sunbirds); Dicaeidae (flowerpeckers)
Stenophagous:	Eelgrass: *Branta bernicla* (brant) Snails: *Rostrhamus sociabilis* (Everglade kite)

Willow ptarmigans may derive some energy from the bacterial production of volatile fatty acids in their ceca. Browsing herbivorous species generally have more extensively developed ceca than graminivorous species, but their role in digestion of cellulose in domestic birds[9] or in bacterial synthesis of vitamins[10] appears doubtful.

Birds require all the essential nutrients in their diets for growth and reproduction. These nutrients have to meet the requirements for energy, water, glucose, essential amino acids, essential fatty acids, vitamins, and minerals. The varieties of birds' diets should be evaluated in terms of their nutrient contents. Unfortunately, we have meager information about this. However, we could draw general conclusions about the qualitative requirements of birds using information about intermediary metabolism and analogies from some other field of biology.

DIGESTIVE ENZYMES IN BIRDS

Definite information about the enzymes present in the digestive tract of birds is confined to a few species, mostly of the domestic type. Some of the relevent data are summarized in Table 2. These enzymes, common to all monogastric animals, use carbohydrates, proteins, and fats as substrates.

Table 2
ENZYMES INVOLVED IN THE DIGESTION OF FOODS

Digestion of Carbohydrate

Amylase:
Duodenum of chicken;[1] small intestine of lapwing (*Vanellus vanellus*),[2] quail (*Coturnix chinesis*),[8] crane (*Grus grus*);[8] pancreas of chicken,[3] domestic pigeon,[4] domestic goose,[5] house sparrow (*Passer domesticus*)[6]

Maltase:
Small intestine of chickens[1] *Coturnix chinesis*,[8] *Grus grus*,[8] king penguin (*Aptenodytes patagonica*),[9] penguin (*Eudyptes crestatus*);[9] skua (*Catharacta skua*),[9] gulls (*Larus dominicanus*)[9]

Sucrase (invertase):
Small intestine of domestic fowl,[7] *Coturnix chinesis*,[8] *Grus grus*,[8] *Catharacta skua*,[9] *Larus dominicanus*;[9] crop of chickens[10]

Lactase:
Intestine of chicken,[7] domestic duck[9]

Palatinase:
Intestine of chicken (hydrolyses [6-O-α-D-glucopyranosyl-D-fructose[7]])

Digestion of Fat

Lipase:
Pancreas of chicken[11]

Digestion of Proteins

Protease:
Proventriculus of chickens;[2] gizzards of owls (*Althena noctua*),[13] kestrel (Falcotin-nunculus),[13] hawks (*Buteo buteo*)[13]

Pepsin
Proventriculus of chickens;[2] gizzards of owls (*Althena noctua*),[13] kestrel (Falcoti-nnunculus),[13] hawks (*Buteo buteo*)[13]

Trypsin:
Duodenum of chickens;[17] pancreas of chickens[5]

Chymotrypsin:
Pancreas of chickens[5]

Enterokinase:
Intestine of chickens,[11] sea gulls[16]

Dipeptidase:
Proventriculus, pancreas, and intestine of domestic fowl[17]

Aminopeptidase:
Proventriculus, pancreas, and intestine of chickens[17]

Carboxypeptidase:
Proventriculus, pancreas, and intestine of chickens[17]

Chitinase and chitobiase
Proventriculus of house sparrow,[18] thrushes (*Trudus merula*),[18] *Leothrix lutea*[18]

REFERENCES

1. Bird, F. H., *Br. Poult. Sci.*, 12, 373, 1971.
2. Bernardi, A. and Schwartz, M. A., *Biochem. Z.*, 262, 175, 1933.
3. Salman, A. J., Dal Borgo, G., Pubols, M. H., and McGinnis, J., *Proc. Soc. Exp. Biol. Med.*, 126, 694, 1967.
4. Dandrifosse, G., *Comp. Biochem. Physiol.*, 34, 299, 1970.
5. Klug, F., *Zentralbl. Physiol.*, 5, 131, 1891.
6. Langendorff, O., *Physiol. Abstr.*, 1, 35, 1879.
7. Siddons, R. C. and Coates, M. E., *Br. J. Nutr.*, 27, 229, 1972.
8. Zoppi, G. and Shmerling, D. H., *Comp. Biochem. Physiol.*, 29, 289, 1969.
9. Kerry, K. R., *Comp. Biochem. Physiol.*, 29, 1015, 1969.
10. Pritchard, P. J., *Comp. Biochem. Physiol.*, 43A, 195, 1972.
11. Lepkovsky, S., Furuta, F., Dimick, M. K., and Yamashina, I., *Poult. Sci.*, 49, 421, 1970.
12. de Réaumer, R. A. F., *C.R. Acad. Sci. Paris*, p. 1152, 1752.

Table 2 (continued)

13.   Tiedemann, F. and Gmelin, L., *Verdauungnach Versuchung*, Vol. 2, Gross, Heidelberg-Leipzig, 1831.
14.   Herpol, C., *Z. Vgl. Physiol.*, 57, 209, 1967.
15.   Herpol, C., *Ann. Biol. Anim. Biochim. Biophys.*, 4, 239, 1964.
16.   Waldschmidt-Letiz, E. and Shinoda, A., *Hoppe-Seyler's Z. Physiol. Chem.*, 177, 301, 1928.
17.   De Rycke, P., *Natuurwet. Tijdschr. Ghent*, 43, 82, 1962.
18.   Jeuniaux, C., *Ann. Soc. Zool. Belg.*, 92, 27, 1962.

MEETING ENERGY REQUIREMENTS OF BIRDS

The energy requirements for all life processes of heterotrophic organisms including birds are furnished by ATP, the biological fuel of life. This implies that all birds have qualitative requirements for readily digestible carbohydrates, lipids, and proteins to satisfy their energy needs.

The energy requirements depend upon the age, sex, reproductive status, and environmental factors. Before migration, most birds tend to become fat. As the metabolizable energy content of fats is higher than that of carbohydrates or proteins, the extra fat helps to meet the additional energy needs of birds during long-distance migration.

WATER REQUIREMENTS

Water is an essential nutrient for all birds, but not all birds have to drink it to meet their requirements. The studies of drinking patterns of desert birds in Australia indicate that a number of species were never observed to drink water.[11] Granivorous species frequented the water supply, but carnivorous and insectivorous birds appeared to be largely independent of water. In some species, the amount of water present in their foods and the water produced as a result of metabolism of dietary fats and carbohydrates in their bodies sufficed for this purpose. The amount of metabolic water obtained per 100 g of food is approximately 107 g for fats, about 40 g for proteins, 60 g for glucose, and 55 g for starch.

Some birds possess a functional nasal salt gland which enables them to concentrate and dispose of the excess salt and thus even utilize sea water. To this group belong the penguins, albatross, boobies, cormorants, flamingoes, gulls, frigate birds, grebes, pelicans, desert partridge, hawks, and eagles.[12]

CARBOHYDRATE REQUIREMENTS OF BIRDS

Starches are the important sources of carbohydrate utilized by granivorous birds who consume cereals, seeds, and tubers. Glucose is the most important monosaccharide essential for all birds and circulates in blood as such. It is transported across the cell membrane and participates in oxidative metabolism in chickens.[13] One mole of glucose yields 39 mol of ATP in metabolism. The essential nature of carbohydrates has been reviewed.[14] Pentosans are partially digested by domestic poultry, but it is doubtful if cellulose plays any important role in the overall nutrition of birds.

Fructose appears to cause heavy mortality of coturnix quail for the first week of life unless glucose is also included in their diets.[15] Fructose increases significantly the specific activity of ketohexokinase in livers of chickens and results in heavier livers with greatest hepatic lipid contents, as compared to dietary glucose, which causes only a slight increase in glucokinase activity.[16] Only about 15% of absorbed fructose is converted to glucose in adult chickens.[17] Carnivorous birds consume very little carbohydrates in their diets, but meet the essential glucose requirement by gluconeogenesis.[17a]

FATS AND ESSENTIAL FATTY ACIDS

Birds can utilize fats and oils because they possess the enzyme lipase in their digestive tract. Chickens appear to utilize fatty acids poorly in comparison to the fatty acid triglycerides, unless an equivalent amount of glycerol and glucose is present. In the process of fatty acid metabolism, acetyl coenzyme A (CoA) and the reduced coenzymes nicotinamide adenine dinucleotide (NADH) and flavin adenine dinucleotide (FADH$_2$) are produced. Oxidation of 1 mol of acetyl CoA gives rise to 12 mol of ATP. Also, the oxidation of NADH and FADH$_2$ through the biological electron transport system further yields 5 mol of ATP per mole of oxygen (O$_2$) utilized. Thus, 1 mol of stearic acid (C$_{18}$H$_{36}$O$_2$) yields 9 mol of acetyl CoA and utilizes 8 mol of O$_2$ or gives rise to 128 mol of ATP, indicating a biological efficiency of about 40%.

Honeyguides (Indicatoridea) of Africa are capable of maintaining themselves on a diet of beeswax for some time. The intestinal microflora of these birds must be capable of digesting beeswax to provide a part of their energy requirements.[18]

As in the case of other monogastric animals, the need of birds for essential fatty acids, especially linoleic acid, must be accepted. A requirement of chickens for linoleic acid has been demonstrated during growth[19] as well as during reproduction.[20] A deficiency of linoleic acid causes poor growth, egg production, and hatchability.

PROTEINS AND AMINO ACIDS

Chickens excrete about 2.2 mg endogenous nitrogen per basal kilocalorie of energy expenditure.[21] This amount of catabolized nitrogen is of the same order as in other species and may be regarded as the maintenance requirement for nitrogen. This is equivalent to about 12.50 mg protein per basal kilocalorie, suggesting the essential nature of proteins for the existence of birds, and is about equal to 250 mg endogenous nitrogen excretion per kilogram of body weight of the chickens or 1.562 g protein per kilogram of body weight. This nitrogen results mostly from digestive secretions and desquamated cells.

In addition to basal requirement, more protein is needed for growth and reproductive functions. The requirement for protein is actually quantitative as well as qualitative as determined by its amino acid composition. The constituent amino acids are grouped into two categories, essential and nonessential. The body tissues cannot synthesize enough essential amino acids to meet their demand, while, on the other hand, nonessential amino acids can be synthesized through the metabolic pathways from the amino groups of essential amino acids if need arises. This does not mean that the birds do not need nonessential amino acids. If these are not supplied, the cost of providing them endogenously would be excessive. The simultaneous presence of nonessential amino acids in the diets of birds spares the essential amino acids for their proper functions. The nonessential amino acids are alanine, aspartic acid, glutamic acid, proline, and serine. A part of the need for nonessential amino acids can be met from ammonium citrate.

Most of the essential amino acids for birds are also required by all monogastric animals: arginine, cystine, histidine, isoleucine, leucine, lysine, methionine, phenylalanine, threonine, tryptophan, tyrosine, and valine. Of these, methionine can usually replace cystine, and phenylalanine can replace tyrosine. Although glycine is regarded as nonessential for other species, birds have a definite requirement for it, as their endogenous synthesis is insufficient to meet this need. Birds excrete uric acid as the end product of nitrogen metabolism, and the uric acid molecule requires a glycine skeleton for its formation.

The main syndrome of any essential amino acid deficiency is poor growth. However, a deficiency of lysine also causes a breakdown in melanin synthesis in pigmented feathers, resulting in the appearance of a white bar. Essential amino acids such as lysine, leucine, and valine appear to be chiefly involved as building blocks of tissue proteins, with a limited amount undergoing oxidative degradation.

The essential amino acids have other functions besides their role as building blocks of tissue proteins. Methionine acts as a methyl donor and is converted to cysteine, which is a constituent of glutathione. Ornithine is derived from arginine and, with glycine, acts as a detoxifying agent in birds. Tyrosine produced as a result of hydroxylation of phenylalanine is a precursor of melanin pigments as well as the hormone adrenaline. Seratonin and nicotinic acid are produced from tryptophan.

Amino acids in excess of a bird's need for body tissue, enzymes, hormones, and other essential factors undergo oxidative degradation and can be used as sources of energy. Some of the amino acids give rise to glucose during their metabolism and are called glucogenic in nature. Most of these belong to the nonessential group, particularly alanine, glycine, serine, aspartic acid, and glutamic acid. The amino acids leucine, lysine, methionine, and tryptophan yield acetoacetic acid in metabolism and are called ketogeneic. This division is not as rigid as may appear from the above statements, but, generally, all the ketogenic amino acids would have to be regarded as essential amino acids for birds.

VITAMINS

The water-soluble vitamins play an important role as coenzymes or prosthetic groups of coenzymes in a large number of biochemical functions, some of which are summarized in Table 3. Even if the essential nature of these vitamins for most of the species of birds has not been established, it is doubtful that birds could survive without their presence in

Table 3
CATALYTIC FUNCTIONS OF SOME WATER-SOLUBLE VITAMINS

Compound	Present or functions in
Thiamin (vitamin B_1)	Thiamin pyrophosphate (TPP) or cocarboxylase; coenzyme of oxidative decarboxylation, transketolation, and transaldolation
Riboflavin (vitamin B_2 or G)	Prosthetic group of flavoprotein flavin mononucleotide (FMN) and flavin adenine dinucleotide (FAD); electron transport
Niacin (nicotinamide)	Nicotinamide adenine dinudeotide (NAD); nicotinamide adenine dinucleotide phosphate (NADP), co-dehydrogenases
Pyridoxine (vitamin B_6)	Pyridoxal phosphate; prosthetic group of transaminases; co-decarboxylases; serine and threonine dehydratases; desulfhydrase; deaminases, amine oxidases, kinureninase
Pantothenic acid	Coenzyme A
Biotin	Carboxylating enzymes: acetyl CoA carboxylase; propionyl CoA carboxylase; pyrurate carboxylase; methylmalonyl-oxaloacetic transcarboxylase
Folic acid	Tetrahydrofolate; carrier of "single" carbon compound such as methyl, methylene, methenyl, hydroxymethyl, formyl
Cobalamin (vitamin B_{12})	Vitamin B_{12} coenzyme; methylmalonyl CoA; methyltransferase
Choline	Phosphalipid metabolism

diets. Some bacteria can synthesize cobalamin (vitamin B_{12}), but unless birds practice coprophagy, dietary vitamin B_{12} would be essential for them.

Ascorbic acid (vitamin C) is nonessential for all birds that have the enzymes D-glucuronic acid reductase and L-gulonolactone oxidase. However, some passerine birds do not have these enzymes and need vitamin C in their diets.[22] Under certain conditions such as stress of high temperature and high production, domestic poultry may require dietary ascorbic acid, as metabolic synthesis may be insufficient to meet the need.

No definite coenzyme functions have been established for the fat-soluble vitamins. However, these are also essential for all animals. The major deficiency symptoms for vitamins in birds are given in Table 4.

Carotenoids are also needed by birds. α-, β-, and γ-Carotene can be converted partially to vitamin A. Certain birds also need xanthophylls in their diets to maintain the yellow and red pigments in their plumage, egg yolks, and shanks. The yellow pigment can be supplied by corn (maize), alfalfa, marigold leaves, or a synthetic pigment such as β-apo-8'-carotenal. The pinkish-red color of flamingoes comes from astaxanthin in their foods; synthetic canthaxanthin in their diets also serves the same purpose.

MINERALS

It is generally believed by biologists that life originated in the ancient seas. Because of their immensity, any changes in temperature, pH, and salt concentration of sea water occurred very slowly over long periods of time. The protoplasm of the early cells shared with the surrounding sea water many of the latter's characteristics. As the earliest animals also appear to have evolved in the same aquatic surroundings, their internal body fluids are similar in many ways to sea water, the cradle of life. Before animals moved to land, evolutionary adaptations enabled them to maintain internally constant fluid environments somewhat resembling the primordial sea, which was rich in sodium, potassium, magnesium, calcium, phosphate, carbonate, chloride, sulfate, and a host of trace elements.

Minerals are essential to birds for deposition in structural elements (skeleton) as well as for other multifunctions such as regulation of osmotic activity and pH, transport of oxygen, energy metabolism, and activation of many enzyme systems. Experimental data are available for domestic poultry (Table 5). The skeleton consists mostly of calcium and phosphorus, two of the elements required in relatively large quantities. The need for calcium further increases for shell formation when the birds are laying eggs. Calcium ion is of importance in the contraction of muscle cell, and is involved in the movement of sodium and potassium ions across biological membranes. Phosphorus is a constituent of the bone minerals and is involved in metabolism of carbohydrates and fats and in the transfer and release of energy.

In general, elements are classified as major or trace elements depending upon their quantitative requirement. The inorganic elements needed in fairly large amounts are calcium, phosphorus, sodium, potassium, magnesium, chlorine, and sulfur. The elements that are needed in small or trace amounts are iron, copper, manganese, zinc, molybdenum, iodine chrominum, selenium, and fluorine.

Based on the above comments as well as information available in a report of the National Academy of Sciences,[2] the qualitative requirements of birds for nutrients are summarized in Table 6.

Table 4
VITAMIN DEFICIENCY SYMPTOMS OF BIRDS

General Deficiency Symptoms

Water-soluble Vitamins

Thiamin (vitamin B_1):
Loss of appetite; general weakness; convulsions; polyneuritis with head drawn over back; death in young and adult chickens and hawks

Riboflavin (vitamin B_2 or G):
Diarrhea; retarded growth; curled toe paralysis with toes curled inward; hypertrophy of brachial and sciatic nerves in chicks; poor hatchability and clubbed down embryos

Niacin:
Inflammation of tongue, mouth cavity and upper part of esophagus; depressed food intake; retarded growth; poor feathering; scaly dermatitis of feet and skin in young chicks; perosis in turkeys (syndromes less severe in turkeys)

Pyridoxine (vitamin B_6):
Poor growth; abnormal excitability; convulsive movements; polyneuritis-like head; death; loss of egg production and poor hatchability

Pantothenic acid:
Retarded growth; poor feathering; granular eyelids sticking together; crusty scabs around vent and corners of mouth; dermatitis of feet

Biotin:
Very severe dermatitis; toes become necrotic and may slough off; mandibular lesions; swollen eyelids stick together; in embryos, micromelia, chondrodystrophy, and parrot beak

Folic acid:
Poor growth; poor feathering; anemia; poor egg production and hatchability; in embryos, bending of tibiotarsus, mandible defects, syndactyl, and hemorrhages

Cobalamine:
Poor growth

Choline:
Poor growth; perosis of legs

Fat-soluble Vitamins

Vitamin A:
Poor growth; emaciation; staggered gait; ruffled plumage; keratinization of the third eyelid; cheesy exudate from eyes and sticky discharge from nose; poor hatchability

Vitamin D_3:
Rickets; retarded growth

Vitamin E:
Nutritional encephalomalacia; prostration with legs outstretched; twisting of head; nutritional myopathy

Vitamin K:
Hemorrhages and death

Table 5
SOME OF THE KNOWN FUNCTIONS OF TRACE ELEMENTS

Mineral	Role
Chromium	Glucose tolerance factor
Cobalt	Constituent of vitamin B_{12} (cobalamin)
Copper	Ceruloplasmin, erythrocuprein, ascorbic acid oxidase, uricase cytochrome oxidase, δ-aminolevulinic acid dehydrase, tyrosinase
Iron	Hemoglobin, myoglobin, transferrin, ferritin, hemosiderin, catalase, cytochromes, succinic dehydrogenase, xanthine oxidase, aconitase, NADH cytochrome reductase
Iodine	Thyroxine
Manganese	Peptidases, deoxyribonuclease, succinic decarboxylase, arginase, kinases
Molybdenum	Xanthine oxidase, aldehyde oxidase
Selenium	Glutathione oxidase
Zinc	Carbonic anhydrase, lactic acid dehydrogenase, glutamic dehydrogenase, alcohol dehydrogenase, peptidases

Table 6
THE QUALITATIVE NUTRIENT REQUIREMENTS
OF BIRDS (AVES)

Amino acids		Vitamins	Minerals
Essential	**Nonessential**	**Vitamins**	**Minerals**
Arginine	Alanine	Retinol (vitamin A)	Calcium
Cystine[a]	Aspartic acid	Calciferol (vitamin D)	Phosphorus
Histidine	Glycine[c]	α-Tocopherol (vitamin E)	Sodium
Isoleucine	Glutamic acid	Menadione (vitamin K)	Potassium
Leucine	Proline	Thiamin (vitamin B_1)	Magnesium
Lysine	Serine[d]	Riboflavin (vitamin B_2 or G)	Manganese
Methionine		Pantothenic acid	Iron
Phenylalanine		Niacin (nicotinamide)	Copper
Threonine		Pyridoxine (vitamin B_6)	Zinc
Tryptophan		Biotin	Selenium
Tyrosine[b]		Folacin (folic acid)	Iodine
Valine		Cobalamin (vitamin B_{12})	Chlorine
		Choline	Fluorine[f]
		Ascorbic acid[e]	Sulfur
			Molybdenum
			Chromium

[a] Replaceable by methionine.
[b] Replaceable by phenylalanine.
[c] Sometimes metabolic synthesis may be insufficient to meet the requirement, so dietary supplementation may be needed.
[d] Replaceable by glycine.
[e] Essential for some passerine birds. Supplementation probably needed for hens under stress of high temperature.
[f] Probably essential but experimental evidence is inconclusive.

REFERENCES

1. Wallace, G. J. and Mahan, H. D., *An Introduction to Ornithology,* Macmillian, New York, 1975.
2. Welty, J. C., *The Life of Birds,* W. B. Saunders, Philadelphia, 1975.
3. Storer, R. W., in *Avian Biology,* Vol. 1, Farner D. S. and King, J. R., Eds., Academic Press, New York, 1971, 149–188.
4. Ziswiler, V. and Farner, D. S., in *Avian Biology,* Vol. 2, Farner, D. S. and King, J. R., Eds., Academic Press, New York, 1972, 343–430.
5. Hill, K. J., in *The Physiology and Biochemistry of the Domestic Fowl,* Vol. 1, Bell, D. J. and Freeman, B. M. Eds., Academic Press, London, 1971, 25–49.
6. Hill, K. J., in *The Physiology and Biochemistry of the Domestic Fowl,* Vol. 1, Bell, D. J. and Freeman, B. M., Eds., Adademic Press, London, 1971, 1–23.
7. McNab, J. M., *World's Poult. Sci. J.,* 29, 251–263, 1973.
8. Moss, R. and Parkinson, J. A., *Br. J. Nutr.,* 27, 285–298, 1972.
9. Mattock, J. G., *Wildfowl,* 22, 107–111, 1971.
10. Coates, M. E., Ford, J. E., and Harrison, G. F., *Br. J. Nutr.,* 22, 493–500, 1968.
11. Fisher, C. D., Lindgren, E., and Dawson, W. R., *Condor,* 74, 111–136, 1972.
12. Shoemaker, V. H., in *Avian Biology,* Vol. 2, Farner, D. S. and King, J. R., Eds., Academic Press, New York, 1972, 527–574.
13. Annison, E. F., Hill, K. J., Shrimpton, D. H., Stringer, D. A., and West, D. E., *Br. Poult. Sci.,* 7, 319–320, 1966.
14. Vohra, P., *World's Poultry Sci. J.,* 23, 20–31, 1967.
15. Roudybush, T. and Vohra, P., unpublished data, 1975.
16. Pearce, J., *Int. J. Biochem.,* 1, 306–312, 1970.
17. Leveille, G. A., Akinbami, T. K., and Ikediobi, C. O., *Proc. Soc. Exp. Biol. Med.,* 135, 483–486, 1970.
17a. Miglioni, R. H., Linder, C., Moura, J. L., and Veiga, J. A. S., *Am. J. Physiol.,* 225, 1389, 1973.
18. Friedmann, H. and Kern, J., *Q. Rev. Biol.,* 31, 19–30, 1956.
19. Machlin, L. J. and Gordon, R. S., *Poult. Sci.,* 39, 1271, 1960.
20. Menge, H., Calvert, C. C., and Denton, C. A., *J. Nutr.,* 86, 115–119, 1965.
21. Leveille, G. A. and Fisher, H., *J. Nutr.,* 66, 441–453, 1958.
22. Chaudhuri, C. R. and Chatterjee, I. B., *Science,* 164, 435–436, 1969.
23. National Academy of Sciences, *Nutrient Requirements of Poultry,* 6th rev. ed., National Academy of Sciences, Washington D.C., 1971.

Cells and Tissues

NUTRIENT REQUIREMENTS OF PLANT TISSUES IN CULTURE FOR GROWTH AND DIFFERENTIATION

I. K. Vasil

INTRODUCTION

The purposes and potentialities of plant tissue culture* were elegantly set forth by the German botanist, Gottlieb Haberlandt, in the introduction to his classical paper, published in 1902.[1] He wrote (as translated by Krikorian and Berquam[2]):

To my knowledge, no systematically organized attempts to culture isolated vegetative cells from higher plants in simple nutrient solutions have been made. Yet the results of such culture experiments should give some interesting insight to the properties and potentialities which the cell as an elementary organism possesses. Moreover, it would provide information about the inter-relationships and complementary influences to which cells within a multicellular whole organism are exposed.

Haberlandt was the first person to seriously attempt the culture of isolated plant cells, but failed largely because of the choice of a wrong and difficult material, and due to the lack of knowledge of the nutrient requirements of isolated plant cells.[3] Later, Kotte,[4,5] a student of Haberlandt, and Robbins[6,7] simultaneously reported growing excised plant roots for several weeks in a nutrient solution. However, the era of modern plant cell culture began with the establishment of continuous cultures of excised tomato roots by White[8] and the cambial tissues of willow and poplar by Gautheret.[9] This was soon followed by descriptions of potentially unlimited growth of the cambial tissues of carrot and tobacco by Gautheret,[10] Nobécourt,[11] and White.[12] Most of these original cell cultures, as well as those of the excised tomato roots by White, are still being maintained through regular subcultures.

Plant tissue cultures are now being widely used to study a variety of problems in the nutrition, growth, differentiation, and reproduction of plants. They have recently also been successfully applied to methods for rapid and mass propagation of plants and for the production of large numbers of haploid plants from anther and microspore cultures; they also have shown considerable potential for use in genetics and plant breeding through their use in experiments on somatic hybridization and parasexual methods of genetic modification.[3,13-17] It must be kept in mind that the continued use of plant tissue cultures for the above and other future novel applications will be possible only if we continue to improve our knowledge of the nutrition of isolated cells in culture and their requirements for organogenesis, and can develop more refined nutrient media which will enable us to grow virtually any living plant cell in vitro and regenerate new plants from it.

Plant tissue culture media are comprised of two major groups of substances: inorganic salts (macro- and micro-nutrients) and organic supplements (sugars and plant growth substances). Complex mixtures of various origin are also used to supplement the above, but in most cases the chemical composition of these mixtures is not known.

INORGANIC NUTRITION

The simple inorganic nutrient solution of Knop,[18] originally developed for the nutrition of whole plants and not for cell cultures, was used by Gautheret[9,10] for his pioneering work on plant cell cultures. White's nutrient solution[19,20] was based on

---

* For the purposes of this review, the term "plant tissue culture" is used in its broadest sense, which includes the culture of protoplasts, cells, tissues, and organs of higher (seed) plants.

Uspenski and Uspenskaia's[21] formula developed for the culture of algae (because of its greater stability over a wide pH range) as well as on the nutrient solution of Trelease and Trelease.[22] Greatly improved growth of a variety of cell cultures was obtained by Hildebrandt et al.[23] and Heller,[24] who modified White's and Gautheret's solutions, respectively. Until the early 1960s, White's nutrient medium in its various modifications and the medium of Heller were the two most common media in use for the cultivation of plant cells. As later work has shown, these media supported optimal growth in many species.[20,25,26]

Various modifications of White's or Heller's medium have been used extensively for the culture of meiotic anthers,[27-29] isolated microsporocytes,[30] and ovaries, ovules, and embryos.[31-33] Similarly, White's medium has proved suitable for the culture of excised roots.[34,35]

Murashige and Skoog[36] found a four- to five-fold increase in growth when White's modified medium was supplemented with an aqueous extract of tobacco leaves. They showed that "this promotion of growth was due mainly though not entirely to inorganic rather than organic constituents in the extract." From these and other detailed experiments, they developed a chemically defined medium that produced optimum growth of tobacco callus tissue. Further addition of common mineral nutrients and organic constituents produced no appreciable change in the rate or amount of growth. The principal features of this medium are greatly increased amounts of inorganic nutrition, particularly the levels of nitrogen (as nitrate and ammonium) and potassium, the presence of iron in a much more stable and chelated form, and the addition of *myo*-inositol. The quantity and form in which nitrogen is present in the nutrient medium are critical for the optimal growth of the tissue, as well as for organogenesis. Nitrogen supplied as nitrate alone will not induce embryogenesis in carrot cultures, although when supplied as ammonium ions in the presence or absence of nitrate it strongly favors embryoid formation.[37-39]

Another completely defined medium, the B5, was described by Gamborg et al.,[40] and has been proven to be especially suitable for suspension cultures of plant cells as well as protoplasts.[41]

The Murashige and Skoog[36] medium (commonly known as the MS medium) — in various minor modifications — is today the most commonly used chemically defined nutrient medium for the cultivation of a variety of plant tissue cultures. The composition of this medium, and of White's medium for comparision and reference, is given in Table 1. A number of other media have been described in the literature and are basically one of the above media with modifications. This has naturally caused serious problems in comparing experimental data published from different laboratories, very often on the same plant species. Recently, a committee acting on behalf of the Plant Division of the Tissue Culture Association (U.S.A.) has recommended standardization of the preparation and composition of nutrient media used for plant cell cultures.[42] Detailed procedures for the preparation of plant tissue culture nutrient media have been described by several authors.[25,26,36,41-43] Commercial preparations of some of the commonly used plant tissue culture media have recently become available, and their use will undoubtedly eliminate or greatly reduce the inconsistencies inherent in the individual preparation of the media.

Table 1
COMPOSITION OF WHITE'S[20]
AND MURASHIGE AND SKOOG'S[36]
NUTRIENT MEDIA COMMONLY USED
IN PLANT TISSUE CULTURES

Composition	White	Murashige and Skoog
Macronutrients (mg/l)		
NH_4NO_3	—	1650
KNO_3	80	1900
$CaCl_2 \cdot 2H_2O$	—	440
$Ca(NO_3)_2 \cdot 4H_2O$	300	—
KCl	65	—
KH_2PO_4	—	170
$NaH_2PO_4 \cdot H_2O$	16.5	—
Na_2SO_4	200	—
$MgSO_4 \cdot 7H_2O$	720	370
Micronutrients (mg/l)		
$MnSO_4 \cdot 4H_2O$	7	22.3
$Fe_2(SO_4)_3$	2.5	—
$FeSO_4 \cdot 7H_2O$	—	27.8
$Na_2 \cdot EDTA$	—	37.3
$ZnSO_4 \cdot 7H_2O$	3	8.6
H_3BO_3	1.5	6.2
KI	0.75	0.83
$CuSO_4 \cdot 5H_2O$	—	0.025
$Na_2MoO_4 \cdot 2H_2O$	—	0.25
$CoCl_2 \cdot 6H_2O$	—	0.025
Organic supplements		
Sucrose	20 g/l	30 g/l
Glycine	3 mg/l	2 mg/l
Nicotinic acid	0.5 mg/l	0.5 mg/l
Thiamine HCl	0.1 mg/l	0.1 mg/l
Pyridoxine HCl	0.1 mg/l	0.5 mg/l
myo-Inositol	—	100 mg/l
Indole-3-acetic acid	—	1—30 mg/l
Kinetin	—	0.04—10 mg/l
Agar	5 g/l	10 g/l
pH	5.5	5.7—5.8

ORGANIC NUTRITION

Plant tissue cultures, with rare exceptions, are not autotrophic. Therefore, it is necessary to provide a suitable source of energy in the nutrient medium. This is most commonly supplied as 2 to 4% sucrose, although other sources of carbon have also been used to a limited extent.[25,43,44]

Many plant tissue cultures possess moderately well-developed chloroplasts and contain large amounts of chlorophyll;[26,45-47] however, photosynthesis and photoautotrophic growth have been demonstrated in very few cases.[48-52] The use of such cell cultures for isolating mutants with improved photosynthetic efficiency is now being actively investigated.[53]

Some of the very early work on the nutrition of plant tissue cultures demonstrated that the addition of autolyzed or nonautolyzed yeast extract, or other similar extracts, to nutrient solutions resulted in improved growth. It was later shown that the beneficial effect of such supplements was due to their amino acid (particularly glycine) and vitamin (especially thiamine) content.[20,43,44] Therefore, most plant cell culture media have routinely contained some vitamins and amino acids.[25,36,42,44] However, detailed investigations have shown that no vitamins other than thiamine and *myo*-inositol markedly affect the growth of tobacco callus tissue on Murashige and Skoog's medium.[54]

Nitsch[55] has described a simple mineral salts-sucrose medium, supplemented with high concentrations of glutamine (800 mg/l) and L-serine (100 mg/l) and very high levels of inositol (5 g/l), for the culture of isolated microspores of *Nicotiana* and *Datura*. Addition of auxins and/or cytokinins to the medium is inhibitory for the desired pattern of growth (production of haploid embryoids and plants).

A very important aspect of organic nutrition is the requirement for plant growth substances in the form of auxins and/or cytokinins. The auxins, indoleacetic acid (IAA), naphthaleneacetic acid (NAA), and 2,4-dichlorophenoxyacetic acid (2,4-D), are widely used.[25,43,44,56] Among these, IAA and NAA not only promote growth but also act synergistically with cytokinins in the control of organ formation. On the other hand, 2,4-D is one of the most potent synthetic auxins known for inducing cell proliferation in vitro,[13,57] but generally inhibits organogenesis. In several instances it has been demonstrated that the gradual or abrupt removal of 2,4-D from the nutrient medium induces organogenesis and embryogenesis,[38,58] although high concentrations of 2,4-D appear to be necessary for the maintenance of many monocotyledonous tissues in culture.[59]

An entirely new class of plant growth substances, the cytokinins, was discovered through plant tissue culture studies.[60-62] Although only a few naturally occurring cytokinins have been isolated and identified so far, a large number of synthetic cytokinins have been prepared and tested for their biological effects.[62] Zeatin and 6-(γ, γ-dimethylallylamino)purine (2iP), both natural cytokinins, are among some of the most active cytokinins known, but for reasons of cost, availability, etc., the commonly used cytokinins are kinetin and benzyladenine. The action of cytokinins on tissue cultures is a function of the amount of IAA available to the tissues in vitro. In general, a high amount of auxin combined with low concentrations of cytokinins favors cell proliferation and root initiation, while low auxin concentrations in the presence of high levels of cytokinins cause shoot formation. Such synergistic action of cytokinins and auxins in organogenesis in vitro is now widely documented.[13,43,60,63] This has greatly aided in developing procedures for the rapid clonal propagation of desirable plant species, particularly many kinds of horticultural and ornamental plants.[13,64]

When cultured on simple nutrient media, without any added auxins and cytokinins, excised immature anthers of tobacco and other species form haploid embryoids directly from the haploid immature microspores.[16,65] The addition of auxins and cytokinins in many cases leads to the formation of haploid callus instead of the direct formation of embryoids. Haploid tissues or plants obtained by this procedure have much potential use in the identification and isolation of mutants, and in plant breeding programs.

A variety of other chemical addenda have been successfully used as supplements to plant tissue culture media, e.g., nucleic acids, purines, pyrimidines, nucleotides, nucleosides, etc.,[28,29,66-69] but their mode of action is far from clear.

COMPLEX MIXTURES OF NATURAL ORIGIN

We have seen that auxins and cytokinins are essential for the growth and differentiation of most plant tissues in culture. However, a variety of complex mixtures of natural origin were used for the same purpose before the role of auxins and cytokinins in plant growth became firmly established. Even now, in many difficult cases where desirable growth and differentiation cannot be obtained by supplementing the media with auxins and cytokinins, these natural complex mixtures, including yeast extract, casein hydrolysate, malt extract, coconut milk (liquid endosperm of coconut), and juices of various fruits such as tomato, orange, watermelon, etc.,[13,25,34,43,44,70] are used with advantage. The beneficial effect of some of these substances is generally related to one or more specific compounds present in them. For example, White[20] showed that the amino acid glycine and the vitamin thiamine accounted for most — if not all — of the growth improvement caused by yeast extract. Similarly, casein hydrolysate stimulates growth by furnishing reduced nitrogen and by acting as a source of specific amino acids.[70] The usefulness of fruit juices is basically due to their high content of plant growth substances, particularly auxins and cytokinins.

One of the most widely used extracts of natural origin in plant tissue cultures has been coconut milk.[71] It was first used by Van Overbeek et al.[72] for the culture of immature embryos of *Datura.* A number of cell-division-inducing substances, including diphenyl-urea,[73] phenylalanine,[75] and others,[70,71,76] have been isolated from coconut milk. None of these are known to completely replace coconut milk in its effect on plant tissue cultures. In many species the optimum growth in vitro by cell proliferation is still possible only by supplementing White's medium[26] or the tobacco medium of Hildebrandt et al.[23,25] with 25 to 50% coconut milk (v/v), the Murashige and Skoog[36] medium notwithstanding.[26,34,43,71]

The once prevalent use of complex and chemically undefined mixtures in plant tissue culture media has been on a steady decline with the availability of more and more potent plant growth substances and with our improved understanding of requirements for growth and differentiation. However, there are still a great many species of plants in which it is either very difficult, or as yet impossible, to obtain organogenesis and plant formation in vitro.

Kao and Michayluk[77] recently described a very complex medium which, in addition to the normal component of mineral salts, contains 14 vitamins, auxins and cytokinins, organic acids, 10 sugars and sugar alcohols, 21 amino acids, 6 nucleic acid bases, casein hydrolysate, and coconut water. The authors reported the formation of a mass of cells from single isolated protoplasts or cells of *Vicia hajastana* cultured in the above medium. This again highlights the fact that our understanding of the nutrient requirements of plant tissues in culture is far from complete.

Several physical and physiological factors also affect the growth and differentiation of plant tissues in culture. The physiological condition of the tissue at the time of excision can be critical to its behavior in vitro.[17,78,79] The physical and physiological conditions — pH of the nutrient medium, quality and quantity of light, semisolid or liquid nutrient media, filter paper rafts, etc. — under which the tissues are cultured all profoundly affect their growth and behavior.[25,43,44,63,80]

REFERENCES

1. **Haberlandt, G.,** Kulturversuche mit isolierten Pflanzenzellen, *Mat. Nat. Kais. Akad. Wiss.* (Wien), 111, 69–92, 1902.
2. **Krikorian, A. D. and Berquam, D. L.,** Plant cell and tissue cultures: the role of Haberlandt, *Bot. Rev.,* 35, 59–88, 1969.
3. **Vasil, I. K. and Vasil, V.,** Totipotency and embryogenesis in plant cell and tissue cultures, *In Vitro,* 8, 117–127, 1972.
4. **Kotte, W.,** Wurzelmeristem in Gewebekultur, *Ber. Dtsch. Bot. Ges.,* 40, 269–272, 1922.
5. **Kotte, W.,** Kulturversuche mit isolierten Wurzelspitzen, *Beitr. Allg. Bot.,* 2, 413–434, 1922.
6. **Robbins, W. J.,** Cultivation of excised root tips and stem tips under sterile conditions, *Bot. Gaz.,* 73, 376–390, 1922.
7. **Robbins, W. J.,** Effect of autolyzed yeast and peptone on growth of excised corn root tips in the dark, *Bot. Gaz.,* 74, 59–79, 1922.
8. **White, P. R.,** Potentially unlimited growth of excised tomato root tips in a liquid medium, *Plant Physiol.,* 9, 585–600, 1934.
9. **Gautheret, R. J.,** Cultur du tissu cambial, *C. R. Acad. Sci.,* 198, 2195–2196, 1934.
10. **Gautheret, R. J.,** Sur la possibilité de réaliser la culture indéfinie des tissus de tubercules de carotte, *C. R. Acad. Sci.,* 208, 118–120, 1939.
11. **Nobécourt, P.,** Sur la perennité et l'augmentation de volume des cultures de tissus végétaux, *C. R. Soc. Biol.,* 130, 1270, 1939.
12. **White, P. R.,** Potentially unlimited growth of excised plant callus in an artificial medium, *Am. J. Bot.,* 26, 59–64, 1939.
13. **Murashige, T.,** Plant propagation through tissue cultures, *Ann. Rev. Plant Physiol.,* 25, 135–166, 1974.
14. **Street, H. E., Ed.,** *Plant Tissue and Cell Culture,* University of California Press, Berkeley, 1973.
15. **Street, H. E., Ed.,** *Tissue Culture and Plant Science 1974,* Academic Press, New York, 1974.
16. **Vasil, I. K. and Nitsch, C.,** Experimental production of pollen haploids and their uses, *Z. Pflanzenphysiol.,* 76, 191–212, 1975.
17. **Vasil, I. K.,** The progress, problems, and prospects of plant protoplast research, *Adv. Agron.,* 28, 119–160, 1976.
18. **Knop, W.,** Quantitative Untersuchungen über den Ernährungsprozess der Pflanzen, *Landwirtsch. Vers. Stn.,* 7, 93–107, 1865.
19. **White, P. R.,** Plant tissue cultures: a preliminary report of results obtained in the culturing of certain plant meristems, *Arch. Exp. Zellforsch.,* 12, 602–620, 1932.
20. **White, P. R.,** *The Cultivation of Animal and Plant Cells,* 2nd ed., Ronald Press, New York, 1963.
21. **Uspenski, E. E. and Uspenskaia, W. J.,** Reinkultur und ungeschlechtliche Fortpflanzung der *Volvox minor* und *Volvox globator* in einer synthetischen Nährlösung, *Z. Bot.,* 17, 273–308, 1925.
22. **Trelease, S. and Trelease, H. M.,** Physiologically balanced culture solutions with stable hydrogen-ion concentration, *Science,* 78, 438–439, 1933.
23. **Hildebrandt, A. C., Riker, A. J., and Duggar, W. M.,** The influence of composition of the medium on growth in vitro of excised tobacco and sunflower tissue cultures, *Am. J. Bot.,* 33, 591–597, 1946.
24. **Heller, R.,** Recherches sur la nutrition minérale des tissus végétaux cultivés in vitro, *Ann. Sci. Nat. Bot. Biol. Vég.,* 14, 1–223, 1953.
25. **Hildebrandt, A. C.,** Tissue and single cell cultures of higher plants as a basic experimental method, in *Modern Methods of Plant Analysis,* Vol. 5, Paech, K., Tracey, M. V., and Linskens, H. F., Eds., Springer-Verlag, Berlin, 1962, 383–421.
26. **Vasil, I. K. and Hildebrandt, A. C.,** Growth and chlorophyll production in plant callus tissues grown in vitro, *Planta,* 68, 69–82, 1966.
27. **Vasil, I. K.,** Effect of kinetin and gibberellic acid on excised anthers of *Allium cepa, Science,* 126, 1294–1295, 1957.
28a. **Vasil, I. K.,** Nucleic acids and survival of excised anthers in vitro, *Science,* 129, 1487–1488, 1959.
28b. **Vasil, I. K.,** Plants: haploid tissue cultures, in *Tissue Culture – Methods and Applications,* Kruse, P. F., Jr. and Patterson, M. K., Jr., Eds., Academic Press, New York, 1973, 157–161.
29. **Vasil, I. K.,** Cultivation of excised anthers in vitro – effect of nucleic acids, *J. Exp. Bot.,* 10, 399–408, 1959.
30. **Ito, M. and Stern, H.,** Studies of meiosis in vitro. I. In vitro culture of meiotic cells, *Dev. Biol.,* 16, 36–53, 1967.

31. Nitsch, J. P., The in vitro culture of flowers and fruits, in *Plant Tissue and Organ Culture — A Symposium,* Maheshwari, P. and Rangaswamy, N. S., Eds., International Society for Plant Morphology, University of Delhi, India, 1963, 198–214.

32. Maheshwari, P., Ed., *Recent Advances in the Embryology of Angiosperms,* International Society for Plant Morphology, University of Delhi, India, 1963.

33. Maheshwari, P. and Rangaswamy, N. S., Eds., *Plant Tissue and Organ Culture — A Symposium,* International Society for Plant Morphology, University of Delhi, India, 1963.

34. Street, H. E., Growth in organized and unorganized systems. Knowledge gained by culture of organs and tissue explants, in *Plant Physiology — A Treatise,* Vol. VB, Steward, F. C., Ed., Academic Press, New York, 1969, 3–224.

35. Street, H. E., Roots, in *Tissue Culture — Methods and Applications,* Kruse, P. F., Jr. and Patterson, M. K., Jr., Eds., Academic Press, New York, 1973, 173–178.

36. Murashige, T. and Skoog, F., A revised medium for rapid growth and bioassays with tobacco tissue cultures, *Physiol. Plant.,* 15, 473–497, 1962.

37. Halperin, W. and Wetherell, W. F., Ammonium requirement for embryogenesis in vitro, *Nature,* 205, 519–520, 1965.

38. Halperin, W., Embryos from somatic plant cells, *Symp. Int. Soc. Cell Biol.,* 9, 169–191, 1970.

39. Tazawa, M. and Reinert, J., Extracellular and intracellular chemical environments in relation to embryogenesis in vitro, *Protoplasma,* 68, 157–173, 1969.

40. Gamborg, O. L., Miller, R. A., and Ojima, K., Nutrient requirements of suspension cultures of soybean root cells, *Exp. Cell Res.,* 50, 151–158, 1968.

41. Gamborg, O. L. and Wetter, L. R., Eds., *Plant Tissue Culture Methods.,* National Research Council, Ottawa, Canada, 1975.

42. Gamborg, O. L., Murashige, T., Thorpe, T. A., and Vasil, I. K., Plant tissue culture media, *In Vitro,* 12, 473–478, 1976.

43. Butenko, R. G., *Plant Tissue Culture and Plant Morphogenesis,* Israel Program for Scientific Translations, Jerusalem, 1968 (English translation).

44. Gautheret, R. J., The nutrition of plant tissue cultures, *Annu. Rev. Plant Physiol.,* 6, 433–484, 1955.

45. Laetsch, W. M. and Stetler, D. A., Chloroplast structure and function in cultured tobacco tissue, *Am. J. Bot.,* 52, 798–804, 1965.

46. Sunderland, N. and Wells, B., Plastid structure and development in green callus tissues of *Oxalis dispar, Ann. Bot.,* 32, 327–346, 1968.

47. Edelman, J. and Hanson, A. D., Sucrose suppression of chlorophyll synthesis in carrot callus cultures, *Planta,* 98, 150–156, 1971.

48. Bergmann, L., Wachstum grüner Suspensionkulturen von *Nicotiana tabacum* var. "Samsun" mit CO_2 als Kohlenstoffquelle, *Planta,* 74, 243–249, 1967.

49. Corduan, G., Autotrophe Gewebekulturen von *Ruta graveolens* und deren $^{14}CO_2$-Markierungsprodukte, *Planta,* 91, 291–301, 1970.

50. Chandler, M. T., De Marsac, N. T., and De Kouchkovsky, Y., Photosynthetic growth of tobacco cells in liquid suspension, *Can. J. Bot.,* 50, 2265–2270, 1972.

51. Hanson, A. D. and Edelman, J., Photosynthesis by carrot tissue cultures, *Planta,* 102, 11–25, 1972.

52. Berlyn, M. B. and Zelitch, I., Photoautotrophic growth and photosynthesis in tobacco callus cells, *Plant Physiol.,* 56, 752–756, 1975.

53. Zelitch, I., Improving the efficiency of photosynthesis, *Science,* 188, 626–633, 1975.

54. Linsmaier, E. M. and Skoog, F., Organic growth factor requirements of tobacco tissue cultures, *Physiol. Plant.,* 18, 100–127, 1965.

55. Nitsch, C., La culture de pollen isolé sur milieu synthétique, *C. R. Acad. Sci. Ser. D.,* 278, 1031–1034, 1974.

56. Thimann, K. V., The natural plant hormones, in *Plant Physiology — A Treatise,* Vol. VIB, Steward, F. C., Ed., Academic Press, New York, 1972, 3–332.

57. Vasil, I. K., Morphogenetic, histochemical, and ultrastructural effects of plant growth substances in vitro, *Biochem. Physiol. Pflanz.,* 164, 58–71, 1973.

58. Green, C. E. and Phillips, R. L., Plant regeneration from tissue cultures of maize, *Crop Sci.,* 15, 417–421, 1975.

59. Schenk, R. U. and Hildebrandt, A. C., Medium and techniques for induction and growth of monocotyledonous and dicotyledonous plant cell cultures, *Can. J. Bot.,* 50, 199–204, 1972.

60. Skoog, F. and Miller, C. O., Chemical regulation of growth and organ formation in plant tissues cultured in vitro, *Symp. Soc. Exp. Biol.,* 11, 118–131, 1957.

61. Fox, J. E., The cytokinins, in *Physiology of Plant Growth and Development,* Wilkins, M. B., Ed., McGraw-Hill, New York, 1969, 85–123.

62. **Skoog, F. and Schmitz, R. Y.,** Cytokinins, in *Plant Physiology – A Treatise,* Vol. VIB, Steward, F. C., Ed., Academic Press, New York, 1972, 181–213.
63. **Vasil, I. K. and Hildebrandt, A. C.,** Variations of morphogenetic behaviour in plant tissue cultures. I. *Cichorium endivia, Am. J. Bot.,* 53, 860–869, 1966.
64. **Vasil, I. K.,** Some alternatives to sex in plants through cell and tissue cultures, *In Vitro,* 12, 287, 1976.
65. **Nitsch, J. P. and Nitsch, C.,** Haploid plants from pollen grains, *Science,* 163, 85–87, 1969.
66. **Skoog, F. and Tsui, C.,** Chemical control of growth and bud formation in tobacco stem segments and callus cultured in vitro, *Am. J. Bot.,* 35, 782–787, 1948.
67. **Hildebrandt, A. C., Riker, A. J., and Muir, E.,** Growth in vitro of marigold and tobacco tissue with nucleic acids and related compounds, *Plant Physiol.,* 32, 231–236, 1957.
68. **Vasil, I. K.,** Some new experiments with excised anthers, in *Plant Tissue and Organ Culture – A Symposium,* Maheshwari, P. and Rangaswamy, N. S., Eds., International Society for Plant Morphology, University of Delhi, India, 1963, 230–238.
69. **Vasil, I. K. and Hildebrandt, A. C.,** Variations of morphogenetic behaviour in plant tissue cultures. II. *Petroselinum hortense, Am. J. Bot.,* 53, 869–874, 1966.
70. **Steward, F. C., Mapes, M. O., and Ammirato, P. V.,** Growth and morphogenesis in tissue and free cell cultures, in *Plant Physiology – A Treatise,* Vol. VB, Steward, F. C., Ed., Academic Press, New York, 1969, 329–376.
71. **Steward, F. C.,** *Growth and Organization in Plants,* Addison-Wesley Publishing, Reading, Mass., 1968.
72. **van Overbeek, J., Conklin, M. E., and Blakeslee, A. F.,** Factors in coconut milk essential for growth and development of very young *Datura* embryos, *Science,* 94, 350–351, 1941.
73. **Shantz, E. M. and Steward, F. C.,** The identification of compound A from coconut milk as 1,3-diphenylurea, *J. Am. Chem. Soc.,* 77, 6351–6353, 1955.
74. **Letham, D. S.,** A new cytokinin bioassay and the naturally occurring cytokinin complex, in *Biochemistry and Physiology of Plant Growth Substances,* Wightman, F. and Setterfield, G., Eds., Runge Press, Ottawa, Canada, 1968, 19–31.
75. **Van Staden, J. and Drewes, S. E.,** Identification of cell division inducing compounds from coconut milk, *Physiol. Plant.,* 32, 347–352, 1974.
76. **Van Staden, J.,** Evidence for the presence of cytokinins in malt and yeast extracts, *Physiol. Plant.,* 30, 182–184, 1974.
77. **Kao, K. N. and Michayluk, M. R.,** Nutritional requirements for growth of *Vicia hajastana* cells and protoplasts at a very low population density in liquid media, *Planta,* 126, 105–110, 1975.
78. **Yeoman, M. M. and Mitchell, J. P.,** Changes accompanying the addition of 2,4-D to excised Jerusalem artichoke tuber tissue, *Ann. Bot.,* 34, 799–810, 1970.
79. **Watts, J. W., Motoyoshi, F., and King, J. M.,** Problems associated with the production of stable protoplasts of cells of tobacco mesophyll, *Ann. Bot.,* 38, 667–671, 1974.
80. **Murashige, T.,** Nutrition of plant cells and organs in vitro, *In Vitro,* 9, 81–85, 1973.

QUALITATIVE REQUIREMENTS AND UTILIZATION OF NUTRIENTS: ARTHROPOD CELLS IN CULTURE

M. M. Stanley

INTRODUCTION

In vitro requirements for arthropod cells are still poorly known. Of the few species that have been cultured for substantial periods, all but one are from insects, and these represent only 4 of 26 orders. Cells of only three species, all cockroaches, have been maintained on chemically defined media (those lacking yeast extract, lactalbumin hydrolysate, insect hemolymph, vertebrate sera or serum fractions, or other impure biological substances), and such media have generally supported poor growth. Even these media may contain nonessential substances. Depletion measurements to show actual use of specific nutrients have been carried out on only a limited basis, as have addition and deletion studies.

Synergisms and substitutions complicate many nutritional studies. For example, amino acid utilization by *Culex tarsalis* cells differ in three media.[1] In addition, unrecognized infections and misidentification of culture origins can lead to erroneous conclusions. Microbial infections, perhaps tracing to the original tissue inocula, have been found in some insect lines.[2] Serological, enzymatic, and karyological studies have also established that certain insect cultures have been misidentified or contaminated by other lines.[3] Data from questionable cultures have been excluded from the present tabulation but unrecognized contamination may, of course, exist.

Techniques for establishing lines of insect cells have reached the point that standard procedures can be described.[4,5] Nevertheless, we understand little of the process. Some lines are very likely "transformed," as is true of many vertebrate lines. At best, insect cell lines are more or less diploid. Many are frankly polyploid but the significance of this fact is clouded because polyploidy is also common in vivo. Perhaps equally important, most lines probably originate from only certain cell types. Rigorous selection may yield lines not at all representative of the whole animal or of the typical cell of that species. Finally, culture conditions may induce enzymes seldom produced in the intact organism. Such problems are suggested by the differing nutritional requirements for two lines of *Laspeyresia pomenella*.[6] On the other hand, mycoplasma have been isolated from some cultures of one line, and it is possible that infection accounted for the differences in metabolism. Clearly, generalizations from the data in Table 1 should be made cautiously, and important facts confirmed by independent experiments.

Mineral requirements of cultured arthropod cells are barely known. At a minimum, most media contain Na, K, Ca, Mg, Cl, and SO_4. In chemically defined media, *Periplaneta americana* cells (EPa) grew on the basic six inorganic ions plus manganese and phosphate.[32] *Blaberus discoidalis* primary culture cells tolerated the basic six ions plus phosphate and iron,[24] while *Leucophea maderae* primary culture cells were maintained on the basic six plus phosphate.[28] In some culture systems the sodium:potassium ratio seems important. Excess potassium is toxic for *Heliothis zea*[35] and *Musca domestica* cells.[36] Much of the importance of inorganic ions lies in their direct effect on osmotic pressure. This parameter is critical for the culture of *Drosophila*[37] and many lepidoptera.[38] High potassium levels limit the osmotic tolerance of *H. zea* cells.[35]

From the data available, no conclusions can be drawn concerning the trace element requirements of cultured arthropod cells.

Table 1
NUTRIENTS FOR CULTURED ARTHROPOD CELLS

Nutrient utilization or requirement is indicated as follows: R = required, R̶ = not required; U = utilized; u = poorly utilized, U̶ = not utilized; S = stimulatory or supportive of growth; s = moderately supportive, $ = not supportive; E = essential, E̶ = nonessential.

Species (synonym)	Designation[a]* and/or origin	Sugars and sugar alcohols															Ref.
		Cellobiose	Fructose	Galactose	Glucose	Inositol[d]	Lactose	Maltose	Mannose	Mannitol	Melibiose	Sorbitol	Sorbose	Sucrose	Trehalose	Turanose	
Ticks																	
Rhipicephalus sanguineus	PC				U	U									U		18
Insects																	
Austrocaligua eucalypti (Antheraea eucalypti)	Ae[b,8]		U		U									U[f]	U		19, 20[h]–22
	RML-10[c,9]																23
Blaberus discoidalis	PC	R̶	R̶	R̶		R̶	R̶	R̶	R̶	R̶	R̶	R̶	R̶	$	R̶	R̶	24
Laspeyresia pomenella (Carpocapsa pomenella)	CP-1268[10]		s		S									$			6[h], 25
	CP-169[10]		S		s									$			6[h]
Culex quinquefasciatus	Ref. 11		S		s										S		26
Culex pipiens	Ref. 12																1
Culex tarsalis	Ref. 13																1
Galleria mellonella	PC																27
Leucophaea maderae	PC	R̶		R̶			R̶	R̶	R̶	R̶	R̶	R̶	R̶			R̶	28
Papilio xuthus	Px-58[14]											R̶					29[h]
Periplaneta americana	EPa[15]	R̶	R̶	R̶		R̶	R̶	R̶	R̶	R̶	R̶	R̶	R̶	R̶	R̶	R̶	30[h]–32
Samia cynthia	PC																33
Spodoptera frugiperda	IPRL-21[16]																16
Trichoplusia ni	TN-368[17]	$	S	$	S	$	$	S	S	$	$	$	$	$	S	$	34

*See footnotes at end of table.

Table 1 (continued)
NUTRIENTS FOR CULTURED ARTHROPOD CELLS

Species (synonym)	Designation[a] and/or origin	α-Alanine	β-Alanine	Aspartic acid	Asparagine	Arginine	Cysteine	Cystine	Glutamic acid	Glutamine	Glycine	Histidine	Isoleucine	Leucine	Lysine	Methionine	Phenylalanine	Proline	Serine	Threonine	Tyrosine	Tryptophane	Valine	Ref.
Ticks																								
Rhipicephalus sanguineus	PC			U		Ū		Ū	U	U	Ū	Ū	Ū	U	Ū	U	U	U	Ū	U	u[e]		Ū	18
Insects																								
Austrocaligua eucalypti (Antheraea eucalypti)	Ae[b,8]	Ū	u	U	U	u[e]		U	U	U	Ū	Ū	U[g]	U[g]	u[e]	U	u[e]	u	Ū	Ū	U	Ū	u[e]	19,20[h]—22
	RML-10[c,9]																							23
Blaberus discoidalis	PC		R		R																			24
Laspeyresia pomenella	CP-1268[10]	Ū		U	Ū			U	U		Ū	Ū	Ū	U[25]	Ū	U	U[k]	Ū[j]	Ū	Ū	Ū[j]		Ū	6,[h]25
(Carpocapsa pomenella)	CP-169[10]	Ū		U	Ū			U	U		Ū	Ū	Ū	Ū	Ū	U	U[k]	Ū	Ū	Ū	U		Ū[+]	6[h]
Culex quinquefasciatus	Ref. 11																							26
Culex pipiens	Ref. 12	Ū		U				U[k]	U[k]		Ū			U[k]		U	U[k]	U[k]	Ū	U[k]		U[k]		1
Culex tarsalis	Ref. 13	Ū	U[k]	U				U[k]	U[k]		U[k]					U[k]	U[k]		U[k]	Ū	Ū	U[k]		1
Galleria mellonella	PC																							27
Leucophaea maderae	PC								R															28
Papilio xuthus	Px-58[14]	Ū	E	S	E	S		E	S	E	S	E	E	E	E	E	u	E	E	E	E	S	E	29[h]
Periplaneta americana	EPa[15]	Ū	E	u	E	E	E		U	E	E	E	S	E	S	E	E	E	S	S	E	E	E	30[h]—32
Samia cynthia	PC																							33
Spodoptera frugiperda	IPRL-21[16]																							16
Trichoplusia ni	TN-368[17]																							34

Table 1 (continued)
NUTRIENTS FOR CULTURED ARTHROPOD CELLS

Species (synonym)	Designation[a] and/or origin	Triglycerides	Free fatty acids	Arachidonic acid	Linoleic acid	Linolenic acid	Oleic acid	Palmitic acid	Palmitoleic acid	Stearic acid	Cholesterol	2-Hydroxyecdysone	Farnesol	Ponasterone A	β-Sitosterol	Ref.
Ticks																
Rhipicephalus sanguineus	PC															18
Insects																
Austrocaligua eucalypti (Antheraea eucalypti)	Ae[b,8]	U	U								U	S[i]	S	S[i]	U	19, 20[h]–22
	RML-10[c,9]										U					23
Blaberus discoidalis	PC	R	R								R					24
Laspeyresia pomenella (Carpocapsa pomenella)	CP-1268[10]															6[h], 25
	CP-169[10]															6[h]
Culex quinquefasciatus	Ref. 11															26
Culex pipiens	Ref. 12															1
Culex tarsalis	Ref. 13															1
Galleria mellonella	PC															27
Leucophaea maderae	PC															28
Papilio xuthus	Px-58[14]															29[h]
Periplaneta americana	EPa[15]															30[h]–32
Samia cynthia	PC															33
Spodoptera frugiperda	IPRL-21[16]			U	U	U	U[j]	U	U[j]	U[j]						16
Trichoplusia ni	TN-368[17]				U	U	U[j]	U	U[j]	U[j]						34

Table 1 (continued)

NUTRIENTS FOR CULTURED ARTHROPOD CELLS

Species (synonym)	Designation[a] and/or origin	Vitamins																Ref.
		Ascorbic acid	α-Tocopherol	Biotin	Carnitine	Choline	Cyanocobalamine	Folic acid	Inositol	Menadione	Nicotinamide	Ca-pantothenate	p-Aminobenzoic acid	Pyridoxine	Retinol	Riboflavin	Thiamine	
Ticks																		
Rhipicephalus sanguineus	PC																	18
Insects																		
Austrocaligua eucalypti (Antheraea eucalypti)	Ae[b,8]																	19, 20[h]–22
Blaberus discoidalis	RML-10[c,9]					R̸	R̸											23
Laspeyresia pomenella (Carpocapsa pomenella)	PC				R̸	R̸												24
	CP-1268[10]																	6[h], 25
Culex quinquefasciatus	CP-1691[10]																	6[h]
Culex pipiens	Ref. 11																	26
Culex tarsalis	Ref. 12																	1
Galleria mellonella	Ref. 13																	1
Leucophaea maderae	PC		R̸				R̸			R̸								27
Papilio xuthus	PC														R̸			28
Periplaneta americana	Px-58[15]	Ⅎ	Ⅎ	S	Ⅎ	S	R̸	S	E	Ⅎ	S[l]	E	Ⅎ	S	R̸	E	E	29[h]
Samia cynthia	EPa[15]		E	E	E	E	E	E	E[m]		E	E	E	E		E	E	30[h]–32
Spodoptera frugiperda	PC																	33
	IPRL-21[16]																	16
Trichoplusia ni	TN-368[17]																	34

Table 1 (continued)
NUTRIENTS FOR CULTURED ARTHROPOD CELLS

Species (synonym)	Designation[a] and/or origin	2-Amino-4-hydroxy-pteridine	2-Amino-4-mercapto-pteridine	Fumaric acid	2-Oxo-glutaric acid	Glycerol	Na-β-glycerophosphate	Malic acid	Pyruvic acid	Succinic acid	Uridine	Ref.
Ticks												
Rhipicephalus sanguineus	PC							R	R	R	R	18
Insects												
Austrocaligua eucalypti (*Antheraea eucalypti*)	Ae[b,8]	R										19, 20[h]–22
Blaberus discoidalis	RML-10[c,9]	R	R	R	R	R	R	R	R	R		23
	PC											24
Laspeyresia pomenella (*Carpocapsa pomenella*)	CP-1268[10]										U[25]	6[h], 25
	CP-169[10]											6[h]
Culex quinquefasciatus	Ref. 11											26
Culex pipiens	Ref. 12											1
Culex tarsalis	Ref. 13											1
Galleria mellonella	PC	S	S									27
Leucophaea maderae	PC	R	R	R	R	R	R	R	R	R	R	28
Papilio xuthus	Px-58[14]	E	E		E	E	E	E	E			29[h]
Periplaneta americana	EPa[15]	E	E	E	E	E	E	E	E	E	E	30[h]–32
Samia cynthia	PC			S	S	S	S	S	S	S		33
Spodoptera frugiperda	IPRL-21[16]											16
Trichoplusia ni	TN-368[17]											34

Table 1 (continued)
NUTRIENTS FOR CULTURED ARTHROPOD CELLS

a   Each line is designated by its code if one is available. Entries marked PC are primary cultures in which cell proliferation and/or migration occurred from explants or disaggregated inocula but from which the cells were not transferred. [7] Evaluation of primary cultures can be more difficult than for cell lines.

b   Not adapted to hemolymph-free media.

c   Adapted to hemolymph-free media.

d   In some instances, inositol is utilized in such large quantities that it can be presumed to be a major energy source and must be considered as a sugar; in other cases, minute quantities seem crucial, so it is considered a vitamin.

e   Utilization is followed by synthesis.

f   Sucrose utilization increases greatly after fructose, glucose, and trehalose are partially depleted.

g   Leucine and isoleucine lumped together.

h   Except for glycine and DL-serine, the 1-enantimorphs of amino acids were used in these studies.

i   Effects vary with dosage and medium used.

j   Utilization follows as initial period of synthesis.

k   Utilized in some media but not in others.

l   Nicotinic acid.

m   *m*-Inositol.

REFERENCES

1. Chao, J. and Ball, G. H., in *Proc. 4th Int. Colloq. on Invertebrate Tissue Culture,* Maramorosch, K., Ed., Academic Press, New York, in press.
2. Hirumi, H., in *Invertebrate Tissue Culture: Applications to Fundamental Research,* Maramorosch, K., Ed., Academic Press, New York, 1976, 233–268.
3. Green, A. E., Charney, J., Nichols, W. W., and Coriel, L. L., *In Vitro,* 7, 313–322, 1972.
4. Goodwin, R. H., *In Vitro,* 11, 369–378, 1975.
5. Schneider, I., in *Tissue Culture Methods and Applications,* Kruse, P. F., Jr. and Patterson, M. K., Jr., Eds., Academic Press, New York, 1973, 150–152.
6. Hink, W. F., Richardson, B. L., Schenk, D. K., and Ellis, B. J., in *Proc. 3rd. Int. Colloq. on Invertebrate Tissue Culture,* Řeháček, J., Blaškovič, D., and Hink, W. F., Eds., Slovak Academy of Science, Bratislava, 1973, 195–208.
7. Federoff, S., *J. Natl. Cancer Inst.,* 38, 607–611, 1967.
8. Grace, T. D. C., *Nature,* 195, 788–789, 1962.
9. Yunker, C. E., Vaughn, J. L., and Cory, J., *Science,* 155, 1565–1566, 1967.
10. Hink, W. F. and Ellis, B. J., *Curr. Top. Microbiol. Immunol.,* 55, 19–28, 1971.
11. Hsu, S. H., Mao, W. H., and Cross, J. H., *J. Med. Entomol.,* 7, 703–707, 1970.
12. Chao, J. and Ball, G. H., *In Vitro,* 8, 406, 1973.
13. Chao, J. and Ball, G. H., in *Proc. 4th Int. Colloq. on Invertebrate Tissue Culture,* Maramorosch, K., Ed., Academic Press, New York, in press.
14. Mitsuhashi, J., *Appl. Entomol. Zool.,* 8, 64–72, 1973.
15. Landureau, J. C., *Exp. Cell Res.,* 50, 323–337, 1968.
16. Louloudes, S. J., Vaughn, J. L., and Dougherty, K. A., *In Vitro,* 8, 473–479, 1973.
17. Hink, W. F., *Nature,* 226, 466–467, 1970.
18. Řeháček, J. and Brzostowski, H. W., *J. Insect Physiol.,* 15, 1683–1686, 1969.
19. Clements, A. N. and Grace, T. D. C., *J. Insect Physiol.,* 13, 1327–1332, 1967.
20. Grace, T. D. C. and Brzostowski, H. W., *J. Insect Physiol.,* 12, 625–633, 1966.
21. Gilby, A. R. and McKellar, J. W., *J. Insect Physiol.,* 20, 2219–2224, 1974.
22. Mitsuhashi, J. and Grace, T. D. C., *Appl. Entomol. Zool.,* 5, 182–188, 1970.
23. Vaughn, J. L., Louloudes, S. J., and Dougherty, K., *Curr. Top. Microbial. Immunol.,* 55, 92–97, 1971.
24. Larsen, W. P., *J. Insect Physiol.,* 13, 613–619, 1967.
25. Gallagher, B. M. and Hartig, W. J., *In Vitro,* 12, 165–172, 1976.
26. Parker, R., Edgar, R., and Yunker, C. E., in *Proc. Conf. Insect and Mite Nutrition,* Rodrigues, J. G., Ed., North Holland, Amsterdam, 1972, 397–406.
27. Saska, J., Grzelakowska-Sztabert, B., and Zielinska, Z. M., *J. Insect Physiol.,* 18, 1733–1737, 1972.
28. Marks, E. P., Reinecke, J. P., and Caldwell, J. M., *In Vitro,* 3, 85–92, 1968.
29. Mitsuhashi, J., *J. Insect Physiol.,* 22, 397–402, 1976.
30. Landureau, J. C. and Jollés, P., *Exp. Cell Res.,* 54, 391–398, 1969.
31. Landureau, J. C., *Exp. Cell Res.,* 54, 399–402, 1969.
32. Landureau, J. C., *C. R. Acad. Sci. Paris,* 270, 3288–3291, 1970.
33. Sanborn, R. C. and Haskell, J. A., in *Proc. 11th Int. Congr. Entomol.,* B3, 237–243, 1960.
34. Stockdale, H. and Gardiner, G. R., in *Proc. 4th Int. Colloq. on Invertebrate Tissue Culture,* Maramorosch, K., Ed., Academic Press, New York, in press.
35. Kurtti, T. J., Chaudhary, S. P. S., and Brooks, M. A., *In Vitro,* 11, 274–285, 1975.
36. Eide, P. E., *J. Insect Physiol.,* 21, 1431–1438, 1975.
37. Eschalier, G. and Ohanessian, A., *In Vitro,* 6, 162–172, 1970.
38. Goodwin, R. H., *In Vitro,* 11, 369–378, 1975.

QUALITATIVE REQUIREMENTS AND UTILIZATION OF NUTRIENTS: POIKILOTHERMIC VERTEBRATE CELLS IN CULTURE

M. Balls and M. A. Monnickendam

The main aim in formulating media for cell and organ culture is to provide an in vitro physiological and nutritional environment as similar as possible to that occurring in vivo. The basic, chemically-defined medium consists of a "physiological solution," comprising a mixture of inorganic salts, to which are added glucose, vitamins, and essential amino acids. Some basic media also contain coenzymes, nucleic acid derivatives, and lipid sources. The final culture medium consists of the basic medium plus serum and/or other natural products such as embryo extracts, tissue extracts, or egg ultrafiltrates.

Methods for poikilothermic vertebrate cell and organ culture have almost always involved the use of basic media originally formulated for mammalian and avian tissue culture. Much attention has been devoted to the temperature requirements and tolerances of lower vertebrate cells and tissues in vitro, and to the osmolality of the physiological solution section of the basic medium. Since there is no evidence that the nutritional requirements of cells from poikilotherms differ fundamentally from those of homoiotherms, very little attention has been paid to this aspect.

In this section, we shall briefly review the little that is known about the requirements of poikilothermic vertebrate cells in culture. Further information may be found in other sections of this handbook and in the reviews of Wolf and Quimby,[1] Freed and Mezger-Freed,[2] Monnickendam and Balls,[3] Balls et al.,[4] and Clark.[5]

Culture Temperature

Poikilothermic vertebrate cells can proliferate and function over a much wider range of incubation temperatures than can homoiothermic vertebrate cells. Cells from fish can be grown at 4 to 26°C, according to the temperature range characteristic of the normal environment of the species of origin and to the tolerance of particular cell lines.[1] The most favored temperature for amphibian tissue culture is 25°C, though cell proliferation occurs over the approximate range 10 to 28°C.[2,3] Fish and amphibian cells of certain cell lines and tissues from tropical species can be maintained at up to 37°C.[3,5] Reptile cells have been grown at 15 to 37°C, but Clark et al.[6] found 30°C to be optimal for cell lines derived from the box turtle, *Terrapene carolina*.

Physiological Solutions

Mammalian physiological solutions, altered to more closely resemble the body fluids of the relevant animals, are normally used (Table 1). The addition of NaCl to media for marine teleost cultures,[7] favored by most workers, was found to be unnecessary by Lee and Loh[8] for cells of a line from the marine Omaka (*Caranx mate*), where optimal growth was at the more usual lower concentration. Although amphibian cells are usually cultured in diluted media,[2,3] cells of a permanent aneuploid line from the South African clawed toad, *Xenopus laevis*, incorporated ³H-thymidine and proliferated in media with osmolalities from 90 to 750 mosmol/kg.[9] The optimal level was 190 mosmol/kg, and high growth rates occurred in the range 120 to 320 mosmol/kg. Monnickendam and Balls[10] found that *Amphiuma means* (Congo eel) liver and spleen organ cultures survived better in hypotonic or isotonic media (125 to 230 mosmol/kg) than in hypertonic media (255 to 305 mosmol/kg).

pH and Buffers

The pH of the medium does not appear to be particularly critical, and most cells do

Table 1
MODIFICATIONS IN HOMOIOTHERM PHYSIOLOGICAL SOLUTIONS FOR POIKILOTHERMIC VERTEBRATE CELL AND ORGAN CULTURE

Animal group	Tonicity compared with homoiotherm solutions	Details	Ref.
Fresh-water teleost fish	Hypotonic or isotonic	Usually diluted by 20% with water	1
Marine teleost fish	Hypertonic	NaCl content increased from 0.137 M to 0.2 M	1, 7, 8
Marine elasmobranch fish	Hypertonic	NaCl content increased from 0.137 M to 0.2 M; approximately 2% urea added	1
Amphibians	Hypotonic	Diluted by 20—50% with water	2, 3, 9, 10
Reptiles	Isotonic	Used as for homoiotherms	5, 6

well in the range 7.2 to 7.6, though some cell lines grow well at up to pH 8.0 or down to pH 6.5. The buffers used for poikilotherm tissue culture, however, are very important. Homoiotherm cells and tissues are normally cultured at 37°C in the presence of 5% CO_2 in media containing $NaHCO_3$. It is a mistake to use CO_2 cabinets for poikilotherm tissue culture because of the higher solubility of CO_2 at lower temperatures and because of the lower pCO_2 of lower vertebrate blood. The basic medium Leibovitz L15, where buffering is provided by free-base amino acids, or more conventional media without $NaHCO_3$ and buffered with 15 mM HEPES are therefore particularly suitable for cells and tissues from the lower vertebrates.[1-4]

Basic Media

Published reports show that most of the commercial synthetic media have at some time been used for poikilotherm tissue culture, but that three media have been used far more than any others: Eagle's minimum essential medium (MEM), Eagle's basal medium (BME), and Leibovitz L15 medium (L15). MEM and BME contain only vitamins, amino acids, glucose, and inorganic salts. L15, which does not need buffering, contains vitamins, inorganic salts, much higher concentrations of amino acids (for buffering as well as for nutrition), galactose (in place of glucose), and sodium pyruvate.

MEM is superior to BME for optimal growth of fresh-water fish cells,[1] whereas the growth of many reptilian cell types is better in BME than in MEM.[5] These differences may be attributable to the higher concentrations of amino acids in MEM or to the exclusive presence of biotin in BME and inositol in MEM. L15 and MEM are the most suitable basic media for amphibian cell monolayer culture.

The suitability of a medium for cell monolayer cultures is usually based on cell survival and proliferation, whereas the success of organ culture is based on the retention of normal structure and function in vitro. Fleming et al.[11] maintained *Amphiuma means* liver in organ culture for up to 60 days in media based on MEM, BME, and L15. The activities of the urea cycle enzymes arginase and ornithine transcarbamylase and of the transaminases glutamate oxalacetate transaminase and glutamate pyruvate transaminase were significantly higher in MEM and BME than in L15. Glucose uptake from the medium and tissue glycogen content were higher in MEM than in L15, whereas nitrogenous excretion was higher in L15 than in MEM. Monnickendam and Balls[12] found that *A. means* kidney fragments produced more ammonia in L15 than in MEM, and other experiments showed that amylase production by *Necturus maculosus* (mud puppy) pancreas cultures was much higher in MEM than in L15 during a 9-week culture period.[4] For these reasons, MEM and BME are considered preferable to L15 for amphibian organ culture.

Serum

Serum is almost invariably added to poikilotherm tissue culture media, since it is normally necessary for the sustained vigorous growth of cells in monolayer cultures and for the retention of tissue structure and function in organ culture. Most of the commercially available sera have been used, as well as homologous sera in a few cases, but most is known about bovine serum. Fetal bovine serum at about 10% is recommended for poikilotherm cell and organ culture, as there is evidence that calf serum lacks essential factors or inhibits the growth of cells from fish[8] and amphibians.[13] Some cell lines will grow slowly in medium with 2% serum, but 20 to 30% is necessary for others.[1]

Many attempts have been made to replace the serum requirement by adding hormones (e.g., insulin) or other factors (e.g., spermine), but it appears that a combination of many factors is necessary for normal cell growth.[14] Sooy and Mezger-Freed[15] fractionated serum in order to develop a medium lacking low molecular weight serum components for selecting auxotrophic mutants from *Rana pipiens* cell lines. A macromolecular fraction from gel filtration on Sephadex G-25 failed to duplicate the effects of whole serum. Growth-promoting effects were restored by the addition of purines or purine nucleosides (hypoxanthine, adenosine, guanosine, inosine), but not by pyrimidine nucleosides. Other nucleic acid derivatives (deoxycitidine, thymidine, deoxyadenosine, deoxyguanosine, 5-methyl-cytosine) partially restored the growth rate. Hypoxanthine was the most effective additive, restoring maximum growth at 10^{-5} M when added with the macromolecular serum fraction. Mammalian cell lines grew at their maximum rates in medium containing only the macromolecular fraction, but whether amphibian cells in general require purine in the medium is not known.

The presence of serum may be less essential for poikilotherm organ culture. *A. means* liver survives in serum-free MEM for at least 28 days.[4] It was found that the glycogen level fell slowly and that urea and ammonia production continued at a relatively constant rate. Adrenaline induced glycogenolysis in liver cultures in serum-free MEM, whereas insulin did not induce glycogenesis and albumin production was greatly reduced. Liver cultures derived from male *Xenopus laevis* produced vitellogenin when treated with estradiol after several days in serum-free MEM.[16] Thus, it seems likely that factors contained in serum are required for some tissue functions, but not for others.

Other Natural Additives

A number of other natural materials have been used as nutrients in poikilotherm tissue culture, including bovine or chick embryo extract, lactalbumin hydrolysate, peptone, tryptose phosphate broth, yeast extract, and whole egg ultrafiltrate. Many workers consider that good serum makes such extras unnecessary,[1,3,5] but others still consider them to be essential for the retention of vigorous cell growth[17] or specific cell functions.[18]

Carbohydrate Metabolism

Most basic media contain glucose, but L15 contains galactose, which, in the presence of pyruvate and alpha alanine, can be substituted for glucose for maximum growth of HeLa and human conjunctiva cells.[19] Since L15 has been the basic medium for many poikilotherm tissue culture experiments, it is clear that this substitution can also be made in media for lower vertebrate cells and tissues.

Extensive studies have been carried out on carbohydrate metabolism in *Amphiuma means* liver organ cultures. Glycogenolysis is induced by adrenalin, glucagon, isoprenaline, and other adrenoceptor agonists, and glycogenesis follows the addition of insulin or culture in high glucose medium.[4,20,21] Liver maintained for several weeks in 50% MEM with 10% fetal bovine serum used less than 25% of the available glucose between the weekly medium changes.[4] Glucagon (1.4×10^{-7} M) induced gluconeogenesis from added pyruvate (2.8 m*M*) or alanine (2.8 m*M*). Pyruvate alone did not increase glucose release or tissue glycogen breakdown, but did cause a significant reduction in nitrogenous excretion.[22]

Solursh and Reiter[18] maintained *Xenopus laevis* hepatocytes in monolayer culture for several months. The cells stored glycogen, which increased in amount following treatment with dexamethasone and corticosterone.

Amino Acid Metabolism

The simpler commercial media such as MEM and BME contain the ten essential amino acids plus cytosine, tyrosine, and glutamine, which also appear to be required by cells in vitro. For some organ cultures, fewer amino acids are required, probably due to reutilization of breakdown products from other parts of the tissue.[23] L15 contains much larger amounts of all these amino acids, and, in addition, several amino acids that are absent from MEM and BME. Two of these, alanine and glycine, have been investigated with respect to their deamination and the subsequent effects on nitrogenous excretion in *Xenopus laevis.*[24-26] We have already noted that total nitrogenous excretion is significantly higher in *Amphiuma means* liver cultures in L15 than in MEM, and that ammonia production by kidney cultures was much higher in L15 than in MEM. Further experiments were carried out to investigate the effect of MEM containing high concentrations (10 m*M*) of added alanine, glycine, and glutamic acid on urea and ammonia production by cultured liver fragments.[27] There was no change in the amount of urea released, but ammonia production was significantly higher in media with added alanine or glycine, and significantly lower in medium with added glutamic acid. In *A. means* liver cultures, it appears that urea production proceeds at the maximum rate, but that at high concentrations alanine and glycine are deaminated, with a resulting increase in ammonia production. A high concentration of glutamic acid apparently stimulates its transamination to other amino acids, with a corresponding reduction in ammonia production. Amino acid uptake, deamination, and transamination appear to be controlled, at least in part, by extracellular amino acid concentration.

Antibiotics

There is no evidence to suggest that cells and tissues from poikilothermic vertebrates differ from those from homoiotherms in their sensitivity to antibiotics. The following concentrations are recommended for routine use in fish, amphibian, and reptile media: 100 IU/ml benzylpenicillin, sodium salt; 100 μg/ml streptomycin sulfate; 2 μg/ml amphotericin B (*or* 25 IU/ml nystatin). Contamination has seriously limited the use of amphibian embryonic material or skin for cell or organ cultures. Laskey[28] tested a number of antibiotics for use in the preparation of cultures from highly contaminated material and found the following to be most satisfactory: 70 μg/ml gentamicin sulfate *or* 100 μg/ml α-carboxybenzyl penicillin, disodium salt, *or* 100 μg/ml kanamycin with 100 IU/ml polymyxin E (colistin sulfomethate, sodium salt).

Conclusions

Since media designed for use with mammalian and avian cells can be used for poikilotherm tissue culture with very few modifications, it is unlikely that the nutritional requirements of cells from lower vertebrates differ fundamentally from those of homoiotherms. However, poikilothermic vertebrate cultures have a number of unique and almost totally unexploited advantages over homoiotherm cultures,[1,3-5] and could be valuable tools in the study of numerous biochemical and physiological characteristics of vertebrate cells.

REFERENCES

1. Wolf, K. and Quimby, M. C., in *Fish Physiology,* Vol. 3, Hoar, W. S. and Randall, D. S., Eds., Academic Press, New York, 1969, 253–305.
2. Freed, J. J. and Mezger-Freed, L., *Methods Cell Physiol.,* 4, 19–47, 1970.
3. Monnickendam, M. A. and Balls, M., *Experientia,* 29, 1–17, 1973.
4. Balls, M., Brown, D., and Fleming, N., *Methods Cell Biol.,* 13, 213–238, 1976.
5. Clark, H. F., in *Growth, Nutrition and Metabolism of Cells in Culture,* Vol. 2, Rothblat, G. H. and Cristofalo, V. J., Eds., Academic Press, New York, 1972, 287–325.
6. Clark, H. F., Cohen, M. M., and Karzon, D. T., *Proc. Soc. Exp. Biol. Med.,* 133, 1039–1047, 1970.
7. Clem, M. W., Moewus, L., and Sigel, M. M., *Proc. Soc. Exp. Biol. Med.,* 108, 762–766, 1961.
8. Lee, M. H. and Loh, P. C., *Proc. Soc. Exp. Biol. Med.,* 150, 40–48, 1975.
9. Balls, M. and Worley, R. S., *Exp. Cell Res.,* 76, 333–336, 1973.
10. Monnickendam, M. A. and Balls, M., *J. Cell Sci.,* 11, 799–813, 1973.
11. Fleming, N., Brown, D., and Balls, M., *J. Cell Sci.,* 18, 533–544, 1975.
12. Monnickendam, M. A. and Balls, M., *Comp. Biochem. Physiol. A,* 50, 359–363, 1975.
13. Arthur, M. E., Ph.D. Thesis, University of East Anglia, Engl., 1971.
14. Hosick, H. L. and Nandi, S., *Exp. Cell Res.,* 84, 419–425, 1974.
15. Sooy, L. E. and Mezger-Freed, L., *Exp. Cell Res.,* 60, 482–485, 1970.
16. Waugh, L. J. and Knowland, J., *Proc. Natl. Acad. Sci. U.S.A.,* 72, 3172–3175, 1975.
17. Seto, T. and Rounds, D. E., *Methods Cell Physiol.,* 3, 75–94, 1968.
18. Solursh, M. and Reiter, R. S., *Z. Zellforsch. Mikrosk. Anat.,* 128, 457–469, 1972.
19. Chang, R. S. and Geyer, R. P., *Proc. Soc. Exp. Biol. Med.,* 96, 336–340, 1958.
20. Monnickendam, M. A., Brown, D., and Balls, M., *Comp. Biochem. Physiol. A,* 47, 567–572, 1974.
21. Brown, D., Pryor, J. S., and Balls, M., in *Organ Culture in Biomedical Research,* Balls, M. and Monnickendam, M. A., Eds., Cambridge University Press, London, 1976, 481–501.
22. Brown, D., Fleming, N., and Balls, M., *Gen. Comp. Endocrinol.,* 27, 380–388, 1975.
23. Paul, J., *Cell and Tissue Culture,* 4th ed., E. & S. Livingstone, Edinburgh, 1970.
24. Balinsky, J. B. and Baldwin, E., *Biochem J.,* 82, 187–191, 1962.
25. Unsworth, B. R. and Crook, E. M., *Comp. Biochem. Physiol.,* 28, 831–845, 1967.
26. Balinsky, J. B., in *Comparative Biochemistry of Nitrogen Metabolism,* Vol. 2, Campbell, J. W., Ed., Academic Press, New York, 1970.
27. Fleming, N. and Balls, M., *Experientia,* 32, 169–171, 1976.
28. Laskey, R. A., *J. Cell Sci.,* 7, 653–659, 1970.

QUALITATIVE REQUIREMENTS AND UTILIZATION OF NUTRIENTS: MAMMALIAN AND AVIAN CELLS IN CULTURE

K. Higuchi

INTRODUCTION

The specific nutritional requirements of animal cells in culture have been studied in detail in only a few cell lines.[1] However, culture media formulated on the basis of these limited studies have generally proven satisfactory for cultivation of a wide variety of mammalian and avian cells. It would appear that very little difference in basic nutritional requirements exists among animal cells in vitro. On the other hand, the apparent similarity may only be a reflection of the fact that all cultured cells have been isolated in similar nutritional environments. A significant exception to the uniformity of nutritional requirements among animal cell cultures is the difference between so-called heteroploid cells and normal diploid cells. Many heteroploid cells can be cultured in serum-free chemically defined media, whereas all normal diploid cells apparently require macro-molecular constituents of serum for growth.[2] Undoubtedly, further developments in cell culture research will reveal other differences in nutritional requirements, particularly among normal differentiated cell types. The ill-defined area of animal cell nutrition dealing with factors involved in growth and development of differentiated cells has been reviewed by Rutter et al.[3] and no attempt will be made to cover it here.

AMINO ACIDS

Thirteen amino acids are required for growth and survival of most animal cells in culture.[4,5] There are six or seven additional amino acids ("nonessential" amino acids) often employed in cell culture media because they may be stimulatory for growth.[6] Use of the term "nonessential" must be qualified because certain cell cultures require some of these amino acids for growth (see Table 1). It should also be noted that requirements for certain amino acids are frequently dependent on population density. Eagle and Piez[7] and Eagle et al.[8] have shown that at high cell populations, requirements for amino acids such as cystine, glutamine, serine, and asparagine are often drastically reduced. Because of space limitations, references for the essential amino acids are given only for the two cell lines with which much of animal cell nutritional studies have been conducted (mouse fibroblast, strain L, and human carcinoma cell, strain HeLa). However, it should be remembered that amino acids essential for HeLa and L929 cells are probably required by almost all mammalian and avian cells in culture.

Data for utilization of amino acids are not specifically listed because of the consistent correlation between utilization and nutritional requirement.[9-11] Study of the uptakes of nonessential amino acids is complicated by the fact that animal cells are able to synthesize substantial amounts of these amino acids; therefore, uptake during one phase of growth may be obscured by later excretion.

VITAMINS

Vitamins that have been shown to be required for growth of cultured animal cells are listed in Table 2. Many of these vitamins are required by all cell cultures; on the other

Table 1
AMINO ACID REQUIREMENTS OF CULTURED ANIMAL CELLS

L-Amino acid	Cell culture	Requirement[a]	Ref.
Alanine	L929, HeLa	Ɇ	4, 5
Arginine[b]	L929, HeLa	E	4, 5
Asparagine	L929, HeLa	Ɇ	4, 5
Asparagine	Walker carcinoma 256	E	12
Asparagine	Mouse leukemia	E	13
Asparagine	Jensen sarcoma	E	14
Aspartate	L929, HeLa	Ɇ	4, 5
Cyst(e)ine[b]	L929, HeLa	E	4, 5
Glutamate	L929, HeLa	Ɇ	4, 5
Glutamine[b]	L929, HeLa	E	4, 5
Glutamine	Monkey kidney primary	Ɇ[c]	15
Glycine	L929, HeLa	Ɇ	4, 5
Glycine	Monkey kidney primary	E[d]	15
Histidine[b]	L929, HeLa	E	4, 5
Hydroxyproline	L929, HeLa	Ɇ	4, 5
Isoleucine[b]	L929, HeLa	E	4, 5
Lysine[b]	L929, HeLa	E	4, 5
Methionine[b]	L929, HeLa	E	4, 5
Phenylalanine[b]	L929, HeLa	E	4, 5
Proline	L929, HeLa	Ɇ	4, 5
Proline	Chinese hamster ovary	E[e]	16
Serine	L929, HeLa	Ɇ	4, 5
Serine	Chinese hamster ovary	E[e]	16
Serine	L cell variant	E	17
Threonine[b]	L929, HeLa	E	4, 5
Tryptophan[b]	L929, HeLa	E	4, 5
Tyrosine[b]	L929, HeLa	E	4, 5
Valine[b]	L929, HeLa	E	4, 5

[a] E = essential; Ɇ = not essential.
[b] "Essential" amino acids.
[c] Primary monkey kidney cells utilize glutamic acid interchangeably with glutamine.[15]
[d] Primary monkey kidney cells require either glycine or folinic acid for optimal growth.[15]
[e] For clonal growth.[16]

hand, requirements for compounds such as ascorbic acid, biotin, and vitamin B_{12} have only infrequently been demonstrated. Eagle and co-workers found that eight vitamins (choline, folic acid, inositol, nicotinamide, pantothenate, pyridoxal, riboflavin, and thiamin) were minimal requirements for most mammalian cell lines studied.[18,19] Swim and Parker[20] observed that the same eight vitamins were essential for growth of a strain of human fibroblasts in culture. These workers also found that cofactor forms of the vitamins were active in cell cultures. Sanford et al.[21] noted that none of the lipid-soluble vitamins (A, D, E, and K) had significant effects on the growth of mouse fibroblasts, nor did ascorbic acid, p-aminobenzoic acid, and vitamin B_{12}. However, a specific role for ascorbic acid for growth of mouse plasmacytoma cells has been reported.[22] Studies in chemically defined media have resulted in detection of biotin as a nutritional requirement of animal cells in vitro.[23-25] Similarly, there is strong evidence that vitamin B_{12} is essential for cultured animal cells.[26]

Data on vitamin utilization by cultured cells are limited. Blaker and Pirt[27] found uptake of biotin, choline, inositol, nicotinamide, pantothenate, pyridoxal, riboflavin, and thiamin by L cells during growth in a chemically defined medium. No uptake of folic acid or vitamin B_{12} was demonstrable.

Table 2
VITAMIN REQUIREMENTS OF CULTURED ANIMAL CELLS

Vitamin	Cell culture	Requirement[a]	Ref.
Ascorbic acid	L	R̶	21
Ascorbic acid	Mouse plasmacytoma	R	22
Biotin	L	R	24–26
Biotin	HeLa	R	26
Biotin	Hamster ovary	R	23
Choline	L, HeLa	R	18
Folic acid	L, HeLa	R	18
Inositol	HeLa	R	19
Inositol	L	R̶	19
Nicotinamide	L, HeLa	R	18
Pantothenate	L, HeLa	R	18
Pyridoxal	L, HeLa	R	18
Riboflavin	L, HeLa	R	18
Thiamin	L, HeLa	R	18
Vitamin B$_{12}$	L, HeLa	R(?)[b]	26

[a] R = required; R̶ = not required.

[b] Requirement for vitamin B$_{12}$ by animal cells in culture is not yet generally recognized.

Table 3
INORGANIC ION REQUIREMENTS OF CULTURED ANIMAL CELLS

Ion	Requirement[a]	Cell strains	Ref.
Na$^+$	R	HeLa, L	28
K$^+$	R	HeLa, L	28
Ca^{++}	R	HeLa, L	28
Ca^{++}	R	HeLa	29
Mg^{++}	R	HeLa, L	28
Mg^{++}	R	HeLa	29
Cl$^-$	R	HeLa, L	28
H$_2$PO$_4$$^-$	R	HeLa, L	28
HCO$_3$$^-$	R	HeLa, human conjunctiva	30
Fe^{++}	R	L	31
Fe^{++}	R	Chinese hamster ovary	16
Zn^{++}	R	L	31
Mn^{++}	R̶[b]	L	31
Cu^{++}	R̶[c]	L	31

[a] R = required; R̶ = not required.

[b] Mn^{++} was toxic at levels tested.

[c] Cu^{++} had sparing effect on iron requirement.

INORGANIC IONS

Most of the inorganic ions listed in Table 3 have been shown to be required for growth of animal cells in culture. The minimal inorganic ion requirements of HeLa and L cells in culture were shown by Eagle[28] to be Na$^+$, K$^+$, Ca^{++}, Mg^{++}, Cl$^-$, and H$_2$PO$_4$$^-$. Wyatt also found Mg^{++} and Ca$^+$ to be required by HeLa cells.[29] Bicarbonate was shown to be an essential ion in the nutrition of several human cell cultures by Geyer and Chang.[30] The role of trace metals in the nutrition of animal cells is largely unknown. Ham[16] reported evidence that iron, copper, and zinc ions were beneficial for clonal growth of hamster and human cell lines. Thomas and Johnson[31] treated culture media with a chelating resin and

Table 4
UTILIZATION OF CARBOHYDRATES BY
CULTURED ANIMAL CELLS

	Cell cultures[a]		
Carbohydrate	Normal human intestine	HeLa	Mouse fibroblast strain L
D-Glucose	U	U	U
D-Fructose	U	U	U
D-Galactose	U	U	U[b]
D-Mannose	U	U	U
D-Mannitol	Ψ̶	Ψ̶	Ψ̶
Trehalose	U	U	–
Turanose	U	U	–
Sucrose	Ψ̶	–	Ψ̶
Lactose	Ψ̶	Ψ̶	Ψ̶
D-Ribose	U[b]	U[b]	–

[a] U = utilized for growth; Ψ̶ = not utilized for growth; – not tested.

[b] Utilization of both galactose and ribose was enhanced by the presence of pyruvate in the medium. Growth in D-ribose medium was significantly less than in glucose medium, whereas other utilizable sugars listed above produced cell yields comparable to glucose.

were able to show that iron and zinc were essential for growth of L cells. They also noted that added Mn^{++} was toxic at any level tested, whereas Cu^{++} had a sparing action on iron requirement. Birch and Pirt[32] studied uptakes of PO_4, Mg, and K ions in L cell cultures and showed that their medium contained these elements substantially in excess of requirements for cell growth.

CARBOHYDRATES

Glucose is the principal carbohydrate in animal cell metabolism. It is essential for cell growth and survival in the absence of other utilizable sugars. Eagle et al.[33] examined the ability of a number of other carbohydrates and derivatives to support growth of various animal cell cultures. The data presented in Table 4 are taken from their work. Some of the compounds tested yielded only slight growth and they are omitted from the table.

MISCELLANEOUS FACTORS IN THE
NUTRITION OF ANIMAL CELLS IN CULTURE

In addition to serum proteins and the defined nutrients listed in Tables 1 through 4, various other substances have been reported to be either necessary or stimulatory for growth of certain animal cell cultures. Some of these substances are discussed briefly below.

Lipids — Exogenous lipids derived from serum in the medium are well utilized by animal cells in culture.[34] Holmes and co-workers showed that cholesterol was an essential nutrient for growth of human diploid fibroblasts.[35] A heteroploid porcine kidney cell was also found to require cholesterol for growth.[36] Fatty acids such as oleic and linoleic acids were required for clonal growth of Chinese hamster cells.[39] Lecithin has been reported to be stimulatory for growth of a strain of human heart cells as well as for a strain of monkey kidney cells.[26]

Purines and pyrimidines — Growth of Walker carcinosarcoma 256 cells was stimulated by purines and purine derivatives.[38] Both hypoxanthine and thymidine were required for optimal growth of Chinese hamster cells.[39]

Keto-acids — Keto-acids such as alpha-ketoglutarate, oxaloacetate, and pyruvate were stimulatory for clonal growth of Walker carcinosarcoma 256 cells.[40]

Hormones — Insulin is known to produce remarkable stimulatory effects on the growth of several human cell lines.[26,41] L-Throxine has been reported to enhance growth of human kidney cells[42] and HeLa cells.[43] Serotonin was shown to promote growth of fibroblasts of both human and mouse origins.[44] Hydrocortisone was found to produce limited stimulation of growth of a heteroploid human liver cell in serum-free medium.[26]

Other substances — Putrescine and related compounds were shown to enhance clonal growth of Chinese hamster cells.[45] Pohjanpelto and Raina found putrescine to be a growth factor for cultured human fibroblasts.[46]

REFERENCES

1. Ham, R. G., *In Vitro,* 10, 119–129, 1974.
2. Higuchi, K., *Adv. Appl. Microbiol.,* 16, 111–136, 1973.
3. Rutter, W. J., Pictet, R. L., and Morris, P. W., *Ann. Rev. Biochem.,* 42, 601–646, 1973.
4. Eagle, H., *J. Biol. Chem.,* 214, 839–852, 1955.
5. Eagle, H., *J. Exp. Med.,* 102, 37–48, 1955.
6. Eagle, H., *Science,* 130, 432–437, 1959.
7. Eagle, H. and Piez, K. A., *J. Exp. Med.,* 116, 29–43, 1962.
8. Eagle, H., Washington, C. L., Levy, M., and Cohen, L., *J. Biol. Chem.,* 241, 4994–4999, 1966.
9. Griffiths, J. B. and Pirt, S. J., *Proc. R. Soc. London Ser. B,* 168, 421–438, 1967.
10. Mohberg, J. and Johnson, M. J., *J. Nat. Cancer Inst.,* 31, 611–623, 1963.
11. Griffiths, J. B., *J. Cell Sci.,* 6, 739–749, 1970.
12. Neuman, R. E. and McCoy, T. A., *Science,* 124, 124–125, 1956.
13. Haley, E. E., Fischer, G. A., and Welch, A. D., *Cancer Res.,* 21, 532–536, 1961.
14. McCoy, T. A., Maxwell, M., and Kruse, P. F., *Cancer Res.,* 19, 591–595, 1959.
15. Eagle, H., Freeman, A. E., and Levy, M., *J. Exp. Med.,* 107, 643–652, 1958.
16. Ham, R. G., *Exp. Cell Res.,* 29, 515–526, 1963.
17. Higuchi, K., *J. Cell Physiol.,* 75, 65–72, 1970.
18. Eagle, H., *J. Exp. Med.,* 102, 595–600, 1955.
19. Eagle, H., Oyama, V. I., Levy, M., and Freeman, A. E., *J. Biol. Chem.,* 226, 191–206, 1957.
20. Swim, H. E. and Parker, R. F., *Arch. Biochem. Biophys.,* 78, 46–53, 1958.
21. Sanford, K. K., Dupree, L. T., and Covalesky, A. B., *Exp. Cell Res.,* 31, 345–375, 1963.
22. Park, E. H., Bergsagel, D. E., and McCulloch, E. A., *Science,* 174, 720–722, 1971.
23. Ham, R. G., *Nat. Cancer Inst. Monogr.,* 26, 19–20, 1967.
24. Haggerty, D. F. and Sato, G., *Biochem. Biophys. Res. Commun.,* 34, 812–813, 1969.
25. Higuchi, K., *Fed. Proc. Abstr.,* 29, 627, 1969.
26. Higuchi, K. and Robinson, R. C., *In Vitro,* 9, 114–121, 1973.
27. Blaker, G. J. and Pirt, S. J., *J. Cell Sci.,* 8, 701–708, 1971.
28. Eagle, H., *Arch. Biochem. Biophys.,* 61, 356–366, 1956.
29. Wyatt, H. V., *Exp. Cell Res.,* 23, 97–107, 1961.
30. Geyer, R. P. and Chang, R. S., *Arch. Biochem. Biophys.,* 73, 500–506, 1958.
31. Thomas, J. A. and Johnson, M. J., *J. Nat. Cancer Inst.,* 39, 337–345, 1967.
32. Birch, J. R. and Pirt, S. J., *J. Cell Sci.,* 8, 693–700, 1971.
33. Eagle, H., Barban, S., Levy, M., and Schulze, H. O., *J. Biol. Chem.,* 233, 551–558, 1958.
34. Bailey, J. M., *Proc. Soc. Exp. Biol. Med.,* 115, 747, 1964.
35. Holmes, R., Helms, J., and Mercer, G., *J. Cell Biol.,* 42, 262–271, 1969.
36. Higuchi, K., *In Vitro,* 6, 239, 1970.
37. Ham, R. G., *Science,* 140, 802–803, 1963.
38. Neuman, R. E. and Tytell, A. A., *Exp. Cell Res.,* 15, 637–639, 1958.
39. Ham, R. G., *Exp. Cell Res.,* 28, 489–500, 1962.
40. Neuman, R. E. and McCoy, T. A., *Proc. Soc. Exp. Biol. Med.,* 98, 303–306, 1958.
41. Lieberman, I. and Ove, P., *J. Biol. Chem.,* 234, 2754–2758, 1959.
42. Siegel, E. and Tobias, C. A., *Nature,* 212, 1318–1321, 1961.
43. Bartfeld, H. and Siegel, S. M., *Exp. Cell Res.,* 49, 25–30, 1968.
44. Boucek, R. J. and Alvarez, T. R., *Science,* 167, 898–899, 1970.
45. Ham, R. G., *Biochem. Biophys. Res. Commun.,* 14, 34–38, 1964.
46. Pohjanpelto, P. and Raina, A., *Nature (London) New Biol.,* 235, 247–249, 1972.

QUALITATIVE REQUIREMENTS AND UTILIZATION OF NUTRIENTS: MAMMALIAN GAMETES AND EMBRYOS

R. B. L. Gwatkin

Data on the nutritional requirements of mammalian gametes and embryos are very limited. Mouse follicular oocytes have been shown to utilize pyruvate, oxaloacetate, and, to some extent, lactate, but not phosphoenolpyruvate or glucose, for maturation to Metaphase II.[1] On the other hand, hamster oocytes require 12 amino acids and appear to meet their energy requirements from one or more of these amino acids by gluconeogenesis.[5] The number of amino acids needed for hamster oocyte maturation increases as the follicle enlarges.[4] Oocyte maturation in both species requires oxygen.[2,5]

Despite numerous studies of in vitro fertilization, few have been concerned with nutritional requirements. The fertilization of rat eggs in a Krebs-Ringer bicarbonate solution, supplemented with crystalline bovine serum albumin, requires the addition of pyruvate and is stimulated by the further addition of oxaloacetate and lactate.[12]

During cleavage of the early mouse embryo, oxygen is utilized[17] and carbon dioxide is incorporated into protein.[19] The mouse zygote utilizes pyruvate, but not phosphoenolpyruvate, lactate, or glucose.[18] As cleavage proceeds further, the number of utilizable sugars increases, so that by midcleavage significant amounts of glucose are utilized.[15] Amino acids begin to be utilized late in cleavage,[14,20,28] and by the blastocyst stage six are required.[16] The rabbit embryo shows a requirement for amino acids at an earlier stage than does the mouse embryo, four being needed at the two-cell stage.[23] As cleavage proceeds to blastocyst, ten amino acids are required.[27]

In the table below, the following symbols are used:

R = required.
Ʀ = not required.
U = utilized.
Ʋ = not utilized.
u = poorly utilized.
S = stimulatory.
s = moderately supportive.

Process and species	Acetate	Fructose	Glucose	Phosphoenolpyruvate	Pyruvate	Lactate	Oxaloacetate	O₂	CO₂	Ca²⁺	Alanine	Arginine	Asparagine	Aspartate	Cysteine	Cystine	Glutamate	Glutamine	Glycine	Histidine	Isoleucine	Leucine	Lysine	Methionine	Phenylalanine	Proline	Serine	Threonine	Tyrosine	Tryptophane	Valine	Ref.
Maturation of Oocytes																																
Mouse (*Mus musculus*)	U		U	U	U	u	U	R			R	R	R	R	R	R	R	R	R	R	R	U	R	R	R	R	R	R	R	R	R	1, 2, 6
Golden hamster (*Mesocricetus auratus*)																																3
From small follicles			S		S	S	U	R			R	R	R	R	R	R	R	R	R	R	R	R	R	R	R	R	R	R	R	R	R	4, 5
From large follicles					S	S																										7
Pig (*Sus scrofa*)		S	S															S					S	S	S		S					8
Rabbit (*Oryctolagus cuniculus*)																																
Fertilization																																
Mouse (*M. musculus*)			S		R					R																						9, 10, 13
Rat (*Rattus norvegicus*)							S			R																						12, 13
Guinea pig (*Cavia porcellus*)										R¹																						11
Preimplantation Development																																
Mouse (*M. musculus*)																																14, 15, 17—20, 28
Zygote			U	U	U	U	U	U	U		u	u	u	u		u	u	u	u	u		u	u	u								14, 15, 17, 19—21, 28
2-Cell		U	U	U	U	U		U	U		u	u	u	u		u	u	u	u	u	s	u	u	u								14, 15, 17, 19, 20, 22, 28, 29
4-Cell to morula			U?						U	R		U²		U²		U²	U²	U²	U²	U²		U²		U²								
Blastocyst			U		R		U	R	U		R	R	R	R	R	R	R	R	R	R	R	R	S	S	S	R	R	R	S	S	s	15—17, 19
Rabbit (*O. cuniculus*)																																
Zygote											R	R	R	R	R	R	R	R	R	R	R	R	R	R	R	R	R	R	R	R	R	23
2-Cell											R	R	R	R	R	R	R	R	R	R	R	R	R	R	R	R	R	R	R	R	R	23
4-Cell to morula											R	R	R	R	R	R	R	R	R	R	R	R	R	R	R	R	R	R	R?	R	R	23
Blastocyst	U	U	U		U			R			R	R	R	R	R	R	R	R	R	R	R	R	R	R	R	R	R	R	R	R	R	24—27

REFERENCES

1. Biggers, J. D., Whittingham, D. G., and Donahue, R. P., *Proc. Natl. Acad. Sci. U.S.A.,* 58, 560–567, 1967.
2. Haidri, A. A., Miller, I. M., and Gwatkin, R. B. L., *Reprod. Fertil.,* 26, 409–411, 1971.
3. Gwatkin, R. B. L. and Haidri, A. A., *Exp. Cell Res.,* 76, 1–7, 1973.
4. Haidri, A. A. and Gwatkin, R. B. L., *J. Reprod. Fertil.,* 35, 173–176, 1973.
5. Gwatkin, R. B. L. and Haidri, A. A., *J. Reprod. Fertil.,* 37, 127–129, 1974.
6. Cross, P. C. and Brinster, R. L., *Exp. Cell Res.,* 86, 43–46, 1974.
7. Tsafriri, A. and Channing, C. P., *J. Reprod. Fertil.,* 43, 149–152, 1975.
8. Bae, I.-H. and Foote, R. H., *Exp. Cell Res.,* 91, 113–118, 1975.
9. Hoppe, P. C. and Whitten, W. K., *J. Reprod. Fertil.,* 39, 433–436, 1974.
10. Iwamatsu, T. and Chang, M. C., *J. Reprod. Fertil.,* 26, 197–208, 1971.
11. Yanagimachi, R. and Usui, N., *Exp. Cell Res.,* 89, 161–174, 1974.
12. Tsunoda, Y. and Chang, M. C., *J. Exp. Zool.,* 193, 79–86, 1975.
13. Miyamoto, H. and Ishibashi, T., *J. Reprod. Fertil.,* 45, 523–526, 1975.
14. Brinster, R. L., *J. Reprod. Fertil.,* 27, 329–338, 1971.
15. Brinster, R. L., *Exp. Cell Res.,* 47, 271–277, 1967.
16. Gwatkin, R. B. L., *J. Cell. Physiol.,* 68, 335–344, 1966.
17. Mills, R. M. and Brinster, R. L., *Exp. Cell Res.,* 47, 337–344, 1967.
18. Whittingham, D. G., *Biol. Reprod.,* 1, 381–386, 1969.
19. Graves, C. N. and Biggers, J. D., *Nature,* 167, 1506–1508, 1970.
20. Monesi, V. and Salfi, V., *Exp. Cell Res.,* 46, 632–635, 1967.
21. Brinster, R. L., *J. Exp. Zool.,* 158, 59–68, 1965.
22. Borland, R. M. and Tasca, R. J., *Dev. Biol.,* 30, 169–182, 1974.
23. Daniel, J. C. and Olson, J. D., *J. Reprod. Fertil.,* 15, 453–455, 1968.
24. Daniel, J. C., *J. Reprod. Fertil.,* 17, 187–190, 1968.
25. Huff, R. L. and Eik-Nes, K. B., *J. Reprod. Fertil.,* 11, 57–63, 1966.
26. Mounib, M. S. and Chang, M. C., *Exp. Cell Res.,* 38, 201–215, 1965.
27. Daniel, J. C. and Krishnan, R. S., *J. Cell. Physiol.,* 70, 155–160, 1968.
28. Mintz, B., *J. Exp. Zool.,* 157, 85–100, 1964.
29. Whitten, W. K., *Adv. Biosci.,* 6, 129–139, 1970.

Index

INDEX

Z

CRC PUBLICATIONS OF RELATED INTEREST

CRC HANDBOOKS:

CRC FENAROLI'S HANDBOOK OF FLAVOR INGREDIENTS
Edited, Translated, and Revised by: **Thomas E. Furia** and **Nicolo Bellanca,** Dyanpol, Palo Alto, California.
This two-volume Handbook is an update of the First Edition and includes comprehensive review chapters by recognized experts in the field. It is extensively indexed for ease of use.

CRC HANDBOOK OF BIOCHEMISTRY AND MOLECULAR BIOLOGY, 3rd Edition
Edited by **Gerald D. Fasman, Ph.D.,** Brandeis University.
This Handbook provides the most up-to-date information in biochemistry and molecular biology, reflecting the wealth of new information that has become available since 1970. It consists of eight volumes and a cumulative series index.

CRC HANDBOOK OF CHEMISTRY AND PHYSICS
Edited by **Robert C. Weast, Ph.D.,** Consolidated Natural Gas Service Co., Inc.
This Handbook is the definitive reference for chemistry and physics and maintains the tradition that has earned it the reputation as the best scientific reference in the world.

CRC HANDBOOK OF FOOD ADDITIVES, 2nd Edition
Edited by **Thomas E. Furia,** Dynapol, Palo Alto, California.
Nearly 1000 pages of pertinent food additive information is offered, reflecting the important changes that have occurred in recent years in the area of food additives.

CRC UNISCIENCE PUBLICATIONS:

MAN, FOOD, AND NUTRITION
Edited by **Miloslav Rechcigl, Jr., M.N.S., Ph.D., F.A.A.A.S., F.A.I.C., F.W.A.S.,** Agency for International Development, U.S. Department of State.
This interdisciplinary treatise offers a comprehensive and integrated critical review of the nature and the scope of the world food problem. It presents strategies and discusses various technological approaches to overcoming world hunger and malnutrition.

PHOSPHATES AS FOOD INGREDIENTS
By **R. H. Ellinger, Ph.D.,** Regulatory Compliance, Kraft Foods Co., Chicago.
This book is a food technologist's guide to the numerous useful applications of phosphates in food processing.

TOXICITY OF PURE FOODS
By **Eldon M. Boyd, Ph.D.** (deceased), Queen's University, Kingston, Ontario. Edited by **Carl E. Boyd, M.D.,** Health and Welfare, Canada.
A systematic study of the toxicity of pure foods is described in this volume.

WORLD FOOD PROBLEM
Edited by **Miloslav Rechcigl, Jr., M.N.S., Ph.D., F.A.A.A.S., F.A.I.C., F.W.A.S.,** Agency for International Development, U.S. Department of State.
This is a comprehensive and up-to-date bibliography on all important facets of the world food problem, encompassing such areas as the availability of natural resources and the present future sources of energy.

CRC PUBLICATIONS OF RELATED INTEREST (continued)

CRC MONOTOPIC REPRINTS:

FLEXIBLE PACKAGING OF FOODS
By **Aaron L. Brody, S.B., M.B.A., Ph.D.,** Arthur D. Little, Inc.
The aim of this book is to describe food products employing flexible packaging, the requirements dictating the flexible packaging being used and the current state of flexible packaging in the food industry.

FREEZE-DRYING OF FOODS
By **C. Judson King, B.E., E.M., Sc.D.,** University of California.
This review concentrations on several papers published in recent years, relates them to the rest of the field, gives an evaluation of their findings, and the conclusions they draw.

LABORATORY TESTS FOR THE ASSESSMENT OF NUTRITIONAL STATUS
By **Howerde E. Sauberlich, Ph.D.,** Fitzsimmons Army Medical Center, Denver.
This book considers the development and application of the various biomedical techniques currently available for use in assessing nutritional status.

SOYBEANS AS A FOOD SOURCE, 2nd Edition
By **W. J. Wolf, B.S., Ph.D.,** and **J. C. Cowan, A.B., Ph.D.,** Northern Marketing and Nutrition Research Division, Peoria, Illinois.
This book summarizes the conversion of soybeans to food. Emphasis is given to the protein content of soybeans because of the current high level of interest.

STORAGE, PROCESSING, AND NUTRITIONAL QUALITY OF FRUITS AND VEGETABLES
By **D. K. Salunkhe, B.S., M.S., Ph.D.,** Utah State University.
This book reviews the nutritional value and quality of fruits and vegetables as influenced by chemical treatments, storage and processing conditions.

THE USE OF FUNGI AS FOOD AND IN FOOD PROCESSING, Parts 1 and 2
By **William D. Gray, A.B., Ph.D.,** Northern Illinois University. Edited by **Thomas E. Furia,** Dynapol, Palo Alto, California.
This two-volume work describes how filamentous fungi have been used in the past and present as a food source in various areas of the world.

CRC CRITICAL REVIEW JOURNALS:

CRC CRITICAL REVIEWS IN FOOD SCIENCE AND NUTRITION
Edited by **Thomas E. Furia,** Dynapol, Palo Alto, California.

Please forward inquiries to CRC Press, Inc., 18901 Cranwood Parkway, Cleveland, Ohio 44128.

CRC FORUMS

The CRC Forums represent a new concept in scientific information. Each Forum brings together recognized authorities to discuss topical, scientific and medical issues, while being tape-recorded. Each Forum is moderated by a designated chairperson. Within a few weeks, edited Casette tapes of the discussion, as well as copies of research papers and appropriate tabular data prepared by the participants, are available for purchase. Each Forum economically provides a total information package that allows scientists and researchers to evaluate and study an issue from a range of perspectives. This is accomplished without the expense of attending costly seminars, and the tapes can be used repeatedly at your listening convenience.

CRC FORUM ON OBESITY

Chairperson, JOHN R. K. ROBSON, M.D., D.P.H., Professor of Nutrition, Medical University of South Carolina.

Participants, GEORGE V. MANN, Sc.D., M.D., Professor of Nutrition, School of Medicine, Vanderbilt University, and **DEREK MILLER, Ph.D.,** Senior Lecturer, Department of Nutrition, Queen Elizabeth College, University of London (England). **HENRY A. JORDAN, M.D.,** Behavioral Weight Control Program, University of Pennsylvania.

Obesity is one of the most prevalent conditions in the Western World. While it undoubtedly occurs when the intake of food energy exceeds energy expenditure many aspects of the condition are not clearly understood.

In an effort to assist physicians in dealing with obesity, a CRC Forum has been held to discuss four of its aspects, namely the genetic, metabolic, and dietary basis of obesity; human behavior and weight gain; obesity in clinical practice, and management of obesity.

These topics are the basis of scientific papers prepared by the participants with expertise in these areas. Subsequent discussions and interaction among the members of the Forum cover important subtopics, including the cultural influences of society on diet, as well as individual attitudes towards food. The discussions on the clinical aspects of obesity cover the difficulties in defining obesity and measuring the extent of the problem that the threats of obesity pose to the health of the community. The discussions on management are concerned with the care of the obese in the Family Practitioners Office as well as the problems of the very obese hospitalized patient.

OUTLINE for CRC FORUM ON OBESITY

Side 1.   Introduction and participants statements
          Prevelance of obesity

Side 2.   Genetics
          Thermogenisis, gene, cellular

Side 3.   Environment
          Early feeding, glandular, cultural, and social effects

Side 4.   Implications
          Diabetes, cardio-vascular disease, hypertension, wear and tear, psychology, minor
              irritations, menarche growth

Side 5.   Treatment
          Diet, weight control, total fast, protein modified fast, quality, slimming groups, bizarre
              treatments

Side 6.   Treatment (cont.)
          Drugs – amphetamines, T3, saccharin, bulking agents, thermogenic drugs
          Activity – exercise and energy dissipation
          Surgery – procedures
          Prognosis for treatment

Side 7.   Behavior modification
          Life modification, material and personnel needs
          Prognosis for treatment

Side 8.   Management
          Who to treat
          Criteria for obesity – weight, height, skinfold
          Standard for obesity – relative weight, Quetelets Index, emotional indices
          Preventive measures
          New and established treatment regimes – individualized and holistic approach, diet quality,
              activity and energy dissipation, drugs, role of paramedical personnel
          Future developments – thermogenic drugs
          Warnings – hazards of surgery, starvation, bizarre diets, dangers during pregnancy and
              growth
          Summary

Direct all inquiries to: CRC Forums 18901 Cranwood Parkway Cleveland, Ohio 44128